HEALTH ASPECTS OF THE DISPOSAL OF WASTE CHEMICALS

Pergamon Titles of Related Interest

Bergman et al. ENVIRONMENTAL HAZARD ASSESSMENT OF EFFLUENTS

Cairns BIOLOGICAL MONITORING IN WATER POLLUTION

Cairns MULTISPECIES TOXICITY TESTING

Fiksel BIORISK

Grabow WATER VIROLOGY

Hutzinger et al. CHLORINATED DIOXINS AND RELATED COMPOUNDS

Jenkins MICROPOLLUTANTS IN THE ENVIRONMENT

Jenkins WATER POLLUTION RESEARCH AND CONTROL

Odendaal MINE WATER POLLUTION

Persson et al. TASTE AND ODORS IN WATER AND AQUATIC ORGANISMS

Ricci & Rowe HEALTH AND ENVIRONMENTAL RISK ASSESSMENT

Wiechers PHOSPHATE REMOVAL IN BIOLOGICAL TREATMENT PROCESSES

Related Journals

(Sample copies available on request)

ATMOSPHERIC ENVIRONMENT

CHEMOSPHERE

CURRENT ADVANCES IN ECOLOGICAL SCIENCES

ENVIRONMENT INTERNATIONAL

ENVIRONMENTAL TOXICOLOGY AND CHEMISTRY

JOURNAL OF AEROSOL SCIENCE

MARINE POLLUTION BULLETIN

NUCLEAR AND CHEMICAL WASTE MANAGEMENT

WATER RESEARCH

WATER SCIENCE AND TECHNOLOGY

WATER SUPPLY

HEALTH ASPECTS OF THE DISPOSAL OF WASTE CHEMICALS

Edited by
Joe W. Grisham

A Report of the
Executive Scientific Panel

Joe W. Grisham, M.D.
University of North Carolina
Chairman

Martin Alexander, Ph.D.
Cornell University

Robert E. Anderson, M.D.
University of New Mexico

Arthur D. Bloom, M.D.
Columbia University

Morton Corn, Ph.D.
Johns Hopkins University

John E. Davies, M.D., M.P.H.
University of Miami

John Doull, M.D., Ph.D.
University of Kansas

Virgil H. Freed, Ph.D.
Oregon State University

Eugene V. Perrin, M.D.
Wayne State University

Henry C. Pitot, M.D., Ph.D.
University of Wisconsin

Gabriel L. Plaa, Ph.D.
University of Montreal

Paul V. Roberts, Ph.D.
Stanford University

Reuel A. Stallones, M.D., M.P.H.
University of Texas

PERGAMON PRESS
New York Oxford Toronto Sydney Frankfurt

Pergamon Press Offices:

U.S.A.	Pergamon Press Inc., Maxwell House, Fairview Park, Elmsford, New York 10523, U.S.A.
U.K.	Pergamon Press Ltd., Headington Hill Hall, Oxford OX3 0BW, England
CANADA	Pergamon Press Canada Ltd., Suite 104, 150 Consumers Road, Willowdale, Ontario M2J 1P9, Canada
AUSTRALIA	Pergamon Press (Aust.) Pty. Ltd., P.O. Box 544, Potts Point, NSW 2011, Australia
FEDERAL REPUBLIC OF GERMANY	Pergamon Press GmbH, Hammerweg 6, D-6242 Kronberg-Taunus, Federal Republic of Germany
BRAZIL	Pergamon Editora Ltda., Rua Eça de Queiros, 346, CEP 04011, São Paulo, Brazil
JAPAN	Pergamon Press Ltd., 8th Floor, Matsuoka Central Building, 1-7-1 Nishishinjuku, Shinjuku, Tokyo 160, Japan
PEOPLE'S REPUBLIC OF CHINA	Pergamon Press, Qianmen Hotel, Beijing, People's Republic of China

---

First printing 1986

Library of Congress Cataloging in Publication Data

Health aspects of the disposal of waste chemicals.

"Prepared under a private contract as the final report of the Executive Scientific Panel (ESP) on Health Aspects of the Disposal of Waste Chemical Chemicals"--Pref.
Includes index.
1. Chemicals--Waste disposal--Hygienic aspects. 2. Chemicals--Waste disposal--Environmental aspects. I. Grisham, Joe W. II. Universities Associated for Research and Education in Pathology. Executive Scientific Panel on Health Aspects of the Disposal of Waste Chemicals. [DNLM: 1. Environmental Exposure. 2. Industrial Waste--adverse effects. WA 788 H434]
RA567.H43 1986 363.7′28 86-1429
ISBN 0-08-033159-9

In order to make this volume available as economically and as rapidly as possible, the authors' typescripts have been reproduced in their original forms. This method unfortunately has its typographical limitations but it is hoped that they in no way distract the reader.

Printed in Great Britain by A. Wheaton & Co. Ltd., Exeter

PREFACE

This manuscript was prepared under a private contract as the Final Report of the Executive Scientific Panel (ESP) on Health Aspects of the Disposal of Waste Chemicals. The Panel, organized by the Board of Directors of Universities Associated for Research and Education in Pathology (UAREP), reviewed the scientific literature on the biological effects of chemicals most frequently associated with uncontrolled disposal and assessed the scientific evidence of association between demonstrated exposure to chemicals from disposal sites and occurrence of human diseases and disorders.

Reports on exposure aspects, health affects and epidemiology were prepared for the ESP by task groups of independent consultants. These reports were integrated into a Preliminary Draft which was peer-reviewed prior to distribution of a Draft Report. After two months of public review the ESP held an open meeting to receive additional data and comments before preparation and delivery of the Final Report to the project sponsors.

This book represents the review by over 40 academic scientists of the salient aspects of human health risk from uncontrolled chemical disposal. It is intended to provide a broad base for interdisciplinary understanding and communication among the diverse professions interested in this problem.

It reviews the current state of knowledge of human health effects that have been attributed to waste chemicals from uncontrolled disposal sites and provides a critical scientific assessment of the investigations of such effects associated with 21 specific sites. The strengths and weaknesses of the approaches to understanding positive associations and establishing causal relationships are analyzed. A synoptic presentation of the current "state-of-the-art" and discussion of practical approaches to assessing risks to human health are provided.

This document is intended to be a source of information on the nature and significance of health effects related to chemical disposal. It is hoped that it will also provide guidance to the general public, various local, state, and federal governmental offices, and to the scientific and industrial communities, on responsible, efficient approaches to resolution of health issues associated with uncontrolled chemical disposal sites.

Joe W. Grisham

ACKNOWLEDGMENTS

The Chairman of the Executive Scientific Panel expresses his sincere appreciation to the members of the Panel, to the workshop participants and to the Task Groups for their skill and dedication in drafting major portions of the report. The Executive Scientific Panel is indebted to its consultants whose peer-review of the Preliminary Draft was most helpful in the development of the Draft Report, to the representatives of the World Health Organization and the United Nations Environment Programme for their assistance in obtaining information on the international aspects of waste chemical health effects, and to the staffs of UAREP and the Life Sciences Research Office of the Federation of American Societies for Experimental Biology for their expert and professional execution of this project. We are especially grateful to Dr. Kenneth Endicott, Executive Officer of UAREP, and to Dr. Kenneth Fisher, Director of LSRO for their leadership as co-Directors and to Mr. Thomas Hill, Associate Project Director, for coordinating the day-to-day operations and for preparing this manuscript for publication. The camera ready copy for this book is the result of the diligent and skillful efforts of Ms. Mable Lam and Ms. Sherry Kloak.

Universities Associated for Research and Education in Pathology gratefully acknowledges the sponsorship of this project by the American Industrial Health Council, Chemical Manufacturers Association, National Agricultural Chemicals Association, National Association of Printing Ink Manufacturers, National Paint and Coatings Association, and the United States Chamber of Commerce, and the key role of Arthur Young and Company and its representative, Mr. Thomas R. Janson, in the management of the contractural relationship between the Sponsors and UAREP.

TABLE OF CONTENTS

1

Introduction and Background

HISTORICAL PERSPECTIVE

Man creates wastes in many ways and the disposal of these products of human activity is an age-old problem. Animal and human waste may contain potentially harmful pathogens and parasites necessitating efforts for their control in both the agricultural and the public health sectors. Combustion of fuels produces potentially toxic gases, aerosols, and solid particles which are discharged into the air. Domestic households discard garbage, solid wastes, and toxic chemicals. Mining, manufacturing, transportation, and service industries all generate by-products and wastes that may be hazardous to human health. Modern agriculture relies on extensive use of pesticides and fertilizers which can contaminate food or water supplies.

In this country, as in most areas of the world, man has traditionally disposed of these wastes in the most expedient manner possible. Throughout North America, and indeed throughout much of the world, liquid wastes and sewage, both treated and untreated, have been discharged into streams, rivers, and oceans, or injected into underground systems; combustibles have been burned and solid wastes have been treated on-site or deposited in the ocean or in landfills, abandoned quarries, mines, or other sites of marginal commerical value.

Following the initial population influx of the eighteenth century, the rate of expansion and industrialization of the United States accelerated during the nineteenth century. Existence of a large labor supply, discovery of abundant natural resources, application of technology to manufacturing, development of an extensive transportation system, creation of markets and capital, as well as favorable government policies, all contributed to the industrialization of the United States. Urbanization was both a factor in, and a result of, these economic changes.

During the past two centuries industrialization of the United States has produced not only the benefits of scientific and technological advances as measured by a general increase in the quality of life, but also an increasing awareness of the hazardous consequences of scientific and technological advances. During the nineteenth and early twentieth centuries and increasingly in recent years, local, state, and federal governments have been called on by an aroused citizenry to provide for sanitation, safe drinking water, worker safety, and a reduction of the morbidity and mortality attributed to the pressures of urbanization and industrialization.

As noted by Corn and Corn (1981), from the early 1800's until the 1930's, local situations resulted in reactions and responses at the local level. Emergency situations precipitated immediate actions, and a crisis mentality has been more evident than not. During this century, citizens as well as local and federal governments have consistently responded to environmental issues with technological remedies, but they have been slow to address social attitudes promoting the generation of waste materials. In a sense, the cliche, "out of sight, out of mind," characterizes this country's longstanding approach to resolving the waste disposal problems of environmental and public health significance.

RECENT DEVELOPMENTS

With World War II and the post-war emergence of the chemical industry in the United States, profound changes occurred to the products of industry. Chemicals such as components of pesticides, drugs, and plastics, as well as products such as synthetic fabrics and packaging materials, began to appear and have since become essential components of a high standard of living which has come to be known as the "American way of life." Advances in analytical chemistry have enabled scientists to detect increasingly smaller and smaller quantities of individual substances. Improved methods of medical science provide new information on interactions of ingested or absorbed chemicals. Chemical engineering has developed techniques for production and manipulation of the exact chemical nature of many substances. New fields of scientific and technical endeavor such as occupational health and safety, environmental engineering, and integrated pest management have emerged.

But just as industrialization and urbanization before them, synthetic chemicals and those derived from nature brought not only benefits and opportunities, but also hazards and risks. As use of chemicals increased, these substances, by-products, and wastes from their production became more widespread. Twenty years ago, public concern was drawn to possible problems of environmental pollution from pesticides. The 1962 book, Silent Spring, by Rachel Carson was perhaps the most widely read of a plethora of articles, documents, and books about pesticides that served to crystallize public opinion concerning the potential problems of chemicals in the environment and food chains. Because use of pesticides is intentional, control of their use could be regulated, thus affording an approach to minimizing or avoiding their environmental and public health impacts.

During this period, other aspects of political and social processes were occurring. One important facet which emerged during the past century is the increasingly dominant role of the Federal Government in regulating interstate commerce. Originally, states regulated such issues as pesticides, foods, drugs, and waste disposal. But a hodge-podge of laws and regulations, benign neglect at various levels of state and local government, and enforcement policies which were inconsistent among the various states led to public support for federal control and regulation. In the specific case of pesticides, the initial response of the Federal Government was a series of studies, panels, and commissions. The 1969 Mrak report entitled "Report of the

Secretary's Commission on Pesticides and Their Relationship to Environmental Health" (USDHEW, 1969) provided a measure of governmental concurrence with the worst fears of those who saw pesticides as chemical threats. Subsequently, public concern was converted into legislative action with the creation of the U.S. Environmental Protection Agency (EPA) in 1970.

Since that time, a primary focus of the EPA has been to weigh the benefits and risks of pesticides in order to develop regulatory policies for this group of chemicals. Public awareness concerning specific pesticides both precipitated and resulted from regulatory actions or the lack thereof. The situation has been continually complicated by strident voices on all sides. Zealous supporters of the chemical industry have decried the public's concerns as unwarranted and have argued that dangers are vastly overstated, often with inadequate supporting data. Equally zealous environmentalists have demanded severe regulatory action in the absence of adequate scientific information about the actual presence or absence of environmental and health hazards.

CONTEMPORARY ISSUES AND DEVELOPMENTS

Public Perceptions and Concerns

Despite the economic uncertainties of the past decades, demand for, and use of products derived from natural and synthetic chemicals have continued to expand since mid-century. There are few aspects of contemporary western civilization that have been immune not only to increasing demands for consumer goods but also to continual substitution of synthetic materials for natural ones. Thus, in part, society's quest for materials and products drives the chemical industry. Nevertheless, public concern for potential adverse effects of chemicals on the environment and on human health is widespread.

The perception that industry wantonly produces pollution, that the environment is being destroyed irreversibly, and that government is hopelessly inept may be overstated, but it is widely held. Limited availability of definitive scholarly reports and repeated news media coverage of events and situations which is more sensational than thorough have influenced public understanding. The public has become highly suspicious of the chemical industry and has little confidence in government efforts to protect it from unknown and unseen chemical threats. The chemical industry, faced with mounting costs of research, development, and regulations, is concerned that its economic base is being eroded. Scientists, who are looked to by all parties for data, are unable to provide enough information that is clear, concise, and unequivocal. Concerned citizens, aware of the potential for environmental and human health hazards, are angered by the unresponsiveness of the system. Government is criticized on all sides for actions and reactions as well as challenged in court at every move. Thus, it is no wonder that the public has concerns that range from serious and well-founded to hysterical when faced with the possible health effects caused by chemical disposal sites. In fact, the very study of issues related to toxic chemicals, use of pesticides, waste disposal practices and problems, and potential hazards contributes to public concern and anxiety. As science and technology approach these issues with greater precision and analytic capabilities, increasingly minute quantities of individual chemicals can be detected. Many of these substances have been present in body tissues and the environment for years; however, our ability to detect them has increased markedly. While scientists can detect and quantify, their ability to understand, let alone predict, the consequences of the presence of trace levels of chemicals has not advanced as rapidly. Thus, new knowledge of the presence of substances brings new questions about effects that cannot as yet be answered. These unknowns bring possibilities of further impacts on human health and the environment, which in turn initiate more research, greater demands for more regulation, and heightened public concern.

With regard to those disciplines that focus on the understanding of possible health effects from chemicals in disposal sites, the situation is acute. The numbers of expert environmental and clinical scientists, engineers, social scientists, and physicians are not adequate to address the magnitude of the research needs. Resources for scientific studies are unavailable or scarce. Scientific studies that are underway are often conducted in a reactionary environment or with an urgency that hinders the critical scientific nature of the investigation. It has become increasingly difficult to separate the various facets of the "hazardous waste problem" because of the interwoven complex of facts, perceptions, consequences, and actions. Thus, the actions and opinions of informed citizens on local problems or national issues range from well-considered active indignation to apathetic inaction. With such a range of opinion and citizen action, it is exceedingly difficult, if not impossible, for health officials, let alone the public, to determine the consequences of voluntary, involuntary, or accidental exposures to hazardous wastes within the time constraints placed on them by the public, corporate, and governmental sectors. Even when scientific data become available, they are often incomplete and thereby become subjects for differing interpretations among expert scientists.

Thus, even the biological, environmental engineering, and social sciences themselves have apparently become recognizable factors in the failure of the U.S. society to assess and manage the "hazardous waste" issue.

Over the last several years, public and governmental attention has been directed to issues related to chemicals associated with water pollution, alteration of air quality, and finally, hazardous wastes and their disposal. Just as Rachel Carson and others brought public attention to pesticides a few years earlier, other authors, as well as leaders in the environmental movement and in the U.S. Congress, pressed for attention to air and water pollution and control of chemical wastes. Each aspect of environmental impact fell under the aegis of EPA as the federal regulator; but the agency, beset by pressures from those to be regulated, those who would regulate, a patchwork of legislative authority, and little historical continuity in regulatory history, was often unequal to the demands placed on its limited staff and budget. Public concern about pollution of air, water, and land grew. Instances of hazardous chemicals in the wrong place at the wrong time, such as the kepone incident in Virginia and the Love Canal situation in New York, together with often well-intentioned but factually incomplete media coverage, have contributed to greater public awareness of potential hazards and risks. The pervasive nature of the hazards, the possibility of uncontrolled risks, and the absence of adequate data and information have all contributed to the increasing public pressures for legislative protection and redress. By the mid-1970's, pressures for federal legislation were overwhelming, and the U.S. Congress responded with new legislation.

Major Legislative Acts

Since 1970, a number of federal statutes have been enacted to respond to a growing public awareness concerning the manufacture, use, potential hazards, and disposal of chemicals, as well as their introduction and control in the atmosphere, water, and soil. Currently, the two laws which primarily govern most aspects of hazardous wastes in the United States are the Resource Conservation and Recovery Act (RCRA) of 1976, as amended, and the Comprehensive Environmental Response, Compensation, and Liability Act (CERCLA) of 1980. The major statutes that are related to or have sections that touch on hazardous wastes and their disposal are summarized in Table 1-1. Each has been amended frequently since passage, primarily to conform with other statutes as they were enacted.

The legislative intent of RCRA was to provide for careful planning and development of management approaches to the future treatment, storage, and disposal of hazardous solid wastes. RCRA also provided federal supervision of municipal solid waste disposal practices as well as those of the private sector.

RCRA provides a comprehensive legislative mandate for a national program on solid waste management (USEPA, 1981). Solid wastes include municipal garbage, industrial wastes, sewage sludge, agricultural refuse, demolition wastes, and mining wastes. Solid wastes also include liquid and gaseous materials in containers.

RCRA includes provisions and programs for the management of non-hazardous wastes and regulation of hazardous wastes. The Act provides for federal regulation of the generation, transport, treatment, storage, and disposal of hazardous wastes. The Act passed by Congress and the regulations implemented by EPA define hazardous waste as solid waste, a solid waste not excluded from regulation, and either: a) a hazardous waste identified and listed by EPA; b) a mixture containing a listed substance; or c) a waste not on the list, but one which exhibits toxicity, persistence in the environment, potential for bioaccumulation, flammability, corrosiveness, reactivity, or other hazardous characteristics. It should be noted that the Act and regulations also specify the following as being excluded from the definition of hazardous wastes: municipal garbage, municipal resource recovery wastes, agricultural residues, certain products of oil and gas exploration and production, residues from mining, and certain chromium-containing wastes, as well as any solid or liquid material in either domestic sewage or irrigation return flows, and industrial point source effluents having discharge permits.

In 1984, RCRA was amended to require EPA to make determinations and promulgate regulations within specified periods of time on disposal of solid and liquid wastes. In addition, the statute was amended to include small-quantity hazardous waste facilities (100 to 1,000 kg/month). Other important provisions of the amendments address permits and inspection procedures, leaking underground storage tanks, deep-well injection of wastes, and exporting hazardous wastes.

It is important to keep in perspective that RCRA covers both non-hazardous and hazardous solid wastes. In regard to hazardous wastes, RCRA, since 1976, has been the primary legislative mandate undergirding federal and state efforts to develop effective plans and regulations related to waste management. In essence, RCRA is the federal statute designed to prevent any environmental and human impacts from hazardous wastes from generation to

TABLE 1–1. Major federal laws containing provisions to control toxic substances and wastes.

Title of Act (and Date Enacted)	Amendments	Further Amendments
Clean Air Act (July 14, 1955)	December 17, 1963	
	October 20, 1965	
	October 15, 1966	November 21, 1967
		December 5, 1969
	December 31, 1970	November 18, 1971
		April 9, 1973
		June 22, 1974
	August 7, 1977	November 16, 1977
		July 2, 1980
		July 17, 1981
		December 21, 1982
Federal Insecticide, Fungicide and Rodenticide Act (June 25, 1947)	August 7, 1959	
	May 12, 1964	
	October 15, 1970	
	December 30, 1970	
	October 21, 1972	
	July 1, 1975	
	October 10, 1975	
	November 28, 1975	
	March 27, 1978	
	September 30, 1978	
	October 17, 1979	
	December 17, 1980	
	December 2, 1983	
Safe Drinking Water Act (December 16, 1974)	November 16, 1977	September 9, 1979
Hazardous Materials Transportation Act (January 3, 1975)	July 19, 1975	
	October 11, 1976	September 30, 1978
		November 30, 1979
Toxic Substances Control Act (October 11, 1976)	December 29, 1981	
Resource Conservation and Recovery Act (October 21, 1976)		
(Solid Waste Disposal Act Amendments of 1980)*	October 15, 1980	December 21, 1982
(Hazardous and Solid Waste Amendments of 1984)	November 8, 1984	
Clean Water Act** (December 27, 1977)	December 16, 1979	
Comprehensive Environmental Response, Compensation, and Liability Act (December 11, 1980)		

*Modified and amended in 1980 to include other provisions and retitled the Solid Waste Disposal Act of 1980 and further amended in 1984. However, "RCRA" is the term most frequently used to refer to the law as amended.

**Previous water pollution legislation was enacted under the Federal Water Pollution Control Act of 1952 and subsequent amendments; major amendments were passed in 1961 and 1972.

disposal, that is "from cradle to grave." As a preventive measure, RCRA focuses upon current and future needs in solid waste management in this country.

However, provisions of the Act as they relate to other federal and state legislative efforts have raised several issues which to date remain unresolved, according to a recent report from the U.S. Congress, Office of Technology Assessment (1983). For example, some hazardous wastes are regulated by statutes other than RCRA, and the Act covers only those substances defined as "hazardous waste" and those locations defined as "regulated waste disposal sites." Further, some states have adopted broader definitions of hazardous waste because they have concluded that RCRA definitions are inadequate or too restrictive. These differences have added to problems associated with defining hazardous wastes and determining quantities of such materials. They have also led to confusion in both governmental agencies and the general public as to the potential for human health effects and the management practices necessary to deal with hazardous wastes (USC/OTA, 1983).

CERCLA, also known as "Superfund," was enacted in 1980 for a five-year period (CERCLA, 1980). The Act provides for remedial action on past problems by giving the Federal Government authority to act in situations where hazardous waste disposal sites are suspected to exist and to pose potential danger to the environment or the public health. CERCLA mandates cleanup of hazardous waste sites and remedial action in response to spills of hazardous substances, and specifies mechanisms to fund such activities. Congress is currently debating reauthorization and modification of CERCLA, and legislative action is expected during 1985.

CERCLA included a provision for the establishment of a Hazardous Substance Response Trust Fund, known as "Superfund," to finance remedial measures. Superfund is financed by a tax on industry (to provide 87.5% of the fund) and appropriations from general tax revenues (to provide the remaining 12.5% of Superfund). The law also requires that companies with financial resources must pay for clean-up of sites if the companies are shown to be directly responsible for the chemical release. In cases where direct responsibility cannot be proven, Superfund resources can be used to bear costs of remedial action. CERCLA established a far-reaching concept of liability that has led to considerable legal challenge by both affected companies in the chemical industry and by representatives of the environmental sector because of exclusionary provisions in the law.

Under CERCLA, EPA has developed a National Contingency Plan (NCP) which deals with flexible but firm approaches to needs for immediate removals, planned removals, and remedial actions in regard to hazardous waste disposal sites. The NCP includes coordinated efforts of federal, state, and local governments, as well as private industry, to address existing problems associated with hazardous substances that have been disposed, the disposal sites, and situations of accidental dispersal. The NCP was meant to ensure effective response and compliance with provisions of CERCLA and the Clean Water Act (USEPA, 1982). The NCP contains provisions for a Hazard Ranking System and guidelines for the types of remedial actions. Since 1982, EPA has attempted to identify and characterize disposal sites throughout the United States and has taken action against responsible parties at a number of disposal sites.

As with RCRA, some differences between CERCLA provisions and definitions and those of state legislation and regulations have contributed to uncertainties and confusion about hazardous substances. For example, CERCLA defines hazardous substances as those substances identified in certain sections of the Federal Water Pollution Control Act, the Clean Air Act, RCRA, the Solid Waste Disposal Act amendments, or the Toxic Substances Control Act. However, CERCLA specifically excludes petroleum and its products (unless listed by other acts), as well as natural and synthetic gases used as fuels. Thus, confusion as to what constitutes a "hazardous substance" has contributed to delays in Superfund actions, as well as problems among state and federal authorities in development and implementation of the National Contingency Plan and a lack of a decision as to "how clean is clean" in the remediation of a site. Indeed, the complexities of the federal and state statutes and regulations are such that even understanding what constitutes a "hazardous substance" within a given chemical disposal site is difficult for most people, except those few whose personal, industrial, governmental, or legal responsibilities or concerns require full-time attention.

In summary, RCRA has provided a legislative mandate to prevent and avoid future problems with hazardous wastes. CERCLA mandates authority and responsibility and remedial actions at abandoned and existing sites in response to current situations. Implementation of both laws has been slow because of the magnitude and complexity of the attendant issues. Since 1976, EPA has made progress in identifying both the numbers and locations of uncontrolled waste disposal sites throughout the United States. However, progress in obtaining data on environmental and human health effects from such sites has been limited (USC/OTA, 1983).

Over the past several decades, a number of federal statutes have addressed matters associated with the manufacture, use, and disposal of natural wastes and hazardous substances (Conservation Foundation, 1984). These portions of the U.S. Code have for the most part been amended to conform to, or be consistent with, the major legislative acts listed in Table 1-1.

Additional information on historical perspectives and evolution of current legislation from various points of view may be found in Brown (1980), the Conservation Foundation (1984), Corn and Corn (1981), Epstein et al. (1982), OTA (USC/OTA, 1983), Peirce and Vesilind (1981), and Pye et al. (1983).

Legal, Political, and Economic Ramifications

Since passage and implementation of RCRA and CERCLA, EPA and other federal agencies, such as the Department of Health and Human Services, as well as many responsible state agencies have instituted a number of studies, plans, and proposals for risk assessment and risk management at disposal sites. Investigations carried out by various agencies, public and private, have identified several thousand chemical disposal sites, and the general assumption is that many more exist. The EPA has identified several hundred "priority" sites which require early attention and, under authority assigned by CERCLA, has initiated remedial action at some of these sites.

In populations living near chemical disposal sites a number of interacting variables influence the degree of public concern and private, corporate, and governmental responses. Certainly, media attention is an important contribution to both concerns and responses, but the impact of mass media has not been well studied (Anonymous, 1983). Typical reports from affected citizens include reference to odors, unsightliness, threats to property values, and, more importantly, fear of adverse health effects. The health effects most feared and most likely to be attributed to a site include cancer, reproductive disorders, congenital defects, and genetic effects. Once the population becomes concerned, there is a tendency by some people to attribute almost any sign or symptom of ill health to living near the site. When this stage is reached, public demand for remedial action reaches a high level and the population becomes skeptical about the validity of scientific efforts to demonstrate a putative cause and effect relationship.

Under these circumstances, public pressure for action creates a host of legal, political, and economic ramifications. Lawsuits are entered at various levels of courts in the judicial system challenging interpretations of regulations, seeking redress and compensation, and appealing the decisions of lower courts. Public officials, caught among varying segments of their constituencies, are faced with political pressures in a series of issues that lack historical or regulatory precedent. Political issues also arise with regard to policy, jurisdiction, and due process. Economic aspects of fault, compensation, resource allocation, and community impact become acute. Costs of remediation of the site may exceed costs of investigating the perceived adverse health effects associated with the site.

One of the greatest problems which government officials, industry, and concerned citizens must face is the availability and reliability of information. Only fragmentary knowledge exists as to the chemicals disposed of at most sites. It is seldom possible to determine the extent to which surrounding populations have been exposed. Identification of health effects, especially those which are chronic, presents many methodologic problems. Lack of information on these concerns has an impact on risk assessment first, but makes risk management all the more tenuous and difficult. Absence of adequate data has an impact on legal, political, and economic effects throughout risk assessment and management.

Basis of Concern for Health Effects

The growth of public concerns related to toxic chemicals in the environment leads to the widespread fear of hidden exposures in which people are not aware of their contact with substances that could possibly lead to irreversible disease processes. While heart disease is the leading cause of mortality in the United States, cancer is the most dreaded. There may be a public perception that if some pesticides are carcinogenic, mutagenic, and/or teratogenic, there must be similar effects for all other organic chemicals.

Science has yet to attain a clear and complete understanding of how cancers are caused, what agents might cause their initiation, what influences the development of various cancers, and how to prevent biological events which lead to development of cancers. Theoretical and practical aspects of carcinogenic risk assessment are formidable challenges to the interpretation of results from contemporary toxicology (NRC, 1983). These uncertainties of science strengthen the public perception noted previously that adverse health effects are the only predictable consequences of exposure to waste substances at disposal sites.

CURRENT SITUATION REGARDING HAZARDOUS WASTES

This section presents a general overview of chemical disposal sites in order to give the reader a historical perspective of past practices, current procedures, and future control methods. Historically, waste material was usually disposed of in the most expeditious way possible, generally by disposal on-site at the industrial facility where the waste was generated or off-site following transportation by truck or rail for relatively short distances. Site selection would be based on various criteria, such as cost of the site, its ownership, ease of disposal, size of the site, and proximity to the source of the waste. Until the passage and implementation of RCRA, long-term environmental and human health impacts were not always considered as site selection criteria. Moreover, most industrial facilities did not analyze the chemical constituents in their wastes unless the wastes were suspected of containing valuable unreacted raw materials, synthetic organic chemicals, solvents, valuable by-products, or trace metals suitable for recycling (Fields and Lindsey, 1975). Thus, there is no historical documentation of the substances disposed of and present in most chemical disposal sites.

Types of Containment

Some waste materials from industrial production and processing may have been suitable for recycling or reuse, but historically most wastes were considered unusable for all practical purposes. One of the least expensive methods of disposition of these wastes in the past had been by disposal on land or by shallow subsurface burial. The physical form of the waste material (i.e., liquids in containers, semi-liquid sludges, bulk solids, etc.) was a key factor in determining the method of land disposal that was used.

The major categories of chemical disposal sites used in the past have been open dumps (waste piles), disposal of containers underground in landfills, storage in bulk containers, and surface impoundments (Pojasek, 1979; Pye et al., 1983). One of the primary differences between a landfill and an open dump is that in a dump, the wastes are exposed for long periods to the weather (precipitation, wind erosion, heating by exposure to direct sunlight); no provisions are made to prevent or to delay the migration of chemicals to soil, water, or air. A landfill is covered with soil, while a secure landfill, as defined and mandated under RCRA regulations, is located appropriately with respect to geology and hydrology, has a protective bottom liner, and is sealed on top with soil or clay on a daily basis. For purposes of this report, the definition of chemical disposal sites covers the full range from open dumps to secure landfills to surface impoundments. Even in a "secure landfill," the underlying liner and the covering materials will not permanently prevent water-soluble compounds from leaching and volatile compounds from evaporating.

Storage in containers is used to segregate wastes which are reactive, corrosive, and/or ignitable (Pojasek, 1979). Such containers include bulk tanks holding thousands of gallons, 55- and 30-gallon drums, and 5- and 1-gallon cans and bottles. Storage in these bulk tanks, drums, bottles, and cans probably will not result in the permanent containment of waste because of the caustic or acidic characteristics of the wastes inside and/or the effects of weathering outside, such as rainwater rusting a steel barrel, as well as the crushing of containers during site compaction.

Disposal pits, ponds, and lagoons are collectively called surface impoundments and receive both liquid wastes and sludges. These waste materials may be further processed to detoxify, neutralize, homogenize, or concentrate the waste with subsequent re-treatment of the liquid at a wastewater treatment facility; the effluent from which is discharged to a receiving watercourse (Suess and Huismans, 1983). Historically, a number of waste lagoons have overflowed into surface waters. Until recently, many lagoons were constructed without a liner, allowing the waste material to leach into the ground and saturate the soil beneath the site.

Nature of Disposal Sites

Some landfill and impoundment sites were located on naturally clayey soil which would retard leakage; however, many sites were located on sandy/-gravelly terrain in abandoned gravel pits or quarries, in areas bordering on wetlands, or even below the seasonal water table. Most sites had no liner installed; however, a few sites were constructed with relatively impervious plastic and/or clay liners, although in some cases the liners were too thin. Liners, including low permeability clays or "chemical resistant" plastics, may eventually break down. Some sites have been abandoned without their owners leaving records and, in some cases, the waste generators contracted with waste-haulers who may have utilized indiscriminate disposal practices at unsuitable sites.

Fred C. Hart Associates, Inc. (1982) studied 929 waste disposal sites throughout the United States by reviewing state and federal files on environmental releases and human health problems, as well as by assessing chemical analysis data for samples collected near each site. The average length of operation for the sites was 7.6 years, with two-thirds having been in operation between 1 and 38 years. Of the 929 sites, 56% were active facilities, 42% were inactive, and the status of the remaining 2% could not be determined based on the information available in the files. The report stated that the types of waste disposal facilities, operating conditions, and environmental settings varied greatly, but that a typical site profile emerged for the 929 sites. Summary data on individual site characteristics were not provided, but the report indicated that the typical site was undesigned, with little information on file to suggest whether adequate operating and maintenance procedures had been routinely employed. Most of the sites were not located on natural clay. They were not constructed with plastic or clay liners to retard leakage nor with leachate or runoff collection systems and/or containment facilities. The inactive sites invariably had not received adequate closure such as capping with clay or dirt, fences around the site, etc. In addition, most of the sites evaluated were located in poor hydrogeologic/environmental settings. For example, in the majority of cases, the facility was located in moderately to highly permeable soils within 10 feet of groundwater and within 100 feet of a surface water body. In many cases, the facility was also located within one-half mile of shallow drinking water wells.

Quantities of Wastes Generated

There are few historical data on the quantities of wastes generated or disposed of in the United States. Recent estimates of the volumes of hazardous wastes relate only to recent practices but are the only data available. Thus, it is difficult to obtain a historical perspective on quantities of waste generated over the last several decades. Nevertheless, recent estimates of the volume of RCRA-regulated hazardous waste produced annually in the United States have increased. For example, the most recent data of Dietz et al. (1984), derived from a comprehensive survey questionnaire, are considerably higher than previous estimates of the volume of hazardous wastes generated.

RCRA-regulated hazardous wastes have been defined by EPA as those which are ignitable, corrosive, reactive, or toxic (USEPA, 1980), and have been estimated to account for 10 to 15% of the total of all industrial wastes generated (Maugh, 1979). The EPA estimated in 1979 that the generation rate of hazardous wastes was then about 35 million tons/year (Maugh, 1979; Pojasek, 1979). The EPA has reported a figure of 41 million wet metric tonnes of hazardous wastes generated in 1980 and until recently had reported that 43 million wet metric tonnes were generated in 1981 (USC/OTA, 1983). (One million metric tonnes equal approximately 1.1 million English [short] tons.) It is acknowledged, however, that these figures do not include the volumes of waste materials not regulated by RCRA (USC/OTA, 1983). RCRA-exempted waste includes mining waste, wastes from energy production, wastes from drilling for petroleum or natural gas, agricultural wastes, and "household wastes" (USEPA, 1980).

OTA (USC/OTA, 1983) summarized EPA estimates of the volume of hazardous waste by state and compared these numbers with results of a survey conducted in 1982 by the Association of State and Territorial Solid Waste Management Officials (ASTSWMO). The total volume for all states estimated by the EPA was 41.2 million metric tonnes; the ASTSWMO estimate was 250 million metric tonnes. The EPA data were the result of survey questionnaires sent to waste generators, transporters, and the owners of treatment, storage, and disposal (TSD) facilities. The data from the ASTSWMO survey were the result of written and telephone responses to a survey questionnaire sent to each state. Seven states and 2 U.S. territories did not respond to the ASTSWMO survey and were estimated to add another 5 million to 25 million metric tonnes annually (for a total "likely range" of 255 million to 275 million metric tonnes). ASTSWMO had requested that the states report data based on their own regulatory definitions of hazardous wastes; as a result, five states (California, Georgia, Indiana, Louisiana, and Texas) reported about 85% of the 250 million metric tonnes reported. Reasons vary for these large volumes: Texas and Louisiana included fly and bottom ash as hazardous waste; of Indiana's estimated 94.9 million metric tonnes, over 97% is accounted for by spent pickle liquor generated by the steel industry (not similarly included by any other state); of Georgia's estimated 38.5 million metric tonnes, over 99.7% represents dilute aqueous solutions neutralized on-site and discharged to sewers and/or receiving waters. Some of the states utilized an inventory system that extrapolated their off-site waste volumes to obtain estimates of on-site storage, while other states utilized only off-site transport manifest data. These estimates do not take into account any materials generated by CERCLA activities (i.e., materials produced by cleanup and remedial action at hazardous waste sites), and it is expected that several million metric tonnes of these "cleanup wastes" will be produced annually during the coming decade (USC/OTA, 1983). Thus, these previous estimates of the volumes of hazardous wastes generated were subject to considerable uncertainty because of the many limitations in the survey methods used.

Dietz et al. (1984) recently published the results of a national survey of both the hazardous waste generators and the treatment, storage, and disposal (TSD) facilities. The questionnaire study was conducted under the sponsorship and direction of the U.S. EPA Office of Solid Waste and included written responses to questionnaires and direct telephone contacts to clarify unclear responses. The survey's most important finding is that an estimated 264 million metric tonnes (71 billion gallons) of RCRA-regulated hazardous wastes were generated during 1981, more than six times the previous EPA estimates of the annual hazardous waste generation. The preliminary draft of this report (Westat, 1983) had estimated the 1981 hazardous waste generation volume to be 150 million metric tonnes, but Dietz et al. (1984) reported that the clarification was due to difficulties in sampling waste generators and that a few major industrial facilities generated and managed a major portion of the amounts of waste.

The estimated values reported by Dietz et al. (1984) were derived from responses from waste generators as well as from operators of the various types of waste management facilities. Information reported in these different questionnaires was analyzed by a statistical weighting and extrapolation procedure to obtain estimated values for the United States. Based on calculations of the sampling error, the 95% confidence estimate of the true quantity of RCRA-regulated hazardous waste generated in the United States in 1981 was reported to be between 35.7 and 106.3 billion gallons (132 to 395 million metric tonnes). Based on the volume of waste reported by respondents to the Waste Generator questionnaire plus the generation volumes reported by the respondents to the TSD Facility questionnaire, Dietz et al. (1984) reported that the recalculated "lower bound" of hazardous waste generation was 44 billion gallons (164 million metric tonnes); this lower bound volume is considerably greater than any previous EPA estimate of the total waste volume generated.

Large portions of the quantities reported are mixtures of hazardous and nonhazardous wastes (e.g., hazardous wastes mixed with industrial process wastes, or small quantities of hazardous wastes dissolved in process water). The survey data did not estimate the actual amounts of such mixtures, nor the concentrations of specific constituents in the mixtures (Dietz et al., 1984). All of these reported mixtures are considered in their entire volume to be hazardous wastes, as regulated under RCRA; moreover, the 264 million metric tonne estimate includes only those wastes which were generated during 1981 and which were managed in treatment, storage, and disposal processes regulated under RCRA.

Dietz et al. (1984) also indicated that the largest quantities of hazardous waste were apparently produced by only a small portion of generators; their statistical procedure indicated that only 1% of the generators (i.e., the largest generators of hazardous waste) accounted for nearly 90% of the total volume of RCRA-regulated hazardous waste generated. The EPA made independent telephone calls to waste generators and TSD management facility operators; in-depth discussions verified that the waste volumes which were reported and utilized in the estimation procedures had been accurately identified as RCRA-regulated hazardous wastes.

Dietz et al. (1984) estimated that there were 14,098 generators of RCRA-regulated hazardous wastes in 1981. Of these waste generators, 8,704 (62%) managed all their wastes by off-site disposal; 3,704 (22%) disposed of wastes both on-site and off-site; and only 2,320 (16%) of the waste generators managed all their wastes by on-site treatment, storage, and disposal. However, of the waste volume generated, 96% was managed on-site and only 4% was shipped off-site. Dietz et al. (1984) attributed this phenomenon to the fact that the larger generators tended to manage their hazardous wastes on-site, while the more numerous smaller generators, for various reasons, shipped their wastes to commercial facilities for treatment, storage, and/or disposal.

The above discussion contains extensive estimates of RCRA-regulated waste generation concerning the volumes of hazardous waste generated in 1981, the patterns of treatment, storage, and disposal, on-site vs. off-site disposal, and the volume generated relative to the size of the waste generator. It should be noted that the 1984 amendments to RCRA include waste facilities generating between 100 and 1,000 kg/month. Thus, future estimates of RCRA-regulated waste volumes will be considerably larger.

The Dietz et al. (1984) survey provides an overview of conditions in 1981; however, it is a single set of estimates. There is no way to utilize these estimates either to project future trends of RCRA-regulated waste generation and management or to estimate the disposal practices in the past which are the purview of CERCLA. Indeed, there are apparently no good estimates of the total volume of chemical wastes which has been generated and disposed historically in the United States. The Conservation Foundation (1984) has reviewed in even greater detail the studies described above concerning RCRA-regulated waste volumes, and the authors did not identify any additional data useful in estimating the total volume of waste materials present in CERCLA-regulated waste sites.

Number of Disposal Sites

The number of RCRA-regulated facilities in use for the management of hazardous waste is also the subject of continued data resolution and refinement. In 1982, Westat Incorporated and the EPA estimated that there were 7,785 facilities in use in 1981 (USC/OTA, 1983), while Dietz et al. (1984) refined the estimate of RCRA sites in use in 1981 downward to 4,818 facilities (Table 1-2). Of the 4,818 TSD facilities listed by type in Table 1-2, 1,495 were for treatment, 4,299 were for storage (for more than 90 days), and 430 were for disposal (total number by category exceeds 4,818 due to more than one type of processing at some facilities).

While these data are useful, this UAREP study is intended to focus on the potential adverse health effects caused by releases from CERCLA-regulated sites rather than from RCRA-regulated TSD facilities. Thus, estimates of the number of uncontrolled and/or abandoned sites regulated under CERCLA are more critical to the scope of this study.

In an effort to characterize the range of the estimated number of sites regulated under CERCLA (Superfund), USC/OTA (1983) reported that two studies conducted in 1979 had provided "national estimates of the number of sites that contain hazardous wastes and that may require cleanup." One source was a telephone survey of state environmental protection officials conducted by the Chemical Manufacturers Association that indicated there were an estimated 4,200 sites, while the second was a report by Fred C. Hart Associates, Inc. (unpublished report; National Hazard Survey Assessment of National Cleanup Costs for Hazardous Waste Problems, December 1978) which estimated that there were 30,000 to 50,000 sites. The OTA (USC/OTA, 1983)

TABLE 1–2. Estimated number of facilities using selected treatment, storage, and disposal processes in 1981.

Process Type		Westat and USEPA Estimates of 1982 (USC/OTA, 1983)	Dietz et al. (1984)
Storage Container		5,652	3,577
Storage Tank		2,280	1,428
Surface Impoundment:		1,096	
	Treatment		410
	Storage		552
	Disposal		116
Treatment Tank		1,951	609
Incinerator		317	240
Landfill		270	199
Waste Pile		312	174
Injection Wells		114	87
Land Treatment		148	70
Other Treatment Processes			392
Other Storage Processes			139
Other Disposal Processes			7
	Total Number of Facilities	7,785*	4,818*

*Sum of process types exceeds reported numbers of facilities due to multiple processing at many facilities.

indicated that both studies had many limitations in the documentation and verification of their survey methods and that both studies were referenced more for their historical perspective than for the accuracy of their estimates.

In February 1983, EPA had approximately 15,000 uncontrolled sites as entries on the Emergency and Remedial Response Information System (ERRIS) (USC/OTA, 1983). In January 1984, the EPA Superfund Task Force reported 16,200 entries on ERRIS and a projection that the number would eventually reach 22,000 sites (Anonymous, 1984a). In December 1984, the ERRIS list contained 19,000 sites, and EPA estimated that the site universe would ultimately reach 22,000 to 25,000 (Anonymous, 1984b). As of March 1982, EPA had designated 115 sites for the interim National Priority List (NPL). In December 1982, EPA released the first complete NPL, containing 418 sites (USC/OTA, 1983). In September 1983, EPA proposed an NPL of 546 final and/or proposed Superfund sites (USEPA/OPA, 1983). In October 1984, EPA (USEPA, 1984) proposed an additional 244 NPL sites and deleted some previously proposed sites, bringing the total number of final and proposed NPL sites to 786 (Figure 1-1). The EPA Superfund Task Force estimated in January 1984 that the number of NPL sites would ultimately be between 1,400 and 2,200 sites (Anonymous, 1984a) and in December 1984, revised these numbers upwards to 1,500 to 2,500 sites, with a projected NPL level of 1,800 sites (Anonymous, 1984b).

As implementation of CERCLA has progressed, EPA has developed policies to promote private cleanup initiatives. Additionally, a non-profit corporation (Clean Sites, Inc.) has been formed to marshal private resources to clean up

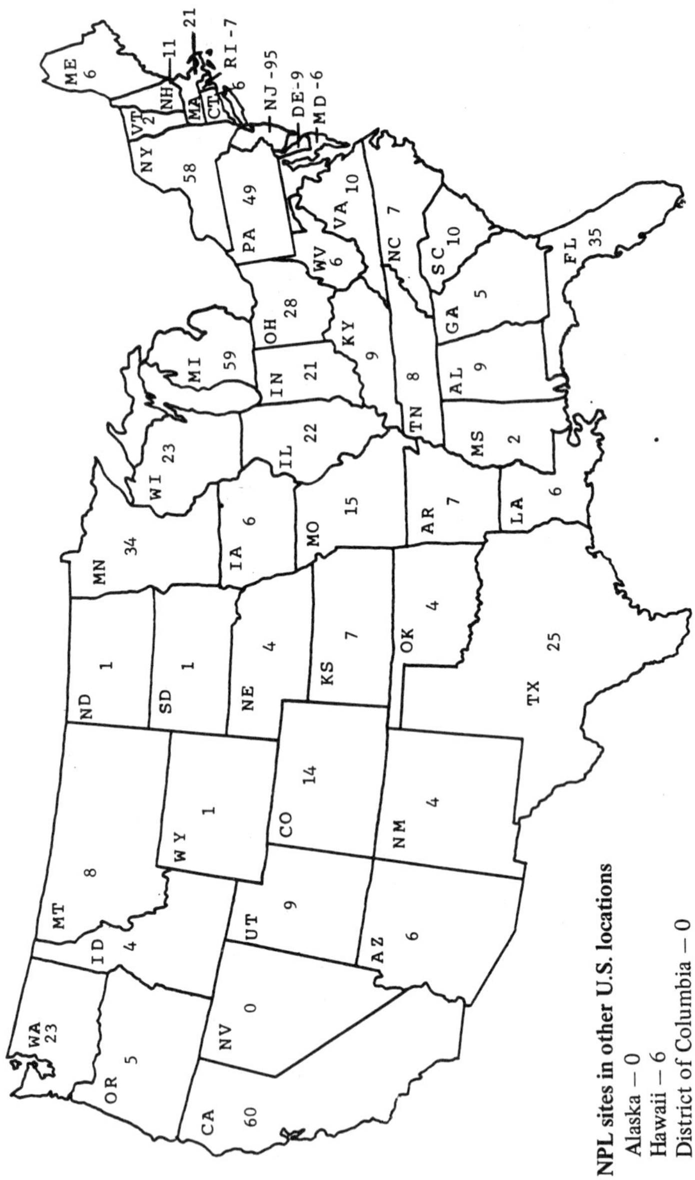

Figure 1–1. The final and proposed National Priority List of 786 (Superfund) sites, as of October 1984.

NPL sites in other U.S. locations

Alaska – 0
Hawaii – 6
District of Columbia – 0
American Samoa – 1
Guam – 1
Northern Mariana Islands – 1
Puerto Rico – 8
Trust Teritories – 1
Virgin Islands – 0

sites. As a result, the EPA Superfund Task Force (Anonymous, 1984b) has estimated that 50% of the total sites will be addressed by private parties, a much higher rate of private cleanup than had previously existed.

A recent report by ASTSWMO (1983) stated that the number of "uncontrolled hazardous waste disposal sites" in the United States in need of cleanup and remedial action is far higher than previous estimates. At that time, the Federal Superfund National Priority List contained 546 sites, which represented only 8% of the 7,113 sites that state solid waste officials reported were in need of cleanup and remedial action. In addition, ASTSWMO (1983) reported that there were at least 10,008 other hazardous waste disposal sites, each of which has the potential to affect public health and the environment. Because ASTSWMO did not receive responses from 7 states or from most of the U.S. Trust territories, it was reported to be nearly certain that the numbers of sites reported were an underestimate of the total number of sites that ASTSWMO indicated require some degree of remedial action.

It is not entirely clear whether the reports cited in the previous paragraphs took into account the waste disposal sites under direct control of federal agencies, such as the Department of Defense, the Department of the Interior, and others. Similarly, these reports often exclude previous and current municipal sanitary landfills and may omit consideration of abandoned private sites and facilities, all of which affect groundwater and possibly public health considerations.

PURPOSE AND SCOPE OF THIS REPORT

The brief introduction and background provided by this chapter only serve to reinforce what the reader already knows: human health effects associated with toxic substances from chemical disposal sites is an exceedingly complex and amorphous topic. What then is the purpose of a report such as this? What does its scope include and exclude? Why was the report prepared by an eminent panel of research scientists?

The Executive Scientific Panel (ESP) has attempted to develop a comprehensive overview of factors, approaches, information, data, observations, and conclusions related to human health effects that may be or have been associated with substances found in chemical disposal sites. The report reviews the uses and limitations of techniques for investigating factors that bear on the identification and establishment of human health effects as well as the available data on substances which produce adverse health effects.

While it is comprehensive, the report is by no means complete. For example, the ESP chose to focus on substances, disposal sites, and potential for adverse health effects generally associated with industrial chemicals and processes. Thus, consideration of municipal solid wastes, wastewater effluents and sewage sludge, mining wastes, transportation accidents, or other incidents involving hazardous materials has been quite limited. Similarly, disposal of radioactive wastes has not been included within the study scope. The ESP acknowledges that there may well be environmental impacts and ecological consequences resulting from the release of chemicals from disposal sites; however, a detailed discussion of these is beyond the scope of this report.

Efforts of many federal, state, and local government agencies, scientific researchers, and private citizens over the past decade have shown that disposal sites vary widely in terms of their geological, hydrological, physical, chemical, and ecological natures. Notwithstanding this general observation on variability, the ESP has become increasingly aware that relatively few chemical disposal sites have been studied extensively over time or with thorough scientific rigor. Indeed, there is no concise and universally accepted definition of what constitutes a chemical disposal site.

The passage and implementation of federal legislation, the intensification of state and local government activities, and the public and private sector concerns with, and actions on, hazardous waste disposal sites have been and continue to be in a dynamic state of change. This dynamic state comprises an evolving mixture of regulatory, remedial, judicial, and scientific approaches. As society grapples with resolution of the complex national issues of the hazardous waste situation and communities require action on local situations, the public and private sectors must reach decisions on courses of action and approaches. Such decisions require more scientific data and conclusions than the scientific community can currently provide; yet, absence or paucity of scientific information by itself is insufficient reason not to address and act on steps toward cleaning up chemical disposal sites and studying their possible impact on human health. In a preliminary study, UAREP (1983) drew several conclusions that oriented the ESP in its selection of topics for this report. These conclusions include the observation that there is large variability in site characteristics. Relatively few chemical disposal sites have been investigated extensively and, for sites where epidemiological studies of exposed populations have been undertaken, results are often equivocal. It became clear that many sites would need to be studied more thoroughly before generalizations could be made with scientific precision. To do this would take years and cost large sums of money. The ESP recognizes that the presently available manpower, financial resources, and facilities of the scientific community are inadequate in terms of society's needs for conclusions.

To focus current thinking on what science does know about human health effects associated with hazardous substances and chemical disposal sites, and to provide a reasoned basis for planning future actions, this report presents a survey of existing knowledge on substances in disposal sites, their human health effects, advantages and limitations of study methods, and factors which have an impact on approaches to human risk assessment. Further, the report identifies potential adverse health effects of those chemicals which have been reported to be present in disposal sites by reviewing available data on health effects of those chemicals derived from studies in industrial situations, epidemiologic studies, and laboratory investigations. Thus, this report addresses the scientific aspects of assessing possible risks of substances in or released from chemical disposal sites. The report does not attempt to deal with risk management of those risks found to be present. For a more complete discussion of the contemporary thinking on the processes involved in and the concepts separating risk assessment and risk management, the reader should consult the National Research Council report on risk evaluation (NRC, 1983).

The report concludes with a synoptic overview and assessment of the current situation in the United States and other countries. Recommendations related to future approaches to scientific studies on chemical disposal sites and their effects on human health are also included. Appendix A contains synopses of information available on eleven selected chemicals, including data on physical and chemical properties, uses, environmental fate, health effects, and current exposure standards. Subsequent appendices identify individuals and organizations who have contributed to this effort.

SUMMARY

Waste is the unusable refuse of human activities. For centuries, disposal of waste has been characterized by removal from the immediate environs where it was generated and deposition in the most expedient and economical alternative environment. Until the industrial revolution, the majority of waste was readily biodegradable; but the advent of mass production of metal and metal alloy products and later development of organic chemical synthesis and synthetic materials have drastically increased the persistence of many waste products. The aesthetic and safety aspects of concentrated accumulation of chemical wastes in particular, have caused public concern of such magnitude that stringent government action was mandated. The result has been a series of environmental protection statutes which have resulted in significiant improvement to the environment. The necessity for continuation of these efforts is apparent in the estimates of the number of uncontrolled and abandoned disposal sites remaining to be eliminated. In the interim, it is important to establish consideration of the public health as the principal criterion for remediation. The purposes of this report are to develop comprehensive overview of factors, approaches, information, data, observations, and conclusions related to human health effects that may be or have been associated with substances found in chemical disposal sites and to review both the available data on substances which produce adverse health effects and the uses and limitations of techniques for investigating factors that bear on identification and establishment of human health effects.

References

Anonymous. (1983) Perceptions of environmental contamination and their implications for health. Hum. Ecol. Health 2:1-2.

Anonymous. (1984a) EPA Superfund Task Force considers removing exemptions to increase tax base. Pestic. Toxic Chem. News 12(12):19-21.

Anonymous. (1984b) EPA's 301 studies project 11.7-billion total Superfund cost. Inside E.P.A. 5:10-13.

Association of State and Territorial Solid Waste Management Officials. (1983) State cleanup programs for hazardous substance sites and spills: a report to the U.S. Environmental Protection Agency. Washington, DC: The Association. 87p.

ASTSWMO SEE Association of State and Territorial Solid Waste Management Officials

Brown, M.H. (1980) Laying waste: the poisoning of America by toxic chemicals. New York: Pantheon Books. 351p.

CERCLA SEE Comprehensive Environmental Response, Compensation, and Liability Act of 1980

Comprehensive Environmental Response, Compensation, and Liability Act of 1980 (Superfund Act). (1980) Pub. L. 96-510, 94 Stat. 2767. 45p. Available from: House or Senate Document Room, U.S. Capitol, Washington, DC.

Conservation Foundation. (1984) State of the environment: an assessment at mid-decade. Washington, DC: The Foundation. 570p.

Corn, J.K.; Corn, M. (1981) The control of health problems related to industrialization. In: Wechsler, H.; Lamont-Havers, R.W.; Cahill, G.F., Jr.; eds. The social context of medical research. Cambridge, MA: Ballinger Publishing Company. p. 193-231.

Dietz, S.; Emmet, M.; DiGaetano, R.; Tuttle, D.; Vincent, C. (1984) National survey of hazardous waste generators and treatment, storage, and disposal facilities regulated under RCRA in 1981. EPA 530/SW-84-005. Prepared by Westat, Inc. Washington, DC: U.S. Environmental Protection Agency, Office of Solid Waste and Emergency Response. 318p. Available from: U.S. Government Printing Office, Washington, DC.

Epstein, S.S.; Brown, L.O.; Pope, C. (1982) Hazardous waste in America. San Francisco: Sierra Club Books. 593p.

Fields, T., Jr.; Lindsey, A.W. (1975) Landfill disposal of hazardous wastes: a review of literature and known approaches. EPA/530/SW-165. Washington, DC: U.S. Government Printing Office. 36p. Available from: Solid Waste Information, U.S. Environmental Protection Agency, Cincinnati, OH.

Fred C. Hart Associates, Inc. (1982) Assessment of hazardous waste mis management damage case histories. 1984 reprint. EPA/530-SW-84-002. Washington, DC: U.S. Environmental Protection Agency, Office of Solid Waste and Emergency Response. 444p. Available from: National Technical Information Service, Springfield, VA.

Maugh, T.H., II. (1979) Toxic waste disposal a growing problem. Science 204:819-823.

National Research Council. Committee on the Institutional Means for Assessment of Risks to Public Health. (1983) Risk assessment in the federal government: managing the process. Washington, DC: National Academy Press.

NRC SEE National Research Council

Peirce, J.J.; Vesilind, P.A.; editors. (1981) Hazardous waste management. Ann Arbor, MI: Ann Arbor Science Publishers, Inc./The Butterworth Group. [now Stoneham, MA: Butterworth Medical and Scientific Publishers]. 190p.

Pojasek, R.B. (1979) Disposing of hazardous chemical wastes. Environ. Sci. Technol. 13:810-814.

Pye, V.I.; Patrick, R.; Quarles, J. (1983) Groundwater contamination in the United States. Philadelphia: University of Pennsylvania Press. 315p.

Suess, M.J.; Huismans, J.W.; editors. (1983) Management of hazardous waste: policy guidelines and code of practice. WHO Regional Publications, European Series No. 14. Copenhagen: World Health Organization, Regional Office for Europe. 101p.

Superfund Act SEE Comprehensive Environmental Response, Compensation, and Liability Act of 1980

UAREP SEE Universities Associated for Research and Education in Pathology, Inc.

Universities Associated for Research and Education in Pathology, Inc. (1983) Report on the feasibility of organizing a study to assess the potential health hazards of waste disposal sites within the United States. 52p. Available from: Universities Associated for Research and Education in Pathology, Inc., Bethesda, MD.

U.S. Congress, Office of Technology Assessment. (1983) Technologies and management strategies for hazardous waste control. Washington, DC: U.S. Congress, Office of Technology Assessment. 407p. Available from: U.S. Government Printing Office, Washington, DC.

USC/OTA SEE U.S. Congress, Office of Technology Assessment

U.S. Department of Health, Education, and Welfare. (1969) Report of the Secretary's Commission on Pesticides and Their Relationship to Environmental Health. Washington, DC: U.S. Government Printing Office. 677p.

USDHEW SEE U.S. Department of Health, Education, and Welfare

U.S. Environmental Protection Agency. (1980) Hazardous waste management system: part 261, identification and listing of hazardous waste. Fed. Regist. 45:33119-33133.

U.S. Environmental Protection Agency. (1981) Resource Conservation and Recovery Act. Public Law 94-580, Oct. 21, 1976: Quiet Communities Act of 1978; Solid Waste Disposal Act Amendments of 1980; Used Oil Recycling Act of 1980; and Comprehensive Environmental Response, Compensation, and Liability Act of 1980. Revised compilation SW-171. [Washington, DC]: The Agency. 70p.

U.S. Environmental Protection Agency. (1982) National oil and hazardous substances contingency plan: final rule. Fed. Regist. 47:31180-31243.

U.S. Environmental Protection Agency. (1984) National Priorities List: 786 current and proposed sites in order of ranking and by state, October 1984. Dec. rev. ed. HW-7.2 Washington, DC: the Agency. 84 p.

U.S. Environmental Protection Agency, Office of Public Affairs. (1983) Environmental news [press release, Sept. 1]. EPA announces new priority Superfund sites. Washington, DC: The Agency.

USEPA SEE U.S. Environmental Protection Agency

USEPA/OPA SEE U.S. Environmental Protection Agency, Office of Public Affairs

Westat, Inc. (1983) National survey of hazardous waste generators and treatment, storage, and disposal facilities regulated under RCRA in 1981: preliminary highlights of findings. 37p. Available from: U.S. Environmental Protection Agency, Washington, DC.

2

Waste Chemicals of Major Concern

The purpose of this chapter is to identify a list of representative chemicals released from chemical disposal sites and to identify the sources of available information concerning potential adverse health effects resulting from exposure to these specific substances. The primary criterion for inclusion of a specific chemical on this report's list of representative chemicals is its reported presence at numerous chemical disposal sites. Two secondary criteria are of equal importance to each other, and are: 1) the potential for the chemical to be transported through the environment, and 2) the potential for the chemical to cause human health effects. A fourth criterion is the existence of a substantial degree of recognition and concern by the scientific community and/or the general public. Information summaries of the representative chemicals are tabulated in Appendix A (Chemical Synopses).

FACTORS IN THE SELECTION OF HAZARDOUS CHEMICALS FOR DETAILED STUDY

Many chemicals have been detected in the mixtures of waste materials which have been deposited at chemical disposal sites. A number of these chemicals have been designated as "hazardous wastes" by the U.S. Environmental Protection Agency (EPA) (USEPA, 1980) as required by the Resource Conservation and Recovery Act. Subsequent additions and deletions have been made to this list. A total of 375 chemicals are listed as Hazardous Constituents of solid waste mixtures in 40 CFR 261, Appendix VIII (USEPA, 1982b), based on their toxicity, ignitability, corrosivity, or reactivity. Subgroups of this listing are contained in 40 CRF 261.33 (USEPA, 1982a). One subgroup of the Hazardous Constituent list includes 107 chemicals designated as Acute Hazardous Wastes because they have been found to be fatal to humans in low doses or, in the absence of data on human toxicity, have been shown to have an oral LD50 (rat) of less than 50 mg/kg, an inhalation LC50 (rat) of less than 2 mg/L, or a dermal LD50 (rabbit) of less than 200 mg/kg, or are otherwise "capable of causing or significantly contributing to an increase in serious irreversible, or incapacitating reversible, illness." Another subgroup of this list includes 233 chemicals listed as Toxic Wastes, "shown in scientific studies to have toxic, carcinogenic, mutagenic, or teratogenic effects on humans or other life forms." In addition, EPA proposed in December 1984 to add 120 compounds to the hazardous constituent list, 28 chemicals to the list of acutely hazardous wastes, and 81 chemicals to the list of toxic hazardous wastes (USEPA, 1984). In January 1985, EPA issued a final rule to modify the list of acute hazardous wastes by adding particular chlorinated-dioxins, -dibenzofurans, and -phenols (USEPA, 1985); formerly only 2,3,7,8-tetrachlorodibenzo-p-dioxin had been listed as a hazardous constituent in 40 CFR 261, Appendix VIII (USEPA, 1982b), and its disposal had been regulated under the Toxic Substances Control Act. For the purposes of this report, these EPA lists are comprised of too many chemicals to allow a detailed review of information on each chemical.

Huismans and Suess (1982) described lists of hazardous compounds which result as wastes from certain industries, wastes containing specific components, or specific waste streams identified by the processes from which they originate. The Council of the European Communities (1978) presented such a list in its Directive on Toxic and Dangerous Waste, shown in Table 2-1. Some of the materials found in this list are too general (for example, "chlorinated solvents") and could potentially contain a large number of compounds. The list in Table 2-1, therefore, is not specific enough for the purposes of this report.

Mitre Corporation (1983) listed the 230 chemicals or groups of chemicals (Table 2-2) that were reported as being present in the environment in the vicinity of the then 546 National Priority List (NPL) sites which had been designated by EPA as mandated under CERCLA (the Superfund Act). One hundred and seventy-three chemicals or groups of chemicals were reported in nearby groundwater, 162 in surface water, and 65 in air. Note that the chemicals and groups of chemicals in Table 2-2 are those which have been reported. Noted below are several reasons why some chemicals appear at only a small number of sites or do not appear in Table 2-2 at all.

- Additional chemicals, quite possibly a large number, may be present at the site, but are not detectable or identifiable with currently utilized analysis techniques. For example, a known compound might go unreported if it is not available in the analyst's computerized gas chromatogram/mass spectrogram library.

- Other chemicals may be present for which specific analyses were not done, and therefore were not detected.

- Some compounds may be present at very low concentrations, may not be of sufficient concentration to be of concern to the analyst, or may not be within detectable limits.

- There may be compounds which are identified, but were not expected to be found at a chemical disposal site, and are not included on the list of identified chemicals reported. It is possible that such chemicals are not reported for several reasons: a) due to low confidence in the accuracy of identification, i.e., the assumption being that the result was in error; b)

TABLE 2–1. List of toxic or dangerous substances and materials (adapted from the Council of the European Communities, 1978).

Acids and/or basic substances used in the surface treatment and finishing of metals
Antimony and compounds
Arsenic and compounds
Asbestos
Beryllium and compounds
Biocides and phytopharmaceutical substances
Cadmium and compounds
Chemical laboratory materials, not identifiable and/or new, with unknown effects on the environment
Chlorinated solvents
Chromium (VI) compounds
Cyanide compounds
Ethers
Isocyanates
Lead and compounds
Mercury and compounds
Metal carbonyls
Organic solvents
Organohalogenated compounds, excluding inert polymeric materials and other substances referred to in this list or covered by other CEC Directives concerning the disposal of toxic or dangerous waste
Peroxides, chlorates, perchlorates and azides
Pharmaceutical compounds
Phenolic compounds
Polycyclic aromatic hydrocarbons (carcinogenic)
Selenium and compounds
Soluble copper compounds
Tarry materials from refining and tar residues from distilling
Tellurium and compounds
Thallium and compounds

> because the chemical in question does not appear on other lists of "hazardous" chemicals; or c) because the chemical is assumed to be innocuous. Moreover, the analytical task in many cases is further complicated by the presence of natural products and/or transformation by-products, often in concentrations (e.g., mg/L) greatly exceeding those of individual hazardous waste constituents which are frequently present in only ug/L concentrations.

In addition to the Mitre list of chemicals found near NPL sites throughout the United States, there are other lists that tabulate chemicals found within or in the vicinity of chemical disposal sites. The Conservation Foundation (1984) reported that EPA had evaluated about 900 sites for possible inclusion on the list for Superfund cleanup, and a total of 444 toxic pollutants had been found at these sites. The 10 substances most commonly reported were lead, trichloroethylene, toluene, benzene, PCBs, chloroform, phenol, arsenic, cadmium, and chromium.

TABLE 2–2. Substances observed and reported in the environment (groundwater, surface water, or air) at the then designated 546 NPL sites. (Excerpted from Mitre, 1983)

(NOS = Not otherwise specified)

U.S. EPA Code	Substance	Number of sites where found: Total	Ground-water	Surface Water	Air
U228	Trichloroethylene	129	110	48	7
U220	Toluene	95	74	41	12
U019	Benzene	94	72	31	17
D008	Lead	93	58	64	7
U044	Chloroform	68	60	22	1
A046	Polychlorinated Biphenyls, NOS	58	32	48	4
U210	Tetrachloroethylene	57	47	17	3
U188	Phenol	55	37	27	3
U226	1,1,1-Trichloroethane	54	45	19	3
D007	Chromium	53	33	33	1
D004	Arsenic	51	36	34	2
C049	Ethylbenzene	47	32	13	7
U079	trans-1,2-Dichlorethylene	46	37	13	1
U239	Xylene	40	30	8	8
C247	Zinc and compounds, NOS	37	21	29	2
D006	Cadmium	37	21	28	2
U080	Dichloromethane	36	26	16	2
U043	Vinyl chloride	34	26	9	4
U151	Mercury	32	22	17	3
P030	Cyanides (soluble salts), NOS	29	17	15	2
W002	Volatile organics, NOS	29	3	1	26
U076	1,1-Dichloroethane	26	19	8	0
U077	1,2-Dichloroethane	25	17	12	1
C178	Copper and compounds, NOS	24	15	16	0
U211	Carbon tetrachloride	22	18	9	2
U037	Chlorobenzene	20	17	7	0
U078	Vinylidene chloride	17	15	3	0
U242	Pentachlorophenol	16	12	8	1
A030	Dichloroethylene, NOS	14	12	2	0
A038	Nickel and compounds, NOS	14	7	9	0
C321	Trichloroethane, NOS	14	12	2	1
U165	Naphthalene	12	9	4	1
W001	Heavy metals, NOS	12	8	5	0
A020	Chromium and compounds, NOS	10	8	7	0
U159	Methyl ethyl ketone	10	4	4	4

TABLE 2–2. Substances observed and reported in the environment (groundwater, surface water, or air) at the then designated 546 NPL sites. (Continued)

U.S. EPA Code	Substance	Number of sites where found: Total	Ground-water	Surface Water	Air
U227	1,1,2-Trichloroethane	10	9	1	1
C151	Iron and compounds, NOS	9	7	3	1
A050	Dioxin, NOS	8	3	6	4
D005	Barium	8	7	3	0
U002	Acetone	8	3	3	3
U022	Benzo[a]pyrene	8	5	3	0
U209	1,1,2,2-Tetrachloroethane	8	8	1	0
A051	Tetrachloroethane, NOS	7	7	2	0
C132	Hydrocarbons, NOS	7	4	4	1
C201	Manganese and compounds, NOS	7	4	4	1
C222	Radon and compounds, NOS	7	1	1	7
C320	Chromium, hexavalent	7	5	3	0
D010	Selenium	7	4	5	0
U036	Chlordane	7	5	4	1
U121	Trichlorofluoromethane	7	6	1	0
U129	Lindane	7	5	5	1
W005	Waste oils/Sludges	7	3	5	0
C349	Dichloroethane, NOS	6	5	0	1
C385	Pyrene	6	4	2	0
U028	bis(2-Ethylhexyl)phthalate	6	4	3	0
C146	Ammonia	5	3	2	0
C307	Methane	5	1	0	4
C311	Phthlates, NOS	5	5	3	0
C324	cis-1,2-Dichloroethylene	5	4	2	0
C332	Phenanthrene	5	3	3	0
U051	Creosote	5	3	3	0
U117	Ethyl ether	5	1	2	2
C013	Anthracene	4	3	2	0
C221	Radium and compounds, NOS	4	2	2	2
C245	Uranium and compounds, NOS	4	2	2	2
C374	Trichlorophenols, NOS	4	2	2	0
U013	Asbestos	4	0	2	2
U061	DDT	4	0	4	0
U120	Benzo[j,k]fluorene	4	3	1	0
U127	Hexachlorobenzene	4	1	3	2
A029	Dichlorobenzene, NOS	3	3	1	0
A082	N-Nitrosodiphenylamine	3	2	2	0
C113	Styrene	3	2	0	2
C319	Chromium, trivalent	3	3	1	0
C334	Acenaphthene	3	3	0	0
C342	Dibromochloromethane	3	2	1	0
C378	Aliphatic hydrocarbons, NOS	3	1	0	2
C470	Trimethylbenzene, NOS	3	2	1	0

TABLE 2–2. Substances observed and reported in the environment (groundwater, surface water, or air) at the then designated 546 NPL sites. (Continued)

U.S. EPA Code	Substance	Number of sites where found: Total	Ground-water	Surface Water	Air
U045	Chloromethane	3	3	0	0
U048	2-Chlorophenol	3	2	1	0
U070	1,2-Dichlorobenzene	3	3	0	0
U101	2,4-Dimethylphenol	3	1	3	0
U135	Hydrogen sulfide	3	0	0	3
W003	Acid, NOS	3	1	2	0
A064	Chlorodifluoromethane	2	2	0	0
A091	1,1,2-Trichloro-1,2,2-trifluoroethane	2	1	0	1
C061	n-Heptane	2	1	1	1
C092	4-Nitrotoluene	2	0	2	0
C098	n-Pentane	2	1	1	0
C156	Aluminum and compounds, NOS	2	1	1	0
C161	Boron and compounds, NOS	2	1	2	0
C170	Chloride (ion)	2	1	1	0
C171	Chlorinated hydrocarbons, NOS	2	2	0	0
C177	Cobalt and compounds	2	0	2	0
C235	Thorium and compounds, NOS	2	1	2	0
C236	Tin and compounds, NOS	2	1	1	0
C266	Ethyl chloride	2	2	0	0
C288	Varsol	2	2	1	0
C300	Methylcyclohexane	2	1	1	0
C306	Bromochloromethane	2	2	0	0
C316	Sodium hydroxide	2	2	0	0
C348	Dimethylaniline	2	1	1	0
C375	Halogenated organics, NOS	2	1	1	0
C379	Olefinic hydrocarbons, NOS	2	0	0	2
C380	Ketones, NOS	2	0	0	2
C484	Dichlorophenol, NOS	2	0	2	0
D011	Silver	2	1	1	0
M001	More than 15 substances listed	2	0	1	1
P015	Beryllium dust, NOS	2	1	2	0
P048	2,4-Dinitrophenol	2	2	0	0
P056	Fluorine	2	1	1	0
P059	Heptachlor	2	2	1	0
U012	Aniline	2	0	2	0
U021	Benzidene	2	2	1	0
U025	bis(2-Chloroethyl)ether	2	2	1	1
U029	Bromomethane	2	2	0	0
U050	Chrysene	2	1	1	0
U081	2,4-Dichlorophenol	2	2	1	0
U102	Dimethyl Phthalate	2	1	1	0
U105	2,4-Dinitrotoluene	2	1	1	0
U106	2,6-Dinitrotoluene	2	1	1	0

TABLE 2–2. Substances observed and reported in the environment (groundwater, surface water, or air) at the then designated 546 NPL sites. (Continued)

U.S. EPA Code	Substance	Number of sites where found: Total	Ground-water	Surface Water	Air
U108	1,4-Dioxane	2	1	1	0
U109	1,2-Diphenylhydrazine	2	1	1	1
U128	Hexachlorobutadiene	2	0	0	2
U130	Hexachlorocyclopentadiene	2	0	1	2
U161	Methyl isobutyl ketone	2	2	0	0
U169	Nitrobenzene	2	0	2	0
U170	4-Nitrophenol	2	2	1	0
U213	Tetrahydrofuran (I)	2	2	0	0
U225	Tribromomethane	2	2	0	0
U233	2,4,5-TP (Silvex)	2	2	0	0
W004	Grease and oil, NOS	2	0	2	0
W010	Waste solvents, NOS	2	2	1	0
W017	Non-volatile organics, NOS	2	1	2	0
W024	Mineral spirits	2	2	1	0
A006	Barium and compounds, NOS	1	0	1	0
A011	Beryllium compounds, NOS	1	0	1	0
A023	DDE	1	0	1	0
A034	Halomethane, NOS	1	0	1	0
A045	Phthalic esters, NOS	1	1	1	0
A054	1,2,4-Trichlorobenzene	1	1	1	0
A059	Bromodichloromethane	1	1	0	0
A066	Coal tars	1	0	1	0
C016	Butadiene	1	0	0	1
C017	Butane	1	1	0	0
C023	Camphor	1	1	1	0
C031	Dicyclopentadiene	1	0	1	0
C054	Ethylene glycol	1	0	1	0
C063	Hexane	1	1	0	0
C067	Isophorone	1	0	1	0
C069	Isopropanol	1	1	0	0
C072	Isopropyl ether	1	1	0	0
C074	Methyl acetate	1	0	1	0
C090	Nitrophenol, NOS	1	1	0	0
C138	Sodium chloride	1	1	0	0
C142	Sulfur dioxide	1	0	0	1
C149	Chlorotoluenes, NOS	1	1	0	0
C153	Titanium and compounds, NOS	1	1	1	0
C196	Hydrochloric acid	1	1	0	0
C200	Magnesium and compounds, NOS	1	1	0	0
C212	Phenolic compounds, NOS	1	0	1	0
C228	Sulfate (ion)	1	1	0	0
C230	Sulfuric acid	1	0	1	0
C246	Vanadium and compounds, NOS	1	1	1	0

TABLE 2–2. Substances observed and reported in the environment (groundwater, surface water, or air) at the then designated 546 NPL sites. (Continued)

U.S. EPA Code	Substance	Number of sites where found: Total	Ground-water	Surface Water	Air
C251	Dibenzofuran	1	0	1	0
C275	Nitroaniline, NOS	1	1	1	0
C325	Nitric acid	1	0	1	0
C326	Octane	1	0	1	0
C336	p-Methyl styrene	1	1	0	0
C337	p-Ethyl toluene	1	0	0	1
C341	Polybrominated biphenyl, NOS	1	0	1	0
C347	Pentachlorobutadiene	1	0	0	1
C350	2-Methylbenzoyl chloride	1	1	0	0
C353	Naphthol, NOS	1	0	1	0
C356	Tetrachlorotoluene	1	1	1	0
C357	Cycloheptane	1	0	1	0
C358	Bromoxylene, NOS	1	1	1	0
C365	Trinitrotoluene (TNT)	1	1	0	0
C368	Nitriles, NOS	1	0	0	1
C376	Esters, NOS	1	0	0	1
C377	Nitrates, NOS	1	1	0	0
C381	Benzothiophene, NOS	1	1	0	0
C382	Propenylbenzene	1	1	0	0
C383	Indene	1	1	0	0
C387	Atrazine	1	1	0	0
C392	Phosphoric acid	1	0	1	0
C395	Tricarboxylic acid, beta-acetoxytributyl	1	1	0	0
C401	Trichlorobenzene, NOS	1	1	1	0
C416	Kelthane	1	1	1	0
C440	Nickel chloride	1	0	1	0
C444	Octachlorodinitrodicyclopentadiene	1	0	0	1
C446	Methylene chloroform	1	1	0	0
C453	Bromobenzene	1	1	0	0
C472	Ammonium, NOS	1	1	0	0
C474	2-Methyl-1,3-dinitrobenzene	1	0	1	0
C475	Trimethyloxabicyclooctane	1	0	1	0
C478	Bromochlorobenzene, NOS	1	1	0	0
C482	Trimethylcyclohexanol	1	1	0	0
C494	Phenol sulfonate	1	0	1	0
C496	Benzyl butyl phthlate	1	1	0	0
C504	RDX	1	1	0	0
C505	Butyl dinitrophenol	1	0	1	0
P003	Acrolein	1	1	0	0
P020	Dinoseb	1	1	0	0
P037	Dieldrin	1	0	1	0
P050	Endosulfan	1	0	1	0
P051	Endrin	1	0	1	0

TABLE 2–2. **Substances observed and reported in the environment (groundwater, surface water, or air) at the then designated 546 NPL sites. (Continued)**

U.S. EPA Code	Substance	Number of sites where found: Total	Ground-water	Surface Water	Air
P067	2-Methylaziridine	1	1	1	0
U018	Benzo[a]anthracene	1	1	1	0
U024	bis(2-Chloroethoxy)methane	1	1	0	0
U038	Chlorobenzilate	1	1	1	0
U039	p-Chloro-m-cresol	1	0	1	0
U052	Cresols, NOS	1	0	1	0
U055	Cumene	1	0	1	0
U056	Cyclohexane	1	0	1	0
U073	3,3'-Dichlorobenzidine	1	0	1	0
U075	Dichlorodifluoromethane	1	1	0	0
U088	Diethyl phthalate	1	0	1	0
U107	Di-n-octyl phthalate	1	1	0	0
U112	Ethyl acetate	1	0	1	0
U131	Hexachloroethane	1	1	0	0
U142	Kepone	1	1	1	0
U158	4,4'-Methylene-bis(2-chloroaniline)	1	1	1	1
U162	Methyl methacrylate	1	0	0	1
U201	Resorcinol	1	0	1	0
W009	#2 Fuel oil	1	1	1	0
W012	Pesticides, NOS	1	1	1	0
W014	Alcohol, NOS	1	0	0	1
W031	Polynuclear aromatic hydrocarbons	1	1	0	0

Other examples of lists of chemicals found at or in the vicinity of chemical disposal sites include the following: Elder et al. (1981) and other research reports have detailed many of the chemicals found at the Love Canal site in Niagara Falls, New York; Clark et al. (1982) listed the range and mean concentrations of thirteen chemicals found in samples of groundwater used for drinking water collected near the Toone site in Hardeman County, Tennessee; the New Jersey Department of Health (New Jersey, 1983) reported air concentrations of chemicals near the GEMS site in Gloucester Township. The Massachusetts Department of Environmental Quality Engineering (Massachusetts, 1981) and the Michigan Department of Natural Resources (Michigan, 1982) have surveyed the chemicals found within disposal sites in their respective states; Levine and Chitwood (1983) reported data obtained from the Florida Department of Environmental Regulation, which had listed chemicals found at 66 disposal sites within the state.

The Mitre list of chemicals found near NPL sites is more extensive in terms of geographical distribution of the sites and therefore includes more types of industrial facilities than would be present in the chemical disposal sites located in any one state. Due to this diversity of industry, the Mitre list, therefore, is likely to be more representative and to include a greater diversity of waste chemicals. In addition, the data tabulated in the Mitre report are from the NPL sites, those sites identified by EPA as being disposal sites causing documented environmental concern.

Various procedures can be used to select lists of chemicals for detailed study. The National Research Council (1984) proposed a priority scheme for the National Toxicology Program to identify chemicals for toxicity testing. In an earlier phase of the study (NRC, 1981), there was a detailed review of various systems developed by either United States or international regulatory agencies to determine priority rankings for toxicity testing. These systems are similar to the procedure utilized in this report for identifying hazardous chemicals for information review; the procedure, described in the beginning of this chapter, is dependent on a chemical being present at disposal sites, its potential for release and migration through the environment, its potential for resulting in adverse human health effects, and its recognition by the scientific community and/or the general public as being a chemical causing concern.

The list of chemicals in Table 2-3 is comprised of selected chemicals found in the environment in the vicinity of chemical disposal sites. The primary criterion for inclusion on the list is a chemical's reported presence at numerous chemical disposal sites. Most, but not all, of the chemicals found at eight or more sites on Mitre's list in Table 2-2 have been included on the list of representative chemicals. Several reasons which may account for those that were omitted are noted below.

- Because neither chemical concentrations nor specific site locations were provided in the Mitre data, there is no indication whether the chemicals listed were found in minute or large quantities or whether they were naturally occurring constituents of the environment at any given site. Iron, for example, was omitted from Table 2-3 because it frequently is found in groundwater from natural sources in locations not affected by chemical disposal sites.

- Some groups of chemicals on the Mitre list were omitted because they were listed as only "generic" categories, for example, "volatile organics" and "heavy metals."

- Other listings were not included because of lack of identification of a specific isomer, for example, "dichloroethylene" and "trichloroethane". As well as omitting the above groups of chemicals from the "selected" list, several additional chemicals were included because they met one of the several other criteria outlined at the beginning of Chapter 2. These chemicals include 1,2-dichlorobenzene, asbestos, DDT, and hexachlorocyclopentadiene.

Those chemicals in Table 2-3 labelled with asterisks were selected for specific and more detailed reviews because they are typical of the types of hazardous chemicals found at chemical disposal sites or because they are compounds for which there exists a substantial degree of recognition and concern by the scientific community and/or the general public. However, it is essential to recognize that Table 2-3 is not meant to be an all-inclusive list of the hundreds of chemicals found at disposal sites. Although a chemical may appear to be fairly representative of its class as far as environmental fate and transport characteristics are concerned, small changes in chemical structure may yield metabolic and pharmacokinetic differences which in turn may account for variation between chemicals with regard to their potential to cause adverse human health effects.

TABLE 2–3. Representative hazardous chemicals found at chemical disposal sites.

	(CAS No.)
Halogenated Aliphatic Compounds:	
Dichloromethane	75-09-2
*Chloroform	67-66-3
*Carbon tetrachloride	56-23-5
1,1-Dichloroethane	75-34-3
1,2-Dichloroethane	107-06-2
*1,1,1-Trichloroethane	71-55-6
1,1,2-Trichloroethane	79-00-5
1,1,2,2-Tetrachloroethane	79-34-5
Vinyl chloride	75-01-4
Vinylidene chloride	75-35-4
trans-1,2-Dichloroethylene	156-60-5
*Trichloroethylene	79-01-6
Tetrachloroethylene	127-18-4
Halogenated Aromatic Compounds:	
Chlorobenzene	108-90-7
1,2-Dichlorobenzene	95-50-1
*Pentachlorophenol	87-86-5
*2,3,7,8-Tetrachlorodibenzo-p-dioxin	1746-01-6
Polychlorinated biphenyls	–
DDT	50-29-3
Non-halogenated Aromatic Compounds:	
*Benzene	71-43-2
Toluene	108-88-3
Xylene	1330-20-7
Ethylbenzene	100-41-4
*Phenol	108-95-2
*Naphthalene	91-20-3
Benzo[a]pyrene	50-32-8
Miscellaneous Organic Compounds:	
Acetone	67-64-1
*Hexachlorocyclopentadiene	77-47-4
*Methyl ethyl ketone	78-93-3
Trace Metals and Other Inorganic Compounds:	
Arsenic	7440-38-2
Asbestos	1332-21-4
Cadmium	7440-43-9
Chromium	7440-47-3
Copper	7440-50-8
Cyanide (free cyanide ion)	57-12-5
Lead	7439-92-1
Mercury	7439-97-6
Nickel	7440-02-0
Zinc	7440-66-6

*Synopses for these chemicals are included in Appendix A of this report.

CHEMICAL SYNOPSES AS AN AID IN ASSESSING RISK

Table 2-4 consists of the general outline format used in preparing detailed synopses of the specific chemicals labelled with asterisks in Table 2-3. The synopses themselves appear in Appendix A of this report. These synopses contain information pertinent to physical and chemical characteristics, as well as information on past and present uses and production volumes, potential transport and fate in the environment, exposure potentials, non-human toxicity, and human health effects. The synopses in Appendix A are not intended to be comprehensive and are only abstracts of the data available from the Toxicology Data Bank (TDB) and supplemental sources (TDB-II was scheduled to be available to the general public through the National Library of Medicine, Specialized Information Services as of November 1984). The reader is cautioned to consult TDB-II or other appropriate references to obtain additional data on these compounds and on chemicals not synopsized in Appendix A. A list of additional pertinent reference material can be found at the end of this chapter.

To obtain an assessment of the risk for a specific chemical at a given disposal site, the data on health effects in Appendix A must be combined with site-specific data on the exposure concentration and duration. (See Chapter 6 for a discussion of the considerations involved in risk assessment.)

SUMMARY

Scientists called upon to assess the potential environmental impact and/or health effects of a chemical disposal site generally find that the materials which have been disposed of at the site have not been adequately documented as to the identities of the specific chemicals, or the sources, quantities, and dates of each deposition. Thus, it is necessary to determine the chemicals which are present at the disposal site and to identify those chemicals which are escaping from the site and their route of entry into the environment. These chemical determinations will usually be qualitative, will almost always indicate a complex mixture of chemicals, and even the qualitative analyses will not be "complete"; that is, some of the chemicals present in the waste mixture will not be identifiable because of lack of specificity or sensitivity of the usual analytical procedures.

Each chemical disposal site is unique, but all have in common the potential for contaminating the environment by releasing toxic substances. The focus of this report is on individual chemicals and, to a lesser degree, mixtures of chemicals. This chapter describes the types of chemicals generally found at chemical disposal sites and identifies some representative chemicals found in the vicinity of a number of sites, suggesting that they have been released from the sites, are migrating through the environment, and that human exposure may be occurring. Any list of chemicals in this report should not be viewed as comprehensive and complete; each site must be treated as a unique entity. A thorough investigation of the health effects in the vicinity of a chemical disposal site should include a detailed analysis of the chemicals present at and escaping from the site, the routes of transport of these chemicals through the environment, the concentrations of the chemicals in the exposure medium, the time the chemicals were in the medium, and an estimate of uptake into the human body.

TABLE 2–4. Format for presentation of data in chemical synopses in Appendix A.

I. **Chemical Identification**
   - A. Chemical name
   - B. CAS Registry number
   - C. Synonyms
   - D. Molecular formula
   - E. Structural formula

II. **Physical and Chemical Properties**
   - A. Molecular weight
   - B. Physical state (color, form)
   - C. Odor
   - D. Melting point
   - E. Boiling point
   - F. Solubilities
     - 1. Water
     - 2. Organic solvents
   - G. Octanol/water partition coefficient (log P)
   - H. Density
   - I. Dissociation constant
   - J. Viscosity
   - K. Surface tension
   - L. Vapor pressure
   - M. Vapor density
   - N. Relative evaporation rate
   - O. Spectral properties

III. **Manufacturing/Use Information**
   - A. U.S. Production volume
   - B. Major uses

IV. **Environmental Fate/Exposure Potential**
   - A. Potential for release from chemical disposal site
     - 1. Water solubility
     - 2. Volatility
     - 3. Adsorption/desorption
   - B. Biological uptake/accumulation
     - 1. Octanol/water partition coefficient (log P)
     - 2. Metabolism
     - 3. Bioaccumulation
   - C. Persistence
     - 1. Chemical transformations
     - 2. Biological transformations
   - D. Measured concentrations
     - 1. Groundwater
     - 2. Surface water
     - 3. Drinking water
     - 4. Air
     - 5. Soil
     - 6. Sediment
     - 7. Food

TABLE 2–4. Format for presentation of data in chemical synopses in Appendix A. (Continued)

E. Human environmental exposure
   1. Probable routes of exposure
   2. Average daily intake
   3. Occupational exposure
   4. Body burden

V. **Health Effects**
   A. Human health effects
      1. Acute effects
         (symptoms and signs, routes of exposure, concentrations x time, etc.)
      2. Chronic effects
         (symptoms and signs, routes of exposure, concentrations x time, etc.)
      3. Carcinogenicity
         (target organs, tumor types, routes of exposure, range of dose, etc.)
      4. Developmental effects
         (fetotoxicity, teratogenicity, birth defects, etc.)
      5. Pharmacokinetics
         (absorption, distribution, metabolism, excretion, etc.)
   B. Non-human mammalian health effects
      1. Acute effects
      2. Chronic effects
      3. Carcinogenicity
      4. Developmental effects
      5. Pharmacokinetics
   C. Non-mammalian effects
      1. Acute effects
      2. Chronic effects
      3. Carcinogenicity
      4. Developmental effects
      5. Pharmacokinetics
   D. Short-term assays
      (DNA damage, mutagenicity, chromosomal anomalies, etc.)

VI. **Exposure Standards and Recommendations**
   A. Air
      1. Immediately Dangerous to Life or Health (IDLH)
      2. Occupational Safety and Health Administration (OSHA) standards [Threshold limit value (TLV), ceiling, peak]
      3. National Institute of Occupational Safety and Health (NIOSH) recommendations [TLV, ceiling, peak]
      4. American Conference of Governmental Industrial Hygienists, Inc. (ACGIH) [TLV, Short-term exposure limit (STEL]
      5. Other
   B. Water
      [Maximum contaminant level (MCL), Water quality criteria (WQC), Suggested no adverse response level (SNARL), etc.]
   C. Other

References

Clark, C.S.; Meyer, C.R.; Gartside, P.S.; Majeti, V.A.; Specker, B.; Balistreri, W.F.; Elia, V.J. (1982) Environmental health survey of drinking water contamination by leachate from a pesticide waste dump in Hardeman County, Tennessee. Arch. Environ. Health 37:9-18,125.

Conservation Foundation. (1984) State of the environment: an assessment at mid-decade. Washington, DC: The Foundation. 570p.

Council of the European Communities. (1978) Council directive of 20 March 1978 on toxic and dangerous waste. Off. J. Eur. Community 84:43-48.

Elder, V.A.; Proctor, B.L.; Hites, R.A. (1981) Organic compounds found near dump sites in Niagara Falls, New York. Environ. Science Technol. 15:1237-1243.

Huismans, J.W.; Suess, M.J.; editors. (1982) Hazardous waste management: policy guidelines and code of practice. European Cooperation on Environmental Health Aspects of the Control of Chemicals: interim document 7. Copenhagen: World Health Organization, Regional Office for Europe. 147p.

Levine, R.S.; Chitwood, D. (1983) Analysis of existing data pertaining to potential public health problems from hazardous wastes in Florida. Prepared for the State of Florida Department of Health and Rehabilitative Services. 177p. Available from: University of Miami, School of Medicine, Miami, FL.

Massachusetts. Special Legislative Commission on Water Supply. (1981) Water quality issues in Massachusetts: chemical contamination, second working paper. 88p. Available from: The Commission, Boston, MA.

Michigan. Department of Natural Resources, Groundwater Quality. (1982) Assessment of groundwater contamination: inventory of sites. [Lansing, MI]: The Department. 242p. Available from: National Technical Information Service, Springfield, VA.

Mitre Corporation. (1983) Computer printouts of National Priorities List data summaries for the 546 final and proposed sites and 881 sites currently on the data base as of September 7, 1983. Prepared for the U.S. Environmental Protection Agency. 94p. Available from: Metrek Division, Mitre Corporation, McLean, VA.

National Research Council. Steering Committee on Identification of Toxic and Potentially Toxic Chemicals for Consideration of the National Toxicology Program. (1981) Strategies to determine needs and priorities for toxicity testing. Volume 1: design. Washington, DC: National Academy Press. 143p.

National Research Council. Steering Committee on Identification of Toxic and Potentially Toxic Chemicals for Consideration of the National Toxicology Program. (1984) Toxicity testing: strategies to determine needs and priorities. Washington, DC: National Academy Press. 382p.

New Jersey. Department of Health, Environmental Health Hazard Evaluation Program. (1983) Environmental air sampling both within and outside of homes located near the Gloucester Environmental Management Services (GEMS) landfill, conducted during winter 1981-82 and summer 1982. In: Summary of health studies conducted at GEMS landfill in Camden County by the New Jersey Department of Health. p. 47-69. Available from: The Program, [Trenton, NJ].

NRC SEE National Research Council

U.S. Environmental Protection Agency. (1980) Hazardous waste management system: part 261, identification and listing of hazardous waste. Fed. Regist. 45:33119-33133.

U.S. Environmental Protection Agency. (1982a) Discarded commercial chemical products, off-specification species, container residues, and spill residues thereof. Code of federal regulations, title 40, part 261.33. Washington, DC: U.S. General Services Administration, Office of the Federal Register. 7p. Available from: U.S. Government Printing Office, Washington, DC.

U.S. Environmental Protection Agency. (1982b) Hazardous constituents. Code of federal regulations, title 40, part 261, app. VIII. Washington, DC: U.S. General Services Administration, Office of the Federal Register. 6p. Available from: U.S. Government Printing Office, Washington, DC.

U.S. Environmental Protection Agency. (1984) Hazardous waste management system; identification and listing of hazardous waste. Fed. Regist. 49:49784-49793.

U.S. Environmental Protection Agency. (1985) Hazardous waste management system; dioxin-containing wastes; final rule. Fed. Regist. 50:1978-2006.

USEPA SEE U.S. Environmental Protection Agency

Additional Pertinent Reference Material

American Conference of Governmental Industrial Hygienists, Inc. (1980) Documentation of the threshold limit values. 4th ed. Cincinnati: The Conference. 486p.

American Conference of Governmental Industrial Hygienists. (1983) TLVs: threshold limit values for chemical substances in the work environment and biological exposure indices with intended changes for 1984-1985. Cincinnati: The Conference. 116p.

Casarett and Doull's toxicology: the basic science of poisons. (1980) 2d ed. Doull, J.; Klaassen, C.D.; Amdur, M.O.; eds. New York: Macmillan Publishing Co. 733p.

CRC handbook of chemistry and physics. (1983) 64th ed. Weast, R.C.; Astle, M.J.; Berger, W.H.; eds. Boca Raton, FL: CRC Press, Inc. 2302p.

Gosselin, R.E., editor. (1984) Clinical toxicology of commercial products. 5th ed. Baltimore, MD: Williams and Wilkins, Co. 1900p.

Hansch, C.; Leo, A.J. (1979) Substituent constants for correlation analysis in chemistry and biology. New York: John Wiley & Sons. 339p.

Hawley, G. G. (1981) Condensed chemical dictionary. 10th ed. New York: Van Nostrand Reinhold Co. 1472p.

Kirk & Othmer Encyclopedia of Chemical Technology. 1978-80. 3rd ed. 12 vols. New York: John Wiley & Sons.

Merck index. (1983) 10th ed. Windholz, M.; Budavari, S.; Blumetti, R.F.; Otterbein, E.S.; eds. Rahway, NJ: Merck & Co., Inc. 2154p.

National Research Council. (1975) Principles for evaluating chemicals in the environment. A report of the Committee for the Working Conference on Principles of Protocols for Evaluating Chemicals in the Environment. Washington, DC: National Academy of Science. 454p.

National Research Council. (1977) Drinking water and health. [Vol. 1]. Washington, DC: National Academy Press. 939p.

National Research Council. (1978) Chloroform, carbon tetrachloride, and other halomethanes: an environmental assessment. Washington, DC.: National Academy of Sciences. 294p.

National Research Council. (1980) Drinking water and health. Vol. 2. Washington, DC: National Academy Press. 393p.

National Research Council. (1980) Drinking water and health. Vol. 3. Washington, DC: National Academy Press. 415p.

National Research Council. (1982) Drinking water and health. Vol. 4. Washington, DC: National Academy Press. 299p.

National Research Council. (1983) Drinking water and health. Vol. 5. Washington, DC: National Academy Press. 157p.

National Research Council. Committee on the Institutional Means for Assessment of Risks to Public Health. (1983) Risk assessment in the federal government: managing the process. Washington, DC: National Academy Press. 189p.

Patty's industrial hygiene and toxicology. (1981) 3rd rev. Vol. 2. Clayton, G.D.; Clayton, F.E.; eds. New York: John Wiley & Sons. 5112p.

Registry of toxic effects of chemical substances. (RTECS). Data base prepared by U.S. Department of Health and Human Services, National Institute for Occupational Safety and Health, Cincinnati, OH, for use on Medlars. Bethesda, MD: U.S. Department of Health and Human Services, National Library of Medicine. Magnetic-tape reels.

Shepard, T.H. (1983) Catalog of teratogenic agents. 4th ed. Baltimore: Johns Hopkins University Press. 410p.

Sittig, M. (1981) Handbook of toxic and hazardous chemicals. Park Ridge, NJ: Noyes Publications. 729p.

Toxicology data bank II. (1984) [Enhanced version of data base]. Bethesda, MD: U.S. Department of Health and Human Services, National Library of Medicine, Specialized Information Services. Magnetic-tape reels.

U.S. Department of Health and Human Services, National Institute for Occupational Safety and Health. Criteria for a recommended standard: occupational exposure to [specific chemicals]. Cincinnati, OH: The Institute.

U.S. Department of Health and Human Services, Public Health Service. (1983) Third annual report on carcinogens: summary. Research Triangle Park, NC: U.S. Department of Health and Human Services, National Toxicology Program Public Information Office. 229p.

U.S. Environmental Protection Agency, Office of Health and Environmental Assessment. Health assessment document[s] [for specific chemicals]. Cincinnati, OH: The Office.

U.S. Environmental Protection Agency, Office of Water Regulations and Standards. Ambient water quality criteria [for specific chemicals]. Washington, DC: The Office. Available from: National Technical Information Service, Springfield, VA.

Verschueren, K. (1983) Handbook of environmental data on organic chemicals. 2nd ed. New York: Van Nostrand Reinhold Company, Inc. 1310p.

World Health Organization, International Agency for Research on Cancer. 1972-1984. IARC monographs on the evaluation of the carcinogenic risk of chemicals to humans. Vols. 1-34 and supplements. Lyon, France: The Agency.

Additional Sources of Information

U.S. Department of Energy, Oak Ridge National Laboratory. Environmental Mutagen Center: Genetox Program. Oak Ridge, TN.

U.S. Department of Energy, Oak Ridge National Laboratory. Environmental Teratology Information Center: Teratology Program. Oak Ridge, TN.

3

Factors Influencing Human Exposure

This chapter discusses the factors important in potential human exposure to chemicals released from chemical disposal sites. For clarity in describing scientific principles, the approach taken in this chapter is to discuss the characteristics of chemicals in one section and the routes of transport in another, rather than to attempt to mesh both concepts in one section. The physical and chemical properties of the compounds disposed of at waste sites are described in terms of their potential for affecting release from the waste disposal sites, migration through the environment, and human exposure. (Specific chemicals found at waste disposal sites are described in Chapter 2.) The major routes of transport and exposure are described, with emphasis on groundwater and surface water, but also including air, soil, and food. The chapter concludes with a discussion of the industrial processing and engineering control methods which are currently available or are being developed to minimize environmental releases from chemical disposal sites.

CHARACTERISTICS OF CHEMICALS WHICH AFFECT RELEASE, TRANSPORT, AND HUMAN EXPOSURE

Diverse classes of chemicals have been deposited in uncontrolled disposal sites (Dietz et al., 1984; Mitre, 1983; USC/OTA, 1983). The chemical and physical characteristics of the individual chemicals determine the degree to which they move within the disposal site, are transported across its boundary, and migrate through the environment. Transformations of chemicals by microbial and chemical degradation reactions are also determined in large part by the chemicals' characteristics, as well as by physical, chemical, and biological factors in the surrounding media (i.e., groundwater, surface water, air, soil). Figure 3-1 illustrates a generalized diagram for the transport of chemicals released from chemical disposal sites. Substances which are released from disposal sites may move through various media in the environment so that humans are potentially exposed via a variety of routes of exposure.

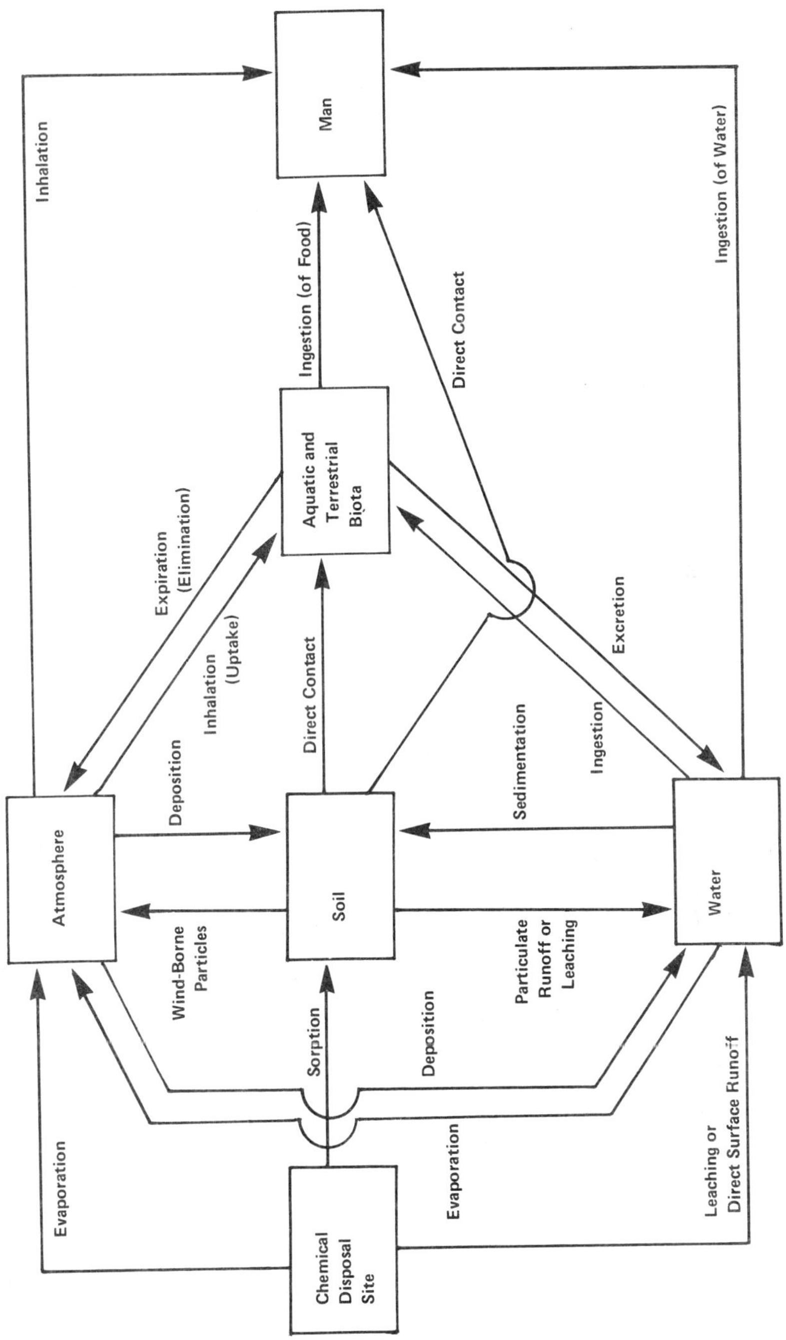

Figure 3-1. Physical and Biological Routes of Transport of Hazardous Substances, their Release from Disposal Sites, and Potential for Human Exposure (adapted from Van Hook, 1978).

Table 3-1 contains a list of certain physical and chemical characteristics of selected compounds found in chemical disposal sites. The characteristics tabulated in Table 3-1 include some of the factors important in controlling the migration and fate of chemicals in the environment. The ranges of the values for each characteristic indicate the diversity in the physical and chemical properties of hazardous chemicals. In very simplified terms, the higher the water solubility value, the more likely that a compound will leach and migrate in groundwater; for low solubility compounds, a low density suggests that a compound will float near the surface of the water table in groundwater, while a high density insoluble compound will tend to sink in the groundwater. The higher a compound's octanol/water partition coefficient (the ratio of the solubility in 1-octanol to that in water), the more hydrophobic the compound, and the more likely its migration will be retarded owing to sorption to soil (and the more readily it would be bioaccumulated into biological tissues); the higher a compound's soil sorption coefficient, the less likely the compound is to migrate. The higher the vapor pressure and the higher the Henry's law constant of a compound, the more likely the compound will volatilize or evaporate from water or soil into the air. If a compound is not readily biodegraded, it will be persistent. These are extremely simplified generalizations and are intended only as a brief overview of the physical and chemical characteristics contained in Table 3-1. The discussions throughout the rest of this section will define and further describe each of these factors, as well as the relationships among these characteristics and the potential for human exposure.

Transport

The extent and rate of chemical movement within a chemical disposal site are functions of the physical containment of the waste materials (e.g., in drums, as semi-liquid sludges, as liquids drained from tanker trailers, as bulk solids dumped onto the ground and/or buried). A low permeability cap over the site will retard percolation of water downward from the surface and thus minimize the leaching of chemicals. In a site without a cap, the uncontained, water-soluble wastes will be subject to movement within the site and to transport across the site boundary.

A chemical, if released across the boundary of the disposal site, may become distributed among the different environmental compartments as illustrated in Figure 3-1. The amount of a given chemical present in each compartment of the environment is determined by both the chemical and physical properties of the chemical as well as the properties of the different compartments. Table 3-2 summarizes reactions and properties of chemicals which affect the transport and fate of chemicals in the environment.

Aqueous migration from the site

The major factors contributing to partitioning of a chemical into the aqueous phase are its water solubility and its latent heat of solution, which is a determinant of the energy required for solubility into water (Freed, 1982). Many organic compounds exhibit hydrophobic characteristics and have water solubilities in the low parts-per-million (ppm) or even parts-per-billion (ppb) range, making exact determinations of solubility in aqueous solutions difficult. Many inorganic salts readily ionize in aqueous systems and become dissolved

TABLE 3–1. Characteristics affecting transport and fate of selected chemicals found at chemical disposal sites. (Information abstracted from the Toxicology Data Bank (TDB II, 1984), unless specified.)

	Water Solubility (mg/L @20°C)	Density (20°C/4°C)	Octanol/Water Partition Coefficient (expressed as the logarithm, log P)	Soil Sorption Coefficient[a]	Vapor Pressure (mm Hg)	Henry's Law Constant[b] ($atm \cdot m^3/mol$ at 25°C)	Biodegradability[c]	
							Oxygenated Water	Anoxic Water
Chloroform	8,200	1.48	1.97[d]	0.94	100 @10°C	0.0034	none	possible
Carbon Tetrachloride	800[d]	1.59	2.64	1.61	91.3 @20°C	0.0302	none	possible
1,1,1-Trichloroethane	4,400	1.34	2.49	1.47	127 @25°C	0.0049	none	possible
Trichloroethylene	1,100	1.46	2.29	1.37	57.8 @20°C	0.0117	none	possible
Chlorobenzene	500	1.10	2.84	1.79	10 @22.2°C	0.0039	possible	none
Pentachlorophenol	14	1.98	5.01	3.75	0.00011 @20°C	0.0001[e]	possible	possible
Benzene	1,780	0.88	2.28	1.36	100 @26.1°C	0.0055	probable	none
Phenol	90,000	1.07	1.46	1.54	0.35 @ 25°C	0.000001	probable	probable
Naphthalene	30	1.15	3.2-3.45	2.11-2.34	1 @53°C	0.00048[e]	possible	none
Hexachlorocyclopentadiene	0.80	1.70	—	—	0.08 @25°C	0.0164	—	—
Methyl Ethyl Ketone	353,000[d]	0.80	0.26[d]	-0.52	70.6 @20°C	0.00009[e]	probable[d]	possible[d]

[a] Calculated following the equation of Chiou et al. (1983): $\log Kom = 0.904 \log P - 0.779$
[b] Shen (1982)
[c] Wilson and McNabb (1983)
[d] Verschueren (1983)
[e] Mackay et al. (1979)

TABLE 3–2. Physical and chemical properties of substances which affect their transport and fate (Freed, 1982).

Physical Chemical Properties and Reactions	Related to
Solubility in water	Leaching, partitioning, mobility in environment
Latent heat of solution	Sorption, leaching, vaporization from surfaces, dissolution
Partition coefficient	Bioaccumulation potential, sorption by organic matter
Hydrolysis	Persistence in environment or biota
Ionization	Route and mechanism of adsorption or uptake, persistence, interaction with other molecular species
Vapor pressure	Atmospheric mobility, rate of vaporization
Photolysis	Persistence, a function of exposure to light based on presence of chemical at soil surface or volatilizing into atmosphere

ions. The thermodynamics of the solubilization energy of the waste mixtures in the soil-water phase found at chemical disposal sites may result in observed solubility values which are different from the pure compounds in distilled water. Moreover, the presence of water-miscible organic solvents can greatly enhance the chemical's solubility, which in turn may increase the potential for leaching.

One of the important processes controlling migration of a chemical out of the waste mass and into groundwater or surface water is the chemical's rate of dissolution into the water moving through the waste, a process termed leaching. Leaching may occur in three directions; the usual direction of water movement within the unsaturated (vadose) zone is downward, but lateral movement of the chemical near the soil surface, and even upward movement and evaporation are sometimes significant (Freed, 1982). Upward movement is a result of mass transfer and capillary action of the water under the influence of evaporation at the surface, and may result in the accumulation of the chemicals at the soil surface. Water moves downward through the soil by gravitational forces and capillary action and is often thought to be in the form of a film. Water arrives at the surface of a disposal site as precipitation, penetrates into the ground, and encounters the chemicals, dissolves them, and carries them downward as the water percolates through the soil. The important factors controlling the rate of leachate production for a specific substance within a chemical disposal site and the leachate movement from the site are:

- water solubility of the substance,
- sorption potential of the substance,
- soil type (including the percentage of organic carbon and of clay), and
- amount and intensity of water penetration.

Substances at chemical disposal sites occur as mixtures, and are likely to leach at different rates. A chemical like chloroform which is highly water soluble and has a low potential for sorption onto soil and/or onto the waste components will be leached rapidly; in contrast, a chemical like pentachlorophenol with low water solubility and high sorption potential would be retarded and have a slower rate of migration (Table 3-1). Sorption can indicate a process of either binding to surfaces (adsorption to soil organic matter or clays, or to waste components) or absorption into patches of high molecular weight organic matter or an unspecified interphase mass transfer from the aqueous to the solid phase, resulting in dissolution from the water. The amount of leachate produced will be a function of the amount of water percolating through and available to dissolve the chemicals, the water solubility of each chemical in the waste mass, and the available contact area on the surface of the waste mass. The temperature of the soil and waste components will also play a role in leaching since temperature affects solubility. If an organic chemical is chemically or biologically transformed into products having higher water solubility, then the rate of leaching will be more rapid, although the products would no longer be migrating as the parent compound.

When percolation is rapid, the bulk movement of the chemicals will be in the direction of water flow. As percolation becomes slower, the distribution of the chemical is determined to a greater extent by molecular diffusion. A dynamic equilibrium develops between the free chemical in solution and the chemical in the solid phase sorbed to the soil. Chemicals tightly sorbed are leached slowly and vice versa. In a sandy aquifer, dissolved sodium ions will behave conservatively and will move at almost the same rate as the groundwater flow, for example, a few inches to a few feet per day (Miller, 1981), and some polar organic molecules such as alcohols may move even faster than groundwater, but many hydrophobic organic chemicals may adsorb onto organic matter and clay components of the soil and migrate very slowly; for example, they may require decades to move even a short distance.

Migration in groundwater

Some chemicals may initially leach from the disposal site as mixtures, but in groundwater transport, differences in their aqueous solubility and soil sorption potential values will influence the rates at which the chemicals migrate. The soil sorption potential for organic compounds onto soil organic matter is inversely proportional to water solubility and is directly proportional to the octanol/water partition coefficient (Kow; frequently expressed as the log of the coefficient, i.e., log P) of the chemical (Chiou et al., 1979; Kenaga and Goring, 1980). For each chemical, the dipole of the molecule and other molecular forces interact with the surface forces of the soil particles. Soil organic matter has a very high retention capacity for nonionic and ionic compounds, and clays also have higher retention capacities for ionic compounds compared with sand, which is relatively nonadsorptive. Organic compounds can be sorbed by relatively weak interaction such as van der Waals forces and dipoles as well as by ionic attraction whereas inorganic species such as metal ions may be chelated through ion exchange and be much more firmly bound.

The layers of soil below the surface of the ground are each comprised of a mixture of various constituents such as organic matter, clays, sand, inorganic salts, pore spaces, and microorganisms. This matrix may have a large surface area with many patches of organic matter containing sites on which sorption can occur. Although the bulk of sorption may be from solution, chemicals present in the vapor state might also be adsorbed, depending on their equilibrium concentrations in the vapor, liquid, and gaseous phases. Adsorption of chemicals from aqueous solution is, in most cases, an exothermic process (Freed, 1982). Usually a lowering in soil temperature causes an increase in the sorption onto soil.

The type of sorption mechanism depends on the nature of the chemical and the type of soil. Inorganic and organic cations may adsorb on the clay portion of soil through cation exchange reactions. Most neutral organic molecules of environmental concern follow a physical type sorption, in which there is initially a formation of a monolayer on the surface followed by a build-up of multilayers of the compound partitioning onto the organic soil phases (Freed, 1982).

The composition of the leachate and its sorption characteristics of exchange and dissolution determine how specific chemicals migrate through the aquifer. Metals or organic compounds previously adsorbed on the soil may be displaced by a new leachate having compounds with a higher sorption potential. The flow of uncontaminated groundwater may also allow desorption of chemicals and alter leaching rates into the groundwater. Components in waste leachate may exceed the sorption and/or exchange capacity of the soil and channeling of high concentrations could occur because of saturation of the available soil sorption sites and spatial variability in soil composition. In certain isolated situations, metals may chemically precipitate out of solution and clog the pore spaces, causing channeling around that zone.

Transport by vaporization

The vapor pressures and Henry's law constants of organic compounds vary over a wide range, from highly volatile substances such as chloroform and vinyl chloride, to moderately volatile compounds such as benzene, to materials of low volatility such as DDT and PCBs (Table 3-1). Although vapor pressure, per se, is an important determinant in the extent of release of a chemical from a disposal site into the atmosphere, other factors must also be considered. For example, when the chemical is bound to the surface of soil particles, vapor pressure expressed simply in terms of vapor-liquid equilibrium cannot be used as an index for vapor transport. Other factors may also affect the release of the chemical, including temperature, soil moisture content, aqueous solubility of the organic compound, its initial concentration in the soil, and the soil pH. For organic compounds dissolved in water, the Henry's law constant of the compound is the important factor in the entry into the atmosphere. The Henry's law constant is proportional to the ratio of vapor pressure and solubility. The vapor pressure of methyl ethyl ketone is about equal to that of trichloroethylene, but the Henry's law constants are vastly different because methyl ethyl ketone is much more soluble in water than trichloroethylene (Table 3-1). Thus, methyl ethyl ketone has a greater tendency to remain in solution, while trichloroethylene is more likely to evaporate from the water into the vapor phase and ultimately enter the atmosphere.

Transport by particles

Organic and inorganic compounds can become sorbed onto soil particles at the chemical disposal site. The log P value of an organic chemical is directly related to its physical sorption onto soil organic matter; for inorganic chemicals, ionic binding and other types of attraction occur. These particles may become airborne by wind erosion (e.g., during remedial action at the site) and any compounds bound to them may be transported from the chemical disposal site via the air. Particles may also be transported as suspended solids in surface waters by soil runoff during rain storms. Some hazardous wastes, such as asbestos, can become dispersed as particulate solids or fibers.

The presence in the atmosphere of suspended dust or other solid aerosols may result in the sorption of some of the chemical vapors onto the particles, and consequently will increase the partition function of the chemical between the atmospheric phase and the other phases of the environment (Freed, 1982). The chemicals sorbed to the airborne particles may then be transported by winds for great distances.

Transformation of Chemicals

Compounds may undergo biological or chemical degradation and be converted to other products. Usually these degradation products are chemically simpler than the parent compound. For example, many organic compounds may ultimately undergo biodegradation to form either carbon dioxide or methane (depending on whether the conditions are aerobic or anaerobic). In some cases degradation or reaction products may be more hazardous than the parent compounds as in the breakdown of trichloroethylene which results in the transient formation of intermediates such as vinyl chloride (McCarty, 1984). Some pairs of compounds may react to form larger molecules. This type of reaction may not be as common as degradation, but little information exists on these possible reactions and products.

Within the disposal site

Chemicals may undergo chemical reactions or biological degradation during storage at the chemical disposal site. The rate and extent of transformation are a function of physical and chemical reaction processes, diffusion, concentration of each chemical, soil temperature, pH, presence of oxygen and redox conditions, etc. It is possible that discarded waste materials in highly concentrated solutions may be toxic to microorganisms and would not be readily biodegraded. Some chemical reactions require energy or elevated temperatures, so even if reactive chemicals come in contact by mixing at the site, these reactions may not occur or may proceed so slowly as to be inconsequential. Other types of reactions are explosive and some waste mixtures are ignitable.

During transport

Chemical degradation

Many factors affect the rate and extent of chemical reactions. The pH and the buffering capacity of the solution may influence the solubility and chemical stability of waste mixtures. The presence of salts in an aqueous solution may cause an ion-association or ion-pair formation. Temperature also affects the reaction rates as well as the solubilities of the compounds.

Surface activity of the chemicals themselves, and of their sorption onto soil sites, will affect the rate and extent of chemical reactions.

The chemical reactions which may occur in groundwater include oxidation, reduction, hydrolysis, polymerization, isomerization, alkylation or dealkylation, esterification, and several types of conjugation reactions with naturally occurring metabolites (Freed, 1982). Many of these reactions can be observed in laboratory experiments designed to study mixtures of discarded wastes; but because information on the rate and extent of the reactions is very limited, it is not known if they occur in nature.

The presence of a hazardous waste leachate can change the physical and chemical characteristics of the surrounding environment. In addition, some reaction products may be as detrimental to groundwater transport as the hazardous chemicals themselves. For example, insoluble hydroxide or carbonate reaction products or carbon dioxide or methane can block the pores in the saturated zone and cause channeling of the flow elsewhere.

Biologic degradation

The primary decomposers active in the biological degradation of those chemicals which occur in and are released from landfills include aerobic and anaerobic bacteria, and probably fungi. Until recent staining and microscopic counting techniques were developed, microbiologists analyzing groundwaters to detect microorganisms had to rely on plate culture and counting techniques. Because few colonies developed from plated groundwater samples, it was assumed that the region below the root zone was largely devoid of life. Recent investigations have shown that there are surprisingly high numbers of bacteria in shallow water-table aquifers (Wilson and McNabb, 1983). However, there is little evidence of the presence of yeasts, fungi, protozoa, or higher animals, so the bacterial assemblages are apparently free of predators.

Biochemical analyses of microbial biomass give good agreement with expected cell numbers based on microscopic counts (Wilson and McNabb, 1983). Bacterial numbers remain relatively constant with increasing depth in the soil column down to depths of 6 m, and sampling during different seasons and from replicate holes at the same site indicate only small differences in cell counts.

New evidence is accumulating to indicate that groundwater microorganisms convert many of the synthetic organic chemicals into inorganic products (Wilson and McNabb, 1983). Some other compounds are transformed only by cometabolism; that is, microbial populations presumably are growing on another substrate while performing this transformation. Some organic molecules are resistant to microbial attack and are persistent. The microbial transformation processes may lead to environmental detoxification by total degradation of the pollutant, to formation of new organic or inorganic toxicants, or sometimes to biosynthesis of persistent products (Alexander, 1981).

The rate and degree of biodegradation appears to be related to environmental factors such as the presence or absence of oxygen (Table 3-3). Chloroethanes and chloroethylenes are not known to be degraded in oxygenated groundwater, but degradation is possible in anoxic groundwater, that is, without dissolved oxygen. Conversely, benzene and substituted benzenes are not broken down or are broken down only slowly under anoxic conditions, but probably are degraded in groundwater containing dissolved oxygen. Compounds present in trace concentrations are usually biodegraded more slowly than when

TABLE 3–3. Potential for biotransformation of selected organic chemicals in water-table aquifers (Wilson and McNabb, 1983).

Class of Compounds	Oxygenated Water >100 μg/L[a]	Oxygenated Water <10 μg/L[a]	Anoxic Water
Halogenated Aliphatic Hydrocarbons			
Trichloroethylene	none	none	possible*
Tetrachloroethylene	none	none	possible*
1,1,1-Trichloroethane	none	none	possible*
Carbon Tetrachloride	none	none	possible*
Chloroform	none	none	possible*
Methylene Chloride	possible	improbable	possible
1,2-Dichloroethane	possible	improbable	possible
Brominated Methanes	improbable	improbable	probable
Chlorobenzenes			
Chlorobenzene	probable	possible	none
1,2-Dichlorobenzene	probable	possible	none
1,4-Dichlorobenzene	probable	possible	none
1,3-Dichlorobenzene	improbable	improbable	none
Alkylbenzenes			
Benzene	probable	possible	none
Toluene	probable	possible	none
Dimethylbenzenes	probable	possible	none
Styrene	probable	possible	none
Phenol and Alkyl Phenols	probable	probable	probable†
Chlorophenols	probable	possible	possible
Aliphatic Hydrocarbons	probable	possible	none
Polynuclear Aromatic Hydrocarbons			
Two and three rings	possible	possible	none
Four or more rings	improbable	improbable	none

[a] Initial concentration of chemical
*Possible, probably incomplete
†Probable, at high concentration

present in higher (but non-toxic) concentrations because reaction rates are a function of concentration. However, Boethling and Alexander (1979) have demonstrated that rates of degradation of trace amount of glucose are even below those that would have been predicted by Michaelis-Menton kinetics. Wilson & McNabb (1983) suggest that optimum concentrations of trace organics may stimulate their own degradation by providing energy for colony growth. Compounds may also induce enzyme activity in a metabolizing species which results in more rapid degradation than would be predicted based on initial conditions.

Photodegradation

A type of reaction particularly important for the destruction of a number of compounds is photochemical breakdown (Chameides and Davis, 1982). These reactions may occur for chemicals on soil surfaces, in surface waters, or in the atmosphere as vapors or sorbed to soil particles. For some compounds, photochemical reactions result in rapid degradation and detoxication, but in others (such as dieldrin), a photochemical product may be formed that has even greater toxicity than the parent compound (Freed, 1982). Photodegradation is only an important process at exposed surfaces and will not occur within the waste masses below the surface of disposal sites.

ROUTES OF TRANSPORT AND EXPOSURE

The mechanisms of release and migration of hazardous chemicals from land disposal sites and surface impoundments include leaching and aqueous transport, volatilization, and wind-borne particulate transport. Ideally, a detailed "materials balance" could be developed for each site. The "materials balance" would be based on the total amount of each hazardous waste chemical deposited at a site, how much of each remains within the site (either unchanged or ultimately degraded), and how much migrates from the site in aqueous solutions via groundwater and surface water, in the vapor phase by volatility, or via soil-adsorbed transport via wind erosion or surface runoff. In fact, the bulk of some chemicals will probably never be transported from the site because the chemicals would break down in transit or, conversely, because they are so persistent (not chemically or biologically degraded) and tightly bound that they would not migrate. However, this type of "materials balance" is not yet feasible because of the major limitations in the availability of data to assess each of the routes of exposure.

Aqueous Transport

Where chemicals have been disposed of on land, there is a potential for their release from the disposal site and subsequent transport through groundwater or in surface waters. The rate at which this occurs depends on the physical and chemical characteristics of the chemicals, the nature of the waste mixtures at the site, their passage across the boundary of the chemical disposal site, and the geological characteristics of the subsurface zone surrounding the site. Chemicals in a disposal site situated adjacent to surface water may run off directly into the water body, while it is possible that chemicals in sites some distance away from surface water may enter the water body via groundwater transport. Groundwater is the source of drinking water for approximately half the population of the United States (Pye et al., 1983) and is rarely treated prior to human consumption except for chlorination of most public water supplies. However, this treatment is for disinfection purposes and would not remove hazardous chemicals; in fact, chlorination in the presence of organic precursors may result in the formation of new chlorinated organic compounds. Ion exchange or activated carbon column treatment would be more effective in the removal of most hazardous chemical compounds.

Groundwater

Critical to the assessment of groundwater transport of hazardous chemicals is an understanding of the nature of geohydrologic flow and the behavior of chemicals in the subsurface environment (Freeze and Cherry, 1979; Miller, 1981). Most soils contain some water in the pore spaces between soil particles, even soils that feel "dry" to the touch. The soil layers above the water table are the "unsaturated zone" (the vadose zone or the zone of aeration above the water table), whereas below a certain depth (variable throughout the country and dependent on many factors), the soil pore spaces are saturated with water (the "saturated zone" of a water-table aquifer). Water generally percolates vertically downward through the unsaturated zone, but usually flows horizontally in the saturated zone. The term "water table" is considered to mean the top of the saturated zone of the groundwater.

Groundwater in the saturated zone occupies almost all the pore spaces and cracks between soil particles and within rocks (Miller, 1981). Unlike the turbulent flow and eddy diffusion mixing of surface-water systems, the flow of groundwater in homogeneous granular deposits is laminar; it moves along distinct, parallel paths through an aquifer with very little mixing (Freeze and Cherry, 1979). Hydrodynamic dispersion allows the gradual spread of a contaminant front outward from its laminar flow channels; in the process, the peak concentration of the contaminant decreases because of dilution. Although clay zones within the ground can retard the rate of flow, almost no component of the groundwater system is impermeable. Given enough time, groundwater will flow through clays; furthermore, contaminants can diffuse through clay even when the rate of aqueous flow is very slow. The diffusion coefficient is reduced by the degree of sorption; thus, the diffusion rate may be reduced by sorption processes.

Groundwater is recharged by rainfall and by inputs from surface waters in the central, upland portions of a land mass. The direction of groundwater flow is determined by the hydraulic gradient. Groundwater moves from regions of high hydrostatic head toward those of lower hydrostatic head. The level of standing water in unpumped wells is an indicator of hydrostatic head. Groundwater discharges into surface water bodies (streams, lakes, oceans) where the hydrostatic heads are the lowest.

Groundwater flow rates in sandy aquifers, such as in southern Long Island, New York, are generally on the order of a few inches to a few feet per day (Miller, 1981). In a fractured rock aquifer, flow rates may be more rapid or much slower, depending on the porosity of the rock and the degree of fracture; thus, an aquifer containing contaminated groundwater may be slowly transporting the accumulations of decades of leachate from a chemical disposal site (Mackay et al., 1984). Therefore, it may take years before the chemicals from a disposal site start appearing in nearby drinking water wells.

Figure 3-2 shows examples of waste disposal sites located above various types of aquifers. Mackay et al. (1984) have reported estimates of the flow rates through various types of aquifers (Table 3-4). Calculations were performed to estimate the transport time of groundwater or a conservative contaminant (i.e., a substance in groundwater which is not chemically or biologically degraded and which is not retarded) moving from a waste disposal site for a distance of 1 kilometer. For the three types of aquifers shown in Figure 3-2, the ranges of calculated transport times are estimated to be as follows: for sand, 10 years to 1,000 years; for unweathered clay, 1,000 years to 100,000 years, but there would be much more rapid flow through fractured clay because of channeling through the fractures; and for fractured rock (karst limestone), 1 year to 10,000 years. These calculations take into account the range of reported porosity values for each type of aquifer, hence the resulting ranges of estimated transport times.

Because of laminar flow, contaminants migrating in a porous aquifer are not diluted by turbulent mixing but rather by hydrodynamic dispersion, which is the

Figure 3–2. Examples of Waste Disposal Sites Located Above Various Aquifer Types

POROUS MEDIA:

Sand: Migration controlled by advection and dispersion

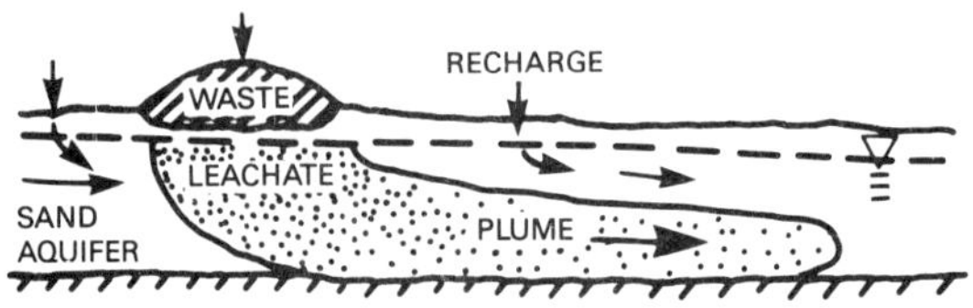

Clay: Migration controlled by diffusion

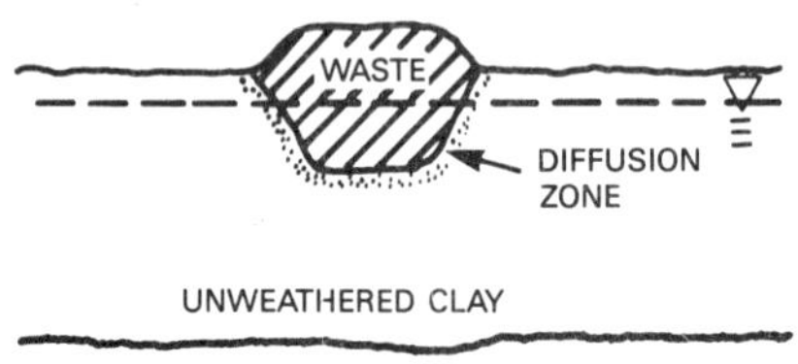

Fractured Rock: Migration controlled by advection in fractures

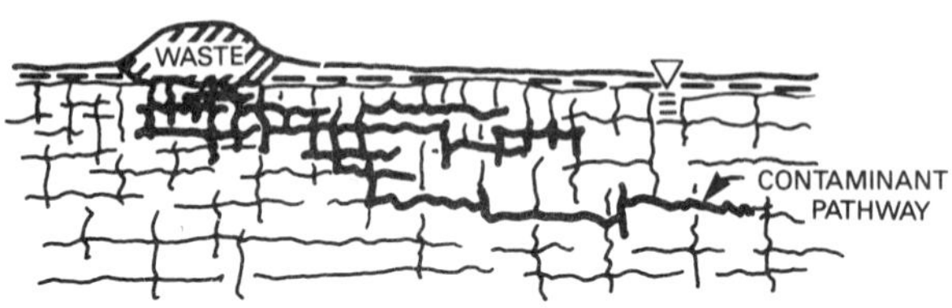

slow process of a plume spreading to occupy an increasing three-dimensional volume as it moves downgradient in the aquifer. This process is similar to smoke rising and drifting downwind while slowly spreading out on an almost windless day. Dispersion can vary considerably within an individual aquifer because of subsurface zones of differential permeability. For example, a pocket of sandy soil and a lens of clay may occur side by side. Because the surface of the ground would look very similar above zones of clay and of sand, it would be difficult to predict the three-dimensional movement of chemicals by only looking at the ground surface. Therefore, the prediction of concentrations of a chemical over time and with distance from the source

TABLE 3–4. **Estimated groundwater flow velocities, travel times, and directions of flow (horizontal or vertical) for aquifers of common granular media.**

	Average Linear Velocity (m/yr)	Travel Time (yr)	Travel Direction and Distance
Aquifers			Lateral transport for 1000 m
Gravel[a]	10 to 10^3	1 to 100	
Sand[a]	1 to 10^2	10 to 1000	
Aquitards			Vertically downward flow for 10 m
Silt[b]	10^{-2} to 1	10 to 1000	
Clay[c]	10^{-4} to 10	1000 to 100,000	

[a]Gravel and sand are assumed to contain less than a few percent silt and clay size fractions. *Source: Mackay et al. (1984)*
[b]Silt is assumed to contain less than 5% clay size fraction and less than 50% sand.
[c]Clay deposit is assumed to contain at least 25% clay size fraction.

would require a detailed understanding of the three-dimensional characteristics of the aquifer.

The shape and size of contaminated plumes in groundwater depend on a number of factors: subsurface geological structures, soil porosities/aquifer permeabilities, localized and regional groundwater flow patterns, chemical characteristics and original concentrations of the hazardous wastes, variations in rainfall, rates of leaching, and withdrawal rates from downgradient wells. In some types of aquifers, plumes spread out widely, while in others, they become elongated. Some plumes have been traced for several miles by sampling the groundwater downgradient from chemical disposal sites (Miller, 1981).

Density of the contaminated wastes is another factor in the formation and movement of a contaminant plume, especially for organic waste mixtures having fractions with low water solubility. The mixtures would percolate downward through the unsaturated zone, probably more rapidly than water-soluble leachates. Once in the groundwater, high-density mixtures would descend to the bedrock ("sinkers"), low-density mixtures would tend to flow near the top of the water table ("floaters"), and intermediate density mixtures which are immiscible with water would flow along the mid-depth of the water column ("mixers"). In each case, some of the mixture could differentially dissolve into the groundwater and move at a different rate from the insoluble and immiscible organic slug.

Figure 3-3 illustrates the movement of a contaminated plume in a typical aquifer with a porous medium (sandy soil) and indicates the effects exerted on the vertical configuration of the plume by the natural flow system. Sampling the mid-depth well would detect the presence of contaminated groundwater, but sampling only the deep and shallow wells would have entirely missed the presence of a contaminated water plume.

The first step in assessing the groundwater quality at a chemical disposal site is a visual survey of the surface of the ground to detect potential sites of discharge to groundwater (Carriere and Canter, 1980; Todd et al., 1976). The next step is a review of the history of the site, including records of spills, leaks, corroding barrels, specific chemicals disposed of at the site, etc., if the information is available (Carriere and Canter, 1980; Todd et al.,

Figure 3–3. Typical Route of Flow of Contaminants in a Water Table Aquifer in a Humid Region (Adapted from Miller, 1981)

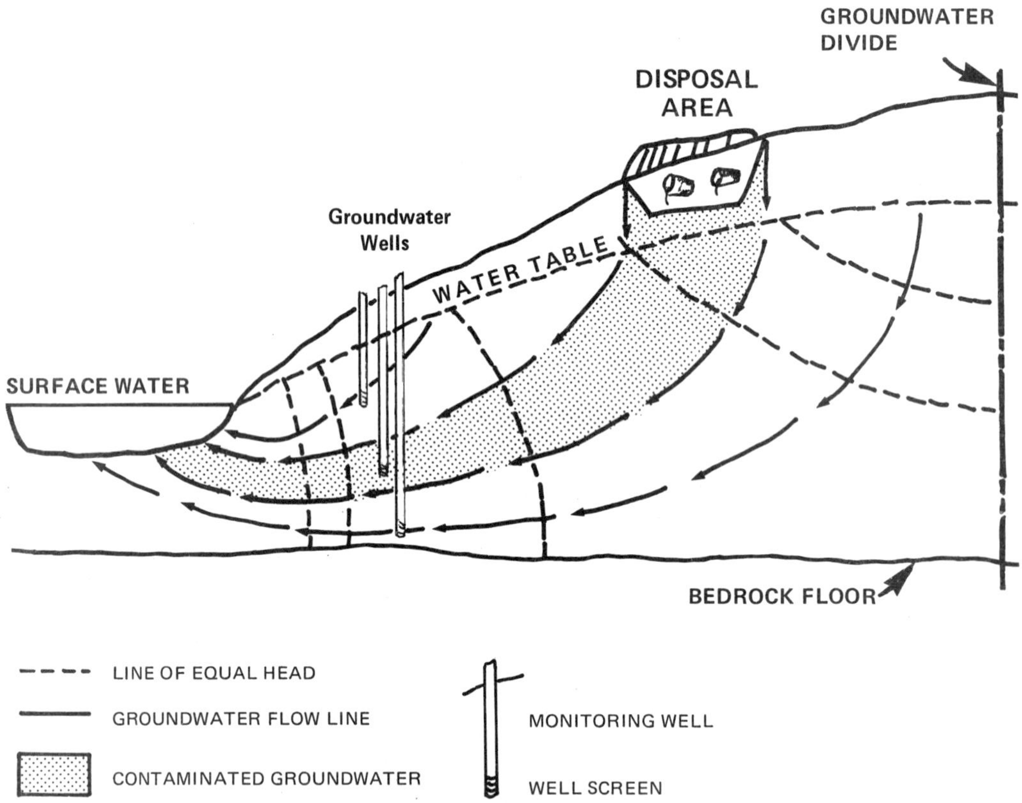

1976). Finally, a hydrogeologic investigation is performed, including the collection of soil borings to determine subsurface soils and geological formations, and wells are installed to sample groundwater and to determine the hydraulic gradient. Because the topographic elevation of the ground surface is not directly related to depth-to-groundwater or to the hydraulic gradient, the initial network of wells must be extensive so as to define both upgradient and downgradient conditions. This investigation will yield information on water table elevations, horizontal and vertical flow patterns, and contamination of the aquifer. The results will provide data for the delineation of pollution sources, evaluation of potential contamination of drinking water wells, and determination of data gaps and the requirement for and location of additional sampling wells.

Miller (1981) reported that this type of groundwater survey at a chemical disposal site could cost from \$50,000 to \$250,000, and possibly require the drilling of dozens of test wells. The survey, however, will only permit determination of the present concentration of contaminants at selected points and a crude estimate of their distribution and rate of movement. In order to predict the future direction and rate of movement of the plume or the concentrations of specific chemicals, development of site-specific mathematical models would be required, necessitating the acquisition of further data on subsurface geology and flow rates. The models may not be expensive, but data acquisition would be, possibly costing up to hundreds of thousands of dollars because of further sampling and the increased number of wells which would have to be dug or drilled.

Surface water

Some chemical disposal sites are located sufficiently near water bodies that hazardous wastes can directly enter the surface water by surface runoff. Figure 3-3 illustrates the principle that groundwater in most aquifers "eventually" enters surface waters, potentially carrying hazardous wastes from sites which are distant from surface waters.

Once hazardous wastes enter the surface water, mixing and dilution occur much more rapidly than in groundwater. Certain hazardous wastes will readily volatilize into the air. Microorganisms may biodegrade contaminants more rapidly in surface water than in groundwater. The aerobic conditions of most surface waters would permit chemical oxidation reactions to proceed more rapidly than in those deep groundwater zones which are anoxic. However, some hazardous wastes are not capable of being readily biodegraded or chemically broken down, for example, certain PCBs and polycyclic aromatic hydrocarbons.

The mixing and dispersion of contaminants in surface waters and their long-distance transport in rivers permit the distribution of waste chemicals to distant locations as compared with the more narrowly-defined geographical distribution of the slowly moving plume in groundwater aquifers. Since many surface waters are utilized as drinking water sources, this wider distribution of surface water contamination may mean that more people are potentially exposed to the contaminants.

Sampling surface waters is easier and less expensive than sampling groundwater because collection of surface water samples is direct and there is no need to dig individual wells at each sampling location. Chemicals present in surface waters can be detected more readily than those in groundwater because the degree of mixing and dispersion within surface waters makes for a more uniform distribution and thus makes their detection more feasible using the current sensitive analytical techniques. The limited mixing of groundwater means that a plume of contaminated water may be missed if only a few wells are sampled.

Other Routes of Exposure

Soil

Many of the hazardous wastes at chemical disposal sites will readily sorb onto soil particles (Chiou et al., 1979; Kenaga and Goring, 1980). The sorption of these compounds is a function of their low solubility in water and high solubility in organic solvents, reflected as a high octanol/water partition coefficient (Kow; log P). This relationship has been utilized to mathematically model the sorption potential of chemicals onto soil organic matter and clay particles.

Humans may be exposed to soil-bound hazardous wastes via direct contact at the chemical disposal site, by inhalation or dermal contact of particles transported from the site via wind erosion, or by exposure to suspended particles in surface waters which escape from sites in surface runoff. In heavily urbanized areas where a population is in close proximity to the disposal sites and where groundwater is not utilized as a drinking water resource, inhalation of wind-borne particles could be a major route of contact with hazardous waste compounds (Johns Hopkins University, 1981). The soil-borne route of transport may be especially important during remedial action such as site excavation. Contaminants sorbed to particles of wind-borne soil may not be detected by organic vapor detectors which are used to determine atmospheric vapors, mists, and gases.

Air

Certain hazardous waste compounds have high vapor pressures and low solubility in water and thus exhibit high Henry's law constants; they will be transferred readily to the vapor phase and ultimately to the atmosphere. Evaporation from surface impoundments is one route of transport into the atmosphere and usually occurs both by the formation of gas bubbles rising to the surface and by single molecules crossing the water-air interface. Compounds can also sublime from the surface of landfills or volatilize from within the ground.

Estimation procedures are available to calculate the rate of evaporation from aqueous solutions (Chiou, 1980; Dilling, 1977; Mackay and Leinonen, 1975). Compounds evaporate more rapidly if they have low molecular weights, low water solubilities, and high vapor pressures. The estimated values of the aqueous evaporation half-life (time necessary for 50% of the dissolved compound to evaporate from shallow depths) range from about 20-30 minutes for the chloromethanes and chloroethanes (Dilling, 1977), to about 5 hours for benzene, to about 3 days for DDT (Mackay and Leinonen, 1975).

Hazardous wastes in landfills can also be transported upwards as gases through the soil and into the atmosphere (Shen, 1981; Thibodeaux, 1981). For buried wastes, the gases must diffuse through air pores in the soil. In landfills where the liquid waste has been co-disposed with refuse, anaerobic degradation will generate additional gases which may accelerate the movement of the hazardous waste vapors upwards at a greater rate than at a chemical waste site without biologically generated gases. Table 3-5 illustrates calculated data for volatilization from three types of disposal sites currently being used for hazardous wastes (California, 1982). Although calculated rates of volatilization for a covered landfill are low, at sites where considerable volumes of waste are disposed, the actual volumes of waste materials which evaporate into the atmosphere may be substantial. On a tons/yr basis, evaporation from southern California waste disposal sites is similar in magnitude to the industrial atmospheric emissions in that area (California, 1982).

Food

Some hazardous wastes may be lost from the soil via uptake by plants and by animals living on or in the soil (Litwin et al., 1983). Movement of water and dissolved hazardous wastes is usually downward but may occur upward,

TABLE 3–5. Estimated time needed for 75% of selected compounds to volatilize in various disposal methods.

Waste Component	Disposal Method		
	Landspreading	Surface Impoundment	Covered Landfill
Phenol	39 days	*	333 years
Tetrachloroethylene	*	12.4 days	*
Benzene	5 days	11.6 days	41 years
Ethanol	16 hours	*	*
Methyl Ethyl Ketone	7 hours	*	*
Ethyl Acetate	*	*	2.6 years
Trichloroethylene	7 hours	11.4 days	2 years
Chloroform	*	11.3 days	11 months
Carbon tetrachloride	*	10.3 days	*
Dichloroethane	3 hours	9.1 days	11.7 months
Dimethylamine	0.7 hours	*	*
Ethylene	*	7.4 days	*
Pentane	*	*	7.8 months
Dichloroethylene	*	*	4.2 months

*Data not available to calculate rate for this substance. *Source: California Air Resources Board (California, 1982)*

especially by capillary action and evaporation as drying of the soil surface takes place (Freed, 1982). Thus, the chemicals may migrate upwards towards the root zone of plants.

Uptake by plants is a function of the availability of the chemical in the soil in the vicinity of the plant roots and of the ability of the plant to absorb the compound. Water solubility is an important factor in transport through the soil, absorption by the roots, and translocation up through the roots, stem, and leaves. The larger the log P value of a compound, the greater is its potential for bioaccumulation in plant tissue (Chiou et al., 1977; Kenaga and Goring, 1980). Accumulation and storage of a chemical within a plant are also dependent on the plant's ability to assimilate and to metabolize the chemical.

Animals may be exposed to hazardous substances by direct contact at the chemical disposal site or indirectly by consuming contaminated plants or other animals. Human exposure could occur via the ingestion of either wild game animals or grazing farm animals which contain bioaccumulated hazardous chemicals. There is also a possibility of human exposure by ingestion of contaminated food crops.

There is a direct correlation between log P and the bioconcentration factor in fish, indicating that hazardous wastes entering surface waters may be absorbed and stored by fish and other aquatic organisms (Kenaga and Goring, 1980). Eating contaminated fish could be a significant source of human exposure to hazardous chemicals, especially if fish comprise a large part of the diet. For example, human consumption of DDT contaminated fish from Indian Creek near Triana, Alabama resulted in blood levels of the chemical which were several times the national mean (Kreiss et al., 1981).

TREATMENT AND CONTROL OF HAZARDOUS WASTES TO LIMIT ENVIRONMENTAL RELEASES

RCRA has mandated that more careful thought and engineering considerations be employed in the selection and design of chemical disposal sites. The disposal of certain waste mixtures and specific hazardous chemicals requires extensive precautions to minimize releases to the environment. For example, liners and caps are to be installed to retard the rate of releases, underground piping systems are to be constructed to intercept and divert leachate for monitoring and subsequent treatment, and safety and health considerations and environmental protection concerns are to be considered in transporting waste materials to the treatment, storage, and disposal (TSD) facility.

Many of the hazardous materials disposed of at RCRA-controlled chemical disposal sites may be prepared or processed in various ways prior to disposal or may be treated in situ to lessen the potential for environmental and health effects. Several of these treatment and control methods are briefly described in this section. Methods of treatment prior to disposal include waste reduction, incineration, and various types of physical, chemical, and biological treatment (Seller and Canter, 1980; USC/OTA, 1983). Treatment methods at the site include chemical stabilization, biological and/or chemical conversion, and encapsulation (Fields and Lindsey, 1975; Suess and Huismans, 1983). Many case studies are included in the references cited, and other management and control methods may be available for specialized wastes, such as marine disposal or underground injection in deep wells. This section is intended as a survey of the most feasible control methods; the salient advantages and/or disadvantages of each are briefly described.

Hazardous Waste Treatment at the Waste Production Site

Waste reduction

- Recovery/Recycling of the Waste: 1) may involve processes in-plant to recover raw materials, by-products, or solvents; 2) can be accomplished at off-site locations by commercial operators treating wastes from a number of facilities to recover materials; and 3) can also include material exchanges whereby the waste materials from one firm might serve as the raw materials for another firm's production or processing.

- Process Modification could involve major restructuring of the piping systems of a facility at great cost or could involve minor operational changes such as adjustments in temperature or pressure of processing (possibly resulting in more efficient production).

- Source Segregation involves the separation of the hazardous constituents from the bulk of the non-hazardous waste to obtain a smaller volume of concentrated hazardous waste.

- End-Product Substitution requires the replacement of a product, the manufacture of which is hazardous-waste intensive, with a new product in which hazardous waste by-products are reduced or eliminated. This option may have a long lead time because changing consumer purchasing practices would be necessary and would take time, and because the construction of new production facilities may be required.

Incineration

The incineration of hazardous waste-containing materials would probably require construction of an on-site combustion chamber facility, or transport of the waste to an off-site incinerator, including at-sea incinerators. Advantages include reduction of waste volume and nearly-complete waste destruction. Disadvantages include energy costs, installation of air pollution control equipment, and the generation of residues, both solid (ashes and small amounts of unburned materials) and gaseous (vapors of unburned combustion products, oxides of sulfur, nitrogen or other elements, and other combustion products). Moreover, some of the residues formed may be extremely toxic, such as dioxins (for example, 2,3,7,8-tetrachlorodibenzo-p-dioxin) and polycyclic aromatic hydrocarbons (for example, benzo[a]pyrene).

Physical, chemical, or biologic treatment

Pretreatment of the hazardous waste at the production facility may reduce the amounts of hazardous constituents generated (Seller and Canter, 1980). Physical treatment includes reverse osmosis, dialysis, evaporation, carbon sorption, ammonia stripping, filtration, sedimentation, and flocculation. Chemical treatment includes ion exchange, neutralization, oxidation, reduction, precipitation, and calcination. Biological treatments include activated sludge, aerated lagoons, trickling filters, waste stabilization ponds, and rotating biological contactors, as well as anaerobic wastewater treatment processes such as sludge digestion. Another new technique which shows some promise for the future involves photochemical treatment of waste materials.

Hazardous Waste Treatment at the Waste Disposal Site

Chemical stabilization

Waste stabilization involves a chemical and/or physical process in which a reaction occurs to fix the waste into a form less susceptible to environmental transport. Adjustment of the pH of some wastes will be the simplest stabilization process. Other such processes include fixation of the waste in a solid matrix (such as plastic, cement or asphalt) or a semi-solid matrix (such as urea-formaldehyde or sulfur foam).

Biologic degradation

Conventional biological treatment involves naturally occurring microorganisms, but many hazardous waste compounds are not degradable by conventional biological means. Biotechnology involves the utilization of bacteria which are selected from nature and acclimated to particular hazardous waste substances to markedly increase the rate of biodegradation. At present, many new biotechnological processes are being studied and developed for possible commercial use.

Chemical degradation

Chemical degradation may be more expensive than incineration or biodegradation. Nevertheless, some chemical procedures may be quite promising for certain wastes. Strong oxidants may be effective in degrading water soluble compounds; chemical dechlorination would break down chlorinated organics, such as some pesticides and possibly PCBs; hydrolysis is effective against organophosphorus and carbamate pesticides. No single procedure for chemical conversion would degrade all hazardous wastes.

Encapsulation

Encapsulation involves the isolation of wastes from the surrounding environment. Chemical stabilization of hazardous wastes is one method of encapsulation; another method involves both impermeable covers over and impervious bottom liners under the chemical disposal sites. Because plastic and clay liners are not completely fail-safe, it may be necessary to install a leachate collection system prior to deposition of any wastes at the disposal site. If any water from the surface percolates through the overlying cap and into the landfill, waste materials may dissolve into the water forming leachate which could percolate through the bottom liner. Periodic chemical analysis of aqueous materials in the leachate collection system would detect any failures in the integrity of the cap and liner system. If any hazardous substances have percolated out of the site, there may be a need for leachate treatment to degrade the hazardous substances prior to any intentional discharges into receiving waters. Current RCRA regulations require monitoring of leachate collection systems for 30 years (USEPA, 1980), although the probability of liner failure and leachate production may continue after closure of the site.

SUMMARY

The chemicals in disposal sites exhibit an enormous diversity of physical and chemical characteristics, and the sites themselves are diverse. The great number of disposal sites, the lack of precautionary practices to retard environmental releases at most of the old sites, and the dearth of documentation regarding quantities and types of chemicals which have been disposed at sites suggest the futility of defining a "typical" site and indicate the enormity of the remediation task. Improvements in chemical disposal practices offer hope that releases of toxic substances can be reduced in the future but it has yet to be proven that modern containment procedures can reduce the rates of release to a negligible level.

Analyses of samples from the vicinity of chemical disposal sites have revealed a variety of chemical waste constituents which have been released and have migrated into the atmosphere, surface water, groundwater, and soil. The greatest potential for health effects is believed to be associated with organic chemical contamination of groundwaters used for public water supplies. Such contamination of aquifers has already reached levels of a significant public health threat in certain areas.

There are a number of different types of preventive/remedial actions to treat contaminated groundwater which is to be utilized as drinking water. The most important is to design the facility to minimize any releases into the groundwater. At many existing chemical disposal sites, reports of groundwater contamination have prompted local and state health officials to close the drinking water wells near those sites. The temporary solution is to transport the required quantities of drinking water to the affected users. Other preventive/remedial actions include constructing a pipeline from another source, or treating the aquifer by filtration through activated carbon filters.

The chemicals reported in drinking-water aquifers in the vicinity of disposal sites include chlorinated solvents, aromatic hydrocarbons, pesticides, trace metals, and polychlorinated biphenyls. A salient feature of many of these chemicals in groundwater is the long-term nature of the contamination. Many chemicals penetrate gradually into the groundwater zone from points of surface or subsurface disposal, are transported very slowly in the direction of the hydraulic gradient, and resist biological and chemical degradation. The time constants for water quality changes in groundwater approach those that characterize oceans and large lakes and, therefore, are much greater than those for streams. There are also time lags for groundwater transport, which are typically measured in decades. This sequencing of time lags seriously complicates the assessment of the potential for health effects from chemical contamination and suggests that present attempts to assess health impacts based on estimated past exposures may be underestimating the ultimate risk stemming from future groundwater contamination. It is likely that groundwater contamination will remain a concern for many years to come, even if efforts to control releases from present and future chemical disposal sites are largely successful.

The understanding of contaminant transport in groundwater has improved greatly in recent years as a result of research in basic principles, laboratory simulation, and, in a limited sense, methods for predicting the movement of some classes of contaminants. Reliable prediction requires detailed knowledge of the physical, chemical, and microbiological characteristics of the subsurface environment; this knowledge is attainable only at enormous cost. In most cases, the available resources are inadequate to provide the data and understanding requisite for meaningful prediction of the movement of contaminated groundwater. An empirical solution to groundwater movement prediction, not without its costs, is to drill a series of wells and frequently sample the groundwater to detect the actual pattern of movement.

The belated recognition of the magnitude and prevalence of groundwater contamination stems, in part, from recent public awareness of growth in production, use, and disposal of chemicals, as well as advances in analytical methodology. Since World War II, there has been a steady increase in U.S. production of synthetic organic chemicals, and presumably also in dispersive use and disposal. Analytical methodology has improved and the development of increasingly sophisticated instrumentation has resulted in the identification of more types of chemicals at much lower detection limits than in the past. Thus, it should come as no surprise that the increases in disposal of chemicals and the ability to identify these chemicals in environmental samples have yielded an increased frequency in the observed cases of groundwater contamination.

It is essential to recognize that chemicals might also escape from disposal sites and migrate in the air, surface waters, and soil as well as bioaccumulate in plants and animals in the vicinity of the sites. Groundwater is the route of transport addressed in this report as the primary cause of concern; however, the other routes of transport and human exposure must also be taken into consideration in assessing the potential health effects resulting from the release of chemicals from disposal sites.

References

Alexander, M. (1981) Biodegradation of chemicals of environmental concern. Science 211:132-138.

Boethling, R.S.; Alexander, M. (1979) Microbial degradation of organic compounds at trace levels. Environ. Sci. Technol. 13:989-991.

California. Air Resources Board. (1982) Assessment of the volatile and toxic organic emissions from hazardous waste disposal in California. 303p. Available from: California Air Resources Board, [Sacramento, CA].

Carriere, G.D.; Canter, L.W. (1980) Planning for ground water quality monitoring. 87p. Available from: National Center for Ground Water Research, Norman, OK.

Chameides, W.L.; Davis, D.D. (1982) Chemistry in the troposphere: special report. Chem. Eng. News (Oct. 4):39-52.

Chiou, C.T. (1980) Evaporation of components from a miscible solution. Environ. Int. 4:15-19.

Chiou, C.T.; Freed, V.H.; Schmedding, D.W.; Kohnert, R.L. (1977) Partition coefficient and bioaccumulation of selected organic chemicals. Environ. Sci. Technol. 11:475-478.

Chiou, C.T.; Kohnert, R.L.; Freed, V.H.; Tonkyn, R.G. (1983) Predictions of evaporative loss rates of solutes in stagnant and turbulent waters in relation to rates of reference materials. Environment Int. 9:13-17.

Chiou, C.T.; Peters, L.F.; Freed, V.H. (1979) A physical concept of soil-water equilibria for nonionic organic compounds. Science 206:831-832.

Dietz, S.; Emmet, M.; DiGaetano, R.; Tuttle, D.; Vincent, C. (1984) National survey of hazardous waste generators and treatment, storage, and disposal facilities regulated under RCRA in 1981. EPA 530/SW-84-005. Prepared by Westat, Inc. Washington, DC: U.S. Environmental Protection Agency, Office of Solid Waste and Emergency Response. 318p. Available from: U.S. Government Printing Office, Washington, DC.

Dilling, W.L. (1977) Interphase transfer processes. II. Evaporation rates of chloromethanes, ethanes, ethylenes, propanes, and propylenes from dilute aqueous solutions: comparisons with theoretical predictions. Environ. Sci. Technol. 11:405-409.

Fields, T., Jr.; Lindsey, A.W. (1975) Landfill disposal of hazardous wastes: a review of literature and known approaches. EPA/530/SW-165. Washington, DC: U.S. Government Printing Office. 36p. Available from: Solid Waste Information, U.S. Environmental Protection Agency, Cincinnati, OH.

Freed, V.H. (1982) Tracing the movement of pollutants through the environment. Paper presented at the Scientists' Forum: an assessment of the scientific knowledge base related to victim compensation, January 27-29, Washington, DC. 30p. Available from: Economics and Environmental Studies Center, The Institute on Man & Science, Rensselaerville, NY.

Freeze, R.A.; Cherry, J.A. (1979) Groundwater. Englewood Cliffs, NJ: Prentice-Hall, Inc. 604p.

Johns Hopkins University. School of Hygiene and Public Health, Hazardous Waste Site Study Group. (1981) Final report: design of environmental health effects studies stemming from the potential for human exposure to toxic waste in Memphis, Tennessee. Prepared for the Tennessee Department of Public Health, Memphis. 249p. Available from: The Department, [Memphis, TN].

Kenaga, E.E.; Goring, C.A.I. (1980) Relationship between water solubility, soil sorption, octanol-water partitioning, and concentration of chemicals in biota. In: Eaton, J.G.; Parrish, P.R.; Hendricks, A.C.; eds. Aquatic toxicology: special technical publication 707. Philadelphia: American Society for Testing and Materials. p. 78-115.

Kreiss, K.; Zack, M.M.; Kimbrough, R.D.; Needham, L.L.; Smrek, A.L.; Jones, B.T. (1981) Cross-sectional study of a community with exceptional exposure to DDT. J. Am. Med. Assoc. 245:1926-1930.

Litwin, Y.J.; Hantzache, N.N.; George, N.A. (1983) Groundwater contamination by pesticides: a California assessment. Prepared for the California State Water Resources Control Board by RAMLIT Associates, Inc., Berkeley. 209p. Available from: The Board, Sacramento, CA.

Mackay, D.; Leinonen, P.J. (1975) Rate of evaporation of low-solubility contaminants from water bodies to atmosphere. Environ. Sci. Technol. 9:1178-1180.

Mackay, D.; Shiu, W.Y.; Sutherland, R.P. (1979) Determination of air-water Henry's law constants for hydrophobic pollutants. Environ. Sci. Technol. 13:333-337.

Mackay, D.M.; Roberts, P.V.; Cherry, J.A. (1984) Basic concepts of organic contaminant transport in groundwater: final draft. 30p. Available from: principal author, Stanford University, Stanford, CA.

McCarty, P.L. (1984) Application of biological transformations in ground water. Paper presented at the Second International Conference on Ground Water Quality Research, March 27, Tulsa, OK. 22p. Available from: the author, Stanford University, Stanford, CA.

Miller, D.W. (1981) Geohydrological surveys at chemical disposal sites. In: Lowrance, W.W., ed. Assessment of health effects at chemical disposal sites: proceedings of a symposium held in New York City on June 1-2, by the Life Sciences and Public Policy Program of The Rockefeller University. New York: The University. p. 23-40.

Mitre Corporation. (1983) Computer printouts of National Priorities List data summaries for the 546 final and proposed sites and 881 sites currently on the data base as of September 7, 1983. Prepared for the U.S. Environmental Protection Agency. 94p. Available from: Metrek Division, Mitre Corporation, McLean, VA.

Pye, V.I.; Patrick, R.; Quarles, J. (1983) Groundwater contamination in the United States. Philadelphia: University of Pennsylvania Press. 315p.

Seller, L.E.; Canter, L.W. (1980) Effects of hazardous waste disposal on ground water quality. 247p. Available from: National Center for Ground Water Research, Norman, OK.

Shen, T.T. (1981) Estimating hazardous air emissions from disposal sites. Pollut. Eng. 13(8):31-34.

Shen, T.T. (1982) Estimation of organic compound emissions from waste lagoons. J. Air Poll. Control Assoc. 32:79-82.

Suess, M.J.; Huismans, J.W.; editors. (1983) Management of hazardous waste: policy guidelines and code of practice. WHO Regional Publications, European Series No. 14. Copenhagen: World Health Organization, Regional Office for Europe. 101p.

TDB II. SEE Toxicology data bank II

Thibodeaux, L.J. (1981) Estimating the air emissions of chemicals from hazardous waste landfills. J. Hazard. Mater. 4:235-244.

Todd, D.K.; Tinlin, R.M.; Schmidt, K.D.; Everett, L.G. (1976) A groundwater quality monitoring methodology. Am. Water Works Assoc. J. 68:586-593.

Toxicology data bank II. (1984) [Enhanced version of data base]. Bethesda, MD: U.S. Department of Health and Human Services, National Library of Medicine, Specialized Information Services. Magnetic disks.

U.S. Congress, Office of Technology Assessment. (1983) Technologies and management strategies for hazardous waste control. Washington, DC: U.S. Congress, Office of Technology Assessment. 407p. Available from: U.S. Government Printing Office, Washington, DC.

USC/OTA SEE U.S. Congress, Office of Technology Assessment

U.S. Environmental Protection Agency. (1980) Hazardous waste management system: part 265, standards for owners and operators of hazardous waste, treatment, storage, and disposal facilities. Fed. Regist. 45:33243.

USEPA SEE U.S. Environmental Protection Agency

Van Hook, R.I. (1978) Transport and transportation pathways of hazardous chemicals from solid waste disposal. Environ. Health Perspect. 27:295-308.

Verschueren, K. (1983) Handbook of environmental data on organic chemicals. 2nd ed. New York: Van Nostrand Reinhold Company, Inc. 1310p.

Wilson, J.T.; McNabb, J.F. (1983) Biological transformation of organic pollutants in groundwater. Trans. Am. Geophys. Union (Eos) 64:505-507. Reprinted by U.S. Environmental Protection Agency as EPA-600/J-83-074. Available from: National Technical Information Service, Springfield, VA.

4

Assessment of Health Effects

ROLE OF LABORATORY TESTS IN HEALTH EVALUATIONS

It is beyond the scope of this report to review the clinical and laboratory methods used by physicians to diagnose and treat illness. We are concerned here only with those clinical and laboratory methods that may be useful in detecting possible health effects in a population possibly exposed to chemicals from a disposal site.

The diagnosis of disease in an individual generally follows a standard decision tree format based on the clinical history and physical examination. The physician then uses additional information including laboratory determinations (e.g., blood and urine tests) and more specialized procedures (e. g., radiological examinations and biopsies) in developing a possible diagnosis. In this setting, highly specific laboratory tests can be of value when they are critically interpreted. A physician's ability to link an illness to an environmental insult depends upon a number of factors including:

- High level of suspicion -- The increased frequency in the occurrence of several instances of an unusual disease or an apparent cluster of a relatively common disorder increases the likelihood for the involvement of environmental factors.
- Exposure history -- The relationship between an environmental exposure and a possible toxin is often exceedingly difficult to determine because individual lifestyles vary greatly and people have a tendency to ignore exposures that have become accepted habits. Highly focused questionnaires are essential to the development of a valid exposure history.

In contrast to the methods used by physicians to develop a diagnosis of disease in individuals, the initial detection of an excess of disease in populations near a chemical disposal site is often best accomplished by other means. Where the history and physical examination is often not practical for large groups of individuals, pertinent information can be gathered more efficiently and effectively with a well constructed questionnaire. Laboratory

HADWC-C*

testing as an initial screening procedure is usually not beneficial (USDHEW-/CCERP, 1978). Preliminary epidemiologic studies frequently find only non-specific complaints such as ambient odors, headache, dizziness, fatigue, loss of libido, or menstrual irregularities. In such instances, careful, detailed clinical and laboratory studies are seldom justified.

Clinical and laboratory studies are rarely included in preliminary studies of people at disposal sites; the use of these highly specific methods is seldom economically feasible until the health effect they detect can be related to the exposure in question. The purpose of preliminary epidemiologic surveys or studies of the chemical content of disposal sites and the extent of chemical migration is to provide leads that permit an intelligent selection of the biomedical methods best suited to study the health effects in affected populations. Perhaps the most helpful leads come from quantitative information pertaining to actual human exposure. For example, if preliminary chemical analyses detect a chemical known to cause liver damage, then clinical and laboratory methods could be selected to focus on this outcome; however, if available exposure data suggest a known neurotoxic chemical, then the clinical and laboratory methods of choice would be quite different.

Detailed clinical and laboratory studies are conducted only if preliminary studies suggest serious health effects might be expected following a long-term exposure to very low levels of toxic chemicals. Public attention has been centered on excess incidence of such outcomes as cancer, genetic defects and developmental anomalies, reproductive failures, and disabling diseases of the nervous system. This chapter offers an overview of the biomedical science involved in the consideration of these concerns and describes the useful application of clinical laboratory methods and other specialized procedures for assessing health effects in populations at chemical disposal sites. Before addressing these tests and procedures, it seems appropriate to examine their basic statistical limitations when applied as population screens.

Limitations of Laboratory Tests for Screening Populations

Clinical laboratory tests are usually performed to increase the accuracy of diagnosis and to monitor therapy. These tests commonly measure an analyte, such as a protein or lipid, or the activity of an enzyme in blood, serum or other body fluid. More complex laboratory tests may analyze the rate of appearance or disappearance in blood or urine of a test substance that has been administered orally or parenterally. These tests are primarily used to measure the functional capability of a specific organ such as the liver or kidneys.

In patients with nonspecific complaints, groups of tests are often used. Theoretically, such an approach should be useful to screen for potential abnormalities in organ or tissue function in populations of individuals who are exposed to environmental pollutants. However, the use of laboratory tests to screen random populations of generally healthy individuals poses profound problems in interpretation of results. Interpretive ambiguities arise largely from statistical considerations, which also reflect the remarkable diversity in normal organ functional capability among truly healthy individuals (Galen and Gambino, 1975; Schichman et al., 1984).

Accuracy and precision of test methods

Until the result is appropriately interpreted, a laboratory test may be of little value in the identification of persons who have an abnormality in the function of an organ or tissue as the result of exposure to an environmental pollutant. Correct interpretation is a complex process which requires knowledge of the accuracy and precision of the test, and the normal range of values found in healthy people. The predictive value of a test depends on its sensitivity and specificity, and on the prevalence of the disease or functional abnormality in the population examined. Powerful methods are required for detecting such effects in populations in the vicinity of chemical disposal sites.

The effectiveness of laboratory tests is a function of the accuracy of the test methods and the precision with which they are conducted. Accuracy is the agreement between the measured quantity and its true value and depends on the analytic methods employed. Precision refers to the agreement between repetitive measurements performed in the same way and is a measure of test reproducibility. It is an important responsibility for the analytical laboratory to assure that both the accuracy and precision of analytical procedures are as high as possible in order to obtain reliable test results. Acceptable analytical laboratories employ quality control programs. Presentation of the basic elements of adequate management of the clinical laboratory is beyond the scope of this discussion on test interpretation, which presumes that any tests are performed with acceptable levels of accuracy and precision.

Assessment of test results

The first step in the interpretation of a laboratory test involves the comparison of the test result with a "normal" range of values. It is often unclear whether the clinical definition of "normal" refers to optimal health (subjective) or simply the absence of clinical disease; on the other hand, the laboratory "normal" values often refer to the most commonly encountered objective values in groups of healthy people. Thus, the clinical and laboratory definitions of "normal" are often not equivalent. The confusion regarding the meaning of "normal," and the fact that there is usually an overlapping range of values between healthy people and those with abnormality of organ functions, has led to the concept of reference interval. This concept has been widely adopted to replace the confusing use of "normal" values.

Reference interval is a range of values found in a specified portion (usually 95%) of a healthy reference population. The range of test values found in healthy people is determined by multiple factors which include the genetic, physiologic, and environmental variables of sex, race, age, weight, diet, nutritional status, exercise, climate, altitude, and the time of day when the sample for analysis is obtained. It is essential, therefore, that the population sample selected for the reference value study be as nearly identical as possible to the population thought to be exposed to substances from a chemical disposal site. Reference intervals obtained from published sources may not always be applicable because differences in the composition of the tested and comparison populations, or differences in the test methods employed, may vary. Furthermore, technical details of specimen collection and handling, and analytical methodology, should be identical. In a reference study it is necessary to determine the form of the distribution of values (e.g., whether Gaussian or not) in order for an appropriate statistical analysis to be performed.

Concepts of sensitivity and specificity of test

In order to determine whether a test reveals organ functional abnormalities or disease, interpretation of a test must take into account diagnostic sensitivity, specificity, prevalence of abnormality, and the predictive value, in addition to whether or not the test result falls within a given reference interval. Sensitivity refers to the fraction of positive test results in individuals who truly have an abnormality of organ function. A sensitivity of 95%, for example, means that 95% of the individuals with the abnormality of organ function will be detected, but that 5% of the individuals who are abnormal will have false-negative test results. Specificity similarly refers to the fraction of negative test results in individuals who truly do not have an abnormality of organ function or a disease. A specificity of 95% means that 95% of the individuals who truly do not have abnormal organ function or disease will have a negative test, but that 5% will have false-positive test results.

Usually the ranges of test values for populations of individuals with and without a particular defect in organ function partially overlap. The sensitivity and specificity of a particular test are defined by the cutoff point chosen arbitrarily to delimit normal from abnormal. Figure 4-1 represents the overlap of test results in populations of individuals with and without abnormalities of an organ function. If, for example, the cutoff point is placed at point X, the test will detect all persons with the abnormality in organ function (100% sensitivity) but it will also falsely indicate many of the individuals in the healthy populations as having the abnormality, that is, many false positives, therefore the test has low specificity. If the cutoff point is placed at point Z, specificity will be 100% but many individuals in the abnormal population will be falsely identified as negative; that is, their abnormality will not be detected. Placing the cutoff at another point (Y) defined arbitrarily or based on the reference interval of the test does not guarantee optimal sensitivity or specificity. In fact, with most test procedures, it is not possible to achieve both high sensitivity and high specificity concurrently. Sensitivity can be increased only by reducing specificity and vice versa. Selection of the optimal cutoff point must be based on the probability of the abnormality in the population tested and the magnitude of the problems that will be associated with false-positive and false-negative errors. If it is deemed necessary to identify all persons in a population who have an abnormality, sensitivity should be maximized; however, this strategy will insure that many unaffected individuals will be falsely assigned to the abnormal category. Such false-positive assignments can have serious psychological and economic implications for the affected individuals, not to mention the possibility and danger of being medically treated for a condition they do not have. Therefore, additional studies accompanied by professional judgement are required.

Prevalence of abnormality and predictive value of tests

The predictive value of a positive test is the probability that the abnormality is present when the test is positive, and is equal to the fraction of individuals with true abnormalities among all patients with positive tests.

Figure 4–1. Frequency Distributions for Hypothetical Test Values[a] (Adapted from Galen and Gambino, 1975)

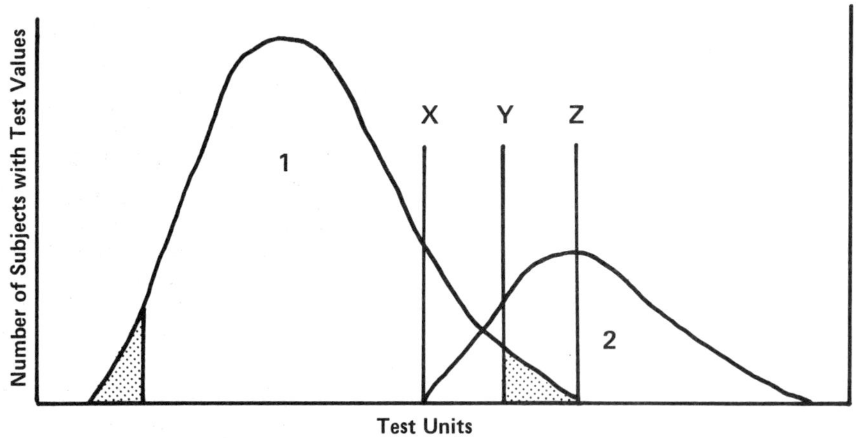

[a]Frequency distributions for a hypothetical test in health (curve 1) and disease (curve 2). Cutoff point X provides 100% sensitivity whereas cutoff point Z provides 100% specificity. Cutoff point Y, corresponding to the upper limit of the defined reference interval, is intermediate in sensitivity and specificity.

Similarly, the predictive value of a negative test is the probability that disease is absent when the test is negative. The predictive value of a test is strongly affected by the prevalence of an abnormality within the tested population (prevalence being defined as the fraction of the test population that presents with the abnormality at the time of testing). The effect of prevalence on a test with 95% sensitivity and 95% specificity is shown in Table 4-1. Only when the prevalence of abnormality in organ function is 50% or above will the predictive value of a positive test be 95%, and when the prevalence of the abnormality is only 0.1%, 98% of the positive test results will be false. Screening tests performed on unselected populations in which the prevalence of abnormality is low will yield overwhelmingly false positive results even when the test has virtually 100% specificity. False positive results of a screening test can be reduced markedly by applying the test to a subpopulation for which the probability of abnormality is increased as the result of specific signs or symptoms. In general, screening tests are not useful unless they are applied to selected subpopulations defined by epidemiological risk factors, historical information, or physical examination. The situation is not improved by employing multiple independent tests. Because reference intervals are usually defined to include 95% of the values from the healthy reference population, there is a 5% chance that a healthy person will have a test value outside that range for a single test. When a series of 'n' independent tests is performed on a healthy population, the probability that all results will be within the various test reference intervals is $P=0.95^n$, and the probability that at least one result will occur outside a reference interval is $P=1-0.95^n$. Thus, the old clinical adage that the more tests that are performed on a healthy person, the more likely is the chance of finding a value outside the normal range when no disease is present. False positive

TABLE 4–1. Effect of prevalence on predictive value when sensitivity and specificity equal 95%.

Prevalence of Disease (%)	Predictive Value of a Positive Test (%)
0.1	1.9
1	16.1
2	27.9
5	50.0
50	95.0
90	99.4

Adapted from Galen, R.S., Gambino, S.R.: Beyond Normality: The Predictive Value and Efficiency of Medical Diagnosis, New York, John Wiley and Sons. 1975, p. 16.

results of statistical origin further complicate studies in which the population has only a low prevalence of abnormality.

Despite these limitations on the use of chemical laboratory analyses for screening random populations, such procedures may be of great value for further identifying the organ that is functionally abnormal in a person or group of persons that may present signs or symptoms suggestive of an early, but nonspecific, disease process.

Reserve capacity as a confounding factor

The testing of individuals and especially groups of individuals is also complicated by the tremendous reserve capacity of most healthy tissues. Thus, a person with only one kidney or with only a portion of the liver can survive without signs or symptoms of dysfunction of that organ. In order to detect subtle abnormalities and avoid the issue of reserve capacity, function tests may be performed during which the host is stressed by imposing an increased workload. For example, renal reserve function can be tested by the capacity to conserve water (and concentrate urine) in the absence of exogenous fluids. Exercise can be utilized to evaluate cardiovascular and pulmonary reserve. The evaluation of functional reserve is generally exceedingly time consuming and not applicable to the screening of populations.

Detecting and Quantitating Chemicals in the Body

Present techniques in analytical chemistry permit both qualitative and quantitative analyses for many exogenous chemicals in the body. Hassan and Pesce (1980) indicated that the selection of samples for analysis is dependent on several factors including:

- the ease of sample collection;
- the concentration of the chemical found in the sample;
- whether the sample contains the parent compound or a metabolite; and
- the chemical properties of the analyte.

The body fluids most readily available for analysis are urine and blood. Other specimens such as hair, breast milk, adipose tissue, saliva, sweat, and semen may also be used depending upon individual circumstances, as personal acceptability of invasive procedures often determines whether a particular target tissue can be analyzed.

Analytical tests for detecting environmental chemicals or their derivatives in tissues, excreta, and body fluids provide the most accurate means to indicate and measure the extent of exposure. These tests offer verifiable information as they are generally both sensitive and specific for the chemical in question. New testing systems that incorporate the study of monoclonal antibodies to identify specific adducts (e.g., benzo[a]pyrene-nucleic acid) further increase sensitivity and specificity by offering a method to measure exposure at the level of the target organ, individual cell type, or even at the site of molecular interaction (Harris et al., 1980). Unfortunately, detailed knowledge of the pharmacodynamics of the chemical involved is seldom available to permit extrapolation between these findings and the diagnosis of an observable biological event or disease process.

One approach is to seek those tests that permit the measurement of total body burden (Gasper and Dauzvardis, 1980). This bypasses many of the uncertainties of exposure extrapolation and offers an opportunity to discuss the biologically relevant dose. Individual exposure to volatile organic compounds can be monitored via breath sampling (Wallace et al., 1982). This method shows promise as it provides data for a biologically integrated measurement that takes into account absorption, transport, biotransformation, and elimination.

The identification of health effects resulting from low-dose exposure is especially difficult. The Centers for Disease Control (CDC) guidelines (USDHHS/CDC, 1984) suggest that the following options be considered in this situation:

- include a large population to increase the chance of detecting a significant increase in the effect;
- look for potential indicators of the health effect rather than the effect itself; and
- use a particularly sensitive test for those health effects, which would not be detected by ordinary clinical tests.

New Probes for Detecting Molecular Lesions Related to Chemical Exposure

The characterization of infectious disease can frequently be determined by the identification of the virus or bacteria involved; however, tests to monitor heart rate or renal clearance have been developed to provide information on disease in the heart and kidney. Similiarly, disease and biological injury as a result of exposure to environmental chemicals require the use of practical clinical tests to measure events directly at the molecular level where xenobiotics interfere with biologic function. Unfortunately, these tests are presently in various stages of development and are not widely available for clinical use as population screens at this time.

Kolber et al. (1983) have compiled a comprehensive review of in vitro methods presently available for the assessment of toxic environmental agents. While in vitro testing can provide important information on the mechanism of action and the potential hazard of chemical agents, continued development is needed to determine the useful application of these tests in assessing the magnitude of human risk. Weinstein and Perera (1982) have recommended the application of advanced laboratory methods in molecular biology to identify, at the biochemical or molecular level, specific exogenous agents and/or host factors that play a role in the evolution of environmental disease and that can be used as epidemiologic endpoints. Much of our present understanding in this area has come from cancer research where the traditional concern has been with biological activation and critical subcellular molecular targets such as DNA, RNA, and protein. These advanced laboratory methods include:

- techniques to assess specific host factors that may influence susceptibility to xenobiotics;
- assays that detect the presence of xenobiotics in human tissues, cells or fluids;
- assays at the cellular level for biologically effective doses of xenobiotic agents; and
- methods to measure early biologic and biochemical responses to xenobiotic agents.

To distinguish these new biochemical tests which have been specifically designed to measure either body burden or critical events at the molecular level, from the clinical tests in current use, they may be grouped together under a new heading and called Early Molecular Biochemical Events tests (EMBE-tests). These include indicators of the biologically active dose of certain chemicals as measured by adduct formation with protein or nucleic acid, urinary excretion of excised DNA, and biologic or biochemical markers in cell targets such as chromosomal abnormalities, point mutations, altered gene expression, reproductive toxicity, and binding of monoclonal antibodies or complementary DNA or RNA probes.

Most of these methods should ultimately be applicable to both laboratory animals and humans, and may be extremely useful in comparative studies by providing a means to develop a database for extrapolation between species. At this time, however, these tests lack sufficient validation to be applicable. Extensive correlation studies are needed to determine the value of such test results in various experimental animal models and human populations. While EMBE-tests may not specifically indicate the outcome, at least they may be able to identify affected individuals within the exposed population. They offer the hope for highly specific methods with relevant endpoints that may enable assessment of risk directly in humans.

Application of Laboratory Tests to Assess Injury to Key Organs

The application of laboratory tests to assess the function of organs, and the perturbation of these functions by toxic chemicals, is illustrated by a brief discussion of the analysis of injury in two key organs -- liver and kidney. The function of these organs is often affected adversely when many chemicals reach toxic levels in the body, since both organs are involved in the metabolism and excretion of exogenous chemicals.

Liver injury

The liver is a primary site for biotransformation of chemicals as well as a target organ for the toxic effects of many chemicals. It is the major site for metabolism of fat soluble chemicals to the more easily excreted water soluble substances.

Zimmerman (1978) has published a comprehensive review of hepatic injury from chemical toxicity, which includes a description of the various alterations of liver function. Liver injury, as mirrored by the accumulation of fatty deposits and hypertrophy of the endoplasmic reticulum, may result from exposure to many chemicals; however, such changes are often nonspecific and may be difficult to detect clinically. For example, cirrhosis is most frequently related to alcohol consumption but can also be caused by occupational exposure to chlorinated hydrocarbons; whereas, granulomas, veno-occlusive disease, or hepatoportal fibrosis can be associated with a number of infectious organisms, therapeutic drugs, and diet derangements, as well as with exposure to toxic chemicals. Kew (1981) has stated that hepatocellular carcinoma, the most common form of liver malignancy, is increasing in prevalence in geographic regions where hepatitis B and aflatoxin exposure are endemic. However, in laboratory animals, hepatocellular carcinoma is frequently caused by halogenated aliphatic, aromatic or polycyclic chemicals (Guzelian, 1983), whereas vinyl chloride, arsenic, and Thorotrast (thorium dioxide, previously used in radio- diagnostic procedures) have been found to be associated with angiosarcoma of the liver in man (Berk et al., 1976). Table 4-2 lists signs of hepatotoxicity caused by some environmental toxicants.

Price and Alberti (1979) compiled a list of biochemical tests of hepatic function. In a recent review of research needs on hepatic injury due to environmental toxicants, Guzelian (1983) updated this list to include more recent technology and expressed concern about the indiscriminate application of batteries of clinical tests because they may provide an unacceptable and misleading excess of false-positive findings. The problem of false-positive test findings was discussed earlier in this chapter. Many of these tests lack either sensitivity or specificity for detecting the effects of environmental chemicals. Guzelian's findings are summarized in Table 4-3 .

In his review, Guzelian (1983) described additional areas of current research emphasis which may cast new light on the relationship between environmental chemical exposure and hepatic injury. He noted that advances in present understanding of the cytochrome P-450 mixed-function oxidase system (responsible for the metabolism of many exogenous chemicals), may permit the continued development of specific, non-invasive tests to measure the activity of this system in the human liver. Two approaches that are being explored are: a) measurement of liver-dependent clearance of model drugs (e.g., antipyrine) (Galizzi et al., 1978), and b) urine analysis for breakdown products of the smooth endoplasmic reticulum itself (e.g., glucaric acid), or for metabolites of endogenous substances (e.g., steroids) (Guzelian et al., 1980; Rosalki et al., 1971).

Recent advances in our understanding of carcinogenesis in an experimental rodent model offer new insight for the sequence of biological events that lead to liver cancer. Farber (1980) has proposed that a multi-step sequence involving initiation, promotion, and proliferation can occur for chemically induced malignancy in hepatocytes. Initiation is a rapid event characterized by the interaction of chemical metabolites with DNA to produce a mutation-like

TABLE 4–2. Summary of syndromes of hepatotoxicity due to environmental agents.

Toxic manifestation	Examples of environmental toxins reported in man
Acute or subacute hepatic necrosis	Chlorinated aliphatics in high doses (CCl_4, tetrachloroethane); mushroom toxins; phosphorus; mycotoxins; high doses of polychlorinated biphenyls (PCB's)
Cholestasis	Methylene diamine (Epping jaundice); dinitrophenol; chromium
Mild fatty metamorphosis	Small doses of chlorinated aliphatics; some organochlorine pesticides
Hypertrophy of endoplasmic reticulum	Polychlorinated biphenyls; some organochlorine pesticides
Veno-occlusive disease	Plant toxins (pyrrolidizine alkaloids); therapeutic liver irradiation
Granuloma	Beryllium
Hepatoportal fibrosis	Vinyl chloride; arsenic
Cirrhosis	Chlorinated aliphatics; aromatics; aflatoxin; arsenic; plant toxins
Hepatocellular carcinoma	Aflatoxin; ? ethanol
Angiosarcoma	Vinyl chloride; arsenic

lesion. Promotion occurs at a later time when this lesion is influenced by a promoting agent to activate the development and proliferation of cancer. Altered (initiated) hepatocytes may never become malignant in the absence of exposure to a promoter. Some chemicals may function as promoters of cancer

TABLE 4–3. Screening tests of liver function.

Test	Abnormality	Sensitive	Liver-specific
Aspartate aminotransferase AST (SGOT)	Hepatocyte necrosis	Yes	No
Alanine aminotransferase ALT (SGPT)	Hepatocyte necrosis	Yes	Yes
Alkaline phosphatase	Biliary obstruction Infiltrative diseases	Yes	No
Bilirubin	Biliary obstruction	No	No
γ-Glutamyl transpeptidase (GGTP)	Biliary obstruction Hepatocellular necrosis Ethanol Microsome induction	Yes	No
Serum bile acids	Biliary obstruction Hepatocellular necrosis Extrahepatic shunts	Yes	Yes
Aminopyrene breath test	Microsome induction Functional liver mass	No	No
Urinary glucaric acid	Microsome induction	No	No
Urinary 6β-hydroxycortisol	Microsome induction	No	No
α-Fetoprotein	Hepatoma	Yes	No

development even though they cannot initiate the process. Proliferation of the activated cancer cell may be quite slow or very rapid.

Farber (1980) has shown that the interaction between chemical metabolites and hepatocytes can induce biochemical changes in those cells affected to convert them into "preneoplastic" cells. These phenotypically abnormal cells form foci or nodules of enzyme-altered hyperplastic cells that subsequently may be evoked by a second chemical stimulus (promoting agent) to develop into hepatocellular carcinoma. Promotion may be induced by chemicals that do not show evidence of toxic injury to the liver, as detected by clinical laboratory tests that are presently available.

New developments in assay techniques have focused on measurements of stable adducts formed between activated carcinogens and genetic material. For example, Poirier et al. (1980) developed a specific radioimmunoassay capable of detecting small quantities of benzo[a]pyrene-deoxyguanosine adducts. Another approach has come from the research efforts of Bennett et al. (1981) in the development of an assay for measuring the urinary excretion of aflatoxin-guanine adducts, found to be primarily of liver origin.

Kidney injury

The kidney filters the blood, excretes the end-products of body metabolism in the form of urine, and regulates the ionic balance of extracellular fluid. The excretory function of the kidney can predispose this organ to a greater degree of exposure to exogenous chemicals than many other organs. The high perfusion rate, active transport capability, and concentrating ability which usually serve to expedite the excretion of waste products may result in greater damage to the kidney when xenobiotic agents are present (Hook, 1980).

Standard clinical laboratory techniques for detecting renal disease include determination of urinary specific gravity, pH, the presence or absence of protein, and levels of blood urea nitrogen (BUN) or serum creatinine. Proteinuria may not become readily apparent until more than two-thirds of renal function is lost. Detailed function studies of glomerular filtration rate using clearance techniques may not be practical, as such studies are cumbersome and the resulting information must be corrected for such things as hydration and osmotic clearance (Wedeen, 1983).

A brief list of chemical pollutants that can increase renal enzyme activities has been compiled by Kluwe and Hook (1980b), and includes:

3,4-benzo[a]pyrene,
chlorinated dibenzodioxins,
dichlorodiphenyltrichloroethane (DDT),
3-methylcholanthrene (3MC),
polybrominated biphenyls (PBB's),
polychlorinated biphenyls (PCB's), and
hexachlorobenzene.

This list is not complete and other chemicals may also produce similar effects. These authors note that chemical inducers or inhibitors of renal metabolism can upset the balance between toxification and detoxification reactions in the kidney, while altering its sensitivity to toxic effects of certain other chemicals. Biosynthesis and degradation of endogenous compounds, such as prostaglandins and 1,25-dihydroxycholecalciferol, may be altered by induced renal enzyme systems.

Some chemicals, such as mercury and chloroform, may directly cause renal injury and produce lesions which may be life-threatening and which show predictable dose-related effects. In their work with PCB's and PBB's, Kluwe and Hook (1980a) have observed that these two groups of chemicals can produce changes in kidney enzymes and yet appear to have no measurable effects on renal function. However, Hook and Serbia (1982) noted that previous exposure to these chemicals resulted in a potentiation of the nephrotoxic response to other chemical agents (i.e., carbon tetrachloride and the therapeutic agent, acetaminophen). In the case of PCB's and PBB's, no functional lesion is produced; however, the bioaccumulation of these compounds may render the individual more susceptible to injury from subsequent exposure to other exogenous chemicals. McCormack et al. (1978) noted that the renal microsomal mixed-function oxidase activity in rats was sharply elevated after three months on a diet containing PBB's. They further observed that renal enzymes were induced in a selective manner; arylhydrocarbon hydroxylase activity was increased several-fold, whereas epoxide hydratase activity was either reduced or remained unaffected. This may be of particular interest as the liver activity for both these enzymes was increased. These findings for PCB's and PBB's indicate that a body burden of these chemicals may upset the balance between the toxification/detoxification reactions by selectively inducing specific microsomal enzyme activities. Thus the response to metabolically activated toxicants may be altered by exposure to even small amounts of these chemicals (Hook and Serbia, 1982).

ASSESSMENT OF MUTAGENIC, CYTOGENETIC, AND CARCINOGENIC EFFECTS

The possibility that chemical exposure might lead to heritable defects is a matter of concern to both the scientific community and the general public. Unfortunately, available human data are inadequate for settling the matter, while the widespread use of short-term mutagenicity tests as a preliminary screen for carcinogenesis continues to focus attention on an increasing number of suspect chemicals. Much of what we know about cancer has been learned through laboratory studies, especially those using mice and rats. In these animals, the incidence of various forms of cancer is influenced by chemicals, irradiation, viruses, diet, selective breeding, etc.

The mechanisms through which these various interventions produce the effect are not fully understood. An attractive hypothesis, supported by a substantial body of evidence, holds that in most instances, something happens to the genetic apparatus of somatic cells which ultimately results in uncontrolled cell proliferation. Based on experimental studies of skin and liver cancer, it is now generally believed that carcinogenesis is a multi-stage process involving initiation, promotion, and progression. In man, it is thought that a similar sequence of events may occur in some tumors. It is extremely difficult to define these factors as each type and possibly each subtype of tumor probably differs in this regard.

Within the scientific community there is a growing tendency to attribute a substantial proportion of human cancers to environmental factors, and more specifically, to chemicals in the environment. There is evidence that more than 20 specific chemicals cause cancer in humans.

There is widespread concern that chemicals migrating from chemical disposal sites and resulting in long-term, low-level exposure of nearby populations may result in increased incidence of human cancer. Epidemiological studies to explore the matter have relied largely on death certificates in which the diagnosis is usually based on clinical diagnoses, and, less frequently, on biopsy findings or autopsy examinations. Occasionally, the records of a local or regional tumor registry are helpful. Population surveys for the general detection of cancer are expensive and often inconclusive. For several forms of cancer, useful screening procedures are available. An example is the use of exfoliative cytology for early detection of uterine, respiratory, or urinary tract cancer. Endoscopy, combined with biopsy, is finding increasing use in the detection of gastrointestinal, respiratory tract, and abdominal cancer. The use of such techniques for population surveys is seldom justified with the exception of special circumstances in high-risk populations.

What is needed is a simple test or battery of tests to detect biological changes which are predictive of cancer in the early stages of the disease. DNA adducts, sister chromatid exchanges (SCE's), chromosomal abnormalities, urinary mutagens, enzyme levels, and monoclonal antibodies offer future possibilities, but have not yet proven to be predictive. Information from laboratory animal studies is useful but poses major problems of interpretation when the results are extrapolated to man.

This discussion of mutagenic, cytogenetic, and carcinogenic effects has as its premise the notion that induced mutation of either the specific locus type or the chromosomal type is undesirable for our species. It is generally accepted that humans are constantly exposed, from numerous sources, to a variety of physical and chemical agents capable of damaging the genetic material of their germ cells or somatic cells. Most of these alterations are appropriately repaired by a group of highly specific enzymes.

The major principles involved in mutagenesis and carcinogenesis testing and their relationships to potential human health effects are summarized below.

Biologic Principles of Mutation

Numerous reviews have been written on the subject of induced mutation (Bloom, 1981; Hollaender, 1971) and there is ample evidence to demonstrate the deleterious effects of these mutations in numerous species, excluding man (ICPEMC, 1983; Neel et al., 1980). Most studies of mutagens in man have come from observations on the genetic effects of ionizing radiation.

Radiation can probably cause cancer in any tissue provided the exposure is of sufficient magnitude and occurs at the appropriate time in the development and maturation of the host. Tumors associated with low doses of external radiation include leukemia, breast cancer, thyroid cancer, and multiple myeloma. By extrapolation from several exposed human populations, it is estimated that 1 to 2 malignant neoplasms are induced per 100,000 persons exposed to one rad of ionizing radiation (NRC, 1980). This indicates that for large populations exposed to even very low doses of ionizing radiation, there may be a significant carcinogenic risk.

A variety of adverse somatic defects have been ascribed to ionizing radiation whose teratogenic effects appear to be related to induced damage in the genetic material of human cells (Brent and Torson, 1972). The observed teratogenic effect is dependent on both the timing and dose of the exposure. For example, if exposure occurs prior to implantation, there is likely to be either loss of the conceptus, or no clinically detectable effect at all. If exposure occurs post-implantation, a variety of organ-specific effects have been described, including defects of the central nervous system such as microcephaly accompanied by mental retardation, ocular defects, alterations in the axial skeleton, and defects in the heart and kidneys.

The present estimate in the general population is that 20 to 40 newborns per 1,000 live births have a significant congenital malformation. In situations in which there is exposure to ionizing radiation in utero, an increase of one child with a malformation per 1,000 live births per rad of exposure is estimated.

The assessment of the transmission to human progeny of clinically significant chromosomal abnormalities or gene mutations is complex and is extrapolated largely from mouse data (NRC, 1972). Two basic methods are generally used in these calculations. The so-called "direct" method assumes that the rates of gene and chromosomal mutations are proportional to the unit dose to the parents. By this method, it is estimated that 20 to 30 affected offspring will result from a one rad exposure to the parents of a million liveborn children. On the other hand, the "doubling dose" method assumes the mutation rate to be proportional, for each defect, to the rate of its spontaneous occurrence. By this method, we arrive at an estimate of 60 to 65 affected children per rad per million liveborn children (UN, 1969).

These extrapolations are from high-dose exposures to low-dose exposures. It is generally accepted that even with the prototype mutagen, ionizing radiation, with which there is extensive experience, an increase in the incidence of genetic disease has never been shown in the offspring of exposed humans. It is thought to be highly unlikely that one will ever be shown because the methods for detecting such increases in disease incidence are simply not sufficiently sensitive to detect small increments against a broad background of "spontaneous" abnormalities.

Evidence of germinal mutations, of either the specific locus or chromosomal types, have been looked for extensively in the progeny of accidentally irradiated human parents, but no radiation-caused mutations have been demonstrated (Neel et al., 1980). These studies have included biochemical genetic approaches, utilizing: 1) multiple loci, expressed either in erythrocytes, in leukocytes, or in serum proteins; and 2) chromosomal rearrangements of the type that would be seen in virtually all cells of the progeny. The latter rearrangements include trisomies, monosomies, deletions, and translocations.

Approaches to Mutagenesis Testing

An important principle in assessing the mutagenicity of chemicals, is that the vast majority (perhaps 90%) of carcinogens are also mutagens (DeSerres and Sheridan, 1983). A lesser but still significant proportion of mutagens are also carcinogens. The basic approach in the genetic evaluation of chemicals has been to assess their mutagenicity first, and then, if the compound is positive, to embark on the more expensive and lengthier carcinogenicity testing in rodents. More than 40 prokaryotic and eukaryotic test systems are available, ranging from tests of single gene mutations in bacterial cells to single gene mutations and chromosomal aberrations in mammalian cells (including human cell lines).

The first level, or Tier 1, approach recommended by the National Academy of Sciences includes tests of reversion at the histidine locus in Salmonella typhimurium (the so-called Ames test), and a specific locus test, such as the thymidine kinase (or TK) locus test in murine cells, in conjunction with a test for chromosomal aberration induction in mammalian cells (Delehanty, 1983). Positivity in two of these three tests is sufficient cause to warrant more extensive study of the compound, including possible carcinogenicity testing. On the basis of the results from long-term testing in laboratory animals, a majority of the mutagenic chemicals defined as positives by these test systems will also become known as carcinogens.

Two criteria must be affirmed before a chemical is labeled a mutagen. First, there must be a dose-response relationship demonstrable in the system under study. While the dose range for this response may be either wide or relatively narrow, an increased mutagenic response must be observed to be within some limit. It is often defined as a linear response, although a variety of mathematical relationships are possible. Secondly, at least two positive tests should be observed in different species. This cross-species positivity is an important criterion that must be met before a chemical is considered to be a mutagen.

The overwhelming preponderance of mutagenicity testing has focused on the effects of single compounds while multiple chemical effects have seldom been studied in these test systems. Research into the genetic effects of multiple chemical exposure at low concentrations can be considered as an area of high priority in the future.

Extrapolations and positivity-negativity

There are many difficulties in extrapolating from these test systems to man, especially in terms of health effects. As indicated above, the carcinogenicity-mutagenicity relationships are only now beginning to be understood, and there is little correlation between in vitro mutagenicity and any other clinical endpoints of a somatic or germ cell genetic type. The induction of specific locus mutations or of chromosomal aberrations or SCE's is not definitely associated with clinical disease. In fact, disease associations are virtually impossible to establish at present, even for strongly positive mutagens.

The question of interpretation of the results of a given mutagenesis assay is complex. The definitions of "positive" and "negative" are generally clear-cut only for a relative minority of tested chemicals. Many chemicals prove to be weakly positive, or neither clearly positive nor negative; false-positives and false-negatives comprise 10-30% of the compounds tested. Approaches to the statistical analyses of these results vary with the test in question. Most mutagenesis assays require the use of positive and negative controls, as well as extensive data to determine the range of normal variation. In addition, there may be considerable subjectivity in some of these assays, and variations of interpretation of test preparations by laboratory personnel must be factored into the statistical analysis.

Mutagen-carcinogen relationships

When activated, at least 22 proto-oncogenes are known that may be associated with cancer in animals or with in vitro transformation of cultured cells. Multiple mechanisms have been hypothesized to explain the possible actions of these oncogenes such as increased expression of the oncogene and/or alteration of the oncogene itself (i.e., mutation). Recent experiments suggest that oncogenes and chemical mutagens may interact.

Investigators have shown that the ras gene from the Harvey murine sarcoma virus may be activated during chemical carcinogenesis (Land et al., 1983). When fifty 1-day-old rats were exposed to a single dose of nitrosomethylurea (NMU), a potent mutagen, it was found that 100% of the rats developed mammary cancer in 6-12 months. The tumor cells of those rats were found to have the modified transforming ras gene. The investigators sequenced the gene and identified a mutation in codon number 12, implying that a structural alteration in the ras gene had led to the malignant transformation. This is the same codon as the ras mutation in eight human lung cancer cell lines. These findings from early experiments on the oncogene effect of chemicals provided valuable information, for there is now increasing evidence suggesting that oncogenes are spontaneously activated or may be induced to be activated by mutation which results in neoplastic transformation. The notion that there are two types of oncogenes, one which, like the myc gene, immortalizes cells and one which, like the ras gene, definitively transforms cells, adds important knowledge to our conceptual understanding of this complex process. The mechanisms by which chemicals induce mutations and cancer will likely be diverse.

Genetic predisposition to cancer

As there are now more than 200 genetic diseases known to be associated with an increased predisposition to neoplastic change, the genetic constitution of the host may increase the likelihood for an individual's susceptibility to malignancy. Many of these disorders are dominant in their mode of inheritance, like neurofibromatosis, with its increased (10%) incidence of intracranial tumors and leukemia. Others are recessive, like the disorders associated with chromosome breakage as in Bloom's or Fanconi's syndromes and xeroderma pigmentosum with their increased risks of leukemias, lymphomas, and solid tumors. Interestingly, there is little definitive biochemistry data available on the mechanisms of DNA repair in human cells even though xeroderma pigmentosum is considered to be the primary disorder of DNA repair (Cleaver, 1983). Deficient DNA repair is considered to be a major cause of increased human sensitivity to mutagens. The 2 major types of DNA repair are excision repair with its enzymatically controlled sequential steps, and post-replication repair with gaps filled in by de novo synthesis. A discussion of the role of DNA repair assays as indicators of genotoxic exposure is presented by Bridges et al. (1982).

The identification of individuals with reduced penetrance of these dominant genes or of individuals who are heterozygous for recessive disorders has been shown to be an important discovery, since all persons are not equally susceptible to the effects of environmental hazards, especially at low doses. Auerbach and Wolman (1976) established the in vitro chemical stress test, in which cells from possible carriers of a gene predisposing to cancer are stressed by exposure to a potent mutagen so that the genetic (genic, chromosomal) effects can be studied. Auerbach developed this test for Fanconi cells enabling distinctions to be made between normal, heterozygous, and homozygous cells. The extent to which mutagenizing experiments activate dormant oncogenes remains unclear.

The work of Haseltine (1983) on DNA base sequence alterations and deletions may lead to the increasing likelihood that in the future, DNA of somatic cells may be examined directly for mutational changes. In the interim, however, the indirect measures of mutagenicity described above will have to be used, as efforts are made to regulate human exposures to environmental hazards. Once a decision is reached on the mutagenicity of a compound, decisions for followup testing may be made more sensibly.

Cytogenetic Monitoring

Cytogenetic monitoring of humans usually involves studies of peripheral blood lymphocytes for chromosome and chromatid type aberrations. For radiation induced genetic injury, the study of complex breaks and rearrangements allows, perhaps, the best biological measures of exposure. In radiation-exposed individuals, there is both an acute effect which is reflected in a dose-response increase in chromosomal aberrations, and a long-term effect which is reflected in the persistence of such aberrations over time (months or years). Studies in numerous occupationally exposed groups have found evidence of chromosome breakage (Bloom, 1972). The use of aberration studies have not been useful in monitoring for chemical exposures, although it has been found that in vitro SCE induction correlates with exposure to mutagenic chemicals better than does chromosomal aberration induction. Studies have shown that SCE's correlate reasonably well with specific locus mutations. It is not clear whether SCE frequencies in vivo will be a sensitive measure of mutagen exposure, since only a few populations, such as petroleum industry workers, have been studied. One interesting observation is that cigarette smoking raises the baseline frequency of SCE's in all groups studied to date (Carrano et al., 1980).

At this time, there is some uncertainty over the clinical significance of findings from in vivo studies of induced aberrations and SCE's. There is a clear and pressing need for an extensive study of the correlations between chromosomal aberrations and disease. Modest attempts at this have been made in the Hiroshima and Nagasaki populations, but there is little data on the clinical significance of cytogenetic findings for these individuals or for others who may have been exposed to chemicals. An important finding is that exposure to clastogens can be associated at the population level rather than at the individual patient level with increases in malignant disease and teratogenic effects.

Carcinogenesis Assay

This discussion of chemically induced carcinogenesis focuses on the identification of human carcinogens and describes the relationship between genetic mutation and cancer. An authoritative review of the science of chemical carcinogenesis has been the subject of two recent reports by the President's Office of Science and Technology Policy (USEOP/OSTP, 1984) and an interdisciplinary panel of scientists (Interdisciplinary Panel on Carcinogenicity, 1984). The OSTP document has been prepared as a guide "to be used by regulatory agencies in assessing cancer risks from chemicals."

Cancer is a disease that probably has plagued man since prehistoric time. The causation of cancer in man has been a subject of concern during the past several centuries. Nearly 300 years ago it was noted that nuns had a higher frequency of breast cancer than did women who bore children. This finding proved to be more puzzling than edifying until the difference was linked to hormonal status. About 200 years ago it was discovered that chimney sweeps, exposed to soot containing wood and coal tars, had a high incidence of scrotal skin cancer. This discovery led to an edict requiring chimney sweeps to bathe weekly and is recognized as one of the first public health measures taken to prevent cancer. About 100 years ago, occupational exposure to chemical aromatic amines was correlated with cancer of the urinary bladder in exposed industrial workers, and sunlight exposure was related to skin cancer. These insights came from early epidemiological investigations of human populations. More recently, it has been shown that the chronic exposure of laboratory animals to certain chemicals can be linked to the development of cancer. Subsequently, long-term tests in several animal species have been used to evaluate chemicals for their ability to induce cancer. The short-term carcinogenicity tests for neoplastic transformation or specific locus mutation in cultured cells, and specific locus mutations in tester strains of bacteria, have been available only during the past twenty years.

The present consensus within the cancer research community is that cancer rarely arises spontaneously (without any known cause), and that the background frequency of "spontaneous" cancer is increased by many inciting agents including biological (viruses), physical (ultraviolet light and x-rays), and chemical agents. Exposure to carcinogenic agents may occur in relation to life style habits, work and the general environment. Individual susceptibility to the induction of cancer by these agents appears to vary with species, age, sex, genetic make-up, diet, and mode of exposure. An in-depth understanding of the mechanisms and development of cancer induction and expression is still lacking. Until this understanding is achieved, the precise role of particular substances and conditions in the causation of cancer will be uncertain.

Present evidence supports the opinion that cancer development is a multi-stage process. The initial stage involves genetic transformation of affected cells, either direct or indirect. Most chemical agents are metabolized by cells to give a combination of active and inactive residues. Active chemical residues may interact with and modify the genomic DNA of the initiated cell. If the damaged DNA is not repaired, it may be transmitted as incorrect information to future generations of daughter cells during normal cell replacement. Under suitable conditions, some specific types of genomic injury (e.g., proto-oncogene) may cause a cell containing incorrect information to be "initiated" and become capable of uncontrolled growth to form a cancer. In addition to the process of initiation, the formation of a cancer also requires the participation of at least two other biological processes, known respectively as promotion and proliferation. Certain chemicals may function as tumor promoters by unknown mechanisms, even though they are incapable of initiating cells.

There is widespread concern that chemicals that have migrated from chemical disposal sites and exposure of nearby populations to long-term, low-level concentrations may cause increased frequencies of a variety of cancers. The most reliable information on the carcinogenic potential of chemicals in man comes from epidemiological studies in human populations. However, as discussed in Chapter 5, epidemiological studies are difficult to perform. Both descriptive epidemiological studies, where cancer frequencies are compared between populations, and correlation studies, where the frequency of cancer in a population is compared with spatial or temporal distribution of possible risk factors, may be used to generate or refine hypotheses, but rarely allow causal inferences to be made. The International Agency for Reseach on Cancer (IARC) has reported that there is evidence to suggest that more than twenty chemicals or industrial processes employing chemicals are carcinogenic for man (WHO/IARC, 1982b).

It is often impossible to structure appropriate epidemiological studies in chemically exposed populations to determine directly if chemicals are carcinogenic for man. As a consequence, other procedures have been developed to provide biological evidence of carcinogenicity. Long-term testing of chemical exposure in laboratory animals can provide useful information. Agents found to be carcinogenic to animals are generally considered to be suspect human carcinogens. However, evaluation of animal carcinogenicity tests may be difficult when the agent tested produces only benign tumors or merely shortens the time to appearance of neoplasms that occur naturally in the test species. Further difficulties in extrapolating from animal studies to man stem from the customary use of the maximum tolerated dose in animal studies, and from the marked cross-species differences in chemical metabolism, DNA repair, and frequency of spontaneous tumors. IARC has long held that "no attempt can be made to interpret animal data directly in terms of human risk, since no objective criteria are available to do so" (WHO/IARC, 1972; 1973a,b; 1974a,b,c,d; 1975a,b; 1976a,b,c; 1977a,b,c; 1978). Nevertheless, in reviewing the world's literature, IARC has concluded that more than 140 chemicals are definitely carcinogenic in laboratory animals and that "it is reasonable to regard [them] for practical purposes as if they were carcinogenic for man" (WHO/IARC, 1982a).

As previously discussed, short-term tests in animal or microbial cell cultures can be used to determine the effects of chemicals on the production of DNA damage and repair, point mutations, chromosomal aberrations (animal cells), or neoplastic transformation (animal cells). These tests are useful to screen chemicals for potential carcinogenicity, provide independent evidence to support the carcinogenicity of a chemical from animal studies, and/or yield information on the cellular mechanisms of cancer development. However, while both the short-term and the long-term testing methods may be used to characterize the nature of the carcinogenic risk of chemical exposure to human populations, neither can be used to supplant data from epidemiological studies in terms of their ability to quantitate the actual human risk (WHO/IARC, 1972; 1973a,b; 1974a,b,c,d; 1975a,b; 1976a,b,c; 1977a,b,c; 1978). Only appropriately structured epidemiological studies of human populations exposed to measured levels of a chemical can provide this information. In the absence of epidemiologic data, quantitative assessment of risk for humans is attempted by extrapolation of animal data to man. Although information derived from experimental laboratory studies may arouse strong suspicion that a particular chemical may be carcinogenic in man, there is, to date, no known mathematical procedure that allows the definitive extrapolation of data from high-dose studies in animals to quantification of the actual risk associated with low-dose exposures in man.

ASSESSMENT OF EFFECTS ON REPRODUCTION AND FETAL AND NEONATAL GROWTH AND DEVELOPMENT

Both physical and chemical factors in the environment are known to induce a variety of adverse physiologic effects which may intervene during the process of reproduction. The mechanisms through which these factors affect mammalian reproduction are not fully understood and are further complicated by the normal physiologic diversity associated with the process. Environmental interventions may occur at any combination or permutation of three biologically sensitive points: the production of gametes (ova and sperm), fertilization and implantation within the uterus, and growth and development at the embryonic and fetal stage. The effects brought about by a specific environmental insult may be quite dissimilar; for example, the germ cells, the uterus, the blastocyst, the embryo, and the feto-placental unit will each respond to an environmental agent in a manner unique to the insult and often with a different response dependent on either the amount or the timing of the exposure.

Adverse reproductive outcomes are not an uncommon phenomenon throughout the animal kingdom, and numerous studies suggest that multiple factors are frequently involved. The normal background incidence for spontaneously occurring adverse reproductive effects is not known, but is probably much higher than previously suspected. Current evidence suggests that exposure to environmental agents can be related to at least several adverse reproductive outcomes. It has been known since ancient times, for example, that exposure to a chemical such as lead may result in an adverse reproductive effect; in contrast, the effects of radiation were not understood until the modern age. Concern for adverse reproductive effects resulting from chemical exposure was heightened with the unfortunate Thalidomide incident of the 1960's. Since that time numerous chemicals have been identified as mutagenic, teratogenic, or embryo-/fetotoxic agents.

Public awareness that inadvertent exposure to chemical agents migrating from disposal sites may predispose individuals to reproductive failures or congenital abnormalities has prompted several epidemiologic studies to detect such outcomes in exposed populations. Traditionally, studies of reproductive outcomes rely on information from existing registries or survey interviews. Recall bias within the affected community and limitations and inconsistencies inherent to available registries are contributing elements for the frequently inconclusive findings in population surveys of this nature. Population screening methods to detect physiologic changes are available; however, the stigma surrounding the use of such testing within the community is not well accepted. The use of screening methods, such as semen assay, amniocentesis, neonatal blood sampling and genetic characterization of an aborted or malformed conceptus, are techniques that provide useful information for an individual but may not be practical when applied to groups at disposal sites due to the costs associated with these specialized techniques.

Biologic Principles of Reproductive Dysfunction

A recent March of Dimes Birth Defects Symposium (Bloom, 1981) compiled the following list of potential endpoints that might be studied in association with reproductive failure or dysfunction. It is not clearly understood how these endpoints relate to one another, nor whether changes in one will necessarily be accompanied by changes in another. The list attempts to follow a biologic chronology:

1. Sexual dysfunction: (libido, potency)
2. Sperm abnormalities: (number, motility, shape)
3. Sub-fecundity: (abnormal gonads, abnormal prepubertal development, infertility, amenorrhea, anovulatory cycles, delay in conception)
4. Illness during pregnancy and parturition:

(toxemia, hemorrhage)
5. Early fetal loss (to 28 weeks): (toxemia, hemorrhage)
6. Late fetal loss (after 28 weeks)
7. Intrapartum death
8. Death in first week postpartum
9. Decreased birthweight
10. Change in gestation age at delivery: (prematurity, postmaturity)
11. Altered sex ratio: (in fetal deaths, at birth)
12. Multiple births
13. Birth defects: (major, minor)
14. Chromosomal abnormalities: (in fetal deaths, at amniocentesis, in perinatal deaths, in livebirths)
15. Infant death
16. Childhood morbidity
17. Childhood malignancies
18. Postnatal effects (decreased adult weight, CNS effects, intellect, behavior)
19. Functional defects
20. Age at menopause

Environmental agents seldom have only one effect on a biological system and such factors may influence the reproductive process in any of three modes of action: a) genetically, by changing the genes themselves; b) as a teratogen, by affecting embryonic development during gestation; and c) as a parental or fetal toxin, by causing impairment of reproductive function or abnormal fetal growth (Perrin, 1982a; Warburton, 1981; Wilson and Fraser, 1977).

Genetic changes may occur within the gametes themselves or during the period of embryonic differentiation. Effects that result from a genetic etiology follow a spectrum of pathologic processes that include gene deficiency or abnormality, and chromosomal deletion, excess or rearrangement. Mutation is among the most frequently discussed mechanisms, followed by chromosomal nondisjunction and breakage, mitotic interference, and altered nucleic acid integrity or function. If exposure occurs at a dose of sufficient magnitude prior to implantation, there is likely to be either loss of the conceptus, or no clinically detectable effect at all. Post-implantation effects may result in a variety of organ-specific alterations including microcephaly and/or mental retardation. Although environmental factors may modify the outcome of a conceptus that has inherited a genetic defect, the genetic abnormality remains the predominant contributor to the adverse outcome.

Wilson (1973a,b) has proposed a list of possible mechanisms that may lead to teratogenesis. The list demonstrates the vulnerability of the embryo during development, but does not take into consideration the fact that pathologic lesions may result from either genetic or environmental factors, or that an environmental agent may induce a mechanism but not result in a teratogenic effect (Beckman and Brent, 1984). The teratogenic effects on prenatal development of humans and experimental animals follow a series of basic principles:

- Susceptibility is dependent upon the genetic makeup of the conceptus.

- Susceptibility varies with the stage of development at the time of exposure.

- The chemical agent acts through specific mechanisms on developing tissues and cells to initiate abnormal development.

- Death of the embryo or fetus, malformation, growth retardation, and functional disorders are the final manifestations of abnormal development.

- The excess of the abnormal environmental influence depends upon the nature of the influence itself.

- There is a dose-dependent relationship for most compounds ranging from no effect to a lethal effect.

It is important to note that unlike mutagenesis and carcinogenesis, teratogenesis is a threshold phenomena. Therefore, terata would not be expected to increase at exposure concentrations below threshold, even for proven teratogens. Chemicals such as DDT, dioxins, dieldrin, captan, and carbaryl are known to produce malformations in experimental animals. Some metals, such as lead and mercury, are known teratogens in man (Shepard, 1983). Industrial compounds present as environmental contaminants may produce damage in the infant and child, both directly and by crossing the placenta as a result of maternal exposure; however, these areas have received insufficient study to provide conclusive evidence for adverse effects at the low dosage levels that typify environmental exposure (Perrin, 1982b).

Several authors in Wilson and Fraser's Handbook of Teratology (1977) suggest that one of the most frequently observed signs of abnormal embryogenesis is cell death in excess of that normally associated with development, resulting in too few cells for subsequent morphogenesis or functional maturation. Chemical substances may also interfere with morphogenetic movements; as such, cellular interaction in the form of induction is not always quantifiable morphologically but may lead to abnormal tissue development. Some agents may initially cause cellular depletion at a single site, and others may result in excessive tissue through hyperplasia or failure of normal resolution and resorption (Wilson and Fraser, 1977).

Shepard (1983) has prepared a catalogue of reference sources on the intrauterine tissue response to material ingestion, exposure to medication, and other toxic substances. In general, an agent that interferes with maternal metabolism and also crosses the placental barrier will exert an influence on the growth and development of the fetus. Toxic effects ranging from fetal death to changes in birth weight or behavioral abnormalities have been observed for numerous chemicals. It is important to note that changes in the precursors and substrates needed for biosynthesis, altered energy sources, enzyme inhibition, imbalance of an osmolar nature, altered membrane characteristics, and infectious disease (e.g., rubella or poliomyelitis virus) may also affect the developing fetus and can produce numerous malformations (Perrin, 1982b).

Researchers have found that no one mechanism of action can be used to describe the effects of all chemicals on the reproductive system, and that the unique nature of each chemical requires individual study. A list of chemicals frequently identified in the vicinity of disposal sites and their known embryo- and fetotoxicity is presented in Table 4-4. This table provides a useful guide in directing the attention of epidemiologic research teams to the reproductive outcomes of concern at chemical disposal sites.

Methods Used in Assessing Reproduction

Government regulatory agencies are dependent on information from laboratory animal testing to make assessments of the extent to which chemical exposure may contribute to the etiology of a congenital malformation. The predictive ability of such testing is frequently open to criticism. Variation in placental anatomy and function, as well as metabolic differences between species, can lead to false negative findings, as exemplified by the thalidomide incident (Brent, 1964). The ability to extrapolate dose-effects to one species on the basis of data from another species is not always predictive of the effects observed in man, as can be noted in Table 4-4 .

As early as 1963, a panel of experts recognized that available testing methods could not guarantee safety to the unborn child (Warkany, 1963). The panel offered a series of recommendations for the use of standardized methods in future chemical screening programs. These recommendations take into consideration the selection of appropriate animal models and multiple species testing, the phase of development at the time of testing, and the use of multiple dose levels to ensure a threshold effect. These improvements in basic research have greatly reduced the risks associated with the human consumption of potentially harmful agents that might effect the unborn child.

Standard protocols for teratogenicity testing are frequently expanded to include the study of animals exposed to the test material prior to breeding, during in utero development and lactation, and in some instances into the adult life of the offspring. Some testing protocols call for the study of reproductive performance through multiple generations. Chemicals known to be teratogenic in man have been demonstrated to be teratogenic in one or more commonly studied laboratory species; however, no one species has been found to be consistently more predictive of human teratogenicity than another. Multiple species testing is recognized as a rule, and a teratologic response in any species is considered to be a teratologic concern for man.

It is important to note that not all animal teratogens have been demonstrated to be teratogenic in man. This difference in the toxicological data base may be due to insufficient epidemiologic data, dissimilarities of exposure concentrations or duration, or differences in the selection of endpoints evaluated. Information gained from epidemiology studies and reproductive registries provides the most convincing evidence for teratogenic effects in man.

Methods of Monitoring for Reproductive Effects

The primary sources of reproductive data include records of vital statistics (e.g., birth and death certificates, chromosomal disorder and malformation registries), hospital and physician records, interview data, and laboratory data such as amniocentesis and newborn blood sampling (Lowrance, 1981). A detailed description of the types of malformations and birth defects was compiled by Warkany (1971) and Bergsma (1979). A listing of the common major malformations among newborns was compiled by Bloom (1981).

TABLE 4–4. Embryo- and fetotoxicity of selected chemicals frequently found at uncontrolled disposal sites.

Chemical	Organism	Exposure Level and Duration [day of gestation]	Malformation	Developmental Difference	Lethal
Halogenated Aliphatics					
1,2-Dichloroethane	rat	IH, 100 ppm, 7 hrs/day [6-15]	nr	–	–
		IH, 300 ppm, 7 hrs/day [6-15]	nr	(dams died)	+
		PO in food, 500 ppm x 2 yrs	–	–	–
		PO in water, 1000 mg/kg	–	–	–
Dichloromethane	mouse, rat	IH, 1225 ppm, 7 hrs/day [6-15]	–	–	–
	rat	IH, 4500 ppm, 6 hrs/day, pre [1-17]	–	wt	–
Trichlorethylene	mouse, rat	IH, 300 ppm, 7 hrs/day [6-15]	–	–	–
	mouse	IH, 3000 ppm, 4 hrs x 5 days		(sperm abnormalities)	
	rat	IH, 1800 ppm, 6 hrs x 10 days pre [1-10]	–	os?	–
1,1,1-Trichloroethane	mouse, rat	IH, 875 ppm, 7 hrs/day [6-15]	–	–	–
	mouse	PO in water, 1000 mg/kg x 35 days pre	–	–	–
	rat	IH, 2100 ppm, 6 hrs x 10 days pre [1-20]	–	wt, os	–
Vinyl chloride	man	environment near industry, both parents	?	–	–
		occupational, paternal only	–	–	?
	mouse	IH, 30,000 ppm, 6 hrs x 5 days		(no dominant lethal mutation)	
	rat	IH, 1000 ppm	–	–	–
		IH, 1500 ppm [during gestation]	–	–	+
Halogenated Aromatics					
DDT	bird	environmental (food)		(decreased egg shell thickness)	
	rat	PO in food, 200 ppm	?	(tail abnormalities)	nr
	mouse	PO in food, 7 ppm, chronic		(reduced fertility)	
	rabbit	PO, 50 mg/kg [7-9]	–	+	+
PBB's	mouse, rat	PO in food, 1000 ppm [during gestation]	–	nr	nr
	rat	pre-explant (development in vitro)	+	+	nr
PCB's	man	PO in food	Yushō	nr	nr
	rat	PO, 100 mg/kg [8-14]	–	nr	+
		PO in food, 500 ppm	–	wt	–

Pentachlorophenol	hamster	PO, 1.25 mg/kg	nr	nr	+
	rat	PO, 15 mg/kg [6-15]	nr	nr	+
2,3,7,8 TCDD	mouse	SC, 3 μg/kg [6-15]	+	nr	nr
	rat	PO, 0.125 μg/kg [6-15]		(GI hemorrhage)	+
		SC, 0.5 μg/kg [6-15]	+	–	–
Aromatics					
Benzene	mouse	PO, 0.3 ml/kg	–	nr	+
		SC, 3 ml/kg (single dose)	+	nr	nr
	rat	IH, 670 mg/m^3, continuous	–	nr	+
		IH, 1500 mg/m^3 [9-14]	–	+	nr
		IH, 7000, mg/m^3 [during gestation]	–	?	–
Benzo[a] pyrene	mouse	IP, 300 mg/kg	+	nr	nr
	rat	PO in food, 1000 ppm [during gestation]	–	–	+
Naphthalene (as metabolite 2-naphthol)	rabbit	PO, 1 g /kg [20, 22, 24]	+ (eye)	nr	nr
Phenol	rat	IP, 200 mg/kg [8-10, 11-13]	–	–	–
Toluene	man	occupational (maternal)	–	wt	–
	rat	IH, 1500 mg/m^3	–	+	nr
Xylene	chicken embryo	xylene atmosphere, 1 hr	+	nr	nr
	man	occupational (maternal)	+	nr	nr
	mouse	PO, 3.6 ml/kg [5-14]	+	nr	+
	rat	IH, 1000 mg/m^3 [9-14]	?	–	–
Miscellaneous					
Acetone	rat	PO in food, 0.045% chronic	–	–	–

+	Present	wt	Low fetal weight	IV	Intravenous
–	Not present	DC	Direct contact	PO	Oral
nr	Not reported	IH	Inhalation	SC	Subcutaneous
?	Questionable significance	IM	Intramuscular	pre	Pre-pregnancy
os	Incomplete ossification	IP	Intraperitoneal		

TABLE 4–4. Embryo- and fetotoxicity of selected chemicals frequently found at uncontrolled disposal sites. (Continued)

Chemical	Organism	Exposure Level and Duration [day of gestation]	Malformation	Developmental Difference	Lethal
Miscellaneous (continued)					
Arsenic	chicken embryo	DC, 0.2 mg in medium	+	nr	nr
	hamster	PO, 15 mg/kg [8]	+	nr	nr
	man	PO, maternal	–	–	–
	mouse	IP, 45 mg/kg, single dose [6, 7, 8, 9, 10, 11]	+	nr	nr
	sheep	PO, 0.5 mg/kg [during gestation]	–	nr	nr
Asbestos	mouse	PO in water, 143 μg/ml	–	–	–
Cyanide	chicken embryo	DC, 0.005 m in medium	nr	+	nr
	hamster	IV, 0.126 mM/kg/hr [6-9]	+	nr	nr
Metals, etc.					
Cadmium	hamster	IV, 2 mg/kg [8]	+	–	–
	man	occupational (maternal)	nr	wt	nr
		environmental (maternal)	–	–	–
	rat	IP, 16 μmoles [9, 10 or 11]	+	nr	nr
		SC, 40 μmoles, single dose	nr	nr	+
Chromium (II)	hamster	IV, 8 mg/kg [7, 8 or 9]	+	nr	+
(VI)	chicken embryo	DC, 2.5 mg [8]	nr	+	nr
Lead	chicken embryo	DC, 0.1 mg [4]	+	nr	nr
	hamster	IV, 50 mg/kg [8]	+	nr	nr
	mouse	PO in food, 0.25%	nr	wt	nr
	rat	IV, 35 mg/kg [9, 16]	+	+	nr
	sheep	maternal blood level 34 μg/100 ml		(learning defects)	
Mercury	man	environmental (maternal, food)	+	?	+
	mouse	IM, 30 mg/kg [6-13]	+	–	–
	rat	PO in food, 20 mg/kg	–	–	–
		vaginal, 0.1 mg	+	–	–
Zinc	mouse	IP, 20 mg/kg	+	–	–

(Adapted from Shepard, 1983)

Registries of reproductive outcomes provide a valuable resource for detecting pattern changes in the incidence of congenital disease over a period of time. Since 1970, the Congenital Malformations Surveillance Branch of the Centers for Disease Control (CDC) has maintained a monitoring service for some 150 types of congenital malformations (USDHHS/CDC, 1982). This national registry constitutes a data base of substantial proportions for studying these reproductive outcomes. Increases in the incidence of some malformations, such as renal agenesis, ventricular septal defects, patent ductus arteriosus, and congenital hip dislocations, have been noted, while other malformations such as spina bifida, anencephaly, and rectal atresia have decreased, and Down's syndrome and deformities of the limbs have remained relatively unchanged (Wilson and Brent, 1981).

Determining the normal background rate of adverse reproductive outcomes in the general population is a complex endeavor. Problems of definition and classification, as well as confounding variables such as maternal age, smoking, and drug and alcohol consumption, present obstacles to the environmental epidemiologist. The ideal reproductive registry would contain sufficient data to confirm a causal relationship between a teratologic agent and an adverse reproductive outcome. The data needed would have to follow a series of criteria, originally outlined by Wilson (1973a), and would provide information to determine: 1) an abrupt increase in the incidence of a particular defect, 2) a coincidence of this increase with a known environmental change, 3) known exposure to the environmental change early in pregnancy, yielding characteristically affected infants, and 4) the absence of other factors.

Unfortunately, most registries provide only limited detail regarding etiology. The usefulness of registries to establish etiologic relationships will remain limited unless they can be improved to reveal changes that are consistent in direction, appreciable in magnitude, and can be correlated with known changes in the human environment. Those trends presently noted by the CDC monitoring program are not indicative of any of these changes. An increased incidence of malformations associated with chemical exposure within relatively small population samples is not apparent when compared with the incidence of specific outcomes in much larger samples of the general or national population. A recent example of this observation was demonstrated in a study of variations in the incidence of teratologic effects following the increased use of female sex hormones as contraceptive agents (Wilson and Brent, 1981). In many instances, the data necessary to give a definitive answer are frequently not available for the questions asked. Ironically, it is often impossible to know all of the appropriate questions until after the surveillance registry has demonstrated a positive trend.

The statistical limitations involved in detecting alterations in the occurrence of reproductive outcomes are exceedingly complex. The probability that an effect will be observed is influenced by: a) the frequency of the outcome observed, b) the magnitude of the effect expected, c) the accepted level of statistical significance, and d) the sample size (Nelson et al., 1979; Strobino et al., 1978). The estimated risk for a dysfunction is usually small (less than 2 per 1,000) for exposed versus unexposed individuals. In the case of a doubling of the risk of an event from 1 per 1,000 to 2 per 1,000, an exposed population of 10,000 individuals is needed to ensure a 75% probability of observing a difference that is significant at the 5% level, using a two-tailed test. Population samples of sufficient size are rarely located proximate to disposal sites to permit computations of this order of magnitude.

Techniques for Detecting Reproductive Effects

The severity and reversibility/irreversibility of the effect, demonstration of threshold and dose relationship, as well as the sensitivity of the test selected to measure the toxicologic response, are important considerations in the assessment of human risk from exposure to environmental chemicals. Many tests are available to detect changes in the reproductive process although specific tests may be either currently inappropriate, economically not feasible, or too dependent on sophisticated instrumentation for use in population studies at disposal sites. An excellent review of present methods for assessing risks to the reproductive process in man has been prepared by the EPA and the Department of Energy (USDOE/ORNL, 1982).

Sperm assays

Human sperm assays have been widely used in assessing spermatogenic function in chemically exposed men and animals (Wyrobek et al., 1983). Sperm anomalies can be used as indicators and, in certain cases, dosimeters of induced antispermatogenic effects. In addition, they can reveal spermatotoxic effects as well as possible genetic damage to the gonads. Experience with agents like dibromochloropropane (DBCP), kepone, lead, and organic solvents such as toluene, benzene, and xylene has shown that severe spermatogenic damage occurs at doses that may cause no other apparent signs of chemical toxicity (Dixon, 1978).

Semen physiology can be assessed by evaluating: a) count (density), b) motility, c) morphology (cytology), and d) YYF test (representative of Y-chromosome nondisjunction during spermatogenesis). With the exception of the assay for motility, which is highly sensitive to time and temperature factors, the other assays are technically straightforward methods that do not appear to be affected by post-ejaculation technical factors. Wyrobek (1983) has prepared a recent review of the technical literature for these assays with emphasis on the more reliable techniques.

It is important to note that variability in the characteristics of semen is high even among fertile "healthy" men, and large numbers of test subjects are required to establish differences between exposed and comparison individuals. Factors such as age, smoking, and medication need to be taken into consideration when interpreting the results from such studies (Bloom, 1981).

Ovarian and uterine function

Chemical agents can interrupt female fertility by one or more mechanisms at several different sites in the reproductive pathway. Compounds such as the purine analogs, azathioprine and 6-mercaptopurine, are known to produce ovarian toxicity by blocking some facet of oogenesis (Chen et al., 1981; Mattison et al., 1981). These compounds can also alter sexual maturation while estrogen agonists such as certain halogenated polycyclic hydrocarbons may produce a precocious pseudopuberty (Gellert et al., 1972; Kimbrough, 1974; Kupfer, 1975; McLachlan, 1980). Certain pesticides, polycyclic aromatic hydrocarbons, halogenated aromatic hydrocarbons, and steroid hormone agonists and antagonists are known to interfere with the hypothalamic-hypophyseal-gonadal-uterine axis, thus affecting ovulation and hormonal changes during the menstrual cycle (Mattison and Ross, 1983). Mattison et al.

(1983) have recently presented a review of the female reproductive cycle with emphasis on the mechanism and site of action of adverse effects of selective agents.

Tests for predicting the estrogenicity of environmental chemicals are available as simple screening tests in animals. These tests include evaluation of a) uterine epithelial cellular hypertrophy, b) estrogen receptor-binding analysis, and c) time of vaginal opening in the neonatal rat. Testing for androgenic effects would include a) observations for masculinization of the female, and b) measurements of testosterone levels in body fluids (USDOE/--ORNL, 1982).

Birth and reproductive failures

As recommended by the American Academy of Pediatrics and the World Health Organization, infants weighing 2,500 grams or less at birth are considered to be of low birth weight. These infants are frequently associated with an increased risk of mortality, mental retardation, congenital malformations, and numerous other physical and neurological impairments. Although this group of infants represents a small proportion of all newborns, more than half of all infant deaths (under 1 year), and nearly three-quarters of all neonatal deaths (under 28 days) can be associated with low birth weight (USDHEW/NCHS, 1980). Recording birth weight is perhaps the most sensitive method readily available for monitoring embryonic effects of toxic agents. This is frequently ignored in human studies, even though it is by far the least expensive. Although it is generally recognized that most women have at least one first trimester spontaneous abortion, any significant increase in this number would be of concern. The sampling methods to evaluate events of early pregnancy are difficult to assess however, because generally the only evidence may be: a) an unusually heavy menstrual flow delayed by one week; b) resorption, with no interruption in the cycle save for a menstrual flow delayed by two weeks; c) a known suspected pregnancy followed by the passage of a clot with no visible embryonic material; or d) the passage of a visible early first trimester embryo. These events are usually not examined or recorded by appropriate health professionals. Systematic studies of stillbirths have not been conducted, but may be organized under the leadership of several state departments of public health where arrangements can be made to evaluate changes as regional or subregional events. An increase in pregnancy failure as an independent event is probably a good indicator of an alteration in the environment; however, this health outcome is also commonly associated with sociopsychological events, nutritional deficiencies, chronic infections, and possibly use of substances of abuse such as alcohol and tobacco. Although national statistics for reproductive impairments among married couples indicate an increase in the incidence of infertility during recent years (Mosher and Pratt, 1982), an increase of infertility due to sterility would have to be sizeable to be statistically significant.

Malformations

Specific malformations are relatively uncommon, but malformations as a class are considered to be suspect, especially when observed in clusters. It is estimated that 2-3% of the population has detectable malformations which, even if not major, may require future correction. The incidence would increase significantly if the list of malformations is broadened to include minor variations such as multiple nevi, hernia-hemorrhoid-varicose vein triad, and minor structural malformations. Such variations are presently viewed to be consistent within the range of normal that is expected within any given population. To include metabolic diseases and enzyme deficiencies would expand the number of diseases that could be included in this category. In addition, multiple congenital malformation patterns such as those produced by chromosomal anomalies, in particular the trisomies, may also be included. By the time a child reaches school age, conditions such as marginal mental defects, minimal learning defects, and problems of sight, hearing, and speech may be identified as being independent of acquired disease etiology and be considered as a malformation. The sum of these additional problems could bring the total number of offspring with malformations to about 10-14% and might even include a major portion of the general population. The problems encountered in detecting an increased incidence of a malformation due to exposure to a disposal site may be exemplified by the incident involving Curene-442 (MOCA) in Adrian, Michigan (Majeti and Clark, 1980; Parris et al., 1980), where increases in neural tube defects have been alleged. At such sites, the concentrations of substances encountered in surface and groundwaters, in air, and by direct contact around the disposal site may be such as to produce a subtle defect, for example, altered enzyme function. Subtle defects are more difficult to detect than those such as facial clefts or congenital heart malformations. Clear evidence for the concomitance of prolonged exposure and the presence of the defect would be required to relate a defect to a disposal site exposure. In addition, coexisting medical problems related to genetic variance and postnatal disease would present a complex monitoring dilemma.

A potent source of information might be the categorizing of teratological events as embryolethal, embryotoxic, and fetotoxic. The likelihood of diagnosing embryolethality -- i.e., early reproductive failure, embryo death, resorption, and abortion in the first trimester -- is problematic. Embryotoxicity and fetotoxicity include specific gross malformations, and growth retardation. Functional defects may occur during fetal or embryonic development but are only detected postpartum. Functional defects require careful examination of physical and mental structure and function.

Infancy and childhood

The list of adverse effects to various organ systems as a result of chemical exposure of the developing fetus is extensive and only a few examples will be discussed here. A more detailed exposition of this topic may be found in Riddell's recent textbook, Pathology of Drug-Induced and Toxic Diseases (1982). Environmental agents such as DDT, carbon tetrachloride, copper sulfate, arsenic compounds, and chlorinated hydrocarbons have all been associated with massive necrosis of the liver or a syndrome resembling hepatitis. Maternal exposure to heavy metals (cadmium, mercury, and lead) has resulted in bone changes that are probably secondary to renal toxicity; on the other hand, direct central nervous system effects may be manifested in the developing central nervous system exposed to soluble mercury compounds.

Diethylstilbestrol (DES) offers an important insight for the intricate complexity required in assessing the effects of chemicals on the developing neonate. DES has been found to produce vaginal adenocarcinoma at puberty in less than 1 in 1000 women who were exposed in utero during the first trimester. Vaginal adenosis and cervical eversion in females, and distinct alterations of normal sperm production and maintenance in males have also been observed (McLachlan and Dixon, 1977; Mills and Bongiovanni, 1978). Vaginal adenosis may grossly mimic atypical growth or neoplasia, and the resulting alterations of reproductive function are well known.

At least one nation, Japan, has established a monitoring program to detect early changes that become evident during infancy. The quality of the Japanese reproductive registry has been enhanced to include a physical evaluation for both mother and child. This government-required examination for each birth must be given during the first months after the child leaves the hospital. Registries of this type will provide the first systematic opportunity to evaluate reproductive effects noted during this period.

ASSESSMENT OF IMMUNOLOGIC EFFECTS

Information on the interaction and function of the multiple components of the immune system has accumulated rapidly in the past decade. The two major branches of immune-system function are humoral and cellular immunity. Humoral immunity is characterized by antigen-specific B-cell differentiation that results in the production of specific immunoglobulins. Cell-mediated immunity encompasses the functions of a variety of cells including T-cells, null cells, and cells of the monocyte-macrophage series. Each type of immunity has many functional subsets and interacting control mechanisms and substances (e.g., complement, lymphokines, and immunomodulators).

Immunological testing methods may be used for both (a) surveying populations exposed to xenobiotic chemicals, and (b) assessing possible effects of such chemicals on the immune system by experimental studies in animals. By screening with in vitro systems, the latter serves as a predictive method and the former to outline a panel of clinical immunology tests. Abnormal clinical test results for measurements of immunologic function, if associated with an agent or mixture of agents, would carry with them concern for an individual's health status when they are statistically significant in an epidemiology study.

Spreafico and Vecchi (1984) define immunotoxicology as a subspeciality concerned with the study of immunologic response to a xenobiotic (e.g., hypersensitivity), as well as the consequences of the influence of a xenobiotic on the immune system (e.g., altered T4/T8 ratio). There is, however, no generally accepted definition of an immunotoxic substance. Immunotoxicity and chemically induced immunomodulation are frequently perceived only as states of decreased or suppressed immune function. In fact, chemicals alter immune function by enhancement or suppression of particular activities, by perturbing the balance of mature or intermediate cell types, and by accelerating or arresting maturation.

Lowered resistance to bacterial, viral, fungal, and parasitic diseases, as well as an increased incidence of neoplastic diseases, are well documented risks of chronic immunosuppression in patients receiving immunosuppressive drugs (Allen, 1976; Allison, 1970; Kripke and Borsos, 1974). Patients with primary immunodeficiencies often succumb to infection at a young age (Gatti and Good, 1971). Methods available to detect changes in the competency of the human immune system represent a valuable adjunct to other clinical features and testing in the diagnosis of disease. Conventional clinical chemistries, hematology, urinalysis, and similiar testing are usually required for adequate interpretation of the results of immune system analysis. The consequences and patterns of altered immune function modulation vary widely, making them difficult to assess with a simple yet reliable panel of tests.

Alterations in immune system function may be caused by many agents including infections, drugs, and stress. Various compartments of the immune system may respond to multimodal insults resulting in a complex summation of compensating mechanisms which tend to be individualized due to genetic factors and antigen exposure.

Chemicals of Environmental Concern

The large number and varied nature of chemicals capable of altering immune system responses increase concern about the environmental occurrence of such chemicals. Spreafico and Vecchi (1984) suggested that an immunotoxic substance induces immunologic effects at doses lower than those producing general signs of toxicity, for example, significant and sustained decrease in body weight. Vos (1977) has reviewed the effects of a number of environmental chemicals on the immune response in experimental animals. Studies of accidentally exposed populations in Michigan, Japan, and China, suggest that polybrominated biphenyls (PBB's) and polychlorinated biphenyls (PCB's) may cause temporary immune dysfunction in humans exposed to sufficient concentrations of these chemicals (Bekesi and Selikoff, 1978). Convincing evidence of long-term effects on the immune system has not been found in these population groups using relatively broad endpoints.

Current research has focused on the development of animal models for assessing immune dysfunction in humans (Bradley and Morahan, 1982; Munson et al., 1982). Luster et al. (1978) noted that the immune dysfunctions associated with exposure of humans to PBB's are similar to those observed in rodent models (Fraker and Aust, 1980). PCB's, PBB's, polycyclic aromatic hydrocarbons (PAH), hexachlorobenzene, diethylstilbestrol (DES), pentachlorophenol (PCP), 2,3,7,8-tetrachlorodibenzo-p-dioxin (TCDD), certain organometals, and many heavy metals produce immunologic alterations in rodent studies (Bellanti et al., 1978; Dean et al., 1982b). Still, the list presented in Table 4-5 is incomplete and many chemicals and combinations of chemicals will need to be evaluated in a validated panel of animal-based immunotoxicity tests if their potential effects on the immune system are to be adequately assessed.

Immune System Sensitivity

Dean et al. (1982b) state that evidence of toxicity can be detected by measuring the apparent increased sensitivity of the immune system as a reflection of injury following exposure to toxic chemicals:

- Most cell types of the immune system continually undergo proliferation and differentiation.
- The immune system is a tightly regulated and highly organized network of cellular components. An imbalance of this regulatory network resulting from cellular injury can be expressed as an immune alteration. Immune alterations may be represented either as immune enhancement, which might ultimately result in autoimmunity, contact hypersensitivity or allergy, or as immune

suppression, which might result in decreased resistance to an infectious agent or impaired immunity surveillance.

- The cell types involved in the immune response can easily be removed or isolated and purified, and their function can be assessed in vitro to provide more sensitive endpoints for detection of cellular injury than are currently available by a routine toxicology evaluation.

Evaluating Immune Systems of Experimental Animals after Chemical Exposure

The National Institute of Environmental Health Sciences (NIEHS) is currently evaluating a panel of tests for defining immune alterations resulting from chemical exposure (Dean et al., 1982a; Moore et al., 1982). The in vivo and in vitro approaches for assessment of the immune function are part of the National Toxicology Program's efforts to develop a comprehensive approach to the evaluation of chemicals of environmental concern. Design of protocols for such testing is difficult because many factors that are not usually thought of as part of the immune system influence the results. For example, if the skin of an experimental animal is dried and cracked by a chemical, the probability of infection will be greater, even if cell-mediated and humoral antibody responses are entirely intact. Immunology represents but one of the disciplines to be considered in the totality of natural host defenses that can be elicited, induced, or acquired. The panel of tests under study by NIEHS for assessing immunologic effects and alterations of host resistance involve evaluation of pathotoxicology, host resistance, cell-mediated immunity (e.g., delayed-type hypersensitivity), humoral immunity, macrophage function, and bone marrow cellularity. Experimental laboratory animal protocols include prenatal as well as adult exposures to test chemicals. The procedures selected should be generally reproducible and easily standardized. In addition, numerous other immune function tests that can be employed in a toxicological assessment are available or are being developed (Haber and Pfitzer, 1982).

In the mouse-model test system being utilized by Munson et al. (1982), a 14-day range finding study is employed in conjunction with a 90-day study in which data collected on body and organ weights, hematology, clinical chemistry, urinalysis, and microsomal enzymes provide a background for the analysis of the immune system.

Dean et al. (1982b) and Bradley and Morahan (1982) suggest that alterations in host resistance would provide useful indicators to predict significant immune dysfunction. Previous research has focused on assessing macrophage function as well as on developing tests to evaluate the immunocompetence of T-cells and B-cells in immunoregulation and host defense. The effects of a chemical or a mixture of chemicals on immune system function can be assessed by a variety of in vivo and in vitro techniques. A consensus on the best panel of tests remains to be developed. It is widely agreed that the immune system represents a very complex constellation of cell and molecular interactions. Several investigators have provided data supporting the hypothesis that immune dysfunction may serve as a co-carcinogenic influence in the etiology of certain tumors (Allison, 1970; Kripke and Borsos, 1974; Penn, 1978).

TABLE 4–5. Hazardous chemicals found at chemical disposal sites that may alter immunological parameters.

Chemical (with code)	Immune Disfunction	Chemical	Species	Source
Aromatic Compounds	Decreased mitogen transformation	(a)	mouse	Wierda et al., 1981
(a) Benzene	Decreased antibody production	(a)	mouse	Wierda et al., 1981
(b) Toluene	Decreased HMI[1]	(d)	mouse	La Via and La Via, 1979
(c) Xylene	Impaired host resistance	(d)	rat	Exon and Koller, 1984
(d) Phenol	Increased susceptibility to tuberculosis	(a)	rabbit	WHO/IARC, 1982b
	Decreased IgG[2], IgA[3] serum levels	(a) (b) (c)	human	Lange et al., 1973
	Decreased complement serum levels	(a) (b) (c)	human	Smolik et al., 1973
	Lymphocytopenia	(a)	human	Aksoy, 1981
Halogenated Aliphatic Compounds	Increased mitogen transformation	(g)	mouse	Sharma and Gehring, 1979
(e) Chloroform (trichloromethane)	Decreased HMI	(f) (h) (i)	mouse	Munson et al., 1982
(f) 1,2-Dichloroethane (ethylene dichloride)	Decreased CMI[4]	(i)	mouse	Sanders et al., 1982
(g) Vinyl chloride (monochloroethylene)	Leukocytopenia	(f)	mouse	Munson et al., 1982
(h) Vinylidene chloride (1,1-dichloroethylene)	Impaired host resistance	(e)	mouse	Capel et al., 1979
(i) 1,1,2-Trichloroethylene	Increased embryonic antigen	(g)	human	Page et al., 1976
Halogenated Aromatic Compounds	Impaired host resistance	(j) (n)	mouse	Loose et al., 1978
(j) Hexachlorobenzene (HCB)		(l)	mouse	Thigpen et al., 1975
(k) Pentachlorophenol (PCP)		(l)	mouse	Luster et al., 1980
(l) 2,3,7,8-Tetrachlorodibenzo-p-dioxin (TCDD)		(o)	mouse	Neilson et al., 1973
(m) 2,3,7,8-Tetrachlorodibenzofuran (TCDF)	Increased GVH[5] reaction	(n)	mouse	Carter and Clanoy, 1980
(n) Polychlorinated biphenyls (PCB's)	Thymus atrophy	(l) (n)	mouse	Vos et al., 1980
(o) 2,2-bis(p-Chlorophenyl)-1,1,1-trichloroethane (DDT)		(o)	mouse	Street and Sharma, 1974
	Decreased mitogen transformation	(l)	mouse	Luster et al., 1980
	Increased mitogen transformation	(n)	mouse	Silkworth and Loose, 1979
	Decreased HMI	(j) (n)	mouse	Loose et al., 1978
		(l)	mouse	Thomas and Hinsdill, 1979
		(n)	mouse	Thomas and Hinsdill, 1978
	Decreased CMI	(l)	mouse	Luster et al., 1980
		(l)	mouse	Clark et al., 1983
		(o)	mouse	Street and Sharma, 1975
	Decreased complement activity	(k)	mouse	Anderson and White, 1984
	Increased HMI	(j)	rat	Vos et al, 1979
	Decreased HMI	(o)	rat	Wasserman et al., 1969
	Increased antitumor activity	(n)	rat	Kerkvliet and Kimeldorf, 1977
	Impaired host resistance	(k)	rat	Exon and Koller, 1984
	Decreased antibody production	(o)	chicken	Rao and Glick, 1977
	Impaired host resistance	(o)	chicken	Radhakrishnan et al., 1972

	Thymus atrophy	(m)	guinea pig	Luster et al., 1979
	Decreased CMI	(m)	guinea pig	Luster et al., 1979
	Decreased mitogen transformation	(m)	guinea pig	Luster et al., 1979
	Normal Ig complement levels	(l)	human	Reggiani, 1980
	Increased respiratory infection	(n)	human	Shigematsu et al., 1978
	Decreased CMI	(n)	human	Chang et al., 1980
Trace Metals	Altered host resistance	(p)	mouse	Gainer and Pry, 1972
(p) Arsenic		(r)	mouse	Koller, 1980
(q) Asbestos		(t)	mouse	Cook et al., 1975
(r) Cadmium		(t) (u)	mouse	Gainer, 1977
(s) Chromium		(v)	mouse	Sobocinski et al., 1977
(t) Lead	Accelerated thymus atrophy	(u)	mouse	Hirokawa and Hayashi, 1980
(u) Mercury	Decreased HMI	(r) (t)	mouse	Koller et al., 1976
(v) Zinc		(u)	mouse	Koller et al., 1977
		(p) (t) (u)	mouse	Blakley et al., 1980
	Impaired macrophage function	(q)	mouse	White, 1981
		(t)	mouse	Blakley and Archer, 1981
		(v)	mouse	Karl et al., 1973
	Lymphocyte-macrophage binding	(q)	rat	Miller et al., 1979
	Increased GVH reaction	(u)	rat	Pelletier et al., 1983
	Decreased virus Ab[6] titer	(s)	rat	Figoni and Treagan, 1975
	Decreased ADCC[7]	(q)	human	Kagamimori et al., 1980
	Increased Ig serum levels	(q)	human	Kagan et al., 1977
	Decreased circulating T-cells[8]	(q)	human	Kagan et al., 1977
	Allergic contact dermatitis	(s)	human	Peltonen and Fraki, 1980
		(u)	human	Baer et al., 1973
	Impaired host resistance	(t)	human	Sachs, 1978
		(t)	human	Ewers et al., 1982
	Decreased HMI	(t)	human	Reigart and Graber, 1976
	Glomerulonephritis	(u)	human	Druet et al., 1982
Miscellaneous Organic Compounds	Decreased NK[9] cell activity	(x)	mouse	Luster et al., 1982
(w) Formaldehyde	Allergic contact dermatitis	(y)	guinea pig	Karol et al., 1980
(x) Ethyl carbamate (urethane)	Respiratory tract hypersensitivity	(y)	guinea pig	Karol et al., 1980
(y) Toluene diisocyanate (TDI)	Respiratory tract hypersensitivity	(y)	human	O'Brien et al., 1979
	Allergic contact dermatitis	(w)	human	Bardana, 1980

[1] Humoral mediated immunity
[2] Immunoglobulin G
[3] Immunoglobulin A
[4] Cell mediated immunity
[5] Graft versus host
[6] Antibody
[7] Antibody-dependent cell-mediated cytotoxicity
[8] Thymus derived cells
[9] Natural killer cells

The response of the immune system is sensitive in that it can be rapidly stimulated or down-regulated. It is able to affect normal defense mechanisms when one or more of its components is compromised as might occur in infection, exposure to xenobiotics, or for other reasons. The importance of this compensation ability can be seen by considering the significant role played by antibody-dependent cell-mediated cytotoxicity (ADCC) and natural killer (NK) cells in the control of malignancies (Fudenberg and Whitten, 1984). Measurements of a perturbation need to be cautiously interpreted relative to adverse effects, as the immune system is continually compensating for both permanent and transient imbalances. However, the absence of an abnormality does not indicate the absence of an insult.

Criteria Used for Immunologic Assessement

Pathotoxicology

Pathotoxicity data are largely unattainable for human studies. For animal experiments, assessment of thymus and spleen weights represent a useful parameter in assessing the toxic effects of chemical exposure. However, a body weight to organ weight ratio is somewhat more sensitive than direct weighing or morphological evaluation when only subtle changes have occurred. For example, thymic weight losses of less than 20% are difficult to discern by routine histopathology (Dean et al., 1982b). Caution is suggested in using thymic atrophy as a specific indicator of immunosuppression because similar effects may be produced by general stress or unrelated toxicities.

Bone marrow progenitor cell assays

Cells of the bone marrow differentiate along several lines giving rise to hematopoietic and lymphoid cell populations. Boorman et al. (1982) examined the site of toxic effects on these rapidly proliferating cells when mice were exposed to a panel of chemicals and drugs including TCDD, PBB's, DES, benzo[a]pyrene, and indomethacin. Each chemical was immunotoxic at concentrations below those causing other toxic manifestations. Effects were noted on hematopoietic stem cells and granulocyte-macrophage progenitors. These findings give support for the continued development of various clonal marrow assays to detect, in a very sensitive manner, the effects of environmental contaminants and drugs. Although some investigators may argue against the selection of a "most sensitive" assay system, it should be noted that the hematopoietic stem cells have a finite renewal capacity and it has been suggested that a permanent decrease in their proliferative capacity might be induced by a limited exposure (Botnick et al., 1978). Boorman et al. (1982) have advanced the untested hypothesis that mixtures of chemicals may present a greater risk if one chemical can induce damage to these cells.

Cell-mediated immunity

In vivo assays of cell-mediated immunity include graft versus host (GVH) reaction, rejection of skin grafts, and measurements of delayed hypersensitivity. In vitro assays such as lymphoproliferation, T-cell cytotoxicity, and lymphokine production have not correlated as well with host resistance to infectious agents as has delayed hypersensitivity (Dean et al., 1982b;

MacLean, 1979). As a result, assays of lymphocyte proliferation and function require careful interpretation. Tolerance and suppressor cell activity are additional monitors of immune responsiveness and may represent extremely sensitive indicators of immunotoxicity if appropriately interpreted (Dean et al., 1982b).

Adequate methods for assessing the potential allergenic or hypersensitivity effects of chemicals remain difficult to develop. Luster and Dean (1982) have summarized the state of the art of immunological hypersensitivity resulting from environmental or occupational exposure to chemicals. At present, useful clinical procedures include a detailed clinical history, skin testing, precipitin test for IgG antibodies, radioallergosorbent test for IgE antibodies (RAST test), bronchial provocation, and lymphokine release assay.

Humoral-mediated immunity

The measurement of immunoglobulin levels as a component of immune responsiveness serves as an index in chronic studies when the study period extends beyond the immunoglobulin half-life. However, immunoglobulin levels are considered neither specific nor sensitive indicators of immunotoxicity. Genetic diversity, seasonal variations, and variance in history of antigen exposure limit the interpretation of serum immunoglobulin levels. For these reasons, immunoglobulin levels may be less sensitive or specific indicators of environmental chemical toxicities than the in vitro assay. It is most practical to use human peripheral blood for functional testing of lymphocyte proliferation and antibody production.

Macrophage function assays

In addition to the familiar phagocytic activity of macrophages, numerous macrophage subsets are now known to play important roles in feedback and regulatory mechanisms involving prostaglandins and monokines. The macrophage subsets are, in turn, partially regulated by lymphokine levels. Macrophages are important in cytostasis and cytolysis of neoplastically transformed or virally infected cells. Assessment of macrophage function as part of an experimental animal protocol is essential in a comprehensive program to assess the possible effects of an environmental contaminant on immune system function.

Testing the Immune System of Humans

The following methods are currently available to investigate immune system functions of humans:

- delayed-type hypersensitivity and contact hypersensitivity;
- mixed lymphocyte reactions using peripheral blood lymphocytes;
- peripheral blood monocyte studies including phagocytosis, tumor cell line killing, cytostasis, surface markers, and enzyme profiles;
- in vitro response to thymus-independent antigens and polyclonal activators;

- enumeration and separation of defined lymphocyte subsets;
- blood chemistry and hematology;
- bone marrow work-up;
- complement studies including specific component concentrations and function; and
- function of cytotoxic T-lymphocytes, natural killer cells, and antibody dependent cytotoxic cells.

The following cannot be studied in humans, but these procedures provide valuable information on the immunotoxicity of xenobiotics to laboratory animals:

- in vivo assessment of reticuloendothelial system clearance;
- in vivo antibody production or plaque forming cells (e.g., spleen cells);
- infectious agent assays;
- organ weights; and
- in vitro T-cell dependent, antigen-specific antibody response.

A single test of an individual's immune function lacks sufficient meaning to permit extrapolation or generalization to the genetically and physiologically heterogeneous population. If, after exposure to a xenobiotic substance, an individual has markedly impaired T-cell or B-cell function, macrophage function, or complement level, or otherwise demonstrates abnormal changes in the traditional immune compartment, then there is cause for concern and additional studies are in order. Other non-immunologic criteria would need to be assessed to gain a complete understanding of the biologic and clinical significance of these changes.

ASSESSMENT OF NEUROLOGIC EFFECTS

Background and Principles of Neurotoxicology

The historic record of neurotoxicology includes several examples of injury to workers and the general public following exposure to a chemical that was later found to be neurotoxic. One of the most widely known episodes of neurologic poisoning occurred in the population surrounding Minamata Bay, Japan between 1953 and the 1960's. In this incident the mercury effluent dumped into the bay from an industrial plant entered into the food chain after it had been converted into the more harmful methylmercury form by plankton in the bay (Kurland et al., 1960; Tedeschi, 1982). Widespread neurotoxic poisoning occurred following the addition of tri-o-cresyl phosphate to a drink in the United States, and into cooking oil in Morocco (Smith and Spalding, 1959; Smith et al., 1930). The finding that numerous alkane hydrocarbons (e.g., n-hexane, methyl n-butyl ketone and other compounds) could be metabolically converted into the highly neurotoxic n-diketones was a complete surprise. Numerous examples of neurotoxic injury from this group of compounds have been cited in the scientific literature including: furniture workers in the Bronx, New York (Herskowitz et al., 1971), workers in the laminating industry and sandal manufacturing in Japan (Inoue et al., 1970; Oishi et al., 1964), shoe manufacturing in Italy (DiBosco and Fonzi, 1974), glue sniffers in the U.S. (Towfighi et al., 1976) and Germany (Altenkirch et al., 1977), and workers in a printing plant in Ohio (Allen et al., 1975; Mendell et al., 1974).

It is difficult to determine whether a chemical or mixture of chemicals at a chemical disposal site presents a neurotoxic danger to the public (Schaumburg et al., 1981, 1983). Table 4-6 presents a summary of available information for known neurotoxic chemicals. The table provides information on the anatomical site of chemical injury, the signs and symptoms observed, and an indication of whether or not the mechanism of injury is understood. As illustrated in the table, there are many neurotoxic chemicals for which either the site of action or the mechanism of injury have not been completely determined. Included in this group are chemicals that induce rather vague clinical pictures of neurologic injury that are difficult to characterize and to quantify. There is another group of chemicals (e.g., carbon tetrachloride, tetrachlorobiphenyl, zinc, aluminum, Paraquat, and styrene) for which the neurotoxic potential is questionable or not known.

The difficulty in diagnosing a neurologic syndrome in the absence of a full understanding of the chemical involved and the mechanism of neurologic injury can be illustrated by an episode among chemical workers. After repeated exposure to the organophosphorus pesticide, leptophos, a group of workers developed symptoms and signs characteristic of spinal cord injury (Xintaris et al., 1978). On the basis of these neurologic symptoms, many of the workers were diagnosed as having multiple sclerosis. However, the parallel that exists between the neurologic injury observed in these workers and that seen in laboratory experiments in chickens given leptophos (Abou-Donia et al., 1980), strongly suggests that this compound was related to the injury. The workers have the impression that they have a progressive neurological disease (i.e., multiple sclerosis), whereas, the nerve injuries they experienced may well have been fixed, nonprogressive spinal cord lesions from their exposure to leptophos. The diagnosis may have been further complicated by exposure of the workers to other neurotoxic chemicals including n-hexane.

There are a number of confounding variables to be considered in evaluating chemicals for potential neurotoxicity. A useful approach is to identify the structural components of the nervous system that are primarily affected as the site of injury: neurons, axons, myelin, glial cells, Schwann cells, and blood vessels (Schaumburg et al., 1983). Neurologic injury is frequently defined by these anatomical sites. For example, chemical injuries to neurons, axons, or to myelin are termed respectively: neuronopathies (e.g., trimethyltin and methylmercury), axonopathies (e.g., n-hexane and organophosphorus compounds), and myelinopathies (e.g., triethyltin). Considering the complexity of the nervous system, a wide variety of clinical pictures can result as specific neurotoxins may show a propensity to affect sensory nerves whereas others may affect motor pathways, basal ganglia, cerebellum or scattered cerebral neurons. If antecedent or concurrent disease, such as diabetes mellitus, alcohol abuse or nutritional deficiencies are factored into a clinical diagnosis, it is clear that early neurotoxic injury can be difficult to detect. Age is another factor in the relationship between chemical toxicants and their effects on the nervous system. The nervous system is particularly vulnerable during development, beginning in utero and continuing during the first years of life. The immature nervous system is capable of greater plasticity, with uninjured

TABLE 4–6. Chemicals known to be neurotoxic.

Chemical	Signs and Symptoms		Site of Injury	Mechanism of Injury		
				Known	Suspected	Unknown
Acrylamide	Acute:	encephalopathy	Unknown			X
	Chronic:	peripheral neuropathy	Axon			X
Allphatic Hydrocarbons Metabolized to γ-diketones	Acute:	narcosis	Unknown			X
	Chronic:	peripheral neuropathy	Axon		X	
Arsenic	Chronic:	peripheral neuropathy	Axon			X
Barium	Paralysis		Muscle		X	
Cyanide	Acute:	death	Neuron	X		
	Chronic:	peripheral neuropathy	Axon (?)			X
Chlordecone	Tremor, opsoclonus		Unknown			X
2,4-Dichlorophenoxy Acetic Acid	Peripheral neuropathy		Unknown			X
Lead	Acute:	encephalopathy	Blood vessels			X
	Chronic:	peripheral neuropathy	Axons			X
Manganese	Acute:	encephalopathy	Unknown			X
	Chronic:	Parkinsonism	Basal ganglia neurons		X	
Mercury	Inorganic:	tremor	Unknown			X
	Organic:	ataxia	Cerebellar neurons			X
		peripheral neuropathy	Axon			X

Methanol	Acute: encephalopathy	Unknown			X
	blindness	Retina		X	
	Chronic: blindness	Retina		X	
Organophosphorous Compounds	Acute: cholinergic poisoning	Acetylcholinesterase	X		
	Chronic: ataxia, paralysis	Long tracts of spinal cord		X	
	peripheral neuropathy	Axon		X	
Organotin Compounds					
Trimethyltin	Memory defects, disorientation, seizures	Neurons of limbic system			X
Triethyltin	Encelphalopathy	Myelin			X
Polybrominated Biphenyls	Memory loss	Unknown			X
Tellurium Compounds	Prenatal: hydrocephalus	Unknown			X
	Postnatal: peripheral neuropathy	Schwann cell			X
Thallium	Acute: encephalopathy	Unknown			X
	Chronic: cranial and peripheral neuropathy	Axon			X
Toluene	Acute: narcosis	Unknown			X
	Chronic: ataxia	Cerebellum			X
Trichloroethylene	Cranial neuropathy	Axon			X

neurons taking over the function of those that have been injured or destroyed. Unfortunately, this capacity is lost with age.

As chemical disposal sites frequently contain mixtures of chemicals, studies involving interactions between more than one chemical are of interest. Unfortunately, most of our present knowledge on the neurotoxic effects of chemicals is based on studies of individual compounds and there is frequently insufficient information available to describe the types of neurologic injury that could follow multiple chemical exposure. Altenkirch et al., (1978) observed a potentiating effect of increased neurotoxicity when n-hexane is administered along with 2-butanone (methyl ethyl ketone). It is not clear whether this effect is mediated by enhanced hepatic oxidation of n-hexane or by a separate effect on its distribution and metabolism to 2,5-hexanedione (a known potent neurotoxicant). This information supports the finding that predictions of neurotoxic potential are made more difficult in the presence of chemical mixtures.

The use of animal models for identifying chemical neurotoxic hazards in man is subject to numerous problems of interspecies extrapolation. For example, rodents are not a predictive animal model for studying the delayed neurotoxic effects of organophosphorus compounds that have been observed in man (Aldridge et al., 1969). This is due to the rodents' lack of sensitivity to acute cholinergic poisoning, as well as to the unique and rapid detoxification of this class of compounds by the rodent liver. The chicken has proven to be the best animal model for studying neurotoxic effects of organophosphate exposure. However, with alkanes there are significant differences among species not only in rates of absorption, metabolism, distribution, and excretion, but also in the relative rates of various metabolic pathways.

Despite these difficulties, a few generalizations can be made regarding the clinical detection of neurologic disease following exposure to neurotoxicants. Neurologic damage is generally diffuse and directed toward susceptible subsystems. Those subsystems that show particular vulnerablity to injury include peripheral nerves, long pathways of the spinal cord, the cerebrum and cerebellum. Neurological deficits are symmetrical in distribution so that unilateral lesions are considered to be of non-toxic origin. When the site of neurotoxic action is the distal axon, myelinated axons are more vulnerable than unmyelinated axons. There is a direct relationship between the severity of neurotoxic injury and both the increasing diameter and length of axons (Schaumburg and Spencer, 1979).

In contrast to the long-term, delayed effect observed for most carcinogenic chemicals, the time period between exposure and effect for a potential neurotoxic chemical is short. The neurotoxic effect of n-hexane can be observed after only several months of heavy exposure (Towfighi et al., 1976; Schaumburg and Spencer, 1976). Lipophilic organophosphorous compounds can progressively accumulate within the nervous system in such a way that symptoms of neurotoxicity develop in close temporal relationship to repeated exposures of otherwise no effect levels (Abou-Donia et al., 1980). With single large doses of organophosphorous compounds, the greatest time period between exposure and symptoms of delayed neurotoxicity would be 1 to 2 weeks (Abou-Donia, 1981).

The nervous system has a limited capability to protect itself against toxic injury. For example, individual components such as cytoplasm, axoplasm, and myelin are constantly being renewed. In the peripheral nervous system, demyelination can be followed by remyelination, and axonal degeneration by regeneration. In addition, intracellular molecules, such as reduced glutathione and polyamines, can quench electrophilic metabolites. For example, Graham (1984) has shown that glutathione peroxidase and superoxide dismutase protect the nervous system against injury from products of the partial reduction of oxygen (i.e., superoxide anions, hydroxyl radicals, and hydrogen peroxides). The combined protective and reparative mechanisms of the nervous system, coupled with no precedence among known neurotoxicants for long-term, delayed effects, lead to the important suggestion that "no-effect" levels of exposure can be defined for neurotoxic chemicals. The level of exposure of an individual can frequently be determined by analysis of certain chemicals in body tissues and fluids. For example, hair and urine can be sampled to quantify arsenic and lead exposure, and biopsies of adipose tissue can be assayed for lipophilic compounds.

Screening for Neurotoxicologic Injury

In a recent review on the use of clinical neurology for studies of potential neurologic injury at chemical disposal sites, Schaumburg et al. (1983) noted that the study of neurologic injury depends on the limited tools of physical examination, electrophysiology, and neuropsychology.

Electrophysiological technique is the study method most commonly used to determine neurologic injury. This method permits measurements of sensory and motor nerve conduction velocities of the large diameter nerve fibers as recorded on an oscilliscope from surface electrodes that are attached to a patients extremities (Landrigan et al., 1976). Under field conditions the response variability is frequently high where factors such as temperature are hard to control. The most throughly studied toxic axonopathy, the distal central-peripheral type, is probably the most common response of the nervous system to exogenous toxins. The pathological feature of this condition is axonal degeneration. This effect can be demonstrated by electrophysiologic techniques when the injury is distal to the points where nerve conduction velocities would be determined. Changes in nerve conduction velocity are characteristic of myelinopathies rather than the more common axonopathies. Electrophysiological studies performed in the field are much more difficult to interpret than those studies performed in a modern, well-equipped, electrodiagnostic laboratory that routinely uses averaging techniques and has extensive experience with normal individuals. Many reported field studies appear to have been poorly controlled and are seriously flawed as the study sample size was frequently too small for an accurate statistical analysis.

One benefit of using an established, modern, electrodiagnostic laboratory is the capability to record computer-averaged visual and somatosensory evoked potentials (Arezzo et al., 1979, 1982; Seppalainen and Harkonen, 1976; Seppalainen et al., 1979, 1980, 1981). These techniques permit monitoring portions of those nerve tracts most vulnerable to injury from neurotoxic exposure. In addition, Arezzo and Schaumburg (1980) have reported the use of a modified Optacon instrument as a means for screening sensory loss. The technique has been verified in hospital patients with diabetic and renal axonopathies and used to detect nerve injury in workers at chemical plants (Arezzo et al., 1983). Aminoff (1980) has outlined techniques for detecting changes in sympathetic nervous system function following exposure to toxic agents.

Neuropsychological deficits that have been reported in association with exposure to toxicants include: disturbances in intelligence, memory, and problem solving, as well as alterations in attention, psychomotor functioning, and mood (Damstra, 1978; Landrigan, 1983). Even in the hands of expert neuropsychologists, detection of these effects in a large, previously unstudied population is difficult (Feldman et al., 1980; Friedlander and Hearne, 1980; Lindstrom, 1980; Maroni et al., 1977). The use of a battery of standardized psychological tests can provide objective and quantifiable information; however, the value of such testing is frequently limited not only by the lack of baseline information from the people being tested but also by within- and across-subject variations in measurements, lack of standardization and specificity of the tests, and the potential for bias on the part of both subject and observer. Whenever possible, psychological data should be correlated with biochemical or physiological findings. The recent study of memory performance in individuals exposed to polybrominated biphenyl (PBB's) by Brown et al. (1981) is exemplary of a well designed study of neuropsychological effects following a chemical exposure.

Abnormal behavior may be manifested by people who learn that they are being exposed to chemicals from a chemical disposal site. Determining whether these behaviors result from exposure to toxicants or from stress presents a considerable problem.. In this regard, it should be noted that what is termed "behavioral neurotoxicology" has little to do with what is generally viewed as abnormal behavior in humans; rather, this field is largely confined to experimental studies that focus on the early clinical detection of neurological abnormalities (Mitchell and Tilson, 1982).

SUMMARY

The potential for adverse health effects in populations living near to disposal sites may involve any organ system depending on the specific chemicals contacted, the extent of exposure, the characteristics of the exposed individuals (e.g., age, sex, and genetic make-up), the metabolism of the chemicals involved, and the presence or absence of confounding variables such as other diseases. Health effects of primary concern to exposed populations at disposal sites include cancer, genetic defects, congenital anomalies, reproductive abnormalities, alterations of immunobiological homeostasis, and disorders of the central nervous system.

This chapter presents a theoretical consideration of the use of clinical and laboratory studies to screen for adverse health effects in populations thought to be exposed to environmental chemicals, and discusses those approaches that might prove beneficial in the investigation of potential health effects. In large populations, the study of adverse health effects or a disease outbreak of unknown etiology is usually based on epidemiological techniques. Clinical and laboratory investigations are usually of limited value in identifying an adverse consequence for the low levels of chemical exposure that are frequently associated with disposal sites. However, in situations where an adverse effect has been previously established, a careful clinical evaluation coupled with specific laboratory determinations may be of significant benefit. Detailed studies that include clinical and laboratory investigations are usually undertaken only when there is reason to suspect that a specific health effect will be found in the exposed population.

When applying these clinical methods to reach a diagnosis in the case of the individual patient, the physician takes into consideration the results of a complete history and physical examination along with laboratory tests (e.g., blood and urine tests, x-rays, etc.). Laboratory tests alone are seldom definitive as screening procedures because there is a wide range of biological variation within the population. The range of test values for a healthy population usually overlaps the range of test values from a diseased population so that both "false positives" and "false negatives" are almost inevitable. When only a small fraction of the population-at-risk is adversely affected by chemicals at a disposal site, the vast majority of the laboratory test values will fall within the normal range for the reference population. Factors to be considered in the selection, use, and evaluation of laboratory tests include quality control, sensitivity, specificity, predictability, and reference intervals.

Methods to determine body burden of specific chemicals are well developed and may be of value not only as an early warning system, but also as measurements of exposure. Recent advances in molecular biology offer several promising approaches for detecting biological changes at the biochemical or molecular level that may precede overt clinical manifestations of disease. Although they are in the early stages of development and validation, these methods include the use of monoclonal antibody probes to detect alterations in cell surface receptors, studies of altered gene expression, and chromosomal abnormalities.

The detection of induced heritable disorders in humans has been unsuccessful despite the fact that the possible induction of heritable defects is a major concern to exposed populations. There is genetic evidence that cancer-inducing genes (i.e., oncogenes) interact in a variety of ways to predispose cells to undergo malignant transformations following exposure to chemical mutagens or carcinogens. If the suspect disease is cancer, the epidemiologist generally relies on data from vital statistics or tumor registries rather than on population surveys that include clinical and laboratory studies.

Although some birth defects are evident at birth and easy to detect, others are subtle or may not become evident until years later. Minor fluctuations in the incidence of birth defects in small, exposed populations near disposal sites are difficult to interpret when compared with the naturally occurring background incidence of birth defects in the general population. At present, the most sensitive method for monitoring effects that have occurred in utero is the measurement of birth weight.

Reproductive failures such as sterility or spontaneous abortion are likely to be related to lifestyle as well as to chemical exposure. When related to chemical exposure, these effects are more readily associated with high-dose occupational exposures rather than the low level of exposures which are usually found in proximity to disposal sites. Since reproductive failures are a frequent occurrence in the general population, incidence rates in small groups of exposed individuals near disposal sites would have to be extraordinary high in order to demonstrate a significant increase. Human sperm assays may be of value in some situations.

Several chemicals known to induce neurotoxic effects in humans exposed in industrial settings may also be of concern at disposal sites. In screening populations exposed to these neurotoxins, the investigator must rely heavily on the limited usefulness of physical examinations, electrophysiology, and neuropsychology. Most procedures are expensive and are best carried out in specialized centers, as they are not suited to field use.

Assessment of immunologic effects may hold promise for the early detection of effects due to toxic chemical exposure since subtle changes in immune response may presage overt disease. While a number of tests are already available, no single test or battery of tests can be recommended for population surveys.

References

Abou-Donia, M.B. (1981) Organophosphorous ester-induced delayed neurotoxicity. Ann. Rev. Pharmacol. Toxicol. 21:511-548.

Abou-Donia, M.B.; Graham, D.G.; Ashry, M.A.; Timmons, P.R. (1980) Delayed neurotoxicity of leptophos and related compounds: differential effects of subchronic oral administration of pure, technical grade and degradation products on the hen. Toxicol. Appl. Pharmacol. 53:150-163.

Aksoy, M. (1981) Problems with benzene in Turkey. Reg. Toxicol. Pharmacol. 1:147-155.

Aldridge, W.N.; Barnes, J.M.; Johnson, M.K. (1969) Studies on delayed neurotoxicity produced by some organophosphorus compounds. Ann. N.Y. Acad. Sci. 160:314-322.

Allen, J.C. (1976) Infection complicating neoplastic disease and cytotoxic therapy. In: Allen, J.C., ed. Infection and the compromised host. Baltimore: Williams and Wilkins Company. p. 151-171.

Allen, N.; Mendell, J.R.; Billmaier, D.J.; Fontaine, R.E.; O'Neill, J. (1975) Toxic polyneuropathy due to methyl-n-butyl ketone. Arch. Neurol. 32:209-218.

Allison, A.C. (1970) Potentiation of viral carcinogenesis by immunosuppression. Br. Med. J. 4:419-420.

Altenkirch, H.; Mager, J.; Stoltenburg, G.; Helmbrecht, J. (1977) Toxic polyneuropathies after sniffing a glue thinner. J. Neurol. 214:137-152.

Altenkirch, H.; Stoltenburg, G.; Wagner, H.M. (1978) Experimental studies on hydrocarbon neuropathies induced by methyl-ethyl-ketone (MEK). J. Neurol. 219:159-170.

Aminoff, M.J. (1980) Peripheral sympathetic function in patients with a polyneuropathy. J. Neurol. Sci. 44:213-219.

Anderson, A.C.; White, K.L. (1984) Suppression of mouse complement activity by dioxin contaminants of technical grade pentachlorophenol [Abstract]. Toxicologist 4:159.

Arezzo, J.C.; Legatt, A.D.; Vaughan, H.G., Jr. (1979) Topography and intracranial sources of somatosensory evoked potentials in the monkey. I. Early components. Electroencephalogr. Clin. Neurophysiol. 46:155-172.

Arezzo, J.C.; Schaumburg, H.H. (1980) Use of the "Optacon" as a screening device: a new technique for detecting sensory loss in individuals exposed to neurotoxins. J. Occup. Med. 22:461-464.

Arezzo, J.C.; Schaumburg, H.H.; Petersen, C.A. (1983) Rapid screening for peripheral neuropathy: a field study with the "Optacon". Neurology 33:626-629.

Arezzo, J.C.; Schaumburg, H.H.; Vaughan, H.G.; Spencer, P.S.; Barna, J. (1982) Hind limb somatosensory evoked potentials in the monkey: the effects of distal axonopathy. Ann. Neurol. 12:24-32.

Auerbach, A.D.; Wolman, S.R. (1976) Susceptibility of Fanconi's anaemia fibroblasts to chromosome damage by carcinogens. Nature 261:494-496.

Baer, R.L.; Ramsey, D.L.; Biondi, E. (1973) Most common contact allergens: 1968-1970. Arch. Dermatol. 108:74-78.

Bardana, E.J., Jr. (1980) Formaldehyde: hypersensitivity and irritant reactions at work and in the home. Immunol. Allergy Prac. 2(3):11-23.

Beckman, D.A.; Brent, R.L. (1984) Mechanisms of teratogenesis. Ann. Rev. Pharmacol. 24:483-500.

Bekesi, J.G.; Selikoff, I.J. (1978) Altered immune function in Michigan dairy farmers and chemical workers exposed to polybrominated biphenyls. In: Asher, I.M., ed. Inadvertent modification of the immune response: the effects of foods, drugs and environmental contaminants. Proceedings of the Fourth FDA Science Symposium, held Aug. 28-30, 1978, Annapolis, MD. [Washington, DC]: U.S. Department of Health and Human Services, Food and Drug Administration. p. 210-212. Available from: U.S. Government Printing Office, Washington, DC.

Bellanti, J.A.; Balter, N.J.; Schwartz, S.L.; Gray, I. (1978) Heavy metals: effects of lead, cadmium and chromium in a rodent model. In: Asher, I.M., ed. Inadvertent modification of the immune response: the effects of foods, drugs and environmental contaminants. Proceedings of the Fourth FDA Science Symposium, held Aug. 28-30, Annapolis, MD. Washington, DC: U.S. Department of Health and Human Services, Food and Drug Administration. p. 191-195. Available from: U.S. Government Printing Office, Washington, DC.

Bennett, R.A.; Essigmann, J.M.; Wogan, G.N. (1981) Excretion of an aflatoxin-guanine adduct in the urine of aflatoxin B1-treated rats. Cancer Res. 41:650-654.

Bergsma, D., editor. (1979) Birth defects compendium. 2nd ed. Published for the National Foundation-March of Dimes. New York: Alan R. Liss, Inc. 1183p.

Berk, P.D.; Martin, J.F.; Young, R.S.; Creech, J.; Selikoff, I.J.; Falk, H.; Watanabe, P.; Popper, H.; Thomas, L. (1976) Vinyl chloride-associated liver disease. Ann. Intern. Med. 84:717-731.

Blakely, B.R.; Archer, D.L. (1981) Effect of lead acetate on the immune response in mice. Toxicol. Appl. Pharmacol. 61:18-26.

Blakely, B.R.; Sisodia, C.S.; Mukkur, T.K. (1980) Effect of methylmercury, tetraethyl lead, and sodium arsenite on the humoral immune response in mice. Toxicol. Appl. Pharmacol. 52:245-254.

Bloom, A.D. (1972) Induced chromosomal aberrations in man. In: Harris, H.; Hirschhorn, K.; eds. Advances in human genetics. Vol. 3. New York: Plenum Press. p. 99-172.

Bloom, A.D., editor. (1981) Guidelines for studies of human populations exposed to mutagenic and reproductive hazards: proceedings of conference held Jan. 26-27, in Washington, DC. White Plains, NY: March of Dimes Birth Defects Foundation. 163p.

Boorman, G.A.; Luster, M.I.; Dean, J.H.; Campbell, M.L. (1982) Assessment of myelotoxicity caused by environmental chemicals. Environ. Health Perspect. 43:129-135.

Botnick, L.E.; Hannon, E.C.; Hellman, S. (1978) Multisystems stem cell failure after apparent recovery from alkylating agents. Cancer Res. 38: 1942-1947.

Bradley, S.G.; Morahan, P.S. (1982) Approaches to assessing host resistance. Environ. Health Perspect. 43:61-69.

Brent, R.L. (1964) Drug testing in animals for teratogenic effects: thalidomide in the pregnant rat. J. Pediatr. 64:762-770.

Brent, R.L.; Torson, R.O. (1972) Radiation exposure in pregnancy. Chicago: Year Book Medical Publishers. 48p.

Bridges, B.A.; Butterworth, B.E.; Weinstein, I.B.; editors. (1982) Banbury report 13: indicators of genotoxic exposure. [Cold Spring Harbor, NY]: Cold Spring Harbor Laboratory. 580p.

Brown, G.G.; Preisan, R.C.; Anderson, M.D.; Nixon, R.K.; Isbister, J.L.; Price, H.A. (1981) Memory performance of chemical workers exposed to polybrominated biphenyls. Science 212:1413-1415.

Capel, I.D.; Dorrell, H.M.; Jenner, M.; Pinnock, M.H.; Williams, D.C. (1979) Effect of chloroform ingestion on the growth of some murine tumours. Eur. J. Cancer 15:1485-1490.

Carrano, A.V.; Minkler, J.L.; Stetka, D.G.; Moore, T.H., II. (1980) Variation in the baseline sister chromatid exchange frequency in human lymphocytes. Environ. Mutagen. 2:325-337.

Carter, J.W.; Calnoy, J., Jr. (1980) Acutely administered polychlorinated biphenyls (PCBs) decrease splenic cellularity but increase its ability to cause graft-versus-host reactions in BALB/c mice. Immunopharmacology 2:341-347.

Chang, K.J.; Chen, J.S.; Huang, P.C.; Tung, T.C. (1980) Study of patients with polychlorinated biphenyls poisoning. I. Blood analyses of patients. J. Formosan Med. Assoc. 79:304-313.

Chen, Y.T.; Mattison, D.R.; Feigenbaum, L.; Fukui, H.; Schulman, J.D. (1981) Reduction in oocyte number following prenatal exposure to a diet high in galactose. Science 214:1145-1147.

Clark, D.A.; Sweeney, G.; Safe, S.; Hancock, E.; Kilburn, D.G.; Gauldie, J. (1983) Cellular and genetic basis for suppression of cytotoxic T-cell generation by haloaromatic hydrocarbons. Immunopharmacology 6:143-153.

Cleaver, J.E. (1983) Xeroderma pigmentosum. In: Stanbury, J.B.; Wyngaarden, Y.B.; Fredrickson, D.S.; Goldstein, J.L.; Brown, M.S.; eds. Metabolic basis of inherited disease. 5th ed. New York: McGraw-Hill Book Co. p. 1227-1248.

Cook, J.A.; DiLuzio, N.R.; Hoffman, E.O. (1975) Factors modifying susceptibility to bacterial endotoxin: the effect of lead and cadmium. CRC Crit. Rev. Toxicol. 3(2):201-229.

Damstra, T. (1978) Environmental chemicals and nervous system dysfunction. Yale J. Biol. Med. 51:457-468.

Dean, J.H.; Luster, M.I.; Boorman, G.A. (1982a) Methods and approaches for assessing immunotoxicity: an overview. Environ. Health Perspect. 43:27-29.

Dean, J.H.; Luster, M.I.; Boorman, G.A.; Lauer, L.D. (1982b) Procedures available to examine the immunotoxicity of chemicals and drugs. Pharmacol. Rev. 34:137-148.

Delehanty, J. (1983) Identification and impact of chemical mutagens. CLS Lifelines [National Academy of Science, Commission on Life Science] 9:1-2,8-9.

DeSerres, F.J.; Sheridan, W.; editors. (1983) Utilization of mammalian specific locus studies in hazard evaluation estimation of genetic risk. New York:

Plenum Press. 341p.
DiBosco, M.M.; Fonzi, S. (1974) La polineuropatia cosiddetta da collanti o dei calzaturieri [In Italian]. Rivista Degli Infortuni E Delle Malattie Professionali 61:665-692.
Dixon, L. (1978) Symposium on target organ toxicity: gonads (reproductive and genetic toxicity). Cosponsored by Society of Toxicology and National Institute of Environmental Health Sciences, December 1-4, 1976, Vanderbilt University, Nashville, TN. Environ. Health Perspect. 24.
Druet, P.; Bernard, A.; Hirsh, F.; Weening, J.J.; Gengoux, P.; Mahieu, P.; Birkeland, S. (1982) Immunologically mediated glomerulonephritis by heavy metals. Arch. Toxicol. 50:187-194.
Ewers, U.; Stiller-Winkler, R.; Idel, H. (1982) Serum immunoglobulin, complement C3, and salivary IgA levels in lead workers. Environ. Res. 29:351-357.
Exon, J.H.; Koller, L.D. (1984) Carcinogenicity and immunotoxicity of chlorinated phenolic drinking water contaminants [Abstract]. Toxicologist 4: 157.
Farber, E. (1980) Sequential analysis of liver cancer induction. Biochim. Biophys. Acta 605:149-166.
Feldman, R.G.; Ricks, N.L.; Baker, E.L. (1980) Neuropsychological effects of industrial toxins: a review. Am. J. Industrial Med. 1:211-227.
Figoni, R.; Treagan, L. (1975) Inhibitory effect of nickel and chromium upon antibody response of rats to immunization with T-1 phage. Res. Commun. Chem. Pathol. Pharmacol. 11:335-338.
Fraker, P.J.; Aust, S. (1980) Effect of polybrominated biphenyls on the immune response of BALB/c mice. In: Asher, I.M., ed. Inadvertant modification of the immune response: the effects of foods, drugs, and environmental contaminants. Proceedings of the Fourth FDA Science Symposium, held August 28-30, 1978, Annapolis, MD. [Washington, DC]: U.S. Department of Health and Human Services, Food and Drug Administration. p. 270-271. Available from: U.S. Government Printing Office, Washington, DC.
Friedlander, B.R.; Hearne, F.T. (1980) Epidemiologic considerations in studying neurotoxic disorders. In: Spencer, P.S.; Schaumburg, H.H.; eds. Experimental and clinical neurotoxicology. Baltimore: Williams and Wilkins Company. p. 650-662.
Fudenberg, H.H.; Whitten, H.D. (1984) Immunostimulation: synthetic and biological modulators of immunity. Ann. Rev. Pharmacol. Toxicol. 24:147-174.
Gainer, J.H. (1977) Effects of heavy metals and of deficiency of zinc on mortality rates in mice infected with encephalomyocarditis virus. Am. J. Vet. Med. Res. 38:869-872.
Gainer, J.H.; Pry, T.W. (1972) Effects of arsenicals on viral infection in mice. Am. J. Vet. Med. Res. 33:2299-2307.
Galen, R.S.; Gambino, S.R. (1975) Beyond normality: the predictive value and efficiency of medical diagonsis. New York: John Wiley & Sons. 237p.
Galizzi, J.; Long, R.G.; Billing, B.H.; Sherlock, S. (1978) Assessment of the (^{14}C)aminopyrine breath test in liver disease. Gut 19:40-45.
Gasper, J.R.; Dauzvardis, P.A. (1980) Body burdens--an integrated approach to health risk assessment. Prepared for the National Conference on Hazardous Material Risk Assessment, Disposal and Management, April 25-27, 1979, Miami Beach, FL. Argonne, IL: U.S. Department of Energy, Argonne National Laboratory. 17p. Available from: National Technical

Information Service, Springfield, VA.
Gatti, R.A.; Good, R.A. (1971) Occurrence of malignancy in immunodeficiency disease: a literature review. Cancer 28:89-98.
Gellert, R.J.; Heinrichs, W.L.; Swerdloff, R.S. (1972) DDT homologues: estrogen-like effects on the vagina, uterus and the pituitary of the rat. Endocrinology 91:1095-1100.
Graham, D.G. (1984) Catecholamine toxicity: a proposal for the molecular pathogensis of manganese neurotoxicity and Parkinson's disease. Neurotoxicol. 5:83-96.
Guzelian, P.S. (1981) Environmental toxins and the liver. Pract. Gastroenterol. 5(4):26-30.
Guzelian, P.S. (1983) Research needs for hepatic injury due to environmental agents. Environ. Health Perspect. 48:65-71.
Guzelian, P.S.; Vranian, G.; Boylan, J.J.; Cohn, W.J.; Blanke, R.V. (1980) Liver structure and function in patients poisoned with chlordecone (Kepone). Gastroenterology 78:206-213.
Haber, E.; Pfitzer, E.A.; editors. (1982) Immunological aspects of toxicology. American Society for Pharmacology and Experimental Therapeutics workshop held May 17-20, 1981, in Chicago. Published for the Pharmaceutical Manufacturers Association Foundation, Inc. 151p. Baltimore, MD: Williams and Wilkins Company.
Harada, M. (1978) Congenital Minamata disease: intrauterine methylmercury poisoning. Teratology 18:285:285-288.
Harris, C.C.; Mulvihill, J.J.; Thorgeirsson, S.S.; Minna, J.D. (1980) Individual differences in cancer susceptibility. Ann. Intern. Med. 92:809-825.
Haseltine, W. (1983) Ultraviolet light repair and mutagenesis revisited. Cell 33:13-17.
Hassan, M.; Pesce, A.J. (1980) Analytical toxicology. In: Hanenson, I.B., ed. Quick reference to clinical toxicology. Philadelphia: J.B. Lippincott Co. p. 265-270.
Herskowitz, A.; Ishii, N.; Schaumburg, H. (1971) N-hexane neuropathy: a syndrome occurring as a result of industrial exposure. N. Engl. J. Med. 285:82-85.
Hirokawa, K.; Hayashi, Y. (1980) Acute methyl mercury intoxication in mice: effect on the immune system. Acta Pathol. Jpn. 30:23-32.
Hollaender, A., editor. (1971) Chemical mutagens: principles and methods for their detection. 3rd pr. Vol. 2. New York: Plenum Press. 299p.
Hook, J.B. (1980) Toxic responses of the kidney. In: Casarett and Doull's toxicology: the basic science of poisons. 2nd ed. Doull, J.; Klaassen, C.D.; Amdur, M.O.; eds. New York: Macmillan Publishing Co, Inc. p. 232-245.
Hook, J.B.; Serbia, V.C. (1982) Potentiation of the action of nephrotoxic agents by environmental contaminants. In: Porter, G.A., ed. Nephrotoxic mechanisms of drugs and environmental toxins. New York: Plenum Medical Book Co. p. 345-356.
ICPEMC SEE International Commission for Protection Against Environmental Mutagens and Carcinogens
Inoue, T.; Takeuchi, Y.; Takeuchi, S.; Yamada, S.; Suzuki, H.; Matsushita, T.; Miyagaki, H.; Maeda, K.; Matsumoto, T. (1970) Health survey on vinyl sandal manufacturers with high incidence of "n-hexane" intoxication (Japanese with English abstract). Jpn. J. Indust. Health 12:73-84.
Interdisciplinary Panel on Carcinogenicity. (1984) Criteria for evidence of

chemical carcinogenicity. Science 225:682-687.
International Commission for Protection Against Environmental Mutagens and Carcinogens. (1983) Screening strategy for chemicals that are potential germ-cell mutagens in mammals: committee 1 final report. Mutat. Res. 114:117-177.
Kagamimori, S.; Scott, M.P.; Brown, D.G.; Edwards, R.E.; Wagner, M.M.F. (1980) Effects of chrysotile asbestos on mononuclear cells in vitro. Br. J. Exp. Pathol. 61:55-60.
Kagan, E.; Solomon, A.; Cochrane, J.C.; Kuba, P.; Rocks, P.H.; Webster, I. (1977) Immunological studies of patients with asbestosis. II. Studies of circulating lymphoid cell numbers and humoral immunity. Clin. Exp. Immunol. 28:268-275.
Karl, L.; Chvapil, M; Zukoski, C.F. (1973) Effect of zinc on the viability and phagocytic capacity of peritoneal macrophages. Proc. Soc. Exp. Biol. Med. 142:1123-1127.
Karol, M.H.; Dixon, C.; Brady, M.; Alarie, Y. (1980) Immunologic sensitization and pulmonary hypersensitivity by repeated inhalation of aromatic isocyanates. Toxicol. Appl. Pharmacol. 53:260-263.
Kerkvliet, N.I.; Kimeldorf, D.J. (1977) Antitumor activity of a polychlorinated biphenyl mixture, Aroclor 1254, in rats innoculated with Walker 256 carcinosarcoma cells. J. Nat. Cancer Inst. 59:951-955.
Kew, M.C. (1981) Hepatitis-B virus and hepatocellular carcinoma. Semin. Liver Dis. 1:59-67.
Kimbrough, R.D. (1974) Toxicity of polychlorinated polycyclic compounds and related chemicals. CRC Crit. Rev. Toxicol. 2:445-498.
Kluwe, W.M.; Hook, J.B. (1980a) Effects of environmental chemicals on kidney metabolism and function. Kidney Int. 18:648-655.
Kluwe, W.M.; Hook, J.B. (1980b) Metabolic activation of nephrotoxic haloalkanes. Fed. Proc. 39:3129-3133.
Kolber, A.R.; Wong, T.K.; Grant, L.D.; DeWoskin, R.S.; Hughes, T.J.; editors. (1983) Proceedings of a NATO Advanced Research Institute on "In vitro toxicity testing of environmental agents: current and future possibilities," held September 22-28, 1979, in Monte Carlo, Monaco. Part A: survey of test systems. Published in cooperation with NATO Scientific Affairs Division. New York: Plenum Press. 553p.
Koller, L.D. (1980) Immunotoxicology of heavy metals. Int. J. Immunopharmacol. 2:269-279.
Koller, L.D.; Exon, J.H.; Arbogast, B. (1977) Methylmercury: effect on serum enzymes and humoral antibody. J. Toxicol. Environ. Health 2:1115-1123.
Koller, L.D.; Exon, J.H.; Roan, J.G. (1976) Humoral antibody response in mice after single dose exposure to lead or cadmium. Proc. Soc. Exp. Biol. Med. 151:339-342.
Kripke, M.L.; Borsos, T. (1974) Immunosuppression and carcinogenesis. Isr. J. Med. Sci. 10:888-903.
Kupfer, D. (1975) Effects of pesticides and related compounds on steroid metabolism and function. CRC Crit. Rev. Toxicol. 4:83-124.
Kurland, L.T.; Faro, S.N.; Siedler, H. (1960) Minamata disease: the outbreak of a neurologic disorder in Minamata, Japan, and its relationship to the ingestion of seafood contaminated by mercuric compounds. World Neurol. 1:370-395.
La Via, M.F.; La Via, D.S. (1979) Phenol derivatives are immunodepressive in

mice. Drug Chem. Toxicol. 2:167-177.

Land, H.; Parada, L.F.; Weinberg, R.A. (1983) Cellular oncogenes and multistep carcinogenesis. Science 222:771-777.

Landrigan, P.J. (1983) Toxic exposures and psychiatric disease--lessons from the epidemiology of cancer. Acta Psychiat. Scand. 67 Suppl. 303:6-15.

Landrigan, P.J.; Baker, E.L., Jr.; Feldman, R.G.; Cox, D.H.; Eden, K.V.; Orenstein, W.A.; Mather, J.A.; Yankel, A.J.; Von Lindern, I.H. (1976) Increased lead absorption with anemia and slowed nerve conduction in children near a lead smelter. J. Pediatr. 89:904-910.

Lange, A.; Smolik, R.; Zatonski, W.; Szymanska, J. (1973) Serum immunoglobulin levels in workers exposed to benzene, toluene and xylene. Int. Arch. Arbeitsmed. 31:37-44.

Lindstrom, K. (1980) Changes in psychological performances of solvent-poisoned and solvent-exposed workers. Am. J. Ind. Med. 1:69-84.

Loose, L.D; Silkworth, J.B.; Pittman, K.A.; Benitz, K.F.; Mueller, W. (1978) Impaired host resistance to endotoxin and malaria in polychlorinated biphenyl- and hexachlorobenzene-treated mice. Infect. Immun. 20:30-35.

Lowrance, W.W., editor. (1981) Assessment of health effects at chemical disposal sites: proceedings of a symposium held in New York City on June 1-2, by the Life Sciences and Public Policy Program of The Rockefeller University. New York City: The University. 166p.

Luster, M.I.; Boorman, G.A.; Dean, J.H.; Harris, M.W.; Luebke, R.W.; Padarathsingh, M.L.; Moore, J.A. (1980) Examination of bone marrow, immunologic parameters and host susceptibility following pre- and post-natal exposure to 2,3,7,8-tetrachlorodibenzo-p-dioxin (TCDD). Int. J. Immunopharmacol. 2:301-310.

Luster, M.I.; Dean, J.H. (1982) Immunological hypersensitivity resulting from environmental or occupational exposure to chemicals: a state-of-the-art workshop summary. Fund. Applied Toxicol. 2:327-330.

Luster, M.I.; Dean, J.H.; Boorman, G.A.; Dieter, M.P.; Hayes, H.T. (1982) Immune functions in methyl and ethyl carbamate treated mice. Clin. Exp. Immunol. 50:223-230.

Luster, M.I; Faith, R.E.; Lawson, L.D. (1979) Effects of 2,3,7,8-tetrachlorodibenzofuran (TCDF) on the immune system in guinea pigs. Drug Chem. Toxicol. 2:49-60.

Luster, M.I.; Faith, R.E.; Moore, J.A. (1978) Effects of polybrominated biphenyls (PBB) on immune response in rodents. Environ. Health Perspect. 23:227-232.

MacLean, L.D. (1979) Host resistance in surgical patients. J. Trauma 19:297-304.

Majeti, V.A.; Clark, C.S. (1980) Potential health effects from persistent organics in wastewater and sludges used for land application. Prepared by the University of Cincinnati Medical Center, Department of Environmental Health. EPA-600/1-80-025. Cincinnati, OH: U.S. Environmental Protection Agency, Health Effects Research Laboratory. 51p. Available from: National Technical Information Service, Springfield, VA.

Maroni, M.; Bulgheroni, C.; Cassitto, M.G.; Merluzzi, F.; Gilioli, R.; Foa, V. (1977) Clinical neuropsychological and behavioral study of female workers exposed to 1,1,1,-trichlorethane. Scand. J. Work Environ. Health 3:16-22.

Marshall, E. (1983) House reviews EPA's record on pesticides. Science 219:1220.

Mattison, D.R.; Chang, L.; Thorgeirsson, S.S.; Shiromizu, K. (1981) Effects of

cyclophosphamide, azathioprine and 6-mercaptopurine on oocyte and follicle number in C57BL/6N mice. Res. Commun. Chem. Pathol. Pharmacol. 31: 155-161.
Mattison, D.R.; Nightingale, M.S.; Shiromizu, K. (1983) Effects of toxic substances on female reproduction. Environ. Health Perspect. 48:43-52.
Mattison, D.R.; Ross, G.T. (1983) Oogenesis and ovulation. In: Vouk, V.B.; Sheehan, P.J., eds. Methods for assessing the effects of chemicals on reproductive functions. New York: John Wiley & Sons, Inc. p. 17-22.
McCormack, K.M.; Kluwe, W.M.; Rickert, D.E.; Sanger, V.L.; Hook, J.B. (1978) Renal and hepatic microsomal enzyme stimulation and renal function following three months of dietary exposure to polybrominated biphenyls. Toxicol. Appl. Pharmacol. 44:539-553.
McLachlan, J.A., editor. (1980) Estrogens in the environment: proceedings of the symposium, Sept. 10-12, 1979, Raleigh, NC. New York: Elsevier North-Holland. 427p.
McLachlan, J.A.; Dixon, R.L. (1977) Advances in sex hormone research. Vol. 3. Toxicologic comparisons of experimental and clinical exposures to diethylstilbestrol during gestation. In: Thomas, J.A.; Singhal, R.L.; eds. Regulatory mechanisms affecting gonadal hormone action. Baltimore: University Park Press. p. 309-336.
Mendell, J.R.; Saida, K.; Ganansia, M.F; Jackson, D.B.; Weiss, H.; Gardier, R.W.; Christman, C.; Allen, N.; Couri, D.; O'Neill, J.; Marks, B.; Hetland, L. (1974) Toxic polyneuropathy produced by methyl n-butyl ketone. Science 185:787-789.
Miller, K.; Weintraub, Z.; Kagan, E. (1979) Manifestations of cellular immunity in the rat after prolonged asbestos inhalation. I. Physical interactions between alveolar macrophages and splenic lymphocytes. J. Immunol. 123:1029-1038.
Mills, J.L.; Bongiovanni, A.M. (1978) Effect of prenatal estrogen exposure on male genitalia. Pediatrics 62(Supplement): 1160-1165.
Mitchell, C.L.; Tilson, H.A. (1982) Behavioral toxicology in risk assessment: problems and research needs. CRC Crit. Rev. Toxicol. 10:265-274.
Moore, J.A.; Huff, J.E.; Dean, J.H. (1982) National Toxicology Program and immunological toxicology. Pharmacol. Rev. 34(1):13-16.
Mosher, W.D.; Pratt, W.F. (1982) Reproductive impairments among married couples: United States. Vital Health Stat. Series 23. Prepared by U.S. Department of Health and Human Services 51p. Available from: U.S. Government Printing Office, Washington, DC.
Munson, A.E.; Sanders, V.M.; Douglas, K.A.; Sain, L.E.; Kauffmann, B.M.; White, K.L., Jr. (1982) In vivo assessment of immunotoxicity. Environ. Health Perspect. 43:41-52.
National Research Council. (1972) Effects on populations of exposure to low levels of ionizing radiation. Washington, DC: National Academy of Sciences. 217p.
National Research Council. Committee on the Biological Effects of Ionizing Radiations. (1980) Effects upon populations of exposure to low levels of ionizing radiation. Washington, DC: National Academy Press. 524p.
Neel, J.V.; Satoh, C.; Hamilton, H.B.; Otake, M.; Goriki, K.; Kageoka, T.; Fujita, M.; Neriishi, S.; Asakawa, J. (1980) Search for mutations affecting protein structure in children of atomic bomb survivors: preliminary report. Proc. Nat. Acad. Sci. 77(7):4221-4225.
Neilson, J.T.M.; Forrester, D.J.; Thompson, N.P. (1973) Immunologic studies

on Heligmosomoides polygyrus infection in the mouse: the dynamics of single and multiple infections and the effect of DDT upon acquired resistance. Internat. J. Parasitol. 3:371-378.

Nelson, C.J.; Holson, J.F.; Green, H.G.; Gaylor, D.W. (1979) Retrospective study of the relationship between agricultural use of 2,4,5-T and cleft palate occurrence in Arkansas. Teratology 19:377-383.

NRC SEE National Research Council

O'Brien, I.M.; Harries, M.G.; Burge, P.S.; Pepys, J. (1979) Toluene di-isocyanate-induced asthma. I. Reactions to TDI, MDI and histamine. Clin. Allergy 9:1-6.

Oishi, H.; Mineno, K.; Yamada, M.; Chiba, K.; Shibata, K. (1964) Polyneuropathy caused by an organic solvent n-hexane [In Japanese]. Saigai- Igaku 7:218-222.

Page, M.; Theriault, L.; Delorme, F. (1976) Elevated CEA levels in polyvinyl chloride workers. Biomedicine 25:279.

Parris, G.E.; Diachenko, G.W.; Entz, R.C.; Poppiti, J.A.; Lombardo, P.; Rohrer, T.K.; Hesse, J.L. (1980) Waterborne methylene bis(2-chloroaniline) and 2-chloroaniline contamination around Adrian, Michigan. Bull Environ. Contam. Toxicol. 24:497-503.

Pelletier, L.; Pasquier, R.; Hirsch, F.; Sapin, C.; Druet, P. (1983) Mercury-induced lymphocyte autoreactivity. In: Gibson, G.G.; Hubbard, R.; Parke, D.V.; eds. International Symposium on Immunotoxicology. London: Academic Press. p. 437-441.

Peltonen, L.; Fraki , J. (1980) Prevalence of dichromate sensitivity. Contact Dermatitis 9:190-194.

Penn, I. (1978) Development of cancer in transplantation patients. Adv. Surg. 12:155-191.

Perrin, E.V.D.K. (1982a) Fetus, infant, and young child. In: Riddell, R.H., ed. Pathology of drug-induced and toxic diseases. New York: Churchill Livingstone. p. 19-48.

Perrin, E.V.D.K. (1982b) Teratology, "dysmorphology," and related subjects. In: Hill, R.B.; Terzian, J.A.; eds. Environmental pathology: an evolving field. New York: Alan R. Liss, Inc. p. 129-153.

Poirer, M.C.; Santella, R.; Weinstein, I.B.; Grunberger, D.; Yuspa, S.H. (1980) Quantification of benzo(a)pyrenedeoxyguanosine adducts by radioimmunoassay. Cancer Res. 40:412-416.

Price, C.P.; Alberti, K.G.M.M. (1979) Biochemical assessment of liver function. In: Wright, R.; Alberti, K.G.M.M.; Karran, S.; Millward-Sadler, G.H.; eds. Liver and biliary disease. New York: W.B. Saunders Company. p. 381-416.

Radhakrishnan, C.V.; Thompson, N.P.; Forrester, D.J. (1972) Susceptibility of chickens fed p,p'-DDT to histomoniasis. Bull. Environ. Contam. Toxicol. 8:147-152.

Rao, D.S.V.S.; Glick, B. (1977) Pesticide effects on the immune response and metabolic activity of chicken lymphocytes. Proc. Soc. Exp. Biol. Med. 154:27-29.

Reggiani, G. (1980) Acute human exposure to TCDD in Seveso, Italy. J. Toxicol. Environ. Health 6:27-43.

Reigart, J.R.; Graber, C.D. (1976) Evaluation of the humoral immune response of children with low level lead exposure. Bull. Environ. Contam. Toxicol. 16:112-117.

Riddell, R.H., editor. (1982) Pathology of drug-induced and toxic diseases.

New York: Churchill Livingstone. 682p.
Rosalki, S.B.; Tarlow, D.; Rau, D. (1971) Plasma gamma-glutamyl transpeptidase elevation in patients receiving enzyme-inducing drugs. Lancet 1:376-377.
Sachs, H.K. (1978) Intercurrent infections in lead poisoning. Am. J. Dis. Child. 132:315-316.
Sanders, V.M.; Tucker, A.N.; White, K.L., Jr.; Kauffmann, B.M.; Hallett, P.; Carchman, R.A.; Borzelleca, J.F.; Munson, A.E. (1982) Humoral and cell-mediated immune status in mice exposed to trichloroethylene in the drinking water. Toxicol. Appl. Pharmacol. 62:358-368.
Schaumberg, H.H.; Arezzo, J.C.; Markowitz, L.; Spencer, P.S. (1981) Neurotoxicity assessment at chemical disposal sites. In: Lowrance, W.W., ed. Assessment of health effects at chemical disposal sites: proceedings of a symposium held in New York City on June 1-2, by the Life Sciences and Public Policy Program of The Rockefeller University. New York: The University. p. 82-104.
Schaumburg, H.H.; Markowitz, L.S.; Arezzo, J.C. (1983) Monitoring potential neurotoxic effects of hazardous waste disposal. Environ. Health Perspect. 48:61-64.
Schaumburg, H.H.; Spencer, P.S. (1976) Degeneration in central and peripheral nervous systems produced by pure n-hexane: an experimental study. Brain 99:183-192.
Schaumburg, H.H.; Spencer, P.S. (1979) Toxic models of certain disorders of the nervous system--a teaching monograph. Neurotoxicol. 1:209-220.
Schichman, S.A.; Chapman, J.F.; Cross, R.E.; Parker, N. (1984) Principles of laboratory test interpretation. Bull. Lab. Med. 81:1-8.
Seppalainen, A.M.; Harkonen, H. (1976) Neurophysiological findings among workers occupationally exposed to styrene. Scand. J. Work Environ. Health 3:140-146.
Seppalainen, A.M.; Lindstrom, K.; Martelin, T. (1980) Neurophysiological and psychological picture of solvent poisoning. Am. J. Ind. Med. 1:31-42.
Seppalainen, A.M.; Raitta, C.; Huuskonen, M.S. (1979) N-Hexane induced changes in visual evoked potentials and electroretinograms of industrial workers. Electroencephalogr. Clin. Neurophysiol. 47:492-498.
Seppalainen, A.M.; Savolainen, K.; Kovala, T. (1981) Changes induced by xylene and alcohol in human evoked potentials. Electroencephalogr. Clin. Neurophysiol. 51:148-155.
Sharma, R.P.; Gehring, P.J. (1979) Immunologic effects of vinyl chloride in mice. Ann. N.Y. Acad. Sci. 320:551-563.
Shepard, T.H. (1983) Catalog of teratogenic agents. 4th ed. Baltimore: Johns Hopkins University Press. 410p.
Shigematsu, N.; Ishimaru, S.; Saito, R.; Ikeda, T.; Matsuba, K.; Sugiyama, K.; Masuda, Y. (1978) Respiratory involvement in polychlorinated biphenyls poisoning. Environ. Res. 16:92-100.
Silkworth, J.B.; Loose, L.D. (1979) PCB and HCB induced alteration of lymphocyte blastogenesis [Abstract]. Toxicol. Appl. Pharmacol. 49:A86.
Smith, H.V.; Spalding, J.M.K. (1959) Outbreak of paralysis in Morocco due to ortho-cresyl phosphate poisoning. Lancet 2:1019-1021.
Smith, M.I.; Elvove, E.; Valaer, P.J., Jr.; Frazier, W.H; Mallory, G.E. (1930) Pharmacological and chemical studies of the cause of so-called ginger paralysis: a preliminary report. Public Health Rep. 45:1703-1716.
Smolik, R.; Grzybek-Hryncewicz, K.; Lange, A.; Zatonski, W. (1973) Serum

complement levels in workers exposed to benzene, toluene, and xylene. Int. Arch. Arbeitsmed. 31:243-247.

Sobocinski, P.Z.; Canterbury, W.J., Jr.; Powanda, M.C. (1977) Differential effect of parenteral zinc on the course of various bacterial infections. Proc. Soc. Exp. Biol. Med. 56:334-339.

Spreafico, F.; Vecchi, A. (1984) Immunomodulation by xenobiotics: the open field of immunotoxicology. In: Fudenberg, H.H.; Whittem, H.D.; Ambrogi, F.; eds. Immunomodulation. New frontiers and advances. Proceedings of a symposium on recent advances on immunomodulators, May 14-16, Viareggio, Italy. New York: Plenum Press p. 311-329.

Street, J.C.; Sharma, R.P. (1974) Quantitative aspects of immunosuppression by selected pesticides [Abstract]. Toxicol. Appl. Pharmacol. 29:135.

Street, J.C.; Sharma, R.P. (1975) Alteration of induced cellular and humoral immune responses by pesticides and chemicals of environmental concern: quantitative studies of immunosuppression by DDT, Aroclor 1254, carbaryl, carbofuran, and methylparathion. Toxicol. Appl. Pharmacol. 32:587-602.

Strobino, B.R.; Kline, J.; Stein, Z. (1978) Chemical and physical exposures of parents: effects on human reproduction and offspring. Early Human Dev. 1:371-399.

Tedeschi, L. G. (1982) The Minamata disease. Am J. Forsensic Med. Pathol. 3:335-338.

Thigpen, J.E.; Faith, R.E.; McConnell, E.E.; Moore, J.A. (1975) Increased susceptibility to bacterial infection as a sequela of exposure to 2,3,7,8-tetrachlorodibenzo-p-dioxin. Infect. Immun. 12:1319-1324.

Thomas, P.T.; Hinsdill, R.D. (1978) Effect of polychlorinated biphenyls on the immune responses of Rhesus monkeys and mice. Toxicol. Appl. Pharmacol. 44:41-51.

Thomas, P.T.; Hinsdill, R.D. (1979) Effect of perinatal exposure to tetrachlorodibenzo-p-dioxin on the immune response of young mice. Drug. Chem. Toxicol. 2:77-98.

Towfighi, J.; Gonatas, N.K.; Pleasure, D.; Cooper, H.S.; McCree, L. (1976) Glue sniffer's neuropathy. Neurology 26:238-243.

UN SEE United Nations

United Nations. General Assembly. (1969) Report of the Scientific Committee on the Effects of Atomic Radiation. Official records: 24th session, Supplement 13(A/7613). New York: United Nations. 98p.

U.S. Department of Energy, Oak Ridge National Laboratory. (1982) Assessment of risks to human reproduction and to development of the human conceptus from exposure to environmental substances. Proceedings of U.S. Environmental Protection Agency-Sponsored Conferences: October 1-3, 1980, Atlanta, GA, and December 7-10, 1980, St. Louis, MO. ORNL/EIS-197. EPA-600/9-82-001. Oak Ridge, TN: The Laboratory. 152p.

U.S. Department of Health and Human Services, Centers for Disease Control. (1982) Congenital malformations surveillance report. Atlanta, GA: U.S. Centers for Disease Control. 41p.

U.S. Department of Health and Human Services, Centers for Disease Control. (1984) System for prevention, assessment, and control of exposures and health effects from hazardous sites (S.P.A.C.E. for Health). Prepared by the Center for Environmental Health and the Center for Professional Development and Training. 187p. Available from: Chronic Diseases Division, Center for Environmental Health, Centers for Disease Control, U.S. Department of Health and Human Services, Atlanta, GA.

U.S. Department of Health, Education, and Welfare; Committee to Coordinate Environmental and Related Programs. ([1978]) Report of the Subcommittee on the Potential Health Effects of Toxic Chemical Dumps. [Bethesda, MD]: The Department. 116p.

U.S. Department of Health, Education, and Welfare; National Center for Health Statistics. (1980) Factors associated with low birth weight: United States, 1976. National Vital Statistics System Series 21, No. 37. DHEW publication no. (PHS)80-1915. Washington, DC: The Center.

USDHEW/CCERP SEE U.S. Department of Health, Education, and Welfare; Committee to Coordinate Environmental and Related Programs

USDHEW/NCHS SEE U.S. Department of Health, Education, and Welfare; National Center for Health Statistics

USDHHS/CDC SEE U.S. Department of Health and Human Services, Centers for Disease Control

USDOE/ORNL SEE U.S. Department of Energy, Oak Ridge National Laboratory

USEOP/OSTP SEE U.S. Executive Office of the President, Office of Science and Technology Policy

U.S. Executive Office of the President, Office of Science and Technology Policy. (1984) Chemical carcinogens: review of the science and its associated principles. Fed. Regist. 49:21594-21661.

Vos, J.G. (1977) Immune suppression as related to toxicology. CRC Crit. Rev. Toxicol. 5:67-101.

Vos, J.G.; Faith, R.E.; Luster, M.I. (1980) Immune alterations. In: Kimbrough, R.D., ed. Halogenated biphenyls, terphenyls, naphthalenes, dibenzodioxins and related products. New York: Elsevier/North-Holland Biomedical Press. p. 241-266.

Vos, J.G.; Van Logten, M.J.; Kreeftenberg, J.G.; Kruizinga, W. (1979) Hexachlorobenzene-induced stimulation of the humoral immune response in rats. Ann. N.Y. Acad. Sci. 320:535-550.

Wallace, L.; Zweidinger, R.; Erickson, M.; Cooper, S.; Whitaker, D.; Pellizzari, E. (1982) Monitoring individual exposure: measurements of volatile organic compounds in breathing-zone air, drinking water, and exhaled breath. Environ. Int. 8:269-282.

Warburton, D. (1981) Selection of human reproductive effects for study. In: Lowrance, W.W., ed. Assessment of health effects at chemical disposal sites: proceedings of a symposium held in New York City on June 1-2, by the Life Sciences and Public Policy Program of The Rockefeller University. New York: The University. p. 105-124.

Warkany, J. (1963) Conference on prenatal effects of drugs. Comm. Drug Saf. Bull. June.

Warkany, J. (1971) Congenital malformations: notes and comments. Chicago: Year Book Medical Publishers, Inc. 1309p.

Wassermann, M.; Wassermann, D.; Gershon, Z.; Zellermayer, L. (1969) Effects of organochlorine insecticides on body defense systems. Ann. N.Y. Acad. Sci. 160:393-401.

Wedeen, R. P. (1983) Toxic wastes and kidney disease: research needs. Environ. Health Perspect. 48:73-76.

Weinstein, I.B.; Perera, F.P. (1982) Molecular cancer epidemiology. In: Bridges, B.A.; Butterworth, B.E.; Weinstein, I.B.; eds. Banbury report 13: indicators of genotoxic exposure. Cold Spring Harbor, NY: Cold Spring Harbor Laboratory. p. 3-17.

White, K.L., Jr. (1981) Immunotoxicology of chrysotile asbestos. Richmond, VA: Virginia Commonwealth University. 208p. Dissertation. Available from: University Microfilms International, Ann Arbor, MI.

WHO/IARC SEE World Health Organization, International Agency for Research on Cancer

Wierda, D.; Irons, R.D.; Greenlee, W.F. (1981) Immunotoxicity in C57BL/6 mice exposed to benzene and Aroclor 1254. Toxicol. Appl. Pharmacol. 60:410-417.

Wilson, J.G. (1973a) Environment and birth defects. New York: Academic Press. 305p.

Wilson, J.G. (1973b) Mechanisms of teratogenesis. Am. J. Anat. 136:129-132.

Wilson, J.G.; Brent, R.L. (1981) Are female sex hormones teratogenic? Am. J. Obstet. Gynecol. 141:567-580.

Wilson, J.G.; Fraser, F.C.; editors. (1977) Handbook of teratology. Vol. I. General principles and etiology. New York: Plenum Press. 476p.

World Health Organization, International Agency for Research on Cancer. (1972) IARC monographs on the evaluation of the carcinogenic risk of chemicals to man. Vol. 1. Lyon, France: The Agency. 184p.

World Health Organization, International Agency for Research on Cancer. (1973a) IARC monographs on the evaluation of the carcinogenic risk of chemicals to man. Vol. 2. Some inorganic and organometallic compounds. Lyon, France: The Agency. 178p.

World Health Organization, International Agency for Research on Cancer. (1973b) IARC monographs on the evaluation of the carcinogenic risk of chemicals to man. Vol. 3. Certain polycyclic aromatic hydrocarbons and heterocyclic compounds. Lyon, France: The Agency. 271p.

World Health Organization, International Agency for Research on Cancer. (1974a) IARC monographs on the evaluation of carcinogenic risk of chemicals to man. Vol. 4. Some aromatic amines, hydrazine and related substances, N-nitros ocompounds and miscellaneous alkylating agents. Lyon, France: The Agency. 286p.

World Health Organization, International Agency for Research on Cancer. (1974b) IARC monographs on the evaluation of the carcinogenic risk of chemicals to man. Vol. 5. Some organochlorine pesticides. Lyon, France: The Agency. 241p.

World Health Organization, International Agency for Research on Cancer. (1974c) IARC monographs on the evaluation of the carcinogenic risk of chemicals to man. Vol. 6. Sex hormones. Lyon, France: The Agency. 242p.

World Health Organization, International Agency for Research on Cancer. (1974d) IARC monographs on the evaluation of the carcinogenic risk of chemicals to man. Vol. 7. Some anti-thyroid and related substances, nitrofurans and industrial chemicals. Lyon, France: The Agency. 326p.

World Health Organization, International Agency for Research on Cancer. (1975a) IARC monographs on the evaluation of the carcinogenic risk of chemicals to man. Vol. 8. Some aromatic azo compounds. Lyon, France: The Agency. 357p.

World Health Organization, International Agency for Research on Cancer. (1975b) IARC monographs on the evaluation of the carcinogenic risk of chemicals to man. Vol. 9. Some aziridines, N-, S- & O- mustards and selenium. Lyon, France: The Agency. 268p.

World Health Organization, International Agency for Research on Cancer.

(1976a) IARC monographs on the evaluation of the carcinogenic risk of chemicals to man. Vol. 10. Some naturally occurring substances. Lyon, France: The Agency. 353p.

World Health Organization, International Agency for Research on Cancer. (1976b) IARC monographs on the evaluation of the carcinogenic risk of chemicals to man. Vol. 11. Cadmium, nickel, some epoxides, miscellaneous industrial chemicals and general considerations on volatile anaesthetics. Lyon, France: The Agency. 306p.

World Health Organization, International Agency for Research on Cancer. (1976c) IARC monographs on the evaluation of the carcinogenic risk of chemicals to man. Vol. 12. Some carbamates, thiocarbamates and carbazides. Lyon, France: The Agency. 282p.

World Health Organization, International Agency for Research on Cancer. (1977a) IARC monographs on the evaluation of the carcinogenic risk of chemicals to man. Vol. 13. Some miscellaneous pharmaceutical substances. Lyon, France: The Agency. 255p.

World Health Organization, International Agency for Research on Cancer. (1977b) IARC monographs on the evaluation of the carcinogenic risk of chemicals to man. Vol. 14. Asbestos. Lyon, France: The Agency. 106p.

World Health Organization, International Agency for Research on Cancer. (1977c) IARC monographs on the evaluation of the carcinogenic risk of chemicals to humans. Vol. 15. Some fumigants, the herbicides 2,4-D and 2,4,5-T, chlorinated dibenzodioxins and miscellaneous industrial chemicals. Lyon, France: The Agency. 354p.

World Health Organization, International Agency for Research on Cancer. (1978) IARC monographs on the evaluation of the carcinogenic risk of chemicals to man. Vol. 16. Some aromatic amines and related nitro compounds--hair dyes, colouring agents and miscellaneous industrial chemicals. Lyon, France: The Agency. 400p.

World Health Organization, International Agency for Research on Cancer. (1982a) IARC monographs on the evaluation of the carcinogenic risk of chemicals to humans. Vol. 29. Some industrial chemicals and dyestuffs. Lyon, France: The Agency. 416p.

World Health Organization, International Agency for Research on Cancer. (1982b) IARC monographs on the evaluation of the carcinogenic risk of chemicals to man. Supplement 4 to Vol. 1-29. Chemicals, industrial processes and industries associated with cancer in humans. Lyon, France: The Agency. 292p.

Wyrobek, A.J. (1983) Methods for evaluating the effects of environmental chemicals on human sperm production. Environ. Health Perspect. 48:53-59.

Wyrobek, A.J.; Gordon, L.A.; Burkhart, J.G.; Francis, M.W.; Kapp, R.W., Jr.; Letz, G.; Malling, H.V.; Topham, J.C.; Whorton, M.D. (1983) Evaluation of human sperm as indicators of chemically induced alterations of spermatogenic function. Report of the U.S. Environmental Protection Agency Gene-Tox Program. Mutat. Res. 115:73-148.

Xintaris, C.; Burg, J.R.; Tanaka, S.; Lee, S.T.; Johnson, B.L.; Cottrill, C.A.; Bender, J. (1978) Occupational exposure to Leptophos and other chemicals. DHEW (NIOSH) Pub. no. 78-136. Available from: U.S. Government Printing Office, Washington, DC.

Zimmerman, H.J. (1978) Hepatotoxicity: the adverse effects of drugs and other chemicals on the liver. New York: Appleton-Century-Crofts. 597p.

5

Epidemiology in Exposed Populations

THE ROLE OF EPIDEMIOLOGY IN THE ASSESSMENT OF RISK AT CHEMICAL DISPOSAL SITES

Epidemiology is the scientific discipline whose purpose is to discover and interpret the patterns of occurrence of diseases and related conditions in human populations, to develop quantitative assessments of these conditions, and to relate such assessments of risk to personal and environmental characteristics which determine the degree of the risk. This purpose is achieved by direct measurement of the health status of people and of the variables associated with health and disease. Epidemiology is unique among the health sciences in its ability to address questions of health concern as they arise in actual community settings under the actual circumstances in which people live. Thus, in the present context, epidemiology has the advantage of dealing directly with people and with the exposures which occur in nature, often in combinations of multiple substances, over varying durations, with differing concentrations, and frequently at relatively low doses. The health effects which may be investigated by epidemiologic methods are also diverse and are not limited to those conditions which can be replicated or represented in a laboratory; thus acute and chronic conditions, ranging in manifestation from self-reported complaints to quantitative measures of disease with clinical and laboratory confirmation, may all be encompassed in the application of these methods. As will be described more fully below, such methods include the evaluation of case reports and case series of known or suspected health problems, as well as case-comparison studies, population surveys, community surveillance programs, and cohort studies. From this perspective, the epidemiological study of human health and disease has great utility.

Epidemiological research is especially important in defining the health effects of toxic chemicals arising from chemical disposal sites because the research results are of direct and immediate relevance to human experience, and represent the effects of real exposures in real environments. The relevance of epidemiology to the study of populations at chemical disposal sites is a direct consequence of the types of questions which arise in this context. While issues of many kinds, relevant to a number of disciplines, are raised by the existence of such sites, this report focuses on the issue of human health. As an introduction to this discussion of the applications of epidemiology in the context of chemical disposal sites, several questions with reponses based on epidemiologic approaches are briefly outlined below:

1) Given that a health problem, which might plausibly be attributed to a chemical exposure, clearly exists within some definable population, can evidence be found of a source of chemical exposure as a possible or reasonable explanation?

2) Given that a particular chemical substance is known to be present at a specific site, is there evidence of an excess of health problems within the population thought to be exposed?

3) Given evidence from previous studies that a particular health problem may be caused by exposure to chemicals arising from disposal sites, can this hypothesis be tested in a new investigation at a different site where a similar exposure may occur?

4) What is the actual or potential distribution of concentrations of waste chemicals in relation to human populations, and what health problems might be detected at an early stage, affording preventive or remedial measures to be taken before serious or large-scale harm has occurred?

The epidemiologic nature of these questions is perhaps self-evident, but some comment on them may make the point clearer. Question (1) is stated first because it represents the most urgent situation in which adverse health conditions await explanation. It is difficult to overstate the intensity of concern on the part of the general public and a number of special constituencies when one or, more often, several cases of serious illness are at issue. Explanations may be generated by the public well in advance of formal investigation and may be firmly believed by many potential participants in the investigation. It then becomes difficult to address, in an objective manner, the immediate and pressing issues of the nature of the problem, the plausibility of alternative explanations, and the resolution of community concerns. These are all aspects of the need to obtain concrete information about human health problems in the real situations in which they are perceived to occur.

The first question assumed the reality of a health problem as the starting point in consideration of the possible implication of a chemical disposal site. Question (2), however, merely recognizes the suspicion which may arise out of the knowledge that a source of chemical exposure exists in the community. The question which must then be addressed is whether there is, in fact, any perceptible health problem among potentially exposed persons: Is any health problem observed in greater than expected frequency among such persons? It is important to distinguish between those problems which might be looked for on an a priori basis, given the nature of the substance in question, and those which might also be attributable to that exposure but only by interpretation of the completed investigation. In both the situation where a number of cases have been recognized and the situation of an acute episode of known exposure (as in a spill or other catastrophe), considerable alarm engendered in advance of formal investigation may precipitate community-wide anxiety which demands epidemiologic inquiry if resolution is to be sought.

Question (3) also has implied extensions to several further issues, once an association appears to be firmly established, whether within the particular investigation in which the hypothesis first arose or in consequence of one or more completed studies. These include characterization of any adverse health effect in the individual, the frequency of occurrence of the health effect among the exposed population, and the magnitude of the excess risk which is attributable to the exposure under investigation. Also to be determined are the relevant circumstances of exposure and the variation in risk, if any, among identifiable subgroups of the population.

The last of these questions (4) indicates the potential for application of epidemiologic methods in a truly preventive mode, in that a long-term societal strategy for anticipation of health problems related to chemical disposal sites is contemplated. This question addresses the concept of monitoring and surveillance studies, in contrast to epidemiologic studies in response to specific events, as discussed in a later section. It is implied that a commitment to prevention could, through epidemiologic approaches, accomplish the monitoring of potential exposures by maintenance of an inventory of the composition and condition of at least major sites, in juxtaposition with surveillance of both chemical body burden and health problems which may be sentinel conditions suggestive of chemically related etiology. Further steps in this direction could include, for example, periodic surveys of potentially exposed populations by low cost means to ascertain actual exposure histories. However, the feasibility, in principle, of doing so poses sharply the following question: Is the societal concern about these matters sufficient to warrant the investment in time and resources which could improve the results of epidemiologic investigations when the decision is made that they be carried out? Many commentators have emphasized not only the essential contributions of epidemiologic investigation to addressing the questions at hand but also the limitations inherent in the situations studied; these limitations can be diminished by improving the quality and the detail of available data such as the documentation of individual exposures.

Finally, in this introduction, it remains to be considered whether evidence from experimental and observational investigations of species other than humans may suffice as a basis for inferring causal relationships between exposures to chemicals from disposal sites and adverse health effects. That evidence may be acceptable in some instances, but does not apply to all diseases or to all situations. The need for direct observations of the affected populations, supplemented by appropriate studies through other disciplines, is strongly supported by a broad consideration of what might be termed "epidemiological risk assessment."

Estimation of the risks to human health due to exposure to toxic chemicals released from disposal sites may be based on direct observation, either clinical or epidemiological, or on inferences drawn from indirect evidence, either experimental or observational, of effects on other biological systems. These alternatives offer a choice between undesirable alternatives. If the risk of severe illness is truly large, we would not wish to await the development of illness in exposed persons to attest to the importance of the problem. However, extrapolating human risk from data derived in non-human systems is insecure. Doll (1984) stated that "extrapolation from experimental systems to Man involves many pitfalls, and particularly if attempts are made to extrapolate quantitatively.....evidence in laboratory animals in the absence of human evidence can be only a pointer, albeit an important one, to the presence of a potential hazard."

Laboratory-based evidence of the toxicity of chemicals is often difficult to classify into categories that state clear implications of the potential hazards to man. The International Agency for Research on Cancer (WHO/IARC, 1978) was the first to define a workable system for defining such categories in terms of the carcinogenicity of chemicals. Evidence of carcinogenicity in animals was evaluated and classified as to its potential for carcinogenicity in man based on the strength of the evidence; evidence was classified as being either "sufficient," "limited," or "inadequate." Chemicals with "sufficient" evidence for carcinogenicity are generally regarded, for practical purposes, as if they are carcinogenic to man. However, studies have shown that carcinogens act by different mechanisms and thus, not all evidence of chemical carcinogens, even those with "sufficient" evidence for carcinogenicity in animals, can be extrapolated from one species to another. Although this discussion addresses the issue of the extrapolation of laboratory data to assess risk to humans of carcinogenesis, in particular, it should be apparent that the commentary is applicable, as well, to the full spectrum of health problems which may be of concern in evaluation of the possible effects of exposure to chemicals from disposal sites. For many health effects, both early and delayed responses to chemical exposures, no suitable animal model exists; the only evidence available is that derived from direct observations of exposed people. The use of bioassays in assessing the risk to humans of exposure to toxic chemicals therefore includes the following shortcomings:

- The evidence from laboratory bioassays is indirect and may not be applicable to human experience.
- The methods commonly employed entail extrapolation of effects in animal systems from the large doses used experimentally to the much smaller doses likely to be encountered in environmental exposures.
- For many important adverse health effects, no suitable bioassay system exists.

For these reasons, laboratory observations are useful in the detection of a potential problem, but do not provide sufficient information to develop a reasonable quantitative estimate of the health risks for humans. Epidemiological findings will provide a more direct basis for the assessment of risk whenever human health effects are suspected to result from exposure to environmental contaminants.

Epidemiological risk assessment entails a comprehensive and evaluative review of the entire body of research data concerning the health hazard in question and the development of estimates of risk, together with statements of the uncertainty surrounding those estimates, for different degrees and circumstances of population exposure. If the body of research data is extensive and of good quality, then epidemiological studies can provide a clear and reliable basis for projecting risks from present and future exposures to a toxic chemical. Commonly, however, the data are sparse, and some aspects of most studies are subject to criticism. Most epidemiological risk assessments, therefore, rely heavily on interpretation and judgment. Some of the most difficult situations arise when the data from different studies indicate widely differing statements of potential risk due to exposure. When the economic, social, or political costs of regulation of exposure are high, those situations afford ample basis for disagreement between those affected by exposure and those affected by the regulation of exposure. An equally difficult problem exists in the interpretation of studies in which no unexpected type or frequency of health effect is found, or in which no association can be made between the occurrence of a health effect and exposure to chemicals from a disposal site. When the absolute risks are small, as they commonly are, when the number of persons included in a study is small, when no direct measure to document actual exposure is available, and when the health effects may occur many years after the exposure, the failure to obtain positive results at the time of the study is not convincing evidence that the exposure is innocuous. Even in circumstances unfavorable to the detection of an effect, however, negative studies can help to determine the probable upper limits of risk. It seems, therefore, that experimental toxicological data can forewarn us of potential hazard to man, but that our final decision must rest on a full assessment of our knowledge of the total situation, including that derived from direct human observations, and modified by the needs and demands of society.

EPIDEMIOLOGIC METHODS

Establishment of Cause and Effect Relationships

In the search for a relationship between a factor such as exposure to chemicals and the occurrence of a disease or a biological change, three types of associations may be observed: an artifactual (spurious) association can result from either a chance occurrence or a bias in the study method, referred to as systematic error; an indirect association can occur when the factor of concern and the disease outcome are not directly related to each other but to a common underlying condition, resulting in a noncausal relationship; a true causal association can be recognized by a change in one variable resulting in a change in a second variable (Mausner and Bahn, 1974). It therefore can be shown that the two events, X (exposure to a chemical) and Y (manifestation of a health effect), are causally related if when exposure to the chemical is diminished, the incidence of the health effect is likewise diminished, or vice versa (Lilienfeld, 1957).

Listed below is a further set of criteria which are widely used to assess various aspects of the causal nature of a statistical association, no one of which alone is sufficient for this determination (Lilienfeld and Lilienfeld, 1980; Mausner and Bahn, 1974; USDHEW, 1965).

Consistency of association

Perhaps the single most important factor in establishing a true causal association is the replicability of the observation (Hoover, 1982). If the association persists in testing under diverse circumstances, using different study methods and populations, then it is unlikely that the initial association occurred merely by chance. One of the most difficult problems in the assessment of causal significance is the appearance of conflicting findings in different studies (Stallones, 1984). In practice, the fact that a single, credible study exists in which an expected association did not appear is not always sufficient rationale for eliminating the variable in question from any further consideration of causal association. In addition to the fact that the quality of study methods and procedures, and hence the quality of the evidence, varies from study to study, there may be in operation various interactions between variables. An association may appear in one situation and not in another due to the necessary presence or absence of some other factor.

Strength of association

Since most health effects have many causes and do not have a one-to-one correspondence with a single factor (Lilienfeld and Lilienfeld, 1980), a measure of risk is used to indicate the relative importance of the possible etiological factor in producing the health effect. The greater the proportion of exposed individuals who have the health effect as compared to unexposed individuals with the health effect, the greater the likelihood of causality. In addition, the likelihood of causality increases if a dose-response gradient can be observed.

Temporal relationship of association

By definition of causes and effects, the exposure must predate the onset of the health effect and allow for the necessary induction and/or latent periods. This can be particularly difficult to establish with a chronic disease condition whose development is so gradual that a date of onset cannot be determined. If the beginning date of disease is not known, there is no assurance that an exposure event preceded it.

Specificity of association

though an exposure to a specific agent may have multiple health effects and a specific health effect may result from any one of a number of different exposures, causal significance of an observed association is more confidently assigned if a health effect is specifically related to a defined exposure. This one disease/one cause model is more universally applicable to acute and infectious diseases than to others; chronic diseases commonly have a complex network of antecedent risk correlates. While chemical toxicity is often organ-specific and hence disease-specific, chemical disposal sites usually contain many different chemicals, and therefore the specificity of a health effect may be indeterminate.

Coherence of association

Often referred to as biological plausibility, coherence implies that a cause and effect interpretation for an association does not conflict with what is known of the natural history and biology of the health effect (Rothman, 1982). However, some observations which appear to contradict the available body of knowledge about a disease state may not necessarily be invalid on that basis alone; existing knowledge of the pathological processes of many diseases is, at this time, obscure or incomplete (Stallones, 1984).

Epidemiologic Study Designs

There are a number of types of studies which epidemiologists employ to examine the health status of population groups. The sources of data include vital statistics records, health data registries, medical and hospital records, mailed questionnaires, interviews by telephone or in-person, and physical examinations and diagnostic tests. These various study designs and their modes of data collection are not equally applicable to the study of populations at chemical disposal sites.

Outbreak investigations

A common type of epidemiological study is an outbreak investigation, in which a group of people have been reported to have developed similar symptoms, generally acute, and such reports are clustered temporally. In situations such as gastrointestinal poisoning or infectious disease epidemics, investigators search for some common link among those affected in order to determine the causal factor. An outbreak investigation can be particularly effective in eliciting the nature and source of the exposure when relatively high exposure to a chemical causes acute health effects (for example, the episode in Perham, Minnesota in which 13 people became ill after drinking wellwater contaminated by arsenic from a nearby waste deposit (Feinglass, 1973)).

Surveys

A survey is an epidemiological investigation in which information is collected systematically by examination of the whole or a sample of a population of interest. Often called a cross-sectional or prevalence study, variables of interest are ascertained as they exist in a defined population at one particular time. As a result, the disease prevalence in relation to selected attributes can be estimated but the temporal sequence of cause and effect may not be apparent.

Surveillance

The surveillance or monitoring of a population involves the ongoing collection and analysis of data for the purpose of detecting changes or trends in the health status of the population. Deviation from expected measurements and observations may suggest the need for further investigation or intervention in high-risk populations or subgroups.

Case series

Series of cases of a specified disease often are assembled from clinical experience and serve as the basis for much clinical research. On occasion, epidemiological inferences may be drawn from such a series. However, in the absence of a defined population base, these inferences cannot be stated in quantitative terms.

Case-comparison studies

A case-comparison study compares the frequency of past exposure among persons who are experiencing the disease (cases) with the frequency of past exposure among persons who are not experiencing the disease (comparisons). The past experience of the comparison subjects (e.g., their exposure to chemicals) is used as a baseline from which to evaluate in quantitative terms the experience of the cases. Case series and case-comparison studies are subject to a special bias of ascertainment; that is, the knowledge of disease may affect recall or perception of antecedent events.

Cohort studies

In contrast to the case-comparison study, a cohort study compares the frequency of disease among exposed persons with the frequency of disease among unexposed persons (or alternatively, compares the frequency of disease among groups of persons with varying levels and/or routes of exposure). The characteristics of study subjects are determined and related to the occurrence of disease during a subsequent period of observation. This eliminates the problem of retrospective bias. However, in studies of chemical exposures, prior knowledge of exposure status may lead to bias in perception and reporting of illness.

The advantages and disadvantages of case-comparison and cohort study methods are summarized in Table 5-1 (Bloom, 1981; Lilienfeld and Lilienfeld, 1980; Mausner and Bahn, 1974; Stallones, 1966).

Statistical Procedures and Risk Assessment

Epidemiologic rates and risk estimates

Epidemiologists utilize rates and ratios (such as absolute risk, prevalence rates, incidence rates, relative risk, odds ratio, and attributable risk) to express the population risk of disease due to exposure to some factor (Cornfield and Haenszel, 1960; Mausner and Bahn, 1974; Roht et al., 1982). The absolute risk of having a particular health effect is the ratio of the number of people with the health effect to the total population at risk of developing the health effect. A prevalence rate is an example of an absolute risk and is expressed as:

$$\frac{\text{number of existing cases of a disease}}{\text{population at risk}} \text{ at a point in time}$$

The prevalence of a disease therefore represents the accumulation of cases in a population, and is a resultant of the rate of acquisition and the rate of attrition. An incidence rate is also an example of an absolute risk, and gives some indication of the temporal relationship between exposure and the development of disease. The incidence rate is expressed as:

$$\frac{\text{number of new cases of a disease}}{\text{population at risk}} \text{ during a specified time interval}$$

TABLE 5–1. Case-comparison and cohort studies; summary of advantages and disadvantages.

	Advantages	Disadvantages
Case-comparison Study	Relatively inexpensive Allows smaller sample size Results obtained relatively quickly Suitable for health effects of low incidence	Incomplete information about past events Bias in recall Problem of bias in case ascertainment and selection Selection of valid comparison group is difficult Yields only estimate of relative risk (odds ratio)
Cohort Study	Reduction of sources of bias as exposure is ascertained prior to health effect onset Yields absolute risk as well as relative risk Can explore a spectrum of clinical outcomes	Can be costly Requires large sample size Long follow-up period Problem of attrition Not feasible for rare diseases

To determine how much excess disease is caused by an exposure, other measures of risk must be calculated. The first of these determinations, relative risk, is defined as the ratio of two incidence rates:

$$\frac{\text{incidence rate among exposed}}{\text{incidence rate among nonexposed}}$$

which can be alternatively expressed as:

$$\frac{a}{a+b} \div \frac{c}{c+d} \quad \text{[from Figure 5-1].}$$

Figure 5–1. Classification of a total population by health effects and exposure status.

	Health Effect	No Health Effect	Total
Exposed	a	b	a+b
Not Exposed	c	d	c+d
Total	a+c	b+d	

This analysis of relative risk is directly applicable to cohort studies, but because of the way the study group is assembled in a case-comparison study, groups (a+b) and (c+d) do not represent the total populations exposed and not exposed to the factor; therefore, the incidence rates in a case-comparison study cannot be determined. An estimate of the relative risk can be calculated, however, if the following criteria can be met: (I) the comparison group is representative of the general population, (II) the cases are representative of all cases, and (III) the absolute risk of the disease (health effect) in the population is small (Mausner and Bahn, 1974). If these assumptions are satisfied, the odds ratio can be used as an estimate of the relative risk and is easily derived. If assumption III is true and the proportion of individuals with the health effects is small, then it follows that 'a' is small in comparison to 'b,' and 'c' is small in comparison to 'd.' Therefore, in the preceeding expression based on Figure 5-1, the denominators can be reduced to 'b' and 'd,' respectively, yielding:

$$\frac{a}{b} \div \frac{c}{d} \quad \text{or,} \quad \frac{ad}{bc},$$

which is an approximation or estimate of the relative risk.

The determination of attributable risk is the second method of calculating the excess risk of a health effect due to exposure. Attributable risk is defined as the arithmetic difference between two incidence rates:

attributable risk =
(incidence rate among exposed) - (incidence rate among nonexposed)

This, then, is that portion of the risk among exposed persons that can be attributed to the exposure.

In conclusion, attributable risk and relative risk are useful in different situations. Attributable risk describes the excess rate of a health effect above the rate that would be expected in the absence of the exposure of interest; relative risk is a measure of the strength of association between exposure and health effect and is the more sensitive measure for assessing the likelihood of causality of an exposure in disease. However, once a causal relationship has been established, the measure of attributable risk may give a better idea than relative risk of the impact of an exposure on a population (Hoover, 1982).

APPLICATION OF EPIDEMIOLOGIC METHODS TO CHEMICAL DISPOSAL SITE POPULATIONS

Advances in knowledge of disease causation have resulted in changes in emphasis in the past century from infectious agents, to metabolic and deficiency states, to the current concern with chemical and other environmental hazards. The enactment of environmental protection legislation and the subsequent need for identification and monitoring of environmental chemicals fostered the rapid development of enhanced laboratory analytic capabilities. For some compounds, the detection limits have been lowered by four or more orders of magnitude in the span of two or three years allowing the detection of minute concentrations of chemicals in soil, air, water, and food supplies. These technological advances and the disclosure of an estimated 20,000 uncontrolled chemical disposal sites nationwide have changed the public impression of the environment as largely safe to widely contaminated by toxic chemicals and have fostered the public suspicion that exposure to those chemicals might be linked to the occurrence of previously unexplained illness. State and federal governments responded to the ensuing public demands for investigation by assigning to groups with experience conducting epidemiologic studies the responsibility for interpreting the health consequences of the contaminant levels being reported.

Epidemiologic Investigations Conducted in Response to Specific Events

The majority of waste chemical exposure investigations conducted to date have been responses to specific events. The principal practitioners of this kind of epidemiologic study are public employees with the responsibility of responding to requests for evaluations. The investigative methods employed must be targeted toward problem (hazard) verification and resolution rather than performance of research and hypothesis generation. Research opportunities abound in communities residing near chemical disposal sites, but should be addressed separately; the questions that need to be answered first are those of the client community, in a simple and understandable manner.

In more traditional epidemiological study situations, the appropriateness of proposed investigations is largely controlled through the external peer review system by approval or disapproval of funding; thus marginal investigations are restrained and restricted. Situations in which a community may be adversely affected by a chemical disposal site unconditionally call for a study to be conducted, even though the environment may not be conducive to careful, thoughtful research. Once embarked upon, such studies, by their mere existence, assume an importance and legitimacy that no long series of caveats can change.

Other characteristics of this type of epidemiologic study need to be kept in mind when deciding on appropriate techniques for problem resolution. Although epidemiologic studies often engender controversy after completion, these studies frequently begin amid controversy over already identified conclusions. Individual and community stress and the impact of the news media increase the need to limit such studies to objective measures of outcomes, but objective measures may be rejected by the community. Epidemiological studies should maintain the flexibility to select the most appropriate study population, but these studies have no such flexibility and are bound to the study of a population at a specific site. Even when making imaginative use of the information that is available, these studies frequently have limited power and, because of the public perception that adverse events have occurred, often face the unenviable task of proving a negative.

Impact of risk perceptions on health studies at chemical disposal sites

The difficulties in conducting valid and properly designed epidemiological studies of populations at chemical disposal sites are frequently compounded by differences in perceptions of risk that exist between the community being studied and the scientific professionals and government officials involved. Not all risks of concern to community residents relate specifically to health effects; environmental damage, economic consequences (e.g., property value, jobs), and fear of litigation are also legitimate concerns. The importance of specific risks perceived in the situation depends largely on what people have at stake (Johns Hopkins University, 1981); the more people have to lose, the more important the situation appears to them. Residents of affected areas may be more concerned about their health than non-residents; residents with children may be particularly concerned about long-term and reproductive health effects; homeowners are more concerned about property values than are renters. Some of these matters of concern to residents may not be addressed by planned investigations. Because such investigations often fail to adequately address the concerns of the community, a breakdown of communication and trust between the investigating agencies and community residents may result. Unfortunately, such a breakdown can seriously impair the ability of an investigation to determine with any validity the information it sets out to find (Levine, 1983). A foundation of trust between professionals and citizens needs to be developed early in an investigation to foster a cooperative environment for research.

Placing the risk of chemical exposure in relation to other daily risks may give some sense of proportion to the actual risk faced by individuals. It has been stated that in reality, the hazard to public health from chemical waste sites "is trivial in comparison to the danger from cigarettes" (Abelson, 1983). These kinds of comparisons, however, are small comfort to those who may be exposed. There is an intrinsic difference between the risk of exposure to chemical waste and ordinary daily risks we all face, such as smoking cigarettes and driving automobiles. Most of the latter are familiar and are taken voluntarily, the costs and benefits having been weighed in some fashion. They are personal and individual choices and for the most part involve personal risk. Exposure to chemical waste, for example, through contaminated drinking water, is an involuntary risk that many individuals feel they have not agreed to, even though they may enjoy the technological benefits provided by the growth of chemical and petroleum industries. Such decisions are not in the hands of "ordinary" individuals but are made by others for them; when individual costs are then sustained, they understandably engender anger.

Often, people's concerns about toxic waste can be explained as responses to organizational behavior and uncertainty about risks (Johns Hopkins University, 1981). A community's misconceptions of risk may, in part, be attributable to an uncertain understanding of agencies and their actions or lack of them. Agencies are only at an early stage of learning how to educate and communicate with communities and citizen action groups. A population may be capable of understanding the situation and scientific methods of study being utilized. However, when confronted with inadequately designed or explained studies, the statistical problems in interpretation may not be readily understood, especially when the findings of such studies are not agreed upon by experts. When scientists' answers to questions consist of "we don't know," they are trying to communicate what they have learned in terms of factual information, leaving open the question of interpretation. Their own broad context of understanding, however, can not be communicated to the nonscientific community.

Difficulties in conducting investigations may be compounded by public officials' uncertainty about citizen groups and their motivations. If a community's knowledge about a health threat is assessed inaccurately, public statements made for the benefit of the community may be perceived as being patronizing or concealing the real issues. Without trust and confidence in government and scientific and health professionals, it is difficult to perform research in a scientifically acceptable manner where that research requires the cooperation of the community. People may come to view the researchers as politically and financially involved, whether or not the scientific professionals view themselves in that light (Levine, 1983). Policy officials may, in fact, be reluctant to generate findings that have social or economic consequences with which they are not prepared to deal. Although in reality, scientific researchers and policy making officials do not necessarily share the same philosophies and goals, a public unfamiliar with the workings of such investigations may conclude that they are in league together and that the truth is being withheld in order to further bureaucratic and political goals (Molotch, 1982). As happened at Love Canal (Levine, 1983), people interpret these kinds of responses and ensuing actions as incomprehensible and respond with fear and anger. The vocal and angry reactions of citizen groups confronted with such situations can be viewed as a normal response to a stressful situation (Johns Hopkins University, 1981).

Volatile situations such as the one at Love Canal arise with alarming speed and frequency. Citizens or community action groups trying to get investigators and public officials to listen to them will find it a worthwhile exercise to examine carefully, in advance of such contact, the approach they need to use. Anger and outrage, although understandable and justifiable in many cases, are not particularly persuasive means of getting those in positions of power to listen seriously to community complaints. Accusations and angry demands for action are likely to be met with anger and resistance. Establishing an organized plan to present the community's case in an orderly and rational manner to the proper individuals may be a more effective approach. Anger and outrage are best saved for a situation in which no other recourse is available.

Concerned residents seem to use criteria for assessing costs and benefits different from those of scientific professionals or policy makers. Failure to understand these differences in criteria for assessing risk is a serious detriment to communication between the investigator and the community. Even if community residents are incorrect in their assessment of a hazard, it makes sense to listen to their complaints and perceptions. If citizens believe that they are being taken seriously, they may be more open to listening to a different perspective and interpretation of the facts.

Good rapport between the professional and the community provides the opportunity for explanation of how a particular environmental problem must be examined in light of available resources and remediability (Freudenberg, 1984) as well as opportunity for education in the principles of scientific analysis. A community living near a chemical disposal site may be acutely aware, due to media coverage, of the media's perception of the potential for various adverse health effects. Community residents, upon hearing from each other about the occurrence of various illnesses or adverse reproductive events, do not understand when scientists tell them that no adverse health effects can be related to the chemical waste. Without some explanation of the importance of such factors as sample size, the need for comparison with other populations and with expected rates of disease, the phenomenon of clustering merely by chance, and the criteria for establishing associations between cause and effect, citizens cannot possibly be anything but outraged at the inconclusive results reached by most epidemiological studies of communities near chemical disposal sites. Scientists who are well qualified to conduct epidemiological research may not know how to communicate effectively with the public or how to understand and address its needs. The community needs to be taught what is attainable by an epidemiological assessment and what is not; each step in the decision tree used in the evaluation must be carefully explained and understood. There is a large need for education of communities, investigating agencies, and public media in order for situations to be clearly understood and various needs to be interpreted correctly and addressed adequately.

While many of the vague and nonspecific complaints that residents attribute to being exposed to chemical waste may be impossible to link directly to the exposure itself, the fact that they think they have been harmed is itself a reality. Individuals living near chemical disposal sites often believe themselves at risk for adverse health effects due to chemical exposure and this belief may be stressful. The exact role that stress or increased concern about one's health plays in determining one's health status is unknown; such stresses, however, may be psychosocial determinants of illness onset. It has been reported that anxiety commonly causes physical symptoms that may be manifested long before a person is aware of his anxiety (McGough, 1982). Such symptoms include dizziness, palpitations, weakness, chest and abdominal pains, headache, back pains, and diarrhea. In addition, chronic anxiety, depression or emotional distress may play a part in affecting host susceptibility to a variety of diseases (Martin, 1984); psychosocial determinants of illness have been correlated with illness onset in close temporal sequence to the stressful experience.

The following example provides evidence that stressful living conditions may have some influence on health status. A report of a questionnaire administered to "exposed" and "non-exposed" groups of individuals at the Price Landfill in Atlantic County, New Jersey determined that in the exposed group, the number and frequency of adverse health symptoms were greater than would be expected given the known toxicity of the comparatively low levels of chemicals found in the drinking water (New Jersey, 1983). The report noted that in a similar study of water contamination in another part of New Jersey, the exposed group also showed an increased number of reported symptoms; however, after the water data were reviewed, the exposed group was found to be misclassified in that their water was not, after all, contaminated. The fact that the same increase in reported symptoms among individuals who thought their drinking water was contaminated was found in individuals who do have low level water contamination suggests that stress may be a factor in the etiology of health complaints.

The temptation to undertake a mental health or stress assessment of a community at a disposal site may be strong, however, severe methodological limitations would hinder such an investigation (Johns Hopkins University, 1981). Perhaps more importantly, such a survey could potentially cause harm by arousing additional concerns, possibly having detrimental effects on other aspects of the community health status investigation. In addition, it is extremely difficult to distinguish between stress-related psychological problems and similar problems due to the effects of actual chemical exposure.

Perceived health status as a predictor of mortality has been studied and a significant association has been reported. Aside from sex and age, investigators reported that perceived health was the strongest predictor of mortality; the report, however, was inconclusive because the investigators were unable to control for objective health status (Goldstein et al., 1984). Another investigation did confirm that for individuals over 29 years of age, increasingly poor perceived health was found to be associated with increasing mortality rates (Kaplan and Camacho, 1983). Both these studies found this association to be independent of physical health status.

Factors that prompt investigations

Observation of a potential environmental hazard

The purpose of this section is to describe approaches that an epidemiologist might take when confronted with a request to do something about a chemical disposal site which is not yet known to have caused any adverse health effects. The following is a sequence of possible approaches that might be undertaken:

- The physical study of the site and of the ecosystem at risk of contamination by migration of the waste constituents is likely to be the first study conducted. Methods for the study of the physical characterization of the waste disposal site and the routes of chemical transport are discussed in Chapter 3. These issues relate to the potential for exposure and not to exposure itself.
- If chemicals in the environment appear to have originated from the disposal site, a search may begin for evidence or indicators of exposure among those persons most likely to have been exposed; contaminants may be measured in body tissues or in environmental media to which the people are exposed (e.g., air, water, soil).
- A prevalence survey of the population may be conducted to ascertain whether any unusual types or frequencies of disease are present.
- The results of the prevalence survey may show an apparent disease cluster, indicating a need for a more extensive study of the population; a case-comparison study is often conducted at this point. The study of clusters associated with chemical disposal sites is detailed in a following section.
- The establishment of a surveillance mechanism for long-term study of a population may be necessary to monitor an exposed population or a population-at-risk of exposure. The epidemiologist will have to contribute to the selection of the health effects to be evaluated and of the specific populations to be studied and will need to use the analytic techniques of epidemiology to determine whether or not the population studied is exhibiting an excess frequency of the health effects sought.

Alterations in disease frequency or distribution

A common form of alteration in disease frequency that attracts the attention of the public or a health authority is a cluster of cases -- the reported occurrence of more cases of a particular disease within a specific geographic area and time period than seems reasonable on the basis of the size of the population of the area. Because the incidence of a disease in different populations is represented by a range of frequencies, the incidence of a given disease for most populations will tend to fall toward the middle of what is considered to be the normal range of incidence. The disease frequency in any given population may, however, fall at the very high or very low end of this distribution. If a high disease frequency occurs in an area that contains a chemical disposal site or the drainage from such a site, the site is likely to be a prime suspect as a cause of the cluster. Clusters associated with chemical disposal sites, as distinct from those associated with other forms of chemical exposure, are generally small, typically between 3 and 20 cases. Some apparent clusters are actually based on heterogeneous aggregations of diseases or syndromes, and these are more difficult to investigate than clusters of a single disease or health effect. The most distinctive and easily identified clusters and those which are most likely to lead to successful investigations are clusters of diseases which are otherwise rare in the community.

To investigate a cluster, an operational definition of a case needs to be established and all persons meeting the definition must be identified. By defining the total population-at-risk, the risk of the disease in the population can be determined and, by appropriate comparison, it can likewise be determined whether this risk is actually in excess in the area and time suggested. The most direct approach to make this determination is to compare the observed number of cases with the expected number computed from age- and sex-specific rates derived from a comparison population.

If the cluster involves deaths, national or state death rates may be used to compute expected numbers. If it involves incident cases of cancer, data from an appropriate cancer registry may be used. The comparison of observed and expected values in such a situation presents almost insuperable problems in statistical terms. Considering the enormous numbers of places in which clusters may arise, the multiple time periods involved, the often arbitrary selection of the geographical and temporal boundaries of the cluster, and the hundreds of diseases that may be involved, it is to be expected that many clusters will arise by chance and that many will involve numbers that under other circumstances would be considered statistically significant. While a particular observed vs. expected difference must clearly satisfy the ordinary standards of statistical significance, if a cluster initiated the investigation, much more stringent criteria must be applied.

In a cluster situation in which the age and sex distributions of the population surrounding the cluster are not known, or the disease is one for which no external rates are available, a direct comparison of observed and expected numbers of cases cannot be made. Other statistical procedures (Knox, 1963, 1964; Mantel, 1967) may be used. However, these procedures, like the direct comparison, have the limitation that, if the investigation was initiated by observation of a cluster, an accurate assessment of the probability of the cluster being due to chance cannot be made. That is, a hypothesis cannot be tested by the observation upon which that hypothesis was based. The various types of analyses may, however, lead to the conclusion that an occurrence of a cluster by chance is sufficiently improbable that further investigation is warranted. Such investigation would involve looking for any obvious links between the cases by interviewing them or their surrogates if the actual cases have died or are in some way unable to speak for themselves. Residence (past or present) around the disposal site would be inquired into as would such things as similarities in occupation, place of employment, socioeconomic status, marital status, housing conditions, household composition, residence in relation to any other potential health hazard, and any other factors that the investigator thinks may be of value.

If the disease is one which occurs with some regular frequency in the area, a more formal investigation may be required. For example, if in a small town over a period of a year, 10 cases were observed but 5 were expected, then only 5 of the 10 cases were in excess and these 5 cannot be distinguished from the 5 which would have normally occurred. The 5 natural cases may obscure patterns present among the 5 excess cases and make them difficult to detect by casual inspection.

A research design commonly employed in these circumstances is the case-comparison study, in which comparison subjects similar to the cases in demographic features such as sex and age are selected from the general population. A number of case-comparison studies have been undertaken in the context of chemical disposal site problems, but with little success. On the other hand, case-comparison studies have been highly successful in the investigation of clusters due to infectious agents and chemical contamination unrelated to disposal sites (e.g., occupational exposure) and should not be abandoned as an approach to the investigation of clusters.

Changes in disease patterns other than clustering may draw attention to possible chemical disposal site hazards. For example, changes in the pattern of disease incidence temporally associated with the initiation or the termination of chemical disposal at a site may be a stimulus that prompts an investigation into the meaning of the apparent association. At the present time, such changes in the frequency of disease would only be observed in routine sources of data such as death certificates, hospital records of birth weights, cancer registries, and the relatively few registries of congenital malformation.

In the investigation of clusters and other unusual disease patterns, present indications are that disposal sites are likely to be among the least frequent explanations of such patterns. Investigations prompted by health effects of chemicals arising from disposal sites should therefore address the epidemiology of the health effects in question in a comprehensive way, including consideration of alternative hypotheses regarding etiology.

Guidelines for setting priorities for epidemiologic studies

Guidelines have been developed by several sources as an aid in determining whether sufficient information is available to conduct an adequate epidemiological investigation at a chemical disposal site (USDHHS/CDC, 1984). Although there are always exceptions, the ability to satisfy the following requirements can help to guide epidemiologists and other health professionals when setting priorities for specific projects.

- Identify types, concentrations, and characteristics of the chemicals present in leachates, air, water, or other media that might result in human exposure.

- Develop a relative exposure index that includes consideration of the route of human exposure. (Proximity of residence to a chemical disposal site may be too insensitive an index.)
- Select methods to measure objectively adverse health effects previously attributed to specific chemicals.
- Evaluate the potential for exposure at concentrations that are expected to cause adverse health effects.
- Unless the expected outcomes are agent specific, select comparison subjects from appropriate reference populations.
- Provide power calculations for all proposed studies and carefully explain the design and limitations of the study to state and local officials, representatives of the industry involved, and the population in the vicinity of the site.

Other questions involve assessing the impact of the study results. All studies can pose a variety of issues but, for epidemiologic studies in response to events at chemical disposal sites, the key question to ask is whether the results of a study will stimulate remedial or preventive activity at the site or other similar sites which otherwise would not have been undertaken.

Interpretation of Epidemiologic Data in the Assessment of Health Effects at Chemical Disposal Sites

This section provides an overview of some of the major difficulties with epidemiologic approaches to the evaluation of health effects associated with exposure to chemicals from disposal sites. There are problems inherent both in nature (e.g., latent periods of disease) and in epidemiologic data (e.g., bias). Some of these problems characterize epidemiologic approaches to disease in general whereas others are of special relevance to the epidemiology of the exposures and diseases that may be produced by chemical disposal sites. These difficulties, singly or in combination, can cause an epidemiologic study to yield results that are invalid or unconvincing.

Confounding and bias

Epidemiology is usually an observational, not an experimental, science, an attribute it shares with such sciences as astronomy and population genetics. Nonexperimental sciences such as these are prone to confounding and bias.

Confounding has been defined as the mixing of the effect of the exposure variable with the effects of other (confounding) variables (Ibrahim and Spitzer, 1979). Confounding variables may include age, sex, socioeconomic status, marital status, personal and cultural habits, drug intake, and pre-existing medical conditions which may result in lowered or altered adaptive capacity (Carnow and Conibear, 1982). For example, if an investigation found factory workers to have an excess mortality from lung cancer, it might be concluded that an exposure in the factory caused that excess. However, if the group of factory workers included a particularly high proportion of cigarette smokers, this factor could explain, at least in part, the apparent excess of lung cancer. In this example, the factory exposure is confounded and smoking is the confounding factor. Thus, the relationship between an exposure and a health effect may be obscured by confounding variables that may also contribute to the outcome of interest.

Control of bias is another general problem in epidemiology. Many biases are unique to a particular investigation or setting; however, a few of the major or generic types of bias are analytic, selection, and information bias, as defined below.

Analytic bias means that the data gathered from two or more groups could be compared validly but are not because the statistical or epidemiologic methods used in the analysis are inappropriate.

Selection bias is best described as a consequence of inappropriate comparisons. For example, it might be suggested that, relative to the general population, people living near a chemical disposal site have a high mortality rate due to all causes. However, even a cursory study might reveal that the disposal site is surrounded by one or several communities with a relatively high average age. Thus, the general population is an inappropriate basis for comparison with the chemical disposal site population. While this example uses a transparent problem, it should be appreciated that sources of selection bias can be exceedingly subtle. Since epidemiologic studies often rest importantly upon the comparison of two or more populations, it is crucial to judge the validity of the comparisons made.

Information bias occurs when, although the groups being compared are truly comparable, the information gathered from them is not. For example, if the people living near a chemical disposal site are subjected to special or intensive examination, it would not be surprising to find an apparent excess prevalence of disease conditions of all sorts as compared to a group that has experienced only superficial or routine examinations. Interview-based information on health status is especially prone to a type of information bias termed recall bias (Bloom, 1981); when the primary source of information is interview data, data on exposure may be influenced by the existence of an adverse outcome and, likewise, data on health effects may be influenced by knowledge of a prior exposure. In a case-comparison study, for example, an exposure may be attributed more often to the cases (whether due to over-reporting or to better or more detailed recall) because publicity has made persons with disease more aware of the potential link between chemical exposures and health effects. Similarly, in a cohort study, an undue number of people in the exposed group may report the occurrence of adverse health effects because of their knowledge that they have been exposed to a potential cause of disease. If interviewers are not blinded to the disease or exposure status of the subjects, they may unintentionally contribute to bias by eliciting information more perseveringly from one group than from another. Another difficulty in the use of interview-based data is that the accuracy of the answers elicited is seriously suspect in a population that has become highly politicized, as often occurs among populations who believe they have been exposed to toxic chemicals (Paigen and Highland, 1982; Warburton, 1981). Unless interview-based data can be verified by another source, they may not be valid for analysis.

Selection of comparison populations

Comparing test populations with control populations is the focus of health assessments. In assessing the health status of human populations, however, epidemiologists do not have the luxury of using a true control population

(Hoover, 1982; Lowrance, 1981). Because of variations in individual lifestyles and the multiplicity of ordinary background exposures, valid reference populations for baseline comparison purposes are difficult to identify.

Often, it is difficult to select comparison individuals who are representative of the group from which the study population came. In order to maximize the comparability of two populations, the comparison population should be as similar in composition as possible in terms of factors such as age, race, sex, socio-economic status, geographical area of residence, and personal and cultural habits (USDHHS/CDC, 1984).

There are special problems in the selection of comparisons in studies of chemical disposal sites stemming from the relationship of exposure to area of residence. Solutions to these problems will vary from one instance to another. Since the number of cases is limited and generally small, it is common to select two or three comparison subjects for each case in order to increase the power of the study. Generally, the small increase in power achieved by including more than three comparison subjects is not worth the additional expense and effort (Cornfield and Haenszel, 1960; Ibrahim and Spitzer, 1979). Occasionally, more than one comparison group may be selected to minimize the possibility of error arising from the unrepresentative nature of a single comparison group. Multiple comparison groups selected by different criteria will provide checks on potential biases, and can increase the likelihood of a convincing association by demonstrating consistency (replicability) in the findings.

Considerations regarding exposures of interest

Most epidemiologic studies deal with exposure circumstances very different from those likely to be created by chemical disposal sites. One such difference is that chemical disposal site exposures are likely to occur at very low concentrations yet may remain persistent in the environment and in body tissues for years or decades. Another difference is that disposal site exposures are likely to involve not a single agent but multiple compounds. A crisis situation in which exposure increased rapidly and resulted in an acute effect of high frequency would, of course, facilitate the identification of the causal chemical species by epidemiologic means. These circumstances, however, are more likely in occupational settings or in the case of accidental spills and are not as likely at chemical disposal sites. The characteristics of disposal site exposures, by contrast, are likely to make difficult the establishment of cause and effect relationships.

Perhaps the greatest difficulty in epidemiological studies at chemical disposal sites is the documentation of actual exposures. In order to define accurately the population-at-risk, it would be most useful to develop exposure estimates which are specific for each individual in the population; in this manner, the critical or most heavily exposed subgroup of the population could be identified. An overestimate of the size of the population-at-risk can result if data are lacking on individual exposures and subgroups are identified based solely on their probable exposure. Individuals with health effects but with minimal or no exposure to the chemical in question might then be grouped together erroneously with truly exposed individuals. This misclassification can severely weaken the power of the study; dilution of the association between exposure and health effect will occur and a causal association may be obscured, resulting in an apparently negative study (Landrigan, 1983; Stein et al., 1981). The difficulty in obtaining actual measurements of individual exposures is apparent in reviewing the studies of the health status of populations at chemical disposal sites detailed in Chapter 8; documentation of actual concentrations and durations of exposure is virtually nonexistent. In most of the studies, attempts at exposure measurements are only estimates based on indirect or surrogate indices such as proximity to the disposal site, or air or drinking water contaminant levels derived from chemical analyses of a single or limited number of samples; from data such as these, actual individual exposures can neither be calculated nor verified.

Considerations regarding health effects of interest

Endpoints

Early in the epidemiological evaluation of a chemical disposal site, it is important to consider the nature of the health effect endpoints that might be sought and enumerated. The earliest effect that could occur is not really a health effect at all but rather an indicator of exposure, which may not necessarily cause a health effect. For example, it might be considered appropriate to measure serum, urine, or adipose tissue concentrations of some substances to learn if people potentially exposed to contaminated media arising from the site have elevated body burdens of such chemicals. The results of these tests may provide a better idea of which members of the population have actually been exposed to certain of the waste chemicals. It would at that point be advantageous to undertake the usually more difficult search for an actual effect on the health of the exposed population.

Alternatively, or in addition to the use of an indicator of exposure, attention may be focused on the asymptomatic stages of disease. Such conditions might include mild hypertension, tissue alterations of a preneoplastic nature, or reduced airway compliance. While the detection of these conditions is useful because of their relatively early appearance in the course of disease, it is very difficult to assess their relationship to any particular exposure because comparable information on non-exposed populations is rarely available. Additionally, the endpoint selected may not be a sensitive or specific indicator of a disease process.

Generally, in order to establish whether or not chemicals arising from a disposal site have caused adverse health effects, it is necessary to establish that exposed persons have an excess of one or more diseases. This may involve the examination of large numbers of people, the review of hospital records, or the assessment of death certificates. Each of these sources may be used in any of several ways and each will have strengths and limitations.

Latent period

A major deterrent to success in attempting to link a disease to a putatively causal exposure is a prolonged induction or latent period between the start of exposure and the manifestation of the disease of interest. A disease does not have an absolutely defined latent period but rather a distribution of durations of latent periods; the average length for a given disease due to a given cause may be modified by a number of factors, including characteristics of host (e.g., age) and exposure (e.g., concentration). Relatively shorter latent periods will enhance the ability to establish a cause and effect relationship.

Those health effects which usually have short latent periods are such things as decline in fertility, congenital abnormalities, and decline in pulmonary function. Malignancies and vascular diseases usually have long latencies.

In many instances, neither the start nor the end of the latent period is known with certainty. In the context of chemical disposal sites, exposure does not necessarily begin when disposal at the site begins. It starts when chemicals escape into the environment under conditions in which human contact with the chemicals may occur; that is, chemicals must be released across the site boundary and be transported through the soil, air, water, or food chain to a situation in which exposure to the environment will result in human absorption. In some cases, months to decades may elapse before human exposure begins. The length of this pre-exposure period will be determined by the nature and packaging (containment) of the waste chemicals, the physical characteristics of the site, the attenuation capacity of the surrounding environment (e.g., microbial breakdown and photodegradation), and the route of uptake by human beings.

Implications of a positive outcome of an epidemiologic study

When the results of an epidemiologic study are positive and appear to support a causal link between an exposure and one or more adverse health effects, a number of issues must be considered before any conclusion is reached. Several of these issues are similar to those described below for a negative study; only those unique to the positive case are discussed here.

Although the study may be positive, it may be difficult or impossible to incriminate a particular agent as a causal factor due to a lack of specificity in two regards: a) epidemiological studies at chemical disposal sites involve the study of effects of exposure to a multiplicity of chemicals rather than exposure to a single substance, leading to problems of confounding (Hoover, 1982); and b) similarly, of the multiplicity of health effects that are known, few are sufficiently specific for exposure to particular chemicals that they can be attributed unequivocally to the exposure and not some other cause (USDHHS/CDC, 1984). In addition, exposures sustained from other sources by the population of interest may produce interactions; any one exposure may appear harmless, but exposure to multiple agents may produce disease.

However, if these difficulties are overcome, investigators, the exposed persons, and the public at large will wish to know the magnitude of the risks involved. They will want to know the extent to which risk is increased as a function of various exposure concentrations - not just those experienced near the chemical disposal site. Indeed, attention is often focused on the magnitude of the risk that may be expected in conjunction with extremely low level exposures. The multiple and complex issues of trying to make such extrapolations are dealt with in Chapter 6. These issues notwithstanding, it is almost certainly better to make such dose-response inferences on the basis of the available human evidence than on experimental laboratory results. Information from animals not only involves the problems encountered when the extrapolation must be made over a wide range of doses, but also involves a speculative presumption that man will respond in the same manner as the experimental test species.

Implications of a negative outcome of an epidemiologic study

The findings of a negative study should be evaluated and, if valid, the study results should contribute to the total base of data used in forming an epidemiological assessment of health. The major factors to consider in evaluating the prospect that a negative study is, in fact, a false negative or a valid but unpersuasive negative include the following:

Sample size - The study may be quite small in the sense that very few people have exhibited a health effect of interest and the number expected to exhibit the health effect (based on a valid comparison group) is also small. Thus, one might see an apparent relative risk of 1.0 (that is, the risks for the study group and comparison group are equal) but the confidence limits may be so wide as to make this value uninterpretable. Table 5-2 demonstrates the relationship between sample size requirements and the ability to detect a doubling of outcome frequency. That is, the relative risk ratio of the incidence rate among exposed to the incidence rate among unexposed is 2.0.

As illustrated, the rarer an outcome in an unexposed population, the larger the sample size needed to detect a doubling of the risk of this outcome among the exposed (Hatch et al., 1981). A doubling of a rare outcome is much more likely to arise by chance than a doubling of a more common outcome. A small sample size does not make a study inherently invalid, only perhaps imprecise; such a study that is otherwise perceived to be valid should not be considered a false-negative but rather a true but unpersuasive negative.

Documentation of exposure - The study may be a true negative but unpersuasive, indeed perhaps not relevant, if exposures are deemed to be inordinately low. This might occur if chemicals had not yet been released and transported from the disposal site. Such a study is not really a false negative; it is only irrelevant to the question of whether the site will ever cause adverse health effects. It does, however, speak in the negative to the question: Has the site yet caused the manifestation of disease?

Latency - If a study is done at a time when any health effects due to exposure are still in a period of latency, the study will appear to be falsely negative. While such a study might not be relevant to questions of the ultimate health effects of the chemical disposal site, it may have some value

TABLE 5–2. Relationship of sample size requirements to detect doubling of outcome frequency (adapted from Hatch et al., 1981).

Sample Sizes* Needed for Exposed and Unexposed Groups	Probability of Outcome among Unexposed Group	Probability of Outcome among Exposed Group
22,403	0.001	0.002
2,243	0.01	0.02
197	0.10	0.20
123	0.15	0.30
65	0.25	0.50
41	0.35	0.70
28	0.45	0.90

*Sample sizes were calculated for $\alpha = 0.05$, two-tailed test, at 80% statistical power.

in that it may indicate, depending on where in the distribution of latent period durations the time scale of the study falls, the magnitude of the frequency of the ultimately occurring outcome.

Dilution - A problem that produces a truly false-negative study is dilution of the exposed population as a result of population mobility. People residing near the chemical disposal site may move away and therefore will not be included in any future studies. In addition, the present study may include recent arrivals who are, effectively, non-exposed, or whose previous exposures under other circumstances may add false-positive results to the study, further complicating the interpretation of the data.

Bias - Finally, it is worth mentioning that inappropriate comparisons and analyses can produce false-negative results.

Negative results of epidemiological studies at chemical disposal sites need to be interpreted cautiously. However, Doll and Peto (1981) suggest that if negative human evidence relates to very prolonged and heavy exposure and is consistent in a variety of studies, such evidence may justify the conclusion that for practical purposes, the exposure need not be treated as the cause of human health effects. In practice, such strong negative evidence is not commonly available, but even less conclusive evidence may "help determine priorities between different courses of action" (Doll and Peto, 1981).

Statistical interpretation in the context of chemical disposal sites

Assessing the probability that an observed value is truly different from some expected value is a most difficult and complex issue. Parametric statistical tests of significance that are frequently applied were developed to assist in the evaluation of experimental data, and, in the strictest sense, are applicable only to experimental situations where the requisite assumptions have some chance of being satisfied. Studies of exposure of people to toxic chemicals cannot, of course, be designed experiments; they are, overwhelmingly, observations of situations that exist entirely outside the control of the research scientist. Therefore, the extent to which they approximate true experiments differs widely from situation to situation. An epidemiologist may hope to alleviate these difficulties by careful selection of the basis for computing an expected risk, by stratification or other means to account for confounding, and appropriate use of non-parametric statistics. Given the best efforts, however, the application of hypothesis-testing statistics is not appropriate in observational studies, and improper interpretation of 'p' values and confidence limits may result. In assessing the results of epidemiological studies conducted in non-controlled situations, tests of statistical significance are not appropriate criteria by which to discriminate between the probability that an observed outcome has occurred by chance or by virtue of a causal association with some other factor. In contrast, however, statistical calculations are very convenient means for simultaneously taking into account the observations and the size of the study groups in which those observations were made; in this application they may be viewed as very valuable descriptive summary statistics.

Anticipatory Epidemiologic Research

A great deal of experience has been gained over the past few years in the treatment of population studies at chemical disposal sites; it is time to give more consideration to moving beyond the restrictions posed by the component involving response to an event in chemical disposal site epidemiology. Consideration needs to be given to the establishment of epidemiologic surveillance of populations chosen for their environmental characteristics. The opportunity exists for epidemiologists to design the necessary and appropriate databases rather than adopt and manipulate existing systems (e.g., death certificates, birth certificates, cancer registries, birth defects registries) which were designed for other purposes; it would be helpful to maintain both disease and exposure registries, especially of populations which have been exposed or will continue to be exposed to compounds of known or suspected toxicity. The decision to establish a follow-up registry must be guided, in part, by ethical issues, and scientific value and feasibility (Heath, 1981). The ethical impact of long-term follow-up registries on the individuals being followed needs to be examined; informed consent is essential as is consideration of the problems arising from the fact that active population contact as a means of follow-up (as opposed to passive follow-up techniques such as relying on linkage with other records systems) will burden each individual being followed with the "sustained awareness...that he or she was exposed to a particular hazard and that it may yet affect his or her health" (Heath, 1981). The scientific value and feasibility of a registry relates to its ability to test hypotheses, in particular, those concerning relationships between exposures and outcomes. "The more precise, answerable, relevant, new, productive, extensive, important in the future, and economically feasible the questions these hypotheses address, the more scientifically valuable or appropriate or both the establishment and the maintenance of such registries become" (USDHHS/CDC/CEH, 1984).

Issues which need development include:

- Standardized questionnaires and required minimal information bases which would allow the combining of populations from different sites for comparison and prospective observation.
- Identification of sentinel environmental health events similar to those known for occupational health outcomes, but including a spatial or temporal clustering factor.
- Site specific census and registry systems to identify exposure early, rather than after the outcome has occurred. For example, consideration of birth registries which would gather information at the time the pregnancy is identified rather than the current retrospective gathering of data after an abnormal event has occurred.
- Identification of referent communities to compare prospectively to chemical disposal site communities.

Regulatory agencies are well along in the development of environmental monitoring programs and will continue to identify reactive circumstances requiring investigation by environmental epidemiologists in response to particular events. However, the maturing of this field is at a point where epidemiologists need to move out of a largely reactive mode and integrate epidemiologically useful monitoring activities into those already supported by regulatory requirements.

SUMMARY

Epidemiology is the discovery and interpretation of disease patterns in human communities. In the investigation of health conditions and their possible relationships to disposal sites, epidemiological studies deal directly with populations experiencing actual or perceived exposure to chemicals.

Epidemiological studies use various study methods and sources of data to examine the health status of populations. Vital statistics, birth and death certificates, registries, and medical records may be used to examine mortality rates and the incidence of various disease states and congenital abnormalities. Studies of populations near chemical disposal sites often include a comparison of the exposed population with one or more populations whose make-up is similar in many characteristics (sex, age, socioeconomic status, race, area of residence, etc.) except that the comparison populations are not located near a chemical disposal site. Such comparisons aid in the assessment of whether any illness that exists in the exposed population is in excess of what might be expected and if such an excess may be associated with exposure to the chemicals from the site.

In evaluating cause and effect relationships between health status and chemical exposures, the following criteria are useful. Confidence that an exposure may be causally related to a disease improves as the strengths of these criteria are increased:

- Consistency of association: observations and study findings can be replicated in diverse circumstances, using different study methods and populations.
- Strength of association: incidence of a disease is much greater in those exposed than in those not exposed, and the incidence increases with increased exposure (dose-response relationship).
- Temporal relationship of association: disease appears after exposure occurs, allowing for the necessary induction or latent period.
- Specificity of association: the extent to which one can predict whether an exposure will result in a specific health effect, or conversely, whether a health condition is the result of a specific exposure.
- Coherence (plausibility) of association: the disease in question might be expected to result from the exposure on the basis of existing knowledge.

Because of gaps in existing knowledge and incomplete data, epidemiological studies of populations at chemical disposal sites are seldom able to determine with certainty, according to the preceding criteria, that an exposure and a health effect are related. Information that is rarely available, yet of great importance, in constructing the etiology of a disease includes data on the concentration, duration, and timing of exposure for each individual in the population who is potentially at risk. In the absence of quantitative data on individual exposures, indirect measurements of risk such as proximity of residence to the site are often used. Indirect data are not as specific or sensitive as direct quantitative measurements of exposure and may contribute to an inaccurate definition of the population-at-risk. Thus, existing relationships between exposure and health effects may not be represented accurately when the study results are interpreted.

There are other limitations that may confound or invalidate the results of epidemiological studies, often resulting from equivocal data and leading to inappropriate interpretation of study results. Such limitations include small population sample sizes, long latent periods of disease, and selection of non-specific disease endpoints. In addition, various sources of bias may be present in the study design. A few of the major types of such biases are analytical bias (e.g., inappropriate statistical methods to compare populations), selection bias (e.g., inappropriate population comparisons), information bias (e.g., inadequate blinding of interviewers to the exposure and health status of study subjects), and confounding bias (e.g., multiple unaccounted-for exposures).

Because these studies often take place in an emotionally and politically charged atmosphere, other biases may exist. Many studies are prompted by community fear of adverse health effects because a chemical disposal site is nearby. People living in the vicinity of a chemical waste disposal site are often quite aware of the potential risks they may face, and are likely to attribute any adverse health effect to chemical exposures from the site. The stress of living in an environment charged with anger or fear may be a factor in illnesses that are attributed by the public to exposure to chemicals from disposal sites; however, this hypothesis has received only limited study. It is exceedingly difficult to make individuals in the public sector aware of the discrepancy between what they perceive the risks to be and what available science can state with relative certainty the risks actually are.

Because of the difficulty in conducting an unbiased and unequivocal epidemiological study in response to the concerns of a community residing near a chemical disposal site, monitoring and surveillance activities are the most logical approaches to ascertaining the baseline health status of these populations. Such studies indicate objectively any perturbations of expected morbidity and mortality rates that may occur, but are exceedingly time consuming and costly if monitoring and surveillance are continued over extended time periods.

References

Abelson, P.H. (1983) Waste management. Science 220:1003.

Bloom, A.D., editor. (1981) Guidelines for studies of human populations exposed to mutagenic and reproductive hazards: proceedings of conference held Jan. 26-27, in Washington, DC. White Plains, NY: March of Dimes Birth Defects Foundation. 163p.

Carnow, B.W.; Conibear, S.A. (1982) Clinical environmental and occupational medicine. Paper presented at the Scientists' Forum: An Assessment of the Scientific Knowledge Base Related to Victim Compensation, January 27-29, Washington, DC. 26p. Available from: Economics and Environmental Studies Center, The Institute on Man and Science, Rensselaerville, NY.

Cornfield, J.; Haenszel, W. (1960) Some aspects of retrospective studies. J. Chronic Dis. 11:523-534.
Doll, R. (1984) Purpose of symposium. Summary of paper presented at a symposium on negative epidemiology, held in 1983 in Oxford, England. 21p. Available from: Imperial Cancer Research Fund, University of Oxford.
Doll, R.; Peto, R. (1981) Causes of cancer: quantitative estimates of avoidable risks of cancer in the United States today. J. Natl. Cancer Inst. 66:1191-1308.
Feinglass, E.J. (1973) Arsenic intoxication from well water in the United States. N. Engl. J. Med. 288:828-830.
Freudenberg, N. (1984) Citizen action for environmental health: report on a survey of community organizations. Am. J. Public Health 74:444-448.
Goldstein, M.S.; Siegel, J.M.; Boyer, R. (1984) Predicting changes in perceived health status. Am. J. Public Health 74:611-614.
Hatch, M.; Kline, J.; Stein, Z. (1981) Power considerations in studies of reproductive effects of vinyl chloride and some structural analogs. Environ. Health Perspect. 41:195-201.
Heath, C.W., Jr., chairman. (1981) Report of HHS/CCERP Subcommittee on Use of Long-term Registries for Health Followup of Populations Exposed to Toxic Hazards. 15p. Available from: Centers for Disease Control, U.S. Department of Health and Human Services, Atlanta, GA.
Hoover, R.N. (1982) Identification and quantification of health hazards in populations: the epidemiologic assessment of risks. Paper presented at the Scientists' Forum: An Assessment of the Scientific Knowledge Base Related to Victim Compensation, January 27-29, Washington, DC. 61p. Available from: Economics and Environmental Studies Center, The Institute on Man and Science, Rensselaerville, NY.
Ibrahim, M.A.; Spitzer, W.O. (1979) Case control study: the problem and the prospect. J. Chron. Dis. 32:139-144.
Johns Hopkins University. School of Hygiene and Public Health, Hazardous Waste Site Study Group. (1981) Final report: design of environmental health effects studies stemming from the potential for human exposure to toxic waste in Memphis, Tennessee. Prepared for the Tennessee Department of Public Health, Memphis. 249p. Available from: The Department, [Memphis, TN].
Kaplan, G.A.; Camacho, T. (1983) Perceived health and mortality: a nine-year follow-up of the human population laboratory cohort. Am. J. Epidemiol. 117:292-304.
Knox, E.G. (1963) Detection of low intensity epidemicity: application to cleft lip and palate. Br. J. Prev. Soc. Med. 17:121-127.
Knox, E.G. (1964) Detection of space-time interactions. Br. J. Prev. Soc. Med. 13:25-30.
Landrigan, P.J. (1983) Epidemiologic approaches to persons with exposures to waste chemicals. Environ. Health Perspect. 48:93-97.
Levine, A. (1983) Psychosocial impact of toxic chemical waste dumps. Environ. Health Perspect. 48:15-17.
Lilienfeld, A.M. (1957) Epidemiological methods and inferences in studies of noninfectious diseases. Public Health Rev. 72:51-60.
Lilienfeld, A.M.; Lilienfeld, D. (1980) Foundations of epidemiology. 2nd ed. New York: Oxford University Press. 375p.
Lowrance, W.W., editor. (1981) Assessment of health effects at chemical disposal sites: proceedings of a symposium held in New York City on June

1-2, by the Life Sciences and Public Policy Program of The Rockefeller University. New York City: The University. 166p.
Mantel, N. (1967) The detection of disease clustering and a generalized regression approach. Cancer Res. 27:209-220.
Martin, R.D. (1984) Critical review of the concept of stress in psychosomatic medicine. Perspect. Biol. Med. 27:443-464.
Mausner, J.S.; Bahn, A.K. (1974) Epidemiology: an introductory text. Philadelphia: W.B. Saunders Company. 377p.
McGough, W.E. (1982) Stress and mental illness: clinical psychiatry for the occupational physician. In: Alderman, M.H.; Hanley, M.J.; eds. Clinical medicine for the occupational physician. New York: Marcel Dekker, Inc. p. 221-251.
Molotch, H.L. (1982) Ethnography of a disaster. Science 216:1401-1402.
New Jersey. Department of Health, Environmental Health Hazard Evaluation Program. (1983) Health survey of the population living near the Price Landfill, Egg Harbor Township, Atlantic County. Prepared in cooperation with Atlantic County Health Department. 29p. Available from: The Program, [Trenton, NJ].
Paigen, B.; Highland, J.H. (1982) Methods for assessing health risks in populations living near hazardous waste sites. In: Highland, J., ed. Hazardous waste disposal: assessing the problem. Ann Arbor, MI: Ann Arbor Science Publishers, Inc./The Butterworth Group [now Stoneham, MA: Butterworth Scientific and Medical Publishers]. p. 147-155.
Roht, L.H.; Selwyn, B.J.; Holguin, A.H., Jr.; Christensen, B.L. (1982) Principles of epidemiology: a self-teaching guide. New York: Academic Press, Inc. 514p.
Rothman, K.J. (1982) Causation and causal inference. In: Schottenfeld, D., Fraumeni, J.F.; eds. Cancer epidemiology and prevention. Philadelphia: W.B. Saunders. p. 15-22.
Stallones, R.A. (1966) Prospective epidemiologic studies of cerebrovascular disease: discussion of methods and applications. In: Cerebrovascular disease epidemiology: a workshop. Public Health Monograph No. 76; Public Health Service Publication No. 1441. Washington, DC: U.S. Government Printing Office. p. 51-55.
Stallones, R.A. (1984) Criteria for assessing the causal significance of associations. Excerpt of paper, "Infectious etiology of ischemic heart disease," presented at the 1984 annual meeting of the Society for Epidemiological Research, held June 13-15, in Houston. 3p. Available from: the author, School of Public Health, University of Texas, Houston, TX.
Stein, Z.; Hatch, M.; Kline, J.; Shrout, P.; Warburton, D. (1981) Epidemiologic considerations in assessing health effects at toxic waste sites. In: Lowrance, W.W., ed. Assessment of health effects at chemical disposal sites: proceedings of a symposium held in New York City on June 1-2, by the Life Sciences and Public Policy Program of The Rockefeller University. New York: The University. p. 125-145.
U.S. Department of Health and Human Services, Centers for Disease Control. (1984) System for prevention, assessment, and control of exposures and health effects from hazardous sites (S.P.A.C.E. for health). Prepared by the Center for Environmental Health and the Center for Professional Development and Training. 187p. Available from: Chronic Diseases Division, Center for Environmental Health, U.S. Centers for Disease Control, Atlanta, GA.

U.S. Department of Health and Human Services, Centers for Disease Control, Center for Environmental Health. (1984) Criteria and methods for establishing and maintaining exposure and outcome registries for environmental public health problems. Prepared for the U.S. Public Health Service, Agency for Toxic Substances and Disease Registry. 123p. Available from: [The Agency, Atlanta, GA].

U.S. Department of Health, Education, and Welfare. Surgeon General's Advisory Committee on Smoking and Health. (1965) Smoking and health. Public Health Service Publication No. 1103. [Washington, DC]: U.S. Department of Health, Education, and Welfare, Public Health Service. 387p. Available from: U.S. Government Printing Office, Washington, DC.

USDHEW SEE U.S. Department of Health, Education, and Welfare

USDHHS/CDC SEE U.S. Department of Health and Human Services, Centers for Disease Control

USDHHS/CDC/CEH SEE U.S. Department of Health and Human Services, Centers for Disease Control, Center for Environmental Health

Warburton, D. (1981) Selection of human reproductive effects for study. In: Lowrance, W.W., ed. Assessment of health effects at chemical disposal sites: proceedings of a symposium held in New York City on June 1-2, by the Life Sciences and Public Policy Program of The Rockefeller University. New York: The University. p. 105-124.

WHO/IARC SEE World Health Organization, International Agency for Research on Cancer

World Health Organization, International Agency for Research on Cancer. (1978) IARC monographs on the evaluation of the carcinogenic risk of chemicals to humans. Vol. 17. Some N-nitroso compounds. Lyon, France: The Agency. 365p.

6

Factors in Risk Assessment

Risk assessment is the evaluation of data from sources such as toxicology and epidemiology to predict the effects of public exposure to environmental factors that pose a potential hazard to the health and well-being of a population. The process of risk assessment includes: identification of the source of the threat, be it a chemical in drinking water, ionizing radiation from a medical examination, or any other potential danger to health, and determination of the potential extent or magnitude of the threat and of the probability that the health effect will ensue from the exposure. This chapter focuses on some significant considerations in the development of risk assessments in this progression from source to exposure to effects and explores factors which contribute to the difficulties in making these assessments in the absence of reported human health effects. It does not address the practices and procedures of risk assessment which have been addressed by the National Research Council (1983), but rather it deals with the collection of data to be used in risk assessments.

FACTORS IN ESTIMATING EXPOSURE AND DOSE

In the context of this report, the word "exposure" relates to contact with chemicals by humans. Another use of the word is in the context of assessing the potential for adverse health effects, in which it is desirable to know how much exposure a population has received and by what route. In the latter context, "exposure" is analogous with "dosage" as applied in laboratory toxicology studies; i.e., exposure is a function of (concentration) x (time). For many hazardous chemicals, not only can the concentration vary with time, but also the contact duration is frequently not known. Furthermore, there are other variables which are often unknown such as absorption, detoxification, excretion and pre-existing body burden.

The potential for exposure of man and other organisms is the basis for concern about the escape and migration of chemicals from disposal sites. The potential for deleterious effects of these exposures depends on the toxicity of the chemicals and the duration and concentration of exposure. In order to assess the potential for human health risk from any particular chemical, data are needed on the likely route of exposure, as well as the duration of the exposure and the concentration of the chemical. Exposure to high concentrations may occur long before the detection of hazardous chemicals and/or the observation of adverse health effects. Both the time for manifestation of health effects and the severity of effects produced by a chemical are dependent upon the amount of chemical entering the body in a given time which, in turn, controls the amount of, and the rate at which the chemical reaches the target organs within the body (Corn and Breysee, 1983). The time interval between the disposal of a chemical in a waste disposal site and the manifestation of adverse health effects can be divided into three phases: 1) the time required for release across the site boundary; 2) the time required for transport through the environment to the site where human exposure occurs; and 3) the latency period for the manifestation of human health effects. For some chemicals the three time phases are brief (minutes to hours), while for other chemicals the time phases may entail years or decades. The time interval from initial release from the disposal site to the ultimate onset of adverse health effects may vary widely as shown by the following two "worst case" scenarios.

In one hypothetical situation, the driver of a tank truck might inadvertently drain acid onto a cyanide salt mixture, which would immediately react to form cyanide gas. In a few minutes the gas would be rapidly released into the air; the driver might then inhale significant quantities and become immediately, and possibly fatally, ill.

Another hypothetical scenario of adverse human health effects from a hazardous waste could involve much longer time intervals. A buried drum containing a hazardous chemical waste could, over a period of years, rust through, releasing the chemical into the ground, where it would then leach across the site boundary and percolate into the groundwater. If the chemical were not readily degradable and groundwater flow rates were low, the chemical could migrate in the groundwater for decades before being pumped out of the ground in drinking water. The latency period might be extensive between human exposure and manifestation of disease, such as cancer or chronic liver or kidney damage.

Human exposure may occur by any or all routes: respiratory, dermal, and oral. Dermal exposure may be by direct contact at the site. Respiratory exposures may result from gases or vapors, solid or liquid aerosols or chemicals bound to solid particles. Contaminated groundwater can lead to exposures at great distances from the site. There may be dermal and respiratory exposures during bathing or showering and oral exposures via drinking water or bioaccumulation in the food chain.

Availability of Exposure Data for Each Route

The Mitre Corporation (1983) recently analyzed the data available for substances released from, and observed in the vicinity of the 546 National Priority List (NPL) sites identified as mandated under CERCLA (Superfund). Concentrations of the substances were not reported, but there are listings of the total number of sites at which individual chemicals have been observed. Of the chemicals or chemical mixtures observed at the NPL sites, 173 were reported in groundwater, 162 in surface water, and 65 in air. The chemicals in the environment at the greatest number of sites were trichloroethylene in groundwater at 110 NPL sites, lead in surface water at 64 NPL sites, and benzene in air at 17 NPL sites.

The releases of hazardous substances to the environment from the NPL sites were estimated to expose a maximum of 5,103,000 people at sites with groundwater releases, up to 4,033,000 people at sites with surface water releases, and as many as 803,000 people within a 4 mile radius of sites with air releases (Mitre Corporation, 1983).

Data from the Mitre analysis are valuable for qualitative, but not quantitative, risk assessment because the report lacks exposure concentration data. In the absence of measured concentration data, a quantitative risk assessment can only be performed by using clearly stated and conservative assumptions, in particular, high concentrations are assumed to develop a "worst-case" risk assessment. With the increased emphasis on risk assessments by the United States Environmental Protection Agency (EPA), "worst-case assessments" may be utilized more frequently leaving the more precise quantitative risk assessment to be conducted when actual data become available.

Fred C. Hart Associates (1982) evaluated data available for 929 sites selected by the following criteria: the disposal sites were active or inactive, had been sampled and chemical analysis data were available; the sites were storage facilities; the sites had been scored by the Mitre Corporation, and there was a likelihood of environmental releases of stored chemicals. Concentration data were available for many chemicals released to the environment at some sites. These data are summarized in Table 6-1 (note that contaminant levels are reported in units of mg/L, while the criteria and standards are in mg/L, ug/L, or ng/L). Concentrations of certain chemicals were very high at certain sites. For example, the maximum concentrations observed at any site were: for trichloroethylene in groundwater, 315 mg/L; for dichloromethane in surface water, 250 mg/L; and for benzene in groundwater and surface water, 80 mg/L and 22 mg/L, respectively. The wide ranges for these values suggest that different volumes of chemicals have been disposed at each site and that conditions resulting in chemical release and environmental transport are different for each chemical disposal site. These site-specific factors make it meaningless to estimate "average exposure concentrations" for the population of the United States.

An assessment of potential for health effects at a specific site must be based on the routes of exposure, concentrations, and durations at that site. Ideally, samplings and analyses would have been conducted at a number of stations around that site for an extended time period to evaluate the spatial and temporal variations in the exposure concentrations for a number of chemicals. A complete exposure assessment for the population around a site would include analyses of ambient air, respirable particles, groundwater, surface water, and locally produced food from appropriate locations. The exposure data would then be integrated on the basis of duration for each exposure to determine the long-term exposure concentrations for each chemical at that site. For those chemicals known to exhibit marked temporal variability, it may be more appropriate to perform an assessment of the maximum exposure concentration.

In practice, these ideal exposure data are seldom available and predictions of exposure have to be made based on the following: 1) potential for disposal site release, transport and fate, based on the physical and chemical properties of the waste chemicals and the meteorologic and hydrogeologic character-

TABLE 6–1. Summary of chemical concentration data at hazardous waste sites.

Most Frequently Observed Chemical Category	Contaminant	Contaminant Concentration Range: Groundwater (mg/1)	Contaminant Concentration Range: Surface Water (mg/1)	Contaminant Concentration Range: Soil (mg/1)	Contaminant Concentration Range: Air (mg/1)	USEPA[a] Human Health Criteria 10^{-6}	USEPA[b] Primary Drinking Water Standards	USEPA[c] Drinking Water Standards
VHO's[d]	1,1,1, trichloroethane	trace-12.0	trace-1.8	trace-1.4	1,062.8	NS*	NS	NS
	trichloroethylene	0.0-315.0	trace-7.50	trace-0.10	ND**	2.7 μg/l	NS	NS
	dichloromethane	0.132-19.0	trace-250	0.034	ND	NS	NS	NS
	tetrachloroethylene	0.0-4.6	0.425	trace-20.5	ND	0.8 μg/l	NS	NS
	chlorobenzene	0.177	0.02	ND	ND	NS	NS	NS
	dibromoethane	ND	ND	1.6	ND	NS	NS	NS
VNHO's[e]	benzene	0.001-80	trace-22.0	trace-43.0	ND	0.66 μg/1	NS	NS
	toluene	0.0009-64.8	trace-6.572	trace-64.0	719	NS	NS	NS
	xylene	0.001-18.8	trace-1.70	2.0-3.8	ND	NS	NS	NS
Metals	lead	0.001-810	trace-175	trace-5,750	ND	NS	0.05 mg/l	NS
	manganese	0.012-120	trace-8,900	ND	ND	NS	NS	0.05 mg/l
	chromium	trace-65.5	0.001-10	0.110-31,765	ND	NS	0.05 mg/l	NS
	cadmium	0.02-0.1	0.1-1.0	4.1	ND	NS	NS	NS
	nickel	0.029-5.50	ND	0.076-0.49	ND	NS	NS	NS
	arsenic	0.001-2.4	0.0-16	trace-510	ND	2.2 ng/l	0.05 mg/l	NS
	mercury	ND	0.0-0.001	0.0-734	ND	NS	0.002 mg/l	NS
Acid Compounds	pentachlorophenol	0.0003	ND	0.9-46	ND	NS	NS	NS
	phenol	0.0001-2.3	0.011-6.920	trace-22.5	ND	NS	NS	NS

*No standard

**No data available

[a] U.S. Environmental Protection Agency, Nov. 28, 1980. Water Quality Criteria Documents; Availability. Federal Register, Vol. 45, No. 231.

[b] U.S. Environmental Protection Agency, Dec. 24, 1975. National Interim Primary Drinking Water Regulations. 40 CFR 141; 40 FR 59565, as amended.

[c] U.S. Environmental Protection Agency, July 19, 1979. National Secondary Drinking Water Regulations. 40 CFR 143; 44 FR 42198. Effective Jan. 19, 1981.

[d] Volatile halogenated organics

[e] Volatile non-halogenated organics

Source: Fred C. Hart Associates (1982)

istics at the site; 2) single or multiple grab samples for the exposure route, e.g., collections from a drinking water well and an extrapolation over time; 3) measurements of actual body burdens of the suspect chemicals in the exposed human population. The limitations in the quantitative data on exposure concentration indicate that, at best, only a qualitative assessment of risk will be possible.

The two models used to estimate releases from waste disposal sites which will be discussed are those of Versar (1982) and Donigian et al. (1983). Versar described the following model.

Stage I in the decision process determines the nature and volume of likely wastes from the local sources (including production, commercial processing and consumer use) for each chemical on a mass/year basis. Stage II characterizes the mass/year of each type of waste residue released as well as the total volume and the physical and chemical nature of each waste stream. Stage III determines the types of disposal methods and estimates the percentage of waste entering each disposal method. Stage IV allocates the waste stream to specific waste disposal sites (either on-site or off-site based on the capacity and current operating characteristics of each candidate disposal site). Stage V estimates the environmental releases from the disposal sites based on the design and operating characteristics of each site; identifies the methods for the prediction of releases using site-specific data or appropriate generic data; estimates the volumes and concentrations released from each disposal site or "typical" site; and, extrapolates these release data to obtain regional or national estimates. Monitoring data are collected to verify the predicted releases and to adjust the estimated releases/concentrations. The data utilized in Stages I and II would be obtained from a materials balance for each of the chemicals involved. Versar reported that methods for estimating releases were not available for most types of disposal methods; thus, generic parameter values would have to be used in the absence of chemical- or site-specific data, with an unknown reliability for any estimates based on the assumed generic parameters.

The Donigian manual (Donigian et al., 1983) provides a methodology for rapidly estimating groundwater contamination and changes in concentration with time and distance from a site. The procedures include evaluation of critical contaminants and site characteristics as input to an assessment methodology utilizing a graphical technique (i.e., nomogram) to analyze the fate and movement of chemicals through the unsaturated and saturated soil zones. Guidelines for evaluating critical waste and site characteristics were provided to allow an estimation of the needed nomogram parameters. The procedures are designed for use under emergency response conditions to allow the assessment of potential groundwater contamination to be completed in less than 24 hours. These time constraints would preclude extensive field sampling, laboratory analyses, data search and collection, and sophisticated computer analyses. Sampling and analysis may be initiated, but the results would not be available for verification of the estimated first-cut, order-of-magnitude modeling results. The user must have background and training in hydrogeology, soil science and chemistry as well as the ability to adequately characterize, within the 24-hour time frame, the subsurface media (e.g., heterogeneities, depth to groundwater, soil/aquifer properties, aquifer thickness) through which the contaminants must move.

Hazardous Chemical Exposures from Other Sources

Chemical disposal sites are not the only sources of anthropogenic (i.e., resulting from human activities) environmental releases of hazardous chemicals. Releases may also occur as a result of losses during industrial production, consumer use of products, industrial and/or municipal wastewater discharges, septic tank leachfield effluents, agricultural practices, mine tailings, petroleum exploration wastes, road runoff, spills, and emissions of vapors, gases or mists. For purposes of assessing the potential for adverse health effects resulting from chemical releases from disposal sites, concentrations resulting from sources other than disposal sites must also be taken into consideration. For example, the population around Triana, Alabama, carries a high body burden of DDT and its metabolites as a result of surface water contamination by production wastes from 1947 to 1970 (Kreiss et al., 1981).

Sampling and analysis of volatile organic contaminants in public finished drinking water were conducted by EPA on 945 underground water supply systems throughout the country (Westrick et al., 1983). Half of these were randomly selected by EPA; the other half were nominated by individual states based on indications of potential contamination by nearby industrial facilities, landfills, or other sources of contamination. Tables 6-2a and b summarize the data for the randomly selected wells and Tables 6-3a and b summarize the data for state-nominated wells. Because of the known potential for contamination in the non-random data set (Tables 6-3a and b), the positive data must be interpreted as reflecting a compendium of contaminants from chemical disposal sites and other sources. However, the data for many randomly selected wells (Tables 6-2a and b) may also have been affected by anthropogenic contamination because these wells also contained measurable quantities of certain chemicals. For example, it was surprising that 1,1-dichloroethylene (vinylidene chloride) was detected in a total of 9 of the 466 randomly selected wells because these were not expected to have been contaminated. The results of this study show that the variability in water quality between different groundwater supplies is site-specific, that volatile organic chemicals cannot be quantitatively characterized on a national scale and that the risks for each site must be individually evaluated.

Westrick et al. (1983) also reported that when single wells were resampled at a later date, the pattern of contamination was nearly identical to the original pattern. These observations reinforce the belief that groundwater contamination levels usually change slowly, and consequently once a well is contaminated it may be difficult and very time consuming to restore it to its natural purity.

ECOLOGIC INDICATORS OF EXPOSURE

This report focuses on the potential for adverse human health effects from exposures to hazardous chemicals released from uncontrolled disposal sites. Although a detailed discussion of environmental impacts and ecological consequences of such releases is beyond its scope, this section describes the application of ecological data from the vicinity of the disposal site to the assessment of human health risk.

Plants and animals are used as the laboratory models to obtain data on the relative toxicity and types of effects caused by a particular chemical or

TABLE 6–2a. Summary of occurrence data. Ground Water Supply Survey, December 1980-December 1981. (Supplies serving less than 10,000 persons, random sample of 280 sample sites.)

Parameter	Quantitation Limit*	Occurrences No.	%	Median of Positives	Maximum Value
vinyl chloride	1.0	0	0	–	–
1,1-dichloroethylene	0.2	4	1.4	1.2	6.3
1,1-dichloroethane	0.2	10	3.6	0.51	3.2
cis and/or trans-1,2-dichloroethylene	0.2	3	1.1	0.23	1.7
1,2-dichloroethane	0.5	0	0	–	–
1,1,1-trichloroethane	0.2	12	4.3	0.32	18
carbon tetrachloride	0.2	5	1.8	0.37	16
1,2-dichloropropane	0.2	1	0.4	0.75	0.75
trichloroethylene	0.2	9	3.2	0.88	40
tetrachloroethylene	0.2	13	4.6	0.35	23
benzene	0.5	1	0.4	0.61	0.61
toluene	0.5	4	1.4	0.62	0.85
ethylbenzene	0.5	2	0.7	0.94	1.1
bromobenzene	0.5	3	1.1	1.9	5.8
m-xylene	0.2	6	2.1	0.32	1.5
ortho and para-xylene	0.2	6	2.1	0.34	0.59
p-dichlorobenzene	0.5	2	0.7	0.60	0.68
1,1,2-trichloroethane	0.5	0	0	–	–
1,1,1,2-tetrachloroethane	0.2	0	0	–	–
1,1,2,2-tetrachloroethane	0.5	0	0	–	–
chlorobenzene	0.5	0	0	–	–
1,2-dibromo-3-chloropropane	5.0	1	0.4	5.5	5.5
n-propylbenzene	0.5	0	0	–	–
o-chlorotoluene	0.5	0	0	–	–
p-chlorotoluene	0.5	0	0	–	–
m-dichlorobenzene	0.5	0	0	–	–
o-dichlorobenzene	0.5	0	0	–	–
styrene	0.5	0	0	–	–
isopropylbenzene	0.5	0	0	–	–
chloroform	0.2	104	37.1	1.4	140
bromodichloromethane	0.2	100	35.7	1.4	60
dibromochloromethane	0.5	87	31.1	2.1	52
dichloroiodomethane	1.0	2	0.7	2.8	4.1
bromoform	1.0	44	15.7	2.4	54

*All concentrations are in $\mu g/l$

Source: Westrick et al. (1983)

TABLE 6–2b. Summary of occurrence data. Ground Water Supply Survey, December 1980-December 1981. (Supplies serving more than 10,000 persons, random sample of 186 sample sites.)

Parameter	Quantitation Limit*	Occurrences No.	%	Median of Positives	Maximum Value
vinyl chloride	1.0	1	0.5	1.1	1.1
1,1-dichloroethylene	0.2	5	2.7	0.28	2.2
1,1-dichloroethane	0.2	8	4.3	0.54	1.2
cis and/or trans-1,2-dichloroethylene	0.2	13	7.0	1.1	2.0
1,2-dichloroethane	0.5	3	1.6	0.57	0.95
1,1,1-trichloroethane	0.2	15	8.1	1.0	3.1
carbon tetrachloride	0.2	10	5.4	0.32	2.8
1,2-dichloropropane	0.2	5	2.7	0.96	21
trichloroethylene	0.2	21	11.3	1.0	78
tetrachloroethylene	0.2	21	11.3	0.52	5.9
benzene	0.5	2	1.1	9.0	15
toluene	0.5	2	1.1	2.6	2.9
ethylbenzene	0.5	1	0.5	0.74	0.74
bromobenzene	0.5	1	0.5	1.7	1.7
m-xylene	0.2	2	1.1	0.46	0.61
ortho and para-xylene	0.2	2	1.1	0.59	0.91
p-dichlorobenzene	0.5	3	1.6	0.66	1.3
1,1,2-trichloroethane	0.5	0	0	–	–
1,1,1,2-tetrachloroethane	0.2	0	0	–	–
1,1,2,2-tetrachloroethane	0.5	0	0	–	–
chlorobenzene	0.5	0	0	–	–
1,2-dibromo-3-chloropropane	5.0	0	0	–	–
n-propylbenzene	0.5	0	0	–	–
o-chlorotoluene	0.5	0	0	–	–
p-chlorotoluene	0.5	0	0	–	–
m-dichlorobenzene	0.5	0	0	–	–
o-dichlorobenzene	0.5	0	0	–	–
styrene	0.5	0	0	–	–
isopropylbenzene	0.5	0	0	–	–
chloroform	0.2	106	57.0	1.6	300
bromodichloromethane	0.2	101	54.3	1.6	71
dibromochloromethane	0.5	96	51.6	2.9	59
dichloroiodomethane	1.0	3	1.6	1.8	4.1
bromoform	1.0	57	30.6	3.8	50

*All concentrations are in μg/l

Source: Westrick et al. (1983)

TABLE 6–3a. Summary of occurrence data. Ground Water Supply Survey, December 1980-December 1981. (Supplies serving less than 10,000 persons, random sample of 321 sample sites.)

Parameter	Quantitation Limit*	Occurrences No.	%	Median of Positives	Maximum Value
vinyl chloride	1.0	0	0	–	–
1,1-dichloroethylene	0.2	5	1.6	0.35	3.0
1,1-dichloroethane	0.2	6	1.9	0.62	1.2
cis and/or trans-1,2-dichloroethylene	0.2	11	3.4	1.3	17
1,2-dichloroethane	0.5	3	0.9	2.9	3.4
1,1,1-trichloroethane	0.2	25	7.8	1.2	8.2
carbon tetrachloride	0.2	9	2.8	0.44	15
1,2-dichloropropane	0.2	3	0.9	1.2	1.4
trichloroethylene	0.2	23	7.2	1.2	29
tetrachloroethylene	0.2	27	8.4	0.79	21
benzene	0.5	5	1.6	1.6	12
toluene	0.5	4	1.2	0.67	0.79
ethylbenzene	0.5	3	0.9	0.87	0.95
bromobenzene	0.5	2	0.6	0.97	1.2
m-xylene	0.2	8	2.5	0.38	0.83
ortho and para-xylene	0.2	10	3.1	0.44	2.5
p-dichlorobenzene	0.5	4	1.2	0.74	0.90
1,1,2-trichloroethane	0.5	0	0	–	–
1,1,1,2-tetrachloroethane	0.2	0	0	–	–
1,1,2,2-tetrachloroethane	0.5	0	0	–	–
chlorobenzene	0.5	1	0.3	2.7	2.7
1,2-dibromo-3-chloropropane	5.0	0	0	–	–
n-propylbenzene	0.5	1	0.3	0.98	0.98
o-chlorotoluene	0.5	0	0	–	–
p-chlorotoluene	0.5	0	0	–	–
m-dichlorobenzene	0.5	0	0	–	–
o-dichlorobenzene	0.5	1	0.3	2.2	2.2
styrene	0.5	0	0	–	–
isopropylbenzene	0.5	0	0	–	–
chloroform	0.2	155	48.3	1.6	100
bromodichloromethane	0.2	144	44.9	2.0	49
dibromochloromethane	0.5	135	42.1	3.5	63
dichloroiodomethane	1.0	5	1.6	1.4	4.2
bromoform	1.0	88	27.4	3.7	100

*All concentrations are in $\mu g/l$

Source: Westrick et al. (1983)

TABLE 6–3b. Summary of occurrence data. Ground Water Supply Survey, December 1980-December 1981. (Supplies serving more than 10,000 persons, random sample of 158 sample sites.)

Parameter	Quantitation Limit*	Occurrences No.	%	Median of Positives	Maximum Value
vinyl chloride	1.0	6	3.8	2.7	8.4
1,1-dichloroethylene	0.2	10	6.3	0.34	0.64
1,1-dichloroethane	0.2	17	10.8	0.87	4.2
cis and/or trans-1,2-dichloroethylene	0.2	27	17.1	2.7	120
1,2-dichloroethane	0.5	4	2.5	1.8	9.8
1,1,1-trichloroethane	0.2	26	16.5	0.93	21
carbon tetrachloride	0.2	6	3.8	0.70	9.4
1,2-dichloropropane	0.2	4	2.5	0.70	18
trichloroethylene	0.2	38	24.1	1.5	130
tetrachloroethylene	0.2	18	11.4	0.66	69
benzene	0.5	3	1.9	2.7	12
toluene	0.5	1	0.6	1.5	1.5
ethylbenzene	0.5	0	0	–	–
bromobenzene	0.5	0	0	–	–
m-xylene	0.2	0	0	–	–
ortho and para-xylene	0.2	0	0	–	–
p-dichlorobenzene	0.5	0	0	–	–
1,1,2-trichloroethane	0.5	0	0	–	–
1,1,1,2-tetrachloroethane	0.2	0	0	–	–
1,1,2,2-tetrachloroethane	0.5	0	0	–	–
chlorobenzene	0.5	0	0	–	–
1,2-dibromo-3-chloropropane	5.0	0	0	–	–
n-propylbenzene	0.5	0	0	–	–
o-chlorotoluene	0.5	1	0.6	2.4	2.4
p-chlorotoluene	0.5	0	0	–	
m-dichlorobenzene	0.5	0	0	–	–
o-dichlorobenzene	0.5	1	0.6	2.7	2.7
styrene	0.5	0	0	–	–
isopropylbenzene	0.5	0	0	–	–
chloroform	0.2	100	63.3	2.1	430
bromodichloromethane	0.2	100	63.3	2.2	110
dibromochloromethane	0.5	87	55.1	4.6	51
dichloroiodomethane	1.0	8	5.1	1.2	4.1
bromoform	1.0	60	38.0	5.1	68

*All concentrations are in μg/l

Source: Westrick et al. (1983)

mixture via a specific exposure route. These organisms may also be used as sentinel organisms in the natural environment to serve as "early-warning" indicators to monitor the hazardous materials that may be released from chemical disposal sites, migrate through the environment, and enter into the food chain.

Organisms in the environment can be monitored in a number of ways to assess the presence of toxicants. Individual organisms can be sampled and examined for gross pathology or chemically analyzed to determine contaminant levels in various organs or tissues. Particular species can be selectively sampled, and the collected organisms then evaluated for discrete parameters such as age, body weight, fecundity, and pathology. Aquatic or terrestrial organisms can be sampled by various collecting methods and each taxonomic group identified and enumerated, e.g., the species of algae or the genera of bottom-dwelling invertebrates from a water body. It is a general ecological principle that the more stable an ecological community the greater the numbers of different types of organisms present; conversely, an unstable or stressed ecological community has fewer different types of organisms. Usually only hardy species will survive the exposure to hazardous chemicals, so there will be fewer species but greater numbers of organisms in those species in a stressed ecosystem (Odum, 1959).

The Centers for Disease Control (USDHHS/CDC, 1984) formulated guidelines for environmental reconnaissance of the area around a waste disposal site in order to assess the potential for human exposure to chemicals. The field survey should include detailed observations regarding: the ecosystem in the vicinity of the site; the appearance of the ground and of nearby water bodies (e.g., color, turbidity, unidentified substances); the appearance of surrounding vegetation (e.g., abundance, growth, degree of vigor); and signs of the presence of animals, birds, or insects (e.g., abundance, indications of health or disease state, animal tracks, droppings).

Few studies are available concerning plants and animals as ecological indicators of the release and transport of chemicals from waste disposal sites. Rowley et al. (1983) performed a trapping survey to collect a small mammal, the vole (Microtus pennsylvanicus), in the vicinity of the waste disposal site at Love Canal in Niagara Falls, New York. Samples were collected in the immediate area (I), in an area very close to the disposal site (II), and in a reference area (III) about 1 km from Love Canal. The population densities were lowest at I, highest at III, and intermediate at II; age analyses (based on dry weights of eye lenses) indicated that mean life expectancy was shortest at I (23.6 days), intermediate at II (29.2 days), and longest at III (48.8 days). Survivorship data indicated that there were higher mortality rates at I and II than at III. Adipose tissue samples were pooled from voles collected at sites I and II and were found to contain unreported amounts of chlorinated hydrocarbons which were not present in the pooled samples from site III specimens. Liver and adrenal weights in females and seminal vesicle weights in males were significantly reduced in the voles collected from I compared with those from III. These results suggest that organisms living at, and in the vicinity of, waste disposal sites may be useful indicators of environmental exposure and might have some utility as ecological indicators of chemical migration from waste disposal sites.

Metcalfe et al. (1984) reported that freshwater leeches from an industrially polluted creek contained very high tissue concentrations of chlorophenols compared with most other benthic (bottom-dwelling)invertebrates, fish, and tadpoles. The principal source of chlorophenols entering this stream was a chemical waste disposal site located about 100 meters from the stream. Although the production and disposal of chlorophenol wastes was terminated in about 1969, leeches collected in 1981 at one sampling location had accumulated chlorophenols to levels 140,000 times greater than the average water concentrations. Leeches are annelids which are primitive organisms and may be incapable of metabolizing and/or excreting these compounds.

The development of tumors in fish and other aquatic organisms is another promising method of detecting the release, transport, and environmental contamination by substances from chemical disposal sites (Harshbarger, 1983). The Registry of Tumors in Lower Animals was founded at the Smithsonian Institution in 1966 to provide a diagnostic center, specimen depository, and clearinghouse for information on the study of tumorigenesis in invertebrate and cold-blooded vertebrate animals (Harshbarger, 1977). This ongoing program is funded primarily by the National Cancer Institute. Over the past 20 years, data on tumor sites, types of tumors, and related pathological signs have been collected from numerous locations throughout the United States. Tumor incidence has been found to be an effective indicator of the presence of carcinogens and can be compared with analyses of biochemical, physiological, genetic, or ecological parameters, if these observations are available. In the absence of other measures, fish tumorigenesis by itself has utility as a potential indicator of the presence of carcinogens (USDHHS/NCI, 1984).

Various studies have reported carcinogenesis in aquatic organisms, including the following: three types of neoplastic skin lesions in tiger salamanders (Ambystoma tigrinum) inhabiting a secondary sewage sedimentation lagoon at Reese Air Force Base, Lubbock County, Texas (Rose and Harshbarger, 1977); liver, lip, and dermal neoplasms and the presence of polynuclear aromatic hydrocarbons (PAH's) in brown bullhead (Ictalurus nebulosus) inhabiting the Black River near its junction with Lake Erie in Lorain, Ohio, an area bordered with heavy industry (Baumann et al., 1982); hepatocellular carcinomas and other neoplasms in saugers (Stizostedion canadense) and walleyes (S. vitreum) in Torch Lake, Michigan, which is heavily polluted with copper as well as various other metals from copper mining tailings and may contain residual organic wastes (Black et al., 1982); liver cancers in English sole (Parophrys vetulus) in the Duwamish River estuary and other locations of Puget Sound, Washington, the hepatomas being positively correlated with elevated concentrations of aromatic hydrocarbons and metals found in the sediments (Pierce et al., 1978; Malins et al., 1983);and, liver and skin neoplasms in brown bullhead (I. nebulosus) in the Buffalo River in New York, adjacent to steel, chemical, and dye facilities (Black, 1983).

The skin neoplasms reported from these surveys were frequently located near the mouth; this anatomic location suggests that pollutants may be present in the contaminated sediments and fish may become exposed during feeding (Black, 1983). To test this hypothesis, an extract of Buffalo River sediment was prepared by organic solvent extraction and this material was painted on the skin of a colony of laboratory bullheads weekly for 2 years. Skin changes were noted that gradually progressed to neoplasia. The extracts were also painted on the skin of Swiss mice and skin cancers developed.

The extracted sediments were also administered to bullheads in feeding studies (Black, 1983). Liver changes including early stage neoplasms developed after only 4 months and liver tumors similar to those in the Buffalo River field populations had developed in only 8 months. Thus, the tumor site itself was found to be an indicator of the biotic response. Liver tumors are known to be closely associated with metabolism of carcinogenic substances (Bailey et al., 1984), and the studies of Black (1983) suggest lip neoplasms in bottom-feeding fish are closely associated with direct exposure to the contaminants.

These examples of prevalent tumors in aquatic organisms from polluted sites suggest that this type of tumor data from surveys or actual controlled exposures provide a method of determining chemical releases and potential human exposures. Animal tumor data obtained from indicator species at potentially impacted sites should be compared with data from surveys of control populations, and the results analyzed statistically to determine whether the increased tumor incidence is statistically significant. Moreover, there are gross and cytologic similarities between certain non-neoplastic conditions and neoplasms in fish and other lower animals such as amphibians and reptiles (Harshbarger, 1984). These similarities may result in misinterpretations and misdiagnoses of causation; thus, caution must be used in attributing a cause-and-effect relationship between fish pathology and the presence nearby of a chemical disposal site or other source of contaminants.

In the near future, standardized fish carcinogenesis bioassays may be available which will be inexpensive, reliable, and sensitive. These laboratory bioassays will involve exposing test fish to extracts of sediments or polluted water, which would likely contain mixtures of compounds, or exposing them to pure compounds or known combinations of pure compounds. The results from these bioassays could be used to corroborate observations from surveys of feral species examined by histopathology. At present, it must be remembered that effects detected in surveys of indicator organisms may or may not be relevant to human effects; however, if tumor incidence rates in plant or animal indicator species are in excess of background rates in control populations, these observations may be evidence of the presence of contaminants. Fish may be a unique type of indicator organism because the naturally occurring incidence of tumors is quite low (Bailey et al., 1984; Grizzle et al., 1984). Harshbarger (1977) reported that nearly every type of fish neoplasm found recently had been known prior to 1940, before the increased production of synthetic organic compounds; however, most early reports were isolated cases rather than epizootics. For this reason, Harshbarger (1977) suggested that the frequency of fish neoplasms was increasing substantially in recent years. Thus, in those situations where tumorigenesis is established, prudence would dictate that the source of the contaminants should be determined and the specific chemicals should be identified.

If such chemicals are known carcinogens, emphasis on risk management rather than risk assessment would be the most logical approach. If a mixture of substances is evident and causally related to the increased incidence rates of tumors in indicator species, reasoned judgment suggests that potential risks to other biota, including man, are a strong possibility and should be dealt with appropriately.

POTENTIAL EFFECTS ON TARGET ORGAN SYSTEMS

Observation of health effects in sentinel animals and plants in the immediate vicinity of chemical disposal sites can give some of the earliest clues to the degree of environmental contamination originating from the site. If the site has existed for a long period of time, there may have been ample opportunity for the chemicals to have produced chronic effects or diseases with long latencies. Observations of plants and animals provide some information about the nature of the contaminants and their potential for widespread pollution and toxic effects in the environment. Since these plant and animal effects may be the first indication of problems, investigators may be able to obtain information on the types of contaminants and the nature of their adverse effects. This information, together with information on the chemicals in the site and chemical analyses of the surrounding environmental milieu, forms a powerful risk-assessment tool for the persons attempting to determine what effects could be expected or what target organ systems might be affected. As a basis for further studies, a general list of the target organs and tissues most likely to be affected was obtained from the literature (NRC, 1984), and revised as appropriate for application to chemical disposal site situations. Table 6-4 is a compilation of the most likely routes of chemical exposure associated with specific organ systems, target tissues, and health effects. It is emphasized that these health effects may be induced by a variety of environmental chemical exposures, not just those at chemical disposal sites.

A consideration in organizing Table 6-4 was identifying those organ systems which may most likely be affected by chemical disposal site exposures. Duration and concentration of exposure are important factors in assessing the likelihood that a potential health effect will occur and be detected. Short-term, low-dose exposures may not induce any identifiable health effects, whereas chronic exposures under similar circumstances and short-term high-dose exposures may induce a wide range of effects. Population exposures to disposal site chemicals are usually of much lower concentration than those observed in work environments or accidental environmental exposures. Therefore, it is likely that some effects associated with environmental exposures will not be detected. The reader will recognize that none of these effects are pathognomic for disposal site exposures, thus, complicating the efforts to derive cause and effect evaluations.

The above considerations resulted in a tentative ranking of these organ systems as shown in Table 6-4, with skin, central nervous system (CNS), liver, and reproductive system being the most probable targets. Skin and CNS conditions were judged to be the most likely effects from direct contact with chemicals, whereas hepatic, hematopoietic, renal, reproductive, and CNS effects seemed to be the most likely indicators of chronic low-dose exposure from ingestion. Both CNS and reproductive disorders appeared likely to occur as a result of a number of exposure circumstances, both high- and low-dose, and also from a variety of chemical agents. Lung, gastrointestinal, and cardiovascular disease, although important in occupational exposures, were considered unlikely in the low-dose exposure situations around a chemical disposal site. Although procedures are becoming available for evaluation of alterations in immune response, the immune system was omitted from the table because it was believed that such procedures are not yet in standard practice in the same degree of precision as, for example, hepatic or renal profiles. Similarly, studies of cytogenetic abnormalities of peripheral blood cells were also considered to be still in the developmental phase.

TABLE 6–4. Target organs and potential chemically induced effects in humans.

Target Organ or Tissue	Potential Effect	Probable Routes of Exposure
Skin	Pigment disturbances	ingestion; direct contact
	Contact dermatitis, acute and chronic	direct contact
	Sensitization reactions and urticaria	direct contact
	Folliculitis and acneform dermatoses	ingestion; direct contact
	Miscellaneous reactions	direct contact
	Neoplasia	ingestion; direct contact
Central Nervous System	Behavioral changes	inhalation; ingestion
	Central nervous system depression	inhalation; ingestion
	Peripheral neuropathy	inhalation; ingestion
	Other neurological conditions	inhalation; ingestion
	Neoplasia	inhalation; ingestion
Liver	Acute hepatocellular injury	inhalation; ingestion
	Cholestasis	inhalation; ingestion
	Veno-occlusive disease	inhalation; ingestion
	Subacute necrosis and cirrhosis	inhalation; ingestion
	Cirrhosis	inhalation; ingestion
	Neoplasia	inhalation; ingestion
Reproductive System and Fetus	Germ cell mutation	inhalation; ingestion
	Infertility	inhalation; ingestion
	Embryotoxicity	inhalation; ingestion
	Teratogenesis	inhalation; ingestion
	Spontaneous abortion	inhalation; ingestion
	Congenital malformation	inhalation; ingestion
	Neonatal death	inhalation; ingestion
Hematopoietic and Lymphatic System	Anemia	inhalation; ingestion
	Bone marrow depression	inhalation; ingestion
	Alteration of immune response	inhalation; ingestion; dir. cont.
	Aplastic anemia	inhalation; ingestion
	Leukemia and lymphoma	inhalation; ingestion
Genitourinary System	Acute renal failure	inhalation; ingestion
	Chronic renal disease	inhalation; ingestion
	Neoplasia	inhalation; ingestion
Lung and Respiratory Tract	Asphyxiation	inhalation
	Irritation	inhalation
	Sensitization	inhalation
	Pneumoconiosis	inhalation
	Fibrosis	inhalation
	Edema	inhalation
	Granulomatous disease	inhalation
	Neoplasia	inhalation; ingestion
Gastrointestinal Tract	Esophagitis	inhalation; ingestion
	Ulcer	inhalation; ingestion
	Gastritis, acute or chronic	inhalation; ingestion
	Regional enteritis	inhalation; ingestion
	Colitis	inhalation; ingestion
	Neoplasia	inhalation; ingestion
Cardiovascular System	Hypertension	inhalation; ingestion
	Heart disease	inhalation; ingestion
	Stroke	inhalation; ingestion

FACTORS IN INTERPRETING EXPERIMENTAL MODELS OF TARGET ORGAN EFFECTS

Efforts to predict accurately the human health effects of disposal site chemicals have been prompted by individual concerns for personal health and quality of the environment and by governmental response to public reaction and/or political pressure. Regardless of the motivation, the prediction of health effects in waste site environs is an extremely complex and inexact process.

The first stage of this process is identification of the chemicals of concern and quantitation of these chemicals in the exposure media. The fact that hazardous chemical waste is most frequently an ill-defined mixture of compounds further complicates the definitive assessment of exposure. In those circumstances where the chemical species and the magnitude and duration of human exposure is analogous to occupational or accidental environmental exposures which have already been investigated, predictions of human health effects can be made with some assurance. These latter exposures are usually closely monitored and well-documented and provide the most directly applicable and useful information. Unfortunately, these preferred sources of data are seldom available as comparisons for disposal site-related exposures because the chemical mixtures and the route and duration of exposure cannot adequately be matched. In the absence of human exposure data, research scientists at present must resort to experimental surrogates in laboratory or controlled field studies. Not unexpectedly, the vast majority of health effects data is derived from these sources. This section of the report deals with the uncertainties of extrapolating meaningful information from studies performed in other organisms to estimate the potential responses of human beings.

Successful extrapolation of laboratory animal response to predicted human experience must reconcile four major multifactorial sources of variability: 1) experimental design considerations, 2) environmental factors, 3) dose delivery and scaling, and 4) interspecies differences in morphology, biochemistry and behavior.

The problems of interpreting experimental model response with respect to these sources of variability have been reviewed by Rall (1969), Casarett, (1980), Grice et al. (1984), and Calabrese (1983). Although many of the variables can be controlled through careful study design and good laboratory practice, others are unavoidable. The following discussion will emphasize factors which may have a significant influence on the interpretation of animal data from experiments that are otherwise well designed and appropriately monitored. It will focus on those aspects which arise primarily from the use of other animal species as models for human effects.

Experimental Design Considerations

The goal of effective experimental design is to eliminate or quantify variability between experimental and control groups so that changes in health status following exposure may be ascribed with maximum assurance to the exposure. When the investigator successfully controls randomization of animals, environmental conditions, and laboratory and animal care practices, the experimental variability can be minimized (USEPA, 1979). There are, however, some experimental design considerations which, in the process of reducing error or variability, increase the difficulty of interpreting the results of an experiment with respect to human health effects.

One of the earliest experimental design considerations is the selection of the form and purity of material to be tested. Environmental exposures are usually of a mixture of chemicals which may interact with biological systems to produce synergism, potentiation, or antagonism of a particular effect. The decision to use a pure compound in laboratory experiments may exclude the occurrence of these interactions. Although the choice of a pure compound greatly simplifies laboratory exposure and analysis procedures, competition among mixture components for molecular binding sites on plasma proteins, enzymes and subcellular components cannot be assessed. The absence of information on these interactions among mixture components may lead to misinterpretation of those response data which are a function of unbound plasma concentration or the bioavailability of a single compound at a target cell or enzyme. In cases in which exposure to a single compound results in responses different from those exhibited in humans exposed to mixtures containing the compound, much utility of the model is lost.

Differences in response to a single compound in repeated experiments could occur if different structural or stereo-isomers of the compound are tested. For example, if the configuration of an isomer alters its acid dissociation constant (Ka), the rate of absorption from the gastrointestinal (GI) tract becomes a function of the local pH. Since the proportion of un-ionized (more readily absorbed) compound varies with pH at a given location, isomers would be absorbed in different quantities depending on transit time at the ideal pH and duration and intensity of metabolic activity of local enzymes on the compound administered. A similar problem could occur with an optically active compound if the health effect is a function of reaction with a stereospecific pharmacologic receptor.

The route of exposure in humans and the mode of administration may be difficult to model if the dose is such that the material cannot be easily delivered in food or water, or if it enters the body in different amounts through each of several routes of absorption. For example, some halogenated compounds are not soluble in water at concentrations compatible with administration of moderate to high doses. If the desired dose cannot be given in the food or water, alternative methods such as intubation may change the mechanics of absorption of the material and its distribution during the early phases of exposure. Effective modeling of human absorption under these circumstances may prove tenuous at best.

Selection of the model species can be a significant design consideration with respect to circadian and seasonal changes in enzyme activity and hormone levels. Because the vast majority of chemical testing is done with rodents, the chronobiology of nocturnal species should be considered an important source of interspecies variability. The impact of circadian rhythms on chemical metabolism is discussed in detail by Reinberg (1978).

In some circumstances, the experimental schedule, the latent period for development of effects, and the portion of lifespan to be covered for the species may require that dosing begin in very young animals. Animal organs and their related functions which might not be affected by a toxic agent given at a later stage of maturation, may respond differently from those in adult humans simply because of the early onset of exposure. This is particularly true for those materials which might be expected to have some impact on sexual maturation or function. For example, it is unlikely that rodents exposed from the time they are weaned until they reach sexual maturity are an effective model for the comparable period of development in children.

As animals approach "old age" for their species, the number and severity of diseases unrelated to exposure further complicates effective modeling. Diseases related to the aging process are frequently species-dependent and may confound the assessment of health effects due to exposure. For example, it has been demonstrated that the rate of DNA excision repair in some species is a function of life span (Hart and Setlow, 1974) and, therefore, may be responsible for changes in rates of cancer incidence in animals, rates which are not analogous to those of human response to carcinogenic compounds. A general discussion of problems associated with aging in experimental animals is provided in a review by Grice and Burek (1984).

Environmental Factors

The impact of environmental factors on the interpretation of animal responses and extrapolation to human responses has become a significant concern in recent years. The majority of these factors arise from the necessity to confine large numbers of experimental subjects in convenient and readily identified groups. However, the responses of animals so confined contributes to several sources of variability. Investigators have reported different responses to chemical exposure as a function of number, sex, and social and breeding status of rodents caged together. Both crowding and isolation have profound effects on reproductive behavior, care of litters, and stress-induced disease patterns. The cage type, whether mesh or solid, as well as the type and volume of external noise have also been shown to cause differences in the effects observed in exposure studies. In addition, exposure to chemical compounds present in bedding material may affect the rate of metabolism of administered chemicals in rodents (Casarett, 1980).

Nutritional status and diet have been shown to affect chemical toxicity (Searle, 1976). Conservative risk assessment practice usually dictates that experimental exposures model worst case conditions; however, daily variations in human diets may result in only intermittent exposures, allowing time for repair and recovery from adverse health effects. Attempts to model intermittent exposures, however, would invariably have to take into account the altered dynamics of tissue distribution and metabolism.

Ad libitum food or water access is the most frequently used method of exposure but it may be modified under some circumstances to provide quantitation of the amounts of test substance consumed. In these modified studies animals are fasted prior to dosing to prevent dilution in the stomach and intestines and to encourage complete consumption of the prepared dose in a small quantity of food or water. Under these circumstances, however, a compound which might otherwise be bound to plasma proteins may reach higher free levels in plasma because binding sites are occupied by an excess of circulating fatty acids. For chemicals which have a very high affinity for plasma proteins, and in particular albumin, competition at binding sites could result in displacement of unacceptably large amounts of fatty acids. The simultaneous presence of the test compound and excess free fatty acids could produce a combined effect which is not characteristic of the toxic agent alone.

The specifications for animal diet can also be an important consideration in study design, particularly if the tested material has effects which are nutrient dependent (Newberne, 1976). For example, the current controversy over the influence of dietary fat and fiber on human colorectal cancer raises the issue of whether proportional quantities in laboratory rations would have an analogous impact on development of gastrointestional cancer during lifetime studies in animals (Weisburger and Williams, 1982; Burkitt and Trowell, 1975; Weisburger, 1979).

For those groups of compounds which interact with vitamins, the absence of adequate amounts of requisite vitamins in the laboratory diet may alter the threshold for effects seen in animal models. For example, the response of guinea pigs to chemical exposure may be modified by the interactions of Vitamin C and metabolic pathways. Similarly, the response of numerous species exposed to hydrazines, semicarbazones and azides may vary as a function of the availability of Vitamin B-6.

Finally, for those species which apparently supplement mineral requirements by coprophagy (ingestion of feces), cages with mesh rather than solid bottoms may prevent the normal recycling of required trace elements. This activity may also be necessary to maintain normal levels of stomach flora characteristic to the species.

Quantitative and Qualitative Aspects of Dosing

A basic requirement for appropriate interpretation of animal data is the development of a dose-response (DR) relationship which demonstrates a positive correlation of degree of effect with increasing dose. The establishment of the DR curve defines for any given effect a range of doses between the response threshold (i.e., the lowest dose at which the effect is observed) and its maximum (i.e., the dose at which each animal exhibits the response). For multiple effects there can be multiple DR curves lying between the extremes of no observed adverse effect and lethality. Interpretation of the DR curve with respect to the potential human exposure concentrations requires consideration of the several quantitative and and qualitative elements of uncertainty associated with the extrapolation of data from animal models.

Quantitative aspects

The primary objective of quantitatively defining the DR relationship is to provide a basis for deriving an appropriate proportioning factor or scaling system between species of diverse size and genetic constitution. A number of procedures to accomplish this stage of scaling animal model dose-response relationships to the projected scale of effects in humans have been reviewed by Calabrese (1983). The most frequently used scaling factor involves a linear relationship between mass of dose and body weight. Although this scaling factor is not suitable for all applications, other scaling methods have not proven to be uniform either between species or over a broad range of classes of chemical compounds. Such other scaling functions include:

- The 2/3 root of body weight
- Body surface area
- Relative rate of metabolism
- Molar ratio of target tissue concentration
- Log dose versus log weight regression
- Moles of DNA/cell (in vitro assays of mutagenic compounds)

The data obtained by repeating an experiment with different species using a single test chemical may suggest a preferred scaling method for projecting dose-relationships for human exposure to that particular chemical.

There are several factors which can complicate the interpretation of results. An overlap in DR relationships for two effects of a single chemical at doses above the threshold for each effect may obscure the more sensitive response and impair accurate assessment of the low-dose effect. If both effects are not seen in all species tested, it will be difficult to determine if the effect is masked or simply not present. The interpretation of long-term studies can be influenced by a similar overlap of acute and chronic effects. That is, if the dose is low enough that acute effects (e.g., lethality) do not occur, then the experimental exposure period may not be long enough for chronic effects to develop within the lifespan of the test animal.

A precise quantitative scaling between animals and humans may also be influenced by factors which are sensitive to dosing schedule, dose rate or route of administration. Ad libitum dosing using food or water as the vehicle maintains a fairly even and continuous input of test chemical throughout the active portion of an animal's day. If oral dosing schedules are modified to administer the dose once per day, pharmacokinetics will have a greater influence on peak concentration in plasma or body fluids and on the duration of tissue exposure. For example, if the absorption process is rate-limited at some stage of transport across membranes, a portion of the dose may be metabolized or excreted by other pathways and become unavailable for absorption or may cause confounding effects. Since many larger test animals may eat only once per day, a dose given in food may not be quantitatively absorbed if the concentration at the rate-limiting membrane exceeds the maximum that can be transported during the period that the chemical is present at the membrane interface. Further, a single daily dose may raise the peak concentration at target tissues above the response threshold for secondary effects which would not have been seen with a more evenly distributed exposure. There is also the possibility that metabolizing enzyme systems may become temporarily saturated during peak concentrations, resulting in a shift of circulating proportions of parent compound and metabolite.

Because intermittent inhalation exposures in humans can often be effectively modeled by inhalation exposure in animals, the impact of the above considerations of tissue absorption on the interpretation of inhalation exposure data is greatly reduced. With the possible exception of particulate compounds (e.g., asbestos) which pose additional complexities, inhalation methods for modeling human exposures provide a quantitatively precise basis for scaling.

The different mechanisms for uptake of chemicals at the cellular level may influence quantitative aspects of scaling and hence interpretation of dose-response data. Depending on the method of uptake (e.g., pinocytosis, active transport, facilitated diffusion or passive diffusion), the amount of administered chemical absorbed can be affected by the concentration present and the duration of these concentrations at the site of uptake. Uptake is generally proportional to concentration unless absorption mechanisms approach saturation; at this point, the unabsorbed portion may be metabolized and excreted via other pathways resulting in a decrease in the total absorbed dose at the target tissue and possible effects on alternate tissues.

Qualitative aspects

In addition to purely quantitative considerations for scaling dose from animals to humans, there are other factors of a qualitative nature which can influence the interpretation of animal data.

While the route of exposure and the mode of administration can be designed to simulate some aspects of human exposure, the highest doses given to animals are usually scaled to represent either the worst case conditions of human exposure or the maximum tolerated dose that does not cause lethality. These concentrations may exceed the water solubility of the compound or become too noxious for the animals to accept in either food or water. Either of these conditions will have a significant impact on the actual absorbed dose and may affect nutritional status of the animal as well.

The use of an alternative dosing vehicle other than water or typical dietary constituents may result in a qualitative difference in response and increased difficulties in interpreting the data in the context of human exposure. Simultaneous exposure to the vehicle and test chemical can result in interactions not typical of the human exposure being modeled. Although there might not be direct chemical reactions between vehicle and test chemical, the transport across the gastrointestinal tract membranes and absorption into the bloodstream may be altered if the test chemical is differentially soluble in the vehicle and in the blood. The vehicle's influence, whether chemical or physiological, may cause a reduction in circulatory concentration of the test chemical. The calculated chemical concentration in the blood, based on dose administered with a vehicle, may therefore be an overestimate. It may be difficult to reconcile the disparity between the animal data arising from such experiments and analogous human effects.

Finally, qualitative differences in the actual structure of the chemical species absorbed into the blood may result from differences in gastrointestinal flora between animal species. For example, these differences are likely to be observed between humans which have an essentially sterile upper alimentary tract and rodents which have colonies of stomach flora due to the ingestion of feces.

Interspecies Differences

In designing a study protocol, an investigator can manipulate the variables which affect interpretation of animal data, particularly with respect to extrapolation to humans. However, once the test species is selected, the morphology, biochemistry and behavior which characterize the species are beyond the control of the investigator. This section will review some of the features of interspecies variability which investigators must evaluate in extrapolating animal data to projected human response.

The principal application of scaling functions is to estimate the range of doses which would define the major portions of a human dose-response curve. Uncertainty in this process results from deviation of observed test data from the expected results of a selected scaling function for an individual species. This discussion of interspecies differences will be limited to those factors which would be most likely to result in deviations of human health effects from expected responses derived from animal models. The interpretation of animal data relative to human exposure should not be attempted without careful consideration of such factors.

Morphology

General considerations

When dose is computed directly on the basis of body mass or some function of body mass, both total body fluid and blood volumes of test animals are in proportion to human volumes. If the test compound is evenly distributed in tissue fluids, proportional doses in comparable tissues would be expected across species. However, if the test compound partitions preferentially into a specific organ (e.g., the brain), then disproportionate concentrations would be expected in that tissue based on the disparity of total body mass. For example, a dose of chemical absorbed by the brain, if computed on the total body mass of an elephant from the effective dose data for humans, could result in concentrations up to 40 times that of the expected level because of the comparable size of the brains of the two species. Similarly, if the toxic material is absorbed preferentially in fat deposits, those species with disproportionately large volumes of fatty tissue may retain the compound for a longer period of time but may also reduce the peak concentration at functional target tissues immediately following dosing.

Skin

Studies which use percutaneous absorption as the route of administration for chemical compounds must allow special consideration for variation in the structure and function of the integument between most commonly used laboratory animals and man. Although structural aspects (e.g., thickness, fur, sweat glands) differ considerably among laboratory species and between man and animals, a reasonably good model of human skin has been found in the dorsal skin of mini-pigs. Data based on appropriately scaled doses among two species, one of which uses the skin for thermoregulation and one which does not, may be distorted by a disproportionate degree of vascularity and hence a notable difference in the rate of uptake into the systemic circulation. This effect may be particularly significant when experiments are conducted at ambient temperatures above the comfort range for animals involved.

Gastrointestinal tract and abdominal viscera

In addition to the major difference in structure and function between the monogastric and ruminant animals, there are more subtle differences among the common laboratory species which may account for deviations from expected dose-response relationships based on a fixed scaling factor. For example, the lack of a gallbladder does not appear to be a handicap in a species such as the rat which feeds frequently during its active period. However, study designs which require fasting prior to dosing and administration of the entire daily dose in a single feeding may disrupt normal digestive function by causing much greater quantities of food to be introduced into the gastrointestinal tract in proportion to the quantity of bile acids available for adequate digestion than would normally occur. In addition to a direct effect on nutrition such actions could then lead to changes in local pH in the region of the GI tract where absorption occurs. If the test compound is a weak acid or base, the local pH will have an impact on the rate of absorption and the total amount absorbed. Although the magnitude of these changes may not be significant for a single dose or short-term experiment, over a period of several months to a year or more, they may be responsible for significant deviations of a dose-response relationship from the expected curve.

Transit time from stomach to intestine is a function not only of species but also of nutritional status and the physical form of food. Interpretation of absorption and metabolism data may be difficult in those species which have large populations of stomach flora. Significant portions of an administered dose may be metabolized by stomach bacteria before the chemical reaches the site of absorption in the small intestine. Given the negligible concentrations of flora in the upper gastrointestinal tract of humans, scaling factors which do not anticipate these differences can distort projections of human response.

Respiratory system

Differences in respiratory function parameters among laboratory animal species are reasonably well modeled by appropriate scaling factors based on respiration rate, respiratory minute-volume, and body weight, or estimates of alveolar surface area. Some differences between species, such as the obligate versus non-obligate nasal breathers, may contribute significantly to inter-species variation in response to inhaled contaminants. Interaction of structural characteristics of the nasopharynx with normal breathing and the water solubility of an inhaled chemical compound can result in 20-70% reduction in chemical concentrations reaching the trachea. In aerosol exposure studies, interspecies variation in alveolar retention of inhaled particles of one micron in diameter was observed in guinea pigs, monkeys, and humans. The decrease in alveolar deposition of the large particles in guinea pigs was attributed to increased efficiency of removal in the upper airways.

A final consideration of interspecies differences in respiratory system morphology is related to the presence of submucosal glands and the response of animals to bronchial irritants. Only in the dog, rat and pig are these glands known to be sufficiently well developed to function in a manner analogous to that in humans. Under some circumstances, the presence or absence of a bronchitis-like condition in various species exposed to bronchial irritants might be attributed to inappropriate selection of exposure concentration rather than a fundamental difference in anatomy.

Brain

Disregarding the cognitive function of the human brain which cannot be adequately modeled in animal experiments, many of the basic neurophysiological functions of the brain respond proportionally to an appropriately scaled dose. For many chemicals, however, the scaling factor is based upon the concentration of the test compound in brain tissue at the time of occurrence of the desired endpoint. The time necessary to reach this effective concentration is in turn dependent on the rate of transfer of the test compound from the capillaries into the brain parenchyma. The collective impediments to this transfer are termed the "blood-brain barrier" and are responsible for assuring minimal interference with the internal biochemistry of the brain from extraneous substances. With the exception of lipid soluble materials and a select group of biochemicals for which specific transport mechanisms exist, most other materials are effectively excluded by the mature brain. There are significant differences in the rate of maturation of the blood-brain barrier after birth among species used to model human exposures to central nervous system toxins. The use of a common scaling factor among young animals of different species may therefore be inappropriate.

Biochemistry

Physical-chemical environment

Although the systemic chemical environment is essentially homogeneous among the laboratory mammals commonly used for toxicity testing, there are significant differences between species in the physico-chemical conditions of the systemic environment in the alimentary canal. Variations in pH in analogous regions of the gastrointestinal tract occur among several species. As pH changes so does the ratio of ionized to un-ionized forms of a chemical compound. Because the un-ionized form of a chemical diffuses much more easily across lipid membranes, absorption of weak acids and bases from the stomach and intestine can be profoundly affected by the pH of the absorption environment. Animals such as rabbits, which have a very low stomach pH, may absorb much larger proportions of weak acids in the upper portions of the GI tract than will other species. Because pH tends to increase in the small intestine and reduce the proportion of un-ionized acid there, peak circulating concentrations, as well as the total absorbed dose of an appropriate compound, are likely to be greater in the rabbit than in other test species. Similar but less exaggerated effects may be observed in monkeys, which maintain a distinctly acid environment throughout the small and large intestine. Guinea pigs, on the other hand, have a distinctly alkaline environment in the uppermost segment of the small intestine with a pH which is two units higher than that of the monkey in analogous intestinal segments. A weak acid such as ascorbic acid, given to both monkeys and guinea pigs, would have a 100-fold absorption advantage in the monkey. This difference should be considered when selecting a scaling factor between species in order to minimize deviation of animal data from expected values.

Enzymes

It is generally agreed that the greatest single factor leading to deviation from ideal dose scaling is the metabolic effect of enzymes. Interspecies differences affect both the rates and types of detoxification mechanisms as well as the extent of enterohepatic recycling of detoxified compounds. Intestinal beta-glucuronidase is chiefly responsible for the degradation of that portion of an ingested test chemical which is conjugated with glucuronic acid in the liver and excreted in the bile. Frequently, this enzymatic action releases the initial compound to be reabsorbed for another pass through the liver.

The microsomal enzyme systems are the major detoxifying systems among test species. Experiments with many species would be required to accurately predict the outcome of liver metabolism for most compounds, because of the extensive variety of mechanisms represented. Nevertheless, lack of effective metabolic pathways frequently complicates the interpretation of toxicity data in animals. For example, the pathways for acetylating and detoxifying isonicotinic acid hydrazide in dogs are ineffective, causing dogs to be inordinately sensitive to doses which are nontoxic to monkeys. Scaling such disproportionate responses on the basis of data in a single hypersensitive species would lead to an underestimation of the effective dose for species with superior detoxifying capability.

Enzymes contained in the normal flora of the gastrointestinal tract of test species vary both in composition and concentration. The rabbit is most notable in this respect and has active colonies of bacteria in its stomach. The upper GI tract of humans, on the other hand, is essentially devoid of permanent flora. This physiological difference could result in a considerably different interpretation of the effect of orally administered compounds depending upon the effect of the rabbit's stomach flora on the toxicity of the parent compound.

The complexity of interpreting the effect of enzyme activity on toxicology data, even in a single species, is probably best illustrated by the demonstration by Lower and Bryan (1973) that N-acetylation of aryl amines is simultaneously an activation pathway for induction of liver cancer and a competitive detoxification pathway for aryl amine-induced bladder cancer.

Protein binding

Exogenous chemicals absorbed into the blood are frequently bound to plasma proteins. This binding causes the chemical to be unavailable for diffusion or transport across cell membranes, decreasing the concentration in the extravascular fluid and at receptor sites. Although the binding capacity of various proteins is species-specific, albumin is the primary binding protein for all of the common laboratory test animals. Variation between species is best exemplified by the dog's sensitivity to copper. Poor binding capacity by a specific copper binding protein causes the dog to build up high storage levels of Cu(II) in the liver. Small excesses over normal intake could overload the storage capacity and result in acute poisoning.

In addition to the direct binding ability of plasma proteins, interactions between exogenous chemicals and endogenous fatty acids is an important consideration. Competition between fatty acids and exogenous chemicals for binding sites can result in shifts in concentration of either substance out of proportion to the administered dose. Interspecies comparisons may have to be interpreted in the context of differential binding of parent compound and its metabolites as well as differences in the rates and types of metabolic reactions. Experiments show that for many compounds, the monkey is no better than the rat as a model for predicting human protein binding levels.

Behavior

Chronobiology

Increasing evidence indicates that there are daily "peaks and valleys" in animal sensitivity to a broad range of drugs and other exogenous chemicals. The mechanism for circadian shifts in sensitivity is linked to the kinetics of absorption and excretion, activity of hormones and metabolizing enzymes, membrane transport characteristics, and bioperiodical shifts in receptor responses (Reinberg, 1978). Further, it has been shown in laboratory animals and humans given evenly spaced doses over 24-hours, that therapeutic drugs have periods of maximum and minimum effectiveness.

The common laboratory practice of dosing animals without regard to this periodicity may result in interspecies differences in toxic response which could prove difficult to interpret. It may be inappropriate to dose nocturnal animals when they are least active as an attempt to model human exposures that occur during periods of greatest activity.

Other behavior

Several of the commonly used laboratory animals instinctively and routinely ingest fecal material. This behavior may be an attempt to replenish or supplement mineral elements lost by excretion in the feces or to maintain the balance of flora in the stomach and upper intestine. The interaction of toxic chemicals with normal flora, noted previously, should also be considered in the context of changes in normal alimentation caused by a direct toxic effect on the flora itself. If the animal is unable to re-establish the normal metabolic functions carried out by these organisms, interspecies differences between coprophagic and noncoprophagic species could result.

Finally, inhalation exposure using either a liquid or solid aerosol may be supplemented during grooming by ingestion of test compound deposited on the fur of the animal. The foregoing behavioral characteristics could lead to inaccuracies in extrapolating the dose-response information to humans.

MATHEMATICAL MODELING OF LOW-DOSE EFFECTS

The results of toxicological experiments in animals are frequently used in assessing human cancer risks of an environmental contaminant. The results of experiments carried out at relatively high doses are extrapolated to a low-dose situation using a variety of mathematical models. These methods are applied most frequently to carcinogenicity data but are also applicable for assessing risk of acute and chronic toxicity and mutagenicity. A brief overview of the objectives and assumptions implicit in the various mathematical models will demonstrate the limitations of such data analyses as well as highlight some abuses in interpretation.

Objective of Analysis

Considerable scientific judgment and skill must be exercised in the design of adequate protocols, in the conduct of pathologic appraisals, and in the maintenance of laboratory animals and records all with due consideration of the physiologic, pharmacokinetic, and biochemical properties of the compound under study. The appropriateness of the animal model selected, relevance of corresponding metabolic pathways, and the conduct of adequate experiments to produce valid scientific data for analysis are assumed when applying the following statistical procedures.

Results obtained in animal carcinogenesis studies with relatively high doses of carcinogens are used to draw inferences to the effects on humans exposed to much smaller amounts. Two dose levels of interest to toxicologists and regulatory agencies are the "no observed effect level" that is measured experimentally and a "virtually safe dose" that is calculated in terms of some preselected level of risk (Cornfield, 1977).

Equating the "no observed effect level" to a true "no effect level" is a concept subject to considerable scientific debate, particularly in the case of carcinogenesis. It is assumed that the observed experimental responses can be mathematically converted to an underlying dose-response curve, relating the dose administered to the probable response to non-experimental doses. Various approaches have been developed to predict the shape of the dose-response curve at levels below those that produce an observable effect (Maugh, 1978; Food Safety Council, 1978).

Mantel and Bryan (1961) advanced the concept of a virtually safe dose at which the probability of a response is a near zero lifetime risk of developing cancer. Mantel and Bryan originally suggested a risk of 1 per 100 million. Some regulatory agencies considering certain societal factors have adopted a figure of 1 per 1 million (USDHEW/FDA, 1977) as a level of risk corresponding to an "acceptable" dose level. A number of mathematical models for extrapolating experimental data from animals to acceptable levels of risk for humans give widely different answers using the same data. The ensuing regulatory levels are, in practice, sometimes chosen on the basis of practically attainable low levels.

Examples of Models and Their Limitations

Existing mathematical models include the probit, one hit, Weibull, logit, gamma multi-hit, Armitage-Doll multi-stage, and a simplified statistico-pharmacokinetic model (Food Safety Council, 1978). Refinements of these models that consider time-to-response include the Hartley-Sielken model (Hartley and Sielken, 1977), National Center for Toxicological Research linear extrapolation procedure (Farmer et al., 1982), as well as a model that is multi-stage in dose and Weibull in time-to-tumor or -response (Crump et al., 1981). For practical purposes, many of the existing mathematical models are indistinguishable from each other in their dose-response curves in the experimental dose range, but diverge at very low doses. Care must be taken to give proper weight to background effects.

The probit, logit, and Weibull models describe the characteristics of the study population in terms of its tolerance to the agent. The proportions of the population having tolerances above or below a certain dosage are thus calculated. The distribution of the logarithm of the tolerances is assumed to be normal in the probit model. Cornfield (1977) suggested that the normal distribution may not provide a reliable description of the distribution at values as low as 1 per 1 million. As dose decreases, the probit curve approaches zero more rapidly than curves generated by other models.

The S-shaped dose response curve of the logit model is symmetrical about the 50% response point. Although its shape is similar to that of the probit model, the logit model leads to a lower virtually safe dose estimate, about 1/25th of that predicted by the probit model, when both models are equally descriptive of the observed data. Thus, the logit model is considered more conservative than the probit model.

The one-hit linear model is equivalent to assuming that the dose response curve is linear in the low dose region. This model assumes that the process producing the observed response, such as carcinogenesis, is stochastic; that is, the response is the result of a single discrete event and is not proportional to dose. However, the likelihood of an initiating event occurring increases linearly with dose. The Weibull model is a generalization encompassing nonlinear relationships of hit rate and dose.

The one-hit linear and the probit models were specified by the U.S. Environmental Protection Agency (1976) in its interim guidelines for assessing the health risks of suspected carcinogens. The most frequently cited example of data that fits these models well is the induction of leukemia following ionizing radiation. They predict a linear-dose response down to an induced incidence of about 0.1% (Maugh, 1978; Upton, 1983).

The gamma multi-hit model represents a generalization of the one-hit model (Cornfield et al., 1978) providing a blend of the probit model at high dose levels and the logit model at low dose levels.

Armitage and Doll derived a multi-stage model by assuming that the effect of an external stimulus at certain stages was additive to the effect of an agent at those stages. Models encompassing the concept of additivity would be useful in the treatment of data concerning exposures to mixtures from waste disposal sites. Uncertainty concerning additivity can make orders of magnitude differences in the estimated risk from low-dose exposure (Crump et al., 1976).

If it is assumed that an agent is subjected to both activation and deactivation processes in the body, the dose-response curve predicted by the simplified statistico-pharmacokinetic model will be nearly flat until the administered dose exceeeds the maximum dose that can be deactivated (Cornfield, 1977). The slope of the curve below this dose will be greatly influenced by the rate constants for activation and deactivation. A major limitation in applying this model is the need for details which are seldom available on the pharmacokinetics of the metabolic pathways at various dose levels.

The presence or absence of competing risks might be of particular significance in assessing a virtually safe exposure. Krewski et al. (1983) have recently compared various statistical methods for low-dose extrapolation utilizing experimentally-determined time-to-response data. They considered time to response, the presence of competing risks, background response, latency, and experimental design in their assessment of different extrapolation procedures. The factors by which estimates of risk in the low-dose range differ from actual risks were found to be on the order of 1,000 or more, even in situations where information on the time of occurrence was considered to be precise. They cautioned, therefore, that point estimates of risks in the low-dose region based purely on statistical extrapolation procedures are subject to considerable uncertainty even if precise and detailed information of the nature and time of tumor development at higher doses is available.

Additional Limitations

The data used in mathematical modeling consist of discrete positive and negative responses at each dose level (e.g., a tumor developed or a tumor did not develop). Less satisfactory analyses can be made of continuous "measured response" data (e.g., blood pressure) because certain information must be deleted in converting such data to positive and negative responses. Questions of possible multiple responses to various mixtures, differences in human exposure to these mixtures, methods of combining numerous studies on the same and on different species, and the selection of appropriate control data remain unanswered even after a particular statistical mathematical model is selected. The modeling for carcinogenicity is unique in that the assumption is made that there is no threshold, whereas, all other toxic responses have thresholds.

COMBINED CHEMICAL EXPOSURES

More than five million substances have been described in the chemical literature. In a recent study prepared by the National Research Council (1984) on priority needs in toxicity testing for the National Toxicology Program, it was estimated that, of the 53,000 chemicals of known commercial significance, 86 percent have so little known about their toxicity that not even a partial assessment of their potential health hazard can be made. The variety of possible effects from simultaneous exposure to combinations of chemicals complicates both the design of appropriate studies and the interpretation of their results.

Most chemical waste sites contain mixtures of chemicals which could result in or contribute to, combined environmental exposures. At the time of the Love Canal incident, toxicology data were available for less than half of the nearly 300 chemicals identified in the site by the New York State Department of Health and the U.S. Environmental Protection Agency (Goyer, 1983). Data to fill this information gap are presently being compiled for the individual chemicals from animal studies and occupational exposure data by the National Library of Medicine. It is not anticipated that information regarding health effects from combined exposures will be available from this toxicology data base.

Potential health effects from multiple exposures to toxic substances have been reviewed by the World Health Organization (1981) and the National Research Council (1980). Concern for chemical interaction is frequently noted by regulatory agencies such as the Food and Drug Administration and National Institute of Occupational Safety and Health. Drug manufacturers are frequently required to provide data on new drugs that may interact with other drugs or may be considered to be inducing agents. Employers are being encouraged to educate employees regarding alcohol and drug abuse in the light of potential interactions with chemicals encountered on the job.

In the broadest sense, interactions among environmental contaminants would include physicochemical interactions prior to the exposure of the biological organism and may or may not produce adverse effects. Two known examples of such situations are the photochemical formation of ozone from auto emissions, and the formation of nitrosamines from inorganic nitrite and dialkyl amines. A subcommittee of the World Health Organization (1981) concluded that most of the known reactions between chemicals in the environment would usually render those chemicals either less toxic or less accessible for human exposure.

Complicating Factors of Combined Exposures

The interaction between chemical exposures and factors within the total living environment can have a profound effect on the observed toxic response. Complicating factors such as: social, psychologic, biologic, and iatrogenic status of the individual, or physical agents are all known to effect chemical absorption in man. In addition, each individual or genetic subset of the population may show a unique response to combined, multifactorial exposures. These effects have been demonstrated in numerous epidemiologic studies.

The relationship between alcohol consumption and workplace exposure has been well documented. Ferguson and Vernon (1970), and Windemuller and Ettema (1978) have observed that socially accepted concentrations of ethanol in blood, combined with exposure to trichloroethylene can handicap mental functions to a greater degree than each of these substances separately. Similar relationships have been observed with drug interactions. Cornish et al. (1973) observed a potentiation of toxicity when workers receiving phenobarbital were exposed to carbon tetrachloride or vinyl chloride.

Tobacco smoke contains a variety of known carcinogenic substances. A summary of information linking habitual tobacco smoking and industrial exposure with increased health problems was presented in the Surgeon General's report on Smoking and Health (USDHEW/OSH, 1979). The synergistic effects of tobacco smoke with substances such as cadmium, chloromethyl ether, and aromatic amines were also reported (USDHEW/NIOSH, 1979). Several health studies have shown an increased incidence of mesothelioma in miners and asbestos workers who smoke (McDonald and McDonald, 1973; Selikoff et al., 1968).

Indulski (1983) recognized that during weekends workers may inhale greater quantities of toxic substances in a polluted community environment than they do at their workplace. Wallace et al. (1984) noted quantitative exposure differences between indoor and outdoor air measurements of various pollutants in an urban residential environment. Diet is an additional source of combined chemical exposure. Residues of chemicals such as ethylenedibromide (EDB) have been found in foodstuffs. Nutrition can modify chemical hazards (Crosby, 1969); for example, a low-protein diet increased the susceptibility to toluene, carbon tetrachloride and carbon disulfide, while vitamin deficiencies decreased the induction by organophosphorus compounds of liver microsomal enzymes thereby decreasing the rate of metabolism.

Mechanisms

A toxic interaction is defined as a condition in which two or more chemicals result in a qualitatively or quantitatively altered biologic response relative to that predicted from the action of the individual chemicals (Murphy, 1983). For any exposure, both exogenous and endogenous interactions may result in either: (a) additivity - where the combined effect is the sum of the effects of the individual agents; (b) synergism - where the combined effect is greater than the sum of the effects of the individual agents; (c) potentiation - where one component enhances the effect of another; or (d) antagonism - where the combined effect is less than the sum of the effects of the individual agents (Fairchild, 1983).

Dispersion of an environmental chemical in several exposure media such as air, water, soil, or food, and in multiple environments such as workplace, residential or recreational areas may result in a cumulative exposure that is significantly higher than that from any one media. The presence of chemical hazards in several environmental media may induce multiple stress factors on an organism, which may interact to enhance or inhibit the potential for a health effect (Helmes et al., 1982). In combined exposures, additional factors such as biological half-life, bioavailability, biotransformation and bioactivation may assume greater importance depending on temporal relationships.

Both the sequencing and the frequency of events in a combined exposure may contribute to the mechanism of toxic interaction. The interaction between chemicals and multiple environmental factors can result in an increase in the incidence of some human cancers that is greater than that expected from an exposure to a single carcinogen. The chemical induction of tumors is considered to be a multi-stage phenomenon, requiring either repetitive exposure to a single agent or induction by a single agent followed by promotion from a secondary agent at a later time (Berenblum, 1974). Initiation takes place rapidly and is considered to be essentially irreversible, while promotion may occur months or even years later (Van Duuren et al., 1978).

Murphy (1983) supports a systematic approach by which predictions of potential for toxic interaction among environmental pollutants would be based on knowledge of toxicokinetic and toxicodynamic characteristics for individual chemicals. In an expansion of Veldstra's (1956) original work, he described the key elements for identifying potential interactions between chemicals as:

- Relative affinity (association/dissociation rate constant) of each chemical for sites of action (e.g., target enzymes, neuro-effector site).
- Relative affinity for sites of loss (e.g., enzymatic detoxification, excretion).
- The intrinsic activity of the chemical at the sites of activation.

These principles are based on the concept that there exists for any organism a finite number of sites of action or sites of loss for any biochemical interaction, and that there is a limiting dose range within which synergism or antagonism can be demonstrated. Toxic interactions can be viewed as taking two forms: (1) the quantity of an active form of one or more chemicals available to the pool at critical cellular reactive sites is altered by the presence (or past presence) of one or more other chemicals, or (2) the reactivity of the critical cellular reactive site as activated by exogenous chemicals is altered by the presence (or past presence) of one or more other chemicals which may or may not be capable of eliciting a response (Murphy, 1983).

In the absence of a uniform approach to studying combined exposures, an understanding of the cellular mechanisms and the disposition of individual chemicals may provide information to permit the assessment of the likelihood that combined exposures will result in toxic interactions.

Although, mechanistic toxicology (i.e., the study of biochemical mechanisms that result in toxicity), has contributed greatly to the understanding of disease processes and the manner in which chemicals form reactive species that interact with molecular targets; it has not made a significant impact on safety evaluation. Many mechanistic studies provide information on chemical injury to tissues and organs but leave the biological significance of these processes poorly defined (Witschi and Munro, 1983). An improved understanding of the biologic mechanisms of combined chemical exposures can best be attained from continued studies with laboratory animals.

Testing Methods

In developing a test system to evaluate the adverse health effects that may arise from exposure to mixtures of chemicals and combined exposures, it is important to consider the potential effects of each environmental component. Short-term tests (e.g., Ames test) offer an unlimited opportunity for varying the quantities and combinations of chemical constitutents; however, such tests are limited to detecting only the potential for mutagenicity and seldom provide quantitative information. Tests are available which provide information on the acute, adverse effects of mixtures of chemicals in various organ systems such as the hemopoietic, nervous, and reproductive systems (Neal, 1983). Studies of chemical mixtures in experimental animals, where the route of exposure is identical to the potential human exposure offer the most sensitive and reproducible data for predicting toxicity in man. Both the current methods and significance of animal testing has been reviewed for government agencies (NAS/NAE/NRC, 1975; National Toxicology Program Board of Scientific Counselors, 1984).

Inherent to the problem of test design for evaluating multiple chemical exposures is the question of what constitutes the mixture to be tested. This becomes a complex issue when considering a study of multiple chemicals at a chemical disposal site, as the composition of the mixture may vary in different areas of the site (Kimbrough, 1982), thus requiring the need for testing several combinations of the possible mixtures. Consideration of the heterogeneity, correct chemical mix, and concentrations of individual components is important in selecting the "representative" mixture in studies with laboratory animals.

Chemical mixtures have been identified in many sources of drinking water. Laboratory tests in which rodents were administered material obtained by concentrating drinking water have shown that the additive effects of multiple carcinogens are an important consideration. Kraybill (1980), observed that while the concentration of a single chemical carcinogen in drinking water may occur at the microgram level, the composite concentration of all carcinogens may be at the milligram level. This concentration is considered to be a dose capable of initiating a positive response in both animal models and man assuming simple, linear, additivity of the carcinogenic effects of the components.

Studies have been reported in the literature that indicate possible interactions between several of the chemicals identified at Love Canal (USC/ Senate, 1980). The interactions involved either potentiation of the toxic (non-carcinogenic) effect of one chemical by another or the enhancement of the progression of latent neoplastic change (promotion or co-carcinogenesis). The interaction sites identified in this study were skin, liver, and kidney.

Known Interactions of Combined Environmental Exposures

There are numerous examples of adverse effects caused by chemical interactions that are independent of confounding factors. Animal studies have shown that mixtures of solvents (e.g., CCl_4 and cyclodienes), pesticides (e.g., EPN and Malathion), metals, and irritant gases can modify the adverse health effect of the individual chemical administered alone. The interaction mechanisms vary; additive, synergistic, and antagonistic effects have been observed as well as carcinogenic promotion and co-carcinogenicity. There is a uniqueness of action for most combinations, that suggest appropriate steps be taken to minimize total carcinogen exposure whenever possible. While it is not possible to include a detailed description of the existing literature on this topic, two examples of different classes of chemicals are included.

Solvents

Ikeda (1983) reviewed 50 reports on the combined effects of organic solvents. The studies were limited to a few chlorinated hydrocarbons, aromatics, and alcohols. Although he presented evidence for synergistic, antagonistic and additive effects, he noted that these studies suggest synergistic effects rather than independent effects, and that studies supporting the additiveness hypothesis were limited to findings in short-term animal experiments.

Pesticides

Interaction between the various forms of pesticides is well documented and appears to be as varied as the number of possible combinations. This literature has been recently reviewed by Kaloyanova (1983). The toxic hazard of a formulation is considered to be a combination of the inherent toxicity of the pesticide and all other ingredients. Of particular concern is the likelihood of production of impurities and other contaminants. Lichtenstein et al. (1973) have noted that herbicides can enhance the toxicity of insecticides severalfold. Both synergistic and antagonistic effects can be anticipated between organophosphorus pesticides that are known to be enzyme inducers (Conney and Burns, 1972).

SUMMARY

Assessing human risk from chemical disposal sites requires estimates of the probability of exposure (i.e., physical contact between a chemical and living tissue) and the probability of receiving a toxic dose (i.e., sufficient duration and intensity of exposure to cause an adverse effect). This chapter focuses on some principal factors considered in making these estimates, not on the various processes of risk assessment.

It is well established that the primary route of human exposure to chemicals from disposal sites is ingestion of contaminated water, but inhalation of atmospheric releases and direct dermal contact are frequently significant. In addition, in many localities these exposures are supplemented by environmental exposures from industrial, agricultural, mining, and domestic disposal practices.

A complete assessment of exposure includes the identification of the chemicals in the site, an analysis of their rates of release from the site and transport into an exposure medium, and qualitative and/or quantitative analyses of concentrations in air, water, and food. Although dose is virtually impossible to measure precisely, effective estimates may be obtained from measurements of the duration and intensity of exposure and concentrations in tissue and body fluids.

In practice, estimates of environmental contamination are frequently based on assumed worst case conditions since adequate quantitative data are seldom available. Estimation systems that have been tested include a 5-stage decision tree and materials balance techniques as well as a graphical technique which develops estimates of the rate of movement and the fate of chemicals in the environment through reference to a series of nomograms.

Environmental reconnaissance allows assessment of the changes in the flora and fauna around a disposal site as biological indicators of chemical release and potential human exposure. Specific data on age, weight, population density, reproductive success, analyses of body burdens, histopathology of abnormal tissues, and similar measures provide an opportunity to make both qualitative and quantitative estimates of the degree of biological exposure and the degree to which the food chain may have become contaminated. While no single disease or tumor site may be used as an absolute indicator, the effects seen in animals near the chemical disposal site environment can suggest potential health effects in man. When this information is supplemented by data from laboratory toxicology research, risk assessments become more effective.

Carefully designed and conducted studies of animal models of human target organ systems form the basis for indirect estimates of the human health effect of a quantifiable dose of a toxic chemical. Because of experimental and ethical constraints, dose-response relationships in humans cannot be tested; thus, a proportional or scaling function for estimating human response from animal models is essential for risk assessment. A linear extrapolation of some function of body weight, body surface area, relative rate of metabolism, target tissue concentration, or quantity of DNA per cell is frequently used. Multiple species testing is useful for identifying interspecies differences which, if undetected, might result in misinterpretation of experimental data. When several animal species are used, the effectiveness of the scaling procedure between species may be verified by regression analysis of scaling factor versus dose for a specific endpoint.

Physiological and structural characteristics of the skin, the gastrointestinal tract, the respiratory system, and the central nervous system are the principal sources of potential problems arising from morphological differences between species. Variations in the biochemical characteristics of enzymes, hormones, and certain tissues, as well as variations in behavior among animal species contribute further to uncertainties in the interpretation of dose-response data and in the extrapolation from animal models to humans of the potential for health effects.

An important problem involved in the application of current risk assessment methods is the use of information from relatively high doses of materials in small numbers of test animals to predict what will happen when large numbers of humans are exposed to much smaller concentrations of these materials. Mathematical models are used to extrapolate the risk for experimental animals from the large doses employed in the laboratory to the usually much smaller doses to which people may be exposed. The models allow (a) comparisons of dose-predicting/risk-predicting systems, (b) consideration of time-to-response in the presence of competing risks, and (c) analysis of background response and/or background effects. These models do not assume that a threshold or no-effect level exists. They are most frequently applied to data on carcinogenicity. However, current mathematical modeling lacks precision due in part to failure to accomodate metabolic differences among species. In spite of these shortcomings, extrapolations are used routinely in quantitative risk assessments even though insecure.

The effects of exposure to combinations of chemicals is a major unresolved problem of risk assessment. For the purposes of this study two sources of chemicals are considered: uncontrolled disposal sites and all other sources in the environment. These non-site sources include naturally occurring chemicals, those used in personal and domestic activities, and those encountered in the occupational setting, as well as chemicals already dispersed in the environment from industrial and agricultural activities. Multiple exposures complicate risk assessment beyond simple summation of effects because interaction of a single chemical with certain tissues can, on rare occasions, modify either the chemical or the tissue or both in a manner which leads to potentiation, synergism, or antagonism of the effects of other chemicals present instead of the usual additivity of effects. These interactions may affect target organ or site specificity, reaction kinetics, bioavailability, metabolic pathways and degradation or elimination. Use of a representative chemical to model effects of a mixture would not be appropriate under these circumstances, but the alternative of conducting an experiment using a mixture of chemicals is equally unsatisfactory because the results cannot be extrapolated to any other concentration or proportion of constituents. Experienced scientific judgement is needed to achieve a balance between the risks presented by exposure to a mixture of chemicals in the environment and the extent and nature of toxicological and related investigations to be undertaken.

References

Bailey, G.S.; Hendricks, J.D.; Nixon, J.E.; Pawlowski, N.E. (1984) Sensitivity of rainbow trout and other fish to carcinogens. Drug Metab. Rev. 15:725-750.

Baumann, P.C.; Smith, W.D.; Ribick, M. (1982) Hepatic tumor rates and polynuclear aromatic hydrocarbon levels in two populations of brown bullhead (Ictalurus nebulosus). In: Cooke, M.; Dennis, A.J.; Fisher, G.L.; eds. Polynuclear aromatic hydrocarbons: physical and biological chemistry. Proceedings of the Sixth International Symposium on Polynuclear Aromatic Hydrocarbons, held at Battelle Columbus Laboratories, October 27-29, 1981. Columbus, OH: Battelle Press. p. 93-102.

Berenblum, I. (1974) Carcinogenesis as a biological problem. Amsterdam: Elsevier North-Holland Publishing Co. 376p.

Black, J.J. (1983) Statement of John J. Black, cancer research scientist, Roswell Park Memorial Institute, Buffalo, NY. In: U.S. Congress, House of Representatives. Merchant Marine and Fisheries Committee. Fish cancer epidemics oversight. Hearing before the Subcommittee on Fisheries and Wildlife Conservation and the Environment. 98th Cong., 1st sess. Serial no. 98-40. Washington, DC: The Committee. p. 19-23.

Black, J.J.; Evans, E.D.; Harshbarger, J.C.; Zeigel, R.F. (1982) Epizootic neoplasms in fishes from a lake polluted by copper mining wastes. J. Nat. Cancer. Inst. 69:915-926.

Burkitt, D.P.; Trowell, H.C., editors. (1975) Refined carbohydrate foods and disease: some implications of dietary fibre. New York: Academic Press, Inc. 356p.

Calabrese, E.J. (1983) Principles of animal extrapolation. New York: John Wiley & Sons. 603p.

Casarett SEE Casarett and Doull's toxicology: the basic science of poisons

Casarett and Doull's toxicology: the basic science of poisons. (1980) 2d edition. Doull, J.; Klaassen, C.D.; Amdur, M.O.; eds. New York: Macmillan Publishing Co. 733p.

Conney, A.H.; Burns, J.J. (1972) Metabolic interactions among environmental chemicals and drugs. Science 178:576-586.

Corn, M.; Breysse, P.N. (1983) Review of "Working List of Issues to be Addressed in a Finding of Completeness." Prepared for the Maryland State Hazardous Waste Facilities Siting Board by the Department of Environmen-

tal Health Sciences, School of Hygiene and Public Health, The Johns Hopkins University. 29p. Available from: The Board, Annapolis, MD.
Cornfield, J.; Carlborg, F.W.; Van Ryzin, J. (1978) Setting tolerances on the basis of mathematical treatment of dose-response data extrapolated to low doses. In: Plaa, G.L.; Duncan, W.A.M.; eds. Proceedings of the First International Congress on Toxicology: toxicology as a predictive science. New York: Academic Press. p. 143-164.
Cornfield, J.C. (1977) Carcinogenic risk assessment. Science 198:693-699.
Cornish, H.H.; Ling, B.P.; Barth, M.L. (1973) Phenobarbital and organic solvent toxicity. Am. Ind. Hyg. Assoc. J. 34:487-492.
Crosby, D.G. (1969) The nonmetabolic decomposition of pesticides. Ann. N.Y. Acad. Sci. 160:82-96.
Crump, K.S.; Hoel, D.G.; Langley, C.H.; Peto, R. (1976) Fundamental carcinogenic processes and their implications for low dose risk assessment. Cancer Res. 36:2973-2979.
Crump, K.S.; Howe, R.B.; Masterman, M.D.; Watson, W.W. (1981) RANK 81: A Fortran program for risk assessment using time-to-occurrence dose-response data. Prepared for U.S. Department of Health and Human Services, National Institute of Environmental Health Services. 20p. Available from: the principal author, K.S. Crump and Co., Inc., Ruston, LA.
Donigian, A.S., Jr.; Yo, T.Y.R.; Shanahan, E.W. (1983) Rapid assessment of potential ground-water contamination under emergency response conditions. EPA-600/8-83-030. Prepared by Anderson-Nichols & Co., Inc., Palo Alto, CA. Washington, D.C.: U.S. Environmental Protection Agency, Office of Health and Environmental Assessment. 147p.
Fairchild, E.J. (1983) Concepts of effects of combined exposures to chemicals. In: Andrzejewski, S.W.; Tarkowski, S.; eds. Health effects of combined exposures to chemicals in work and community environments: proceedings of a course [held] October 18-22, 1982, Lodz, Poland. European Cooperation on Environmental Health Aspects of the Control of Chemicals: interim document 11. Copenhagen: World Health Organization, Regional Office for Europe. p. 15-43.
Farmer, J.H.; Kodell, R.L.; Gaylor, D.W. (1982) Estimation and extrapolation of tumor probabilities from a mouse bioassay with survival/sacrifice components. Risk Anal. 2:27-34.
Ferguson, R.K.; Vernon, R.J. (1970) Trichloroethylene in combination with CNS drugs. Effects on visual-motor tests. Arch. Environ. Health 20:462-467.
Food Safety Council. (1978) Proposed system for food safety assessment. Report of the Scientific Committee of the Food Safety Council. Columbia, MD: The Council. 136p.
Fred C. Hart Associates, Inc. (1982) Assessment of hazardous waste mismanagement damage case histories. 1984 reprint. EPA/530-SW-84-002. Washington, DC: U.S. Environmental Protection Agency, Office of Solid Waste and Emergency Response. 444p. Available from: National Technical Information Service, Springfield, VA.
Goyer, R.A. (1983) Introduction and overview. Environ. Health Perspect. 48:1-2.
Grice, H.C.; Arnold, D.L.; Blumenthal, H.; Emmerson, J.L.; Krewski, D.; editors. (1984) Selection of doses in chronic toxicity/carcinogenicity studies. In: Grice, H.C., ed. Current issues in toxicology. New York: Springer-Verlag. p. 3-49.
Grice, H.C.; Burek, J.D; editors. (1984) Age-associated (geriatric) pathology:

its impact on long-term toxicity studies. In: Grice, H.C., ed. Current issues in toxicology. New York: Springer-Verlag. p. 51-117.

Grizzle, J.M.; Melius, P.; Strength, D.R. (1984) Papillomas on fish exposed to chlorinated wastewater effluent. J. Nat. Cancer Inst. 73:1133-1142.

Harshbarger, J. (1983) Statements [and testimony] of John Harshbarger, director of Registry of Tumors in Lower Animals, National Museum of Natural History; and John J. Black, senior cancer research scientist, Roswell Park Memorial Institute. In: U.S. Congress, House of Representatives. Merchant Marine and Fisheries Committee. Fish cancer epidemics oversight. Hearing before the Subcommittee on Fisheries and Wildlife Conservation and the Environment. 98th Cong., 1st session. Serial no. 98-40. Washington, DC: The Committee. p. 5-14.

Harshbarger, J.C. (1977) Role of the Registry of Tumors in Lower Animals in the study of environmental carcinogenesis in aquatic animals. Ann. N.Y. Acad. Sci. 298:280-289.

Harshbarger, J.C. (1984) Pseudoneoplasms in ectothermic animals. In: U.S. Department of Health and Human Services, National Cancer Institute. Use of small fish species in carcinogenicity testing: proceedings of a symposium held Dec. 8-10, 1981, at Lister Hill Center, Bethesda, MD. National Cancer Institute Monograph 65. NIH publication no. 84-2653. Bethesda, MD: The Institute. p. 251-273.

Hart, R.W.; Setlow, R.B. (1974) Correlation between deoxyribonucleic acid excision-repair and life-span in a number of mammalian species. Proc. Nat. Acad. Sci. 71:2169-2173.

Hartley, H.O.; Sielken, R.L., Jr. (1977) Estimation of "safe doses" in carcinogenic experiments. J. Environ. Pathol. Toxicol. 1:241-278.

Helmes, C.T.; Atkinson, D.L.; Jaffer, J.; Sigman, C.C.; Thompson, K.L.; Kelsey, M.I.; Kraybill, H.F.; Munn, J.I. (1982) Evaluation and classification of the potential carcinogenicity of organic air pollutants. J. Environ. Sci. Health A17:321-388.

Ikeda, M. (1983) Interactions and health aspects of exposure to mixtures of organic solvents. In: Andrzejewski, S.W.; Tarkowski, S.; eds. Health effects of combined exposures to chemicals in work and community environments: proceedings of a course [held] October 18-22, 1982, Lodz, Poland. European Cooperation on Environmental Health Aspects of the Control of Chemicals: interim document 11. Copenhagen: World Health Organization, Regional Office for Europe. p. 77-97.

Indulski, J. (1983) Industry-environment of combined hazards. In: Andrzejewski, S.W.; Tarkowski, S.; eds. Health effects of combined exposures to chemicals in work and community environments: proceedings of a course [held] October 18-22, 1982, Lodz, Poland. European Cooperation on Environmental Health Aspects of the Control of Chemicals: interim document 11. Copenhagen: World Health Organization, Regional Office for Europe. p. 1-14.

Kaloyanova, F. (1983) Interactions of pesticides. In: Andrzejewski, S.W.; Tarkowski, S.; eds. Health effects of combined exposures to chemicals in work and community environments: proceedings of a course [held] October 18-22 ,1982, Lodz, Poland. European Cooperation on Environmental Health Aspects of the Control of Chemicals: interim document 11. Copenhagen: World Health Organization, Regional Office for Europe. p. 165-195.

Kimbrough, R.D. (1982) Centers for Disease Control's human health assessment approach involving improperly disposed chemical wastes. In: Highland,

J.H., ed. Hazardous waste disposal: assessing the problem. [Proceedings of a conference sponsored by the Society for Occupational and Environmental Health, held December 1980, Washington, DC.]. Ann Arbor, MI: Ann Arbor Science/The Butterworth Group [now Stoneham, MA: Butterworth Medical and Scientific Publishers]. p. 121-127.

Kraybill, H.F. (1980) Evaluation of public health aspects of carcinogenic/-mutagenic biorefractories in drinking water. Prev. Med. 9:212-218.

Kreiss, K.; Zack, M.M.; Kimbrough, R.D.; Needham, L.L.; Smrek, A.L.; Jones, B.T. (1981) Cross-sectional study of a community with exceptional exposure to DDT. J. Am. Med. Assoc. 245:1926-1930.

Krewski, D.; Crump, K.S.; Farmer, J.; Gaylor, D.W.; Howe, R.; Portier, C.; Salsburg, D.; Sielken, R.L.; Van Ryzin, J. (1983) Comparison of statistical methods for low dose extrapolation utilizing time-to-tumor data. Fundam. Appl. Toxicol. 3:140-160.

Lichtenstein, E.P.; Liang, T.T.; Anderegg, B.N. (1973) Synergism of insecticides by herbicides. Science 181:847-849.

Lower, G.M., Jr.; Bryan, G.T. (1973) Enzymatic N-acetylation of carcinogenic aromatic amines by liver cytosol of species displaying different organ susceptibilities. Biochem. Pharmacol. 22:1581-1588.

Malins, D.C.; Myers, M.S.; Roubal, W.T. (1983) Organic free radicals associated with idiopathic liver lesions of English sole (Parophrys vetulus) from polluted marine environments. Environ. Sci. Technol. 17:679-685.

Mantel, N.; Bryan, W.R. (1961) "Safety" testing of carcinogenic agents. J. Nat. Cancer Inst. 27:455-470.

Maugh, T.H., II. (1978) Chemical carcinogens: how dangerous are low doses? Science 202:37-41.

McDonald, A.D.; McDonald, J.C. (1973) Epidemiologic surveillance of mesothelioma in Canada. Can. Med. Assoc. J. 109:359-362.

Metcalfe, C.D.; Sonstegard, R.A. (1984) Microinjection of carcinogens into rainbow trout embryos: an in vivo carcinogenesis assay. J. Nat. Cancer Inst. 73:1125-1132.

Metcalfe, J.L.; Fox, M.E.; Carey, J.H. (1984) Aquatic leeches (Hirudinea) as bio-indicators of organic chemical contaminants in freshwater ecosystems. Chemosphere 13:143-150.

Mitre Corporation. (1983) Computer printouts of National Priorities List data summaries for the 546 final and proposed sites and 881 sites currently on the data base as of September 7, 1983. Prepared for the U.S. Environmental Protection Agency. 94p. Available from: Metrek Division, Mitre Corporation, McLean, VA.

Murphy, S.D. (1983) General principles in the assessment of toxicity of chemical mixtures. Environ. Health Perspect. 48:141-144.

NAS/NAE/NRC SEE National Academy of Sciences; and National Academy of Engineering; and National Research Council

National Academy of Sciences; and National Academy of Engineering; and National Research Council. (1975) Principles for evaluating chemicals in the environment. A report of the Committee for the Working Conference on Principles of Protocols for Evaluating Chemicals in the Environment. Washington, DC: National Academy of Science. 454p.

National Research Council. Commitee on the Institutional Means for Assessment of Risk to Public Health. (1983) Risk assessment in the federal government: managing the process. Washington, DC: National Academy Press.

National Research Council. Panel on Evaluation of Hazards Associated with Maritime Personnel Exposed to Multiple Cargo Vapors. (1980) Principles of toxicological interactions associated with multiple chemical exposures. Washington, DC: National Academy Press. 197p.

National Research Council. Steering Committee on Identification of Toxic and Potentially Toxic Chemicals for Consideration of the National Toxicology Program. (1984) Toxicity testing: strategies to determine needs and priorities. Washington, DC: National Academy Press. 382p.

National Toxicology Program Board of Scientific Counselors. (1984) Draft report of the Ad Hoc Panel on Chemical Carcinogenesis Testing and Evaluation. 294p. Available from: The Board, Research Triangle Park, NC.

Neal, R. A. (1983) Protocol for testing the toxicity of chemical mixtures. Environ. Health Perspect. 48:137-139.

Newberne, P.M. (1976) Environmental modifiers of susceptibility to carcinogenesis. Cancer Detect. Prev. 1:129-173.

NRC SEE National Research Council

Odum, E.P. (1959) Fundamentals of ecology. 2d ed. 1967 reprint. Philadelphia: W.B. Saunders Company. 546p.

Pierce, K.V.; McCain, B.B.; Wellings, S.R. (1978) Pathology of hepatomas and other liver abnormalities in English sole (Parophrys vetulus) from the Duwamish River estuary, Seattle, Washington. J. Nat. Cancer Inst. 60: 1445-1453.

Rall, D.P. (1969) Difficulties in extrapolating the results of toxicity studies in laboratory animals to man. Environ. Res. 2:360-367.

Reinberg, A. (1978) Clinical chronopharmacology, an experimental basis for chronotherapy. Arzneim Forsch. 28(10a): 1861-1867.

Rose, F.L.; Harshbarger, J.C. (1977) Neoplastic and possibly related skin lesions in neotenic tiger salamanders from a sewage lagoon. Science 196:315-317.

Rowley, M.H.; Christian, J.J.; Basu, D.K.; Pawlikowski, M.A.; Paigen, B. (1983) Use of small mammals (voles) to assess a hazardous waste site at Love Canal, Niagara Falls, New York. Arch. Environ. Contam. Toxicol. 12:383-397.

Searle, C.E., editor. (1976) Chemical Carcinogens. ACS monograph 173. Washington, DC: American Chemical Society. 788p.

Selikoff, I.J.; Hammond, E.C.; Churg, J. (1968) Asbestos exposure, smoking, and neoplasia. J. Am. Med. Assoc. 204:104-112.

Upton, A.C. (1983) Environmental standards for ionizing radiation: theoretical basis for dose-response curves. Environ. Health Perspect. 52:31-39.

U.S. Congress, Senate. Environment and Public Works Committee. (1980) Possible interactions of selected chemicals identified in Love Canal. Appendix E. In: The Committee. Health effects of toxic pollution: a report from the Surgeon General and a brief review of selected environmental contamination incidents with a potential for health effects. Committee print. 96th Cong., 2d sess. Prepared by the Surgeon General and the Congressional Research Service of the Library of Congress. Washington, DC: The Committee. p. 133-138.

USC/SENATE SEE U.S. Congress, Senate

U.S. Department of Health and Human Services, Centers for Disease Control. (1984) System for prevention, assessment, and control of exposures and health effects from hazardous sites (S.P.A.C.E. for Health). Prepared by the Center for Environmental Health and the Center for Professional

Development and Training. 187p. Available from: Chronic Diseases Division, Center for Environmental Health, Centers for Disease Control, U.S. Department of Health and Human Services, Atlanta, GA.

U.S. Department of Health and Human Services, National Cancer Institute. (1984) Use of small fish species in carcinogenicity testing: proceedings of a symposium held Dec. 8-10, 1981, at Lister Hill Center, Bethesda, MD. National Cancer Institute Monograph 65. NIH publication No. 84-2653. Bethesda, MD: The Institute. 409p.

U.S. Department of Health, Education and Welfare; Food and Drug Administration. (1977) Chemical compounds in food-producing animals: criteria and procedures for evaluating assays for carcinogenic residues in edible products of animals. Fed. Regist. 42:10112-10437.

U.S. Department of Health, Education, and Welfare; National Institute for Occupational Safety and Health. (1979) Interaction between smoking and occupational exposures. In: The Department, Office on Smoking and Health. Smoking and health: a report of the Surgeon General. Washington, DC: U.S. Department of Health, Education, and Welfare, Public Health Service. p. 7.1- 7.25.

U.S. Department of Health, Education, and Welfare; Office on Smoking and Health. (1979) Introduction and summary. In: The Office. Smoking and health: a report of the Surgeon-General. DHEW publication no. (PHS)79-50066. Washington, DC: U.S. Department of Health, Education, and Welfare, Public Health Service. p. 1.1-1.35.

USDHEW/FDA SEE U.S. Department of Health, Education, and Welfare; Food and Drug Administration

USDHEW/NIOSH SEE U.S. Department of Health, Education, and Welfare; National Institute for Occupational Safety and Health

USDHEW/OSH SEE U.S. Department of Health, Education, and Welfare; Office of Smoking and Health

USDHHS/CDC SEE U.S. Department of Health and Human Services, Centers for Disease Control

USDHHS/NCI SEE U.S. Department of Health and Human Services, National Cancer Institute

U.S. Environmental Protection Agency. (1976) Health risk and economic impact assessments of suspected carcinogens: interim procedures & guidelines. Fed. Regist. 41:21102-21105.

U.S. Environmental Protection Agency. (1979) Proposed health effects test standards for Toxic Substances Control Act test rules; proposed good laboratory practice standards for health effects; extension of comment period and rescheduling of public meeting. Fed. Regist. 44:44054-44093.

USEPA SEE U.S. Environmental Protection Agency

Van Duuren, B.L.; Smith, A.C.; Melchionne, S.M. (1978) Effect of aging in two-stage carcinogenesis on mouse skin with phorbol myristate acetate as promoting agent. Cancer Res. 38:865-866.

Veldstra, H. (1956) Synergism and potentiation with special references to the combination of structural analogues. Pharmacol. Rev. 8:339-387.

Versar, Inc. (1982) Phase II: methodology for assessing exposures from disposal of toxic substances. Draft final report. Prepared for U.S. Environmental Protection Agency, Exposure Evaluation Division. 164p. Available from: [U.S. Environmental Protection Agency, Office of Toxic Substances, Washington, DC].

Wallace, L.; Pellizzari, E.; Hartwell, T.; Zelon, H.; Sparacino, C.; Whitmore,

R. ([1984]) Analyses of exhaled breath of 355 urban residents for volatile organic compounds. 6p. Available from: U.S. Environmental Protection Agency, Washington, DC.

Weisburger, J.H. (1979) Mechanisms of action of diet as a carcinogen. Cancer 43:1987-1995.

Weisburger, J.H.; Williams, G.M. (1982) Chemical carcinogenesis. In: Holland, J.F.; Frei, E., III.; eds. Cancer medicine. 2d ed. Philadelphia: Lea & Febiger. p. 42-95.

Westrick, J.J.; Mello, J.W.; Thomas, R.F. (1983) Ground water supply survey: summary of volatile organic contaminant occurrence data. 49p. Available from: U.S. Environmental Protection Agency, Office of Drinking Water, Cincinnati, OH.

WHO SEE World Health Organization

Windemuller, F.J.B.; Ettema, J.H. (1978) Effects of combined exposures to trichloroethylene and alcohol on mental capacity. Int. Arch. Occup. Environ. Health 41:77-85.

Witschi, H.R.; Munro, N.B. (1983) Problems of toxicological testing of complex mixtures. [Proceedings of the Third International Congress on Toxicology, San Diego, CA, August 28-September 3, 1983]. Oak Ridge, TN: U.S. Department of Energy, Oak Ridge National Laboratory. 10p. Available from: National Technical Information Service, Springfield, VA.

World Health Organization. (1981) Health effects of combined exposures in the work environment. Report of a WHO Expert Committee. WHO Technical Report Series No. 662. Geneva: The Organization. 76p.

7

Chemical Exposure Health Studies

EFFECTS OF EXPOSURES OF WELL-DEFINED, SPECIFIC POPULATIONS

This chapter identifies human health effects associated with specific chemicals on the basis of case reports, case series, and epidemiologic studies of occupational cohorts. In addition, published and unpublished reports of environmental chemical exposure incidents not directly related to disposal sites were reviewed to determine if the observed health effects were similar to occupational observations and whether such observations could be useful in predicting outcomes at chemical disposal sites.

Effects of Occupational Exposures

The occupational medicine and other human toxicological literature was reviewed for the representative hazardous chemicals in disposal sites identified in Chapter 2 (see Table 2-3). The list of chemicals and their associated health effects which have been reported in the occupational medicine literature are presented in Table 7-1. The vast number of references prohibited critical review of the quality of the data, but most of the references are clinical case reports and case series. More information is available for dermatologic and central nervous system (CNS) effects than for other target organs, and this is reflected in the table. The severity and immediacy of responses were dose-related, and were directly related to route of exposure. In high-dose situations, skin, CNS, liver, kidney, and reproductive outcomes predominated. In moderate exposure situations, hematologic abnormalities were frequent. In low-dose situations, detectable health effects were rarely reported; many exposed individuals were asymptomatic or showed only transient subclinical illness.

It became apparent from the review of the literature that there were shortcomings both in the documentation and quantitation of chemical exposure and in the relationship of such measurements to appropriate health effect endpoints.

TABLE 7–1. Effects of occupational exposures to selected chemicals.

Chemical and Effects	Source
Acetone	
Skin irritant and contact allergen	Pittelkow, 1975; vonKrough and Maibach, 1983
Arsenic and arsenic compounds	
Hyperpigmentation	
Pigment discoloration (brown)	Granstein and Sober, 1981; Adams, 1983
Skin irritant and contact allergen	Pittelkow, 1975; vonKrough and Maibach, 1983
Skin neoplasia	WHO/IARC, 1980
Neurasthenia, irritability, other mild CNS symptoms	
Peripheral neuropathies	Heyman et al., 1956; Feldman et al., 1979
Acute hepatocellular injury	Salaman et al., 1944
Subacute hepatic necrosis with possible cirrhosis	Salaman et al., 1944
Hepatic neoplasia (uncertain)	WHO/IARC, 1979a
Embryotoxicity (suspected)	Clarkson et al., 1983
Teratogenesis (suspected)	Wilson, 1977
Spontaneous abortion/fetal death (suspected)	Clarkson et al., 1983
Developmental disabilities (suspected)	Clarkson et al., 1983
Anemia	
Arsenic Trioxide	
Arsine	Fowler and Weissberg, 1974; Pinto, 1976
Heart disease, direct effect	Axelson et al., 1978; Rosenberg, 1974
Asbestos	
Pneumoconioses	
Fibrosis	
Mesothelioma	
Other lung and respiratory tract neoplasia	
Pharyngeal, esophageal, laryngeal neoplasia	
Stomach and colon neoplasia	
Benzene	
Skin irritant	Pittelkow, 1975; vonKrough and Maibach, 1983
Female infertility (suspected)	Zenz, 1980
Embryotoxicity (suspected)	Zenz, 1980
Teratogenesis (suspected)	Wilson, 1977
CNS depression	Carpenter et al., 1944; Husman and Karli, 1980.
Bone marrow depression	Aksoy et al., 1971; Pagnotto et al., 1979
Aplastic anemia	Aksoy et al., 1972
Leukemia and lymphoma	WHO/IARC, 1982b
Cadmium	
Male infertility (suspected)	Zenz, 1980
Female infertility	Zenz, 1980
Teratogenesis (suspected)	Barr, 1973
Spontaneous abortion/fetal death (suspected)	Clarkson et al., 1983
Neonatal death (suspected)	Chernoff, 1973
Low birth weight (suspected)	USDHEW/NIOSH, 1976
Developmental disabilities (suspected)	Webster, 1978

TABLE 7–1. Effects of occupational exposures to selected chemicals. (Continued)

Chemical and Effects	Source
Cadmium (continued)	
Chronic tubal or interstitial renal disease	Buchet et al., 1980
Kidney and prostate neoplasia (uncertain)	WHO/IARC, 1979c
Pulmonary edema	
Lung and respiratory tract neoplasia (uncertain)	WHO/IARC, 1979c
Hypertension	Schroeder, 1965
Carbon tetrachloride	
Acute hepatocellular injury (hepatitis)	Recknagel, 1967; Poindexter and Greene, 1934; Smetana, 1939
Hepatic cirrhosis	Poindexter and Greene, 1934
Hepatic neoplasia (uncertain)	WHO/IARC, 1979a
Acute renal failure; direct tissue injury	Folland et al., 1976; New et al., 1962
Chromium and chromium compounds	
Pigment discoloration (yellow)	
Bichromate	Granstein and Sober, 1981
Skin irritant	Pittelkow, 1975; von Krough and Maibach, 1983
Skin irritant and contact allergen	
Chromic Acid	Pittelkow, 1975; von Krough and Maibach, 1983
Asthma	
Lung and respiratory tract neoplasia	
Pharyngeal, esophageal, laryngeal neoplasia (uncertain)	WHO/IARC, 1979c
Copper and copper compounds	
Pigment discoloration (green)	
Copper salts	Granstein and Sober, 1981
Anemia	
Copper sulfate	Fairbanks, 1967
Cyanide (free cyanide ion)	
Skin irritant	
Sodium or potassium cyanide	Pittelkow, 1975; von Krough and Maibach, 1983
Asphyxiation	
Hydrogen cyanide	
Asthma	
Isocyanates	
Hypersensitivity pneumonitis	
Toluene diisocyanate	Fink, 1983
DDT	
Peripheral neuropathies	
Ataxic gait	
Tremors	
Subacute hepatic necrosis with possible cirrhosis	Morgan and Lin, 1978
Male infertility	

TABLE 7–1. Effects of occupational exposures to selected chemicals. (Continued)

Chemical and Effects	Source
Dichlorobenzene (unspecified)	
Leukemia and lymphoma (uncertain)	WHO/IARC, 1979c
Dichloromethane	
CNS depression	Moskowitz and Shapiro, 1952
Heart disease; indrect effect	Stewart et al., 1972
Lead and lead compounds	
Pigment discoloration (blue-gray)	
Lead salts	Granstein and Sober, 1981
Neurasthenia, irritability, other mild CNS symptoms	Grandjean et al., 1978; Needleman et al., 1979
Emotional instability	Grandjean et al., 1978; Needleman et al., 1979
Memory impairment	Grandjean et al., 1978; Needleman et al., 1979
Peripheral neuropathies	Araki and Honma, 1976; Baker et al., 1979; Baloh et al., 1979; Feldman et al., 1977; Winegar et al., 1977
Seizures	
Intracranial pressure	
Male infertility	Lancranjan et al., 1975; Zenz, 1980
Female infertility	Zenz, 1980
Embryotoxicity	Clarkson et al., 1983
Teratogenesis (suspected)	Wilson, 1977
Spontaneous abortion/fetal death	Nordstrom et al., 1978b, 1979; Zenz, 1980
Neonatal death	Zenz, 1980
Low birth weight	Fahim et al, 1976; Nordstrom et al., 1978a
Developmental disabilities	Beattie et al., 1975; Moore et al., 1977
Childhood neoplasia (cancer in offspring of exposed individuals)	Kantor et al., 1979
Anemia	Feldman, 1978; Griggs, 1964; Travers et al., 1956
Bone marrow depression	
Chronic tubal or interstitial renal disease	Lilis et al., 1968; Morgan et al., 1966; Weden et al., 1979
Kidney and bladder neoplasia (uncertain)	WHO/IARC, 1979c
Lung and respiratory tract neoplasia (uncertain)	WHO/IARC, 1979c
Stomach neoplasia	
Colon neoplasia (uncertain)	WHO/IARC, 1979c
Hypertension (uncertain)	
Mercury and mercury compounds	
Hyperpigmentation	
Pigment discoloration (black)	
Mercury salts	Granstein and Sober, 1981
Contact allergen	
Phenyl mercuric proprionate	Pittelkow, 1975; von Krough and Maibach, 1983
Skin irritant and contact allergen	
Mercury salts	Pittelkow, 1975; von Krough and Maibach, 1983
Neurasthenia, irritability, other mild CNS symptoms	Miller et al., 1975
Impaired psychomotor function	Miller et al., 1975

TABLE 7–1. Effects of occupational exposures to selected chemicals. (Continued)

Chemical and Effects	Source
Mercury and mercury compounds (continued)	
Peripheral neuropathies	Kurland et al., 1960; Langolf et al., 1978
Ataxic gait	Smith et al., 1970
Myoclonus	Smith et al., 1970
Tremors	Smith et al., 1970
Impaired visual acuity	Smith et al., 1970
Other behavioral changes	Smith et al., 1970
Embryotoxicity (suspected)	Zenz, 1980
Teratogenesis	Harada, 1978
Spontaneous abortion/fetal death (suspected)	Clarkson et al., 1983
Neonatal death (suspected)	Zenz, 1980
Low birth weight (suspected)	Zenz, 1980
Developmental disabilities	Harada, 1978
Acute renal failure	Battigelli, 1960; Friberg et al., 1953; Kazantzis, 1965
Naphthalene	
Anemia	Dawson et al., 1958; Valaes et al., 1963
Nickel and nickel compounds	
Skin erritant and contact allergen	
Nickel salts and oxide	Pittelkow, 1975; von Krough and Maibach, 1983
Headache	
Asthma	
Nickel sulfate	
Lung and respiratory tract neoplasia (uncertain)	WHO/IARC, 1979c
Pharyngeal, esophageal, laryngeal neoplasia (uncertain)	WHO/IARC, 1979c
Phenol	
Skin irritant	Pittelkow, 1975; von Krough and Maibach, 1983
Asthma	
Polychlorinated biphenyls	
Folliculitis and acneform dermatoses	Fischbein et al., 1979; Kuratsune et al., 1972
Skin neoplasia (uncertain)	WHO/IARC, 1979a, 1979b
Acute hepatocellular injury (hepatitis)	
Subacute hepatic necrosis with possible cirrhosis	Strauss, 1944
Male infertility	Schwartz, 1936
Female infertility (suspected)	USDHEW/NIOSH, 1977
Embryotoxicity	Funatsu et al., 1972; Yoshimura, 1974
Teratogenesis	Miller, 1971
Spontaneous abortion/fetal death	Miller, 1971
Neonatal death (suspected)	USDHEW/NIOSH, 1977
Low birth weight	Miller, 1971
Developmental disabilities (suspected)	USDHEW/NIOSH, 1977
Immune suppression (suspected)	

TABLE 7–1. Effects of occupational exposures to selected chemicals. (Continued)

Chemical and Effects	Source
TCDD	
Folliculitis and acneform dermatoses	Pazderova-Vejlupkova et al., 1981
Acute hepatocellular injury (hepatitis)	
Hepatic neoplasia (uncertain)	WHO/IARC, 1979a
Teratogenesis (suspected)	Homberger et al., 1979
Spontaneous abortion/fetal death (suspected)	Homberger et al., 1979
1,1,2,2-Tetrachloroethane	
Acute hepatocellular injury (hepatitis)	Gurney, 1943
Subacute hepatic necrosis with possible cirrhosis	
Tetrachloroethylene	
Impaired psychomotor function	
Teratogenesis (suspected)	Schwetz, 1975
Low birth weight (suspected)	Schwetz et al., 1975
Cardiac toxicity – direct effect	Abedin et al., 1980
Toluene	
Memory impairment	Husman, 1980; Lindstrom, 1973
CNS depression	Carpenter et al., 1944; Husman and Karli, 1980
Ataxic gait	Grabski, 1961; Knox and Nelson, 1966
Subacute hepatic necrosis with possible cirrhosis	Greenburg et al., 1942
Acute renal failure	Taher et al., 1974
Trichloroethane (unspecified)	
CNS depression	Stahl et al., 1969
1,1,1-Trichloroethane	
CNS depression	Caplan et al., 1976
Heart disease; direct effect	WHO/IARC, 1979c
Trichloroethylene	
Skin irritant and contact allergen	Pittelkow, 1975; von Krough and Maibach, 1983
CNS depression	Lloyd et al., 1975
Peripheral neuropathies	Feldman et al., 1970
Miscellaneous neurological conditions	Feldman et al., 1970; Waters et al., 1977
Leukemia and lymphoma (uncertain)	WHO/IARC, 1979c
Acute renal failure	Baerg and Kimberg, 1970; Gutch et al., 1965
Heart disease; direct effect	Kleinfeld and Tabershaw, 1954
Vinyl chloride	
Acroosteolysis, with or without Raynaud's phenomenon; sclerodermoid changes	Wilson et al., 1967; Harris and Adams, 1967
Miscellaneous skin reactions	
CNS neoplasia	WHO/IARC, 1979a
Subacute necrosis with possible cirrhosis	Thomas et al., 1975
Hepatic neoplasia	WHO/IARC, 1979c
Germ cell mutation (suspected)	Infante et al., 1976a

TABLE 7–1. Effects of occupational exposures to selected chemicals. (Continued)

Chemical and Effects	Source
Vinyl chloride (continued)	
Teratogenesis (suspected)	Edmonds et al., 1975, 1978; USDHHS/CDC, 1975
Spontaneous abortion/fetal death (suspected)	Infante et al., 1976a, 1976b
Leukemia and lymphoma (uncertain)	WHO/IARC, 1979c
Lung and respiratory tract neoplasia	
Stomach and colon neoplasia (uncertain)	WHO/IARC, 1979c
Vinylidene chloride	
Lung and respiratory tract neoplasia	
Xylene	
Memory impairment	Husman, 1980; Lindstrom, 1973
CNS depression	Carpenter et al., 1944; Husman and Karli, 1980
Zinc and zinc compounds	
Skin irritant and contact allergen	
Promethazinc	Pittelkow, 1975; von Krough and Maibach, 1983

Effects of Accidental Chemical Contamination Incidents

Appendix 7-I summarizes an extensive, although far from exhaustive, review of health effects reported to have resulted from 28 accidental releases (unrelated to chemical disposal sites) of chemicals into the environment causing contamination of food, drinking water, and other media. The chemicals and circumstances of exposure are described, followed by the study design employed to assess health effects, the method of assessing exposure, and the study findings. Included are 13 incidents involving miscellaneous single chemicals, 5 of which involved polyhalogenated biphenyls; 6 incidents involving organic pesticides; and 9 involving trace metals or organo-metallic compounds.

Predominant health effects were predictable on the basis of data from occupational health studies. In general, as for occupational studies, reported effects were dose-dependent and related to route of exposure. Ingestion was the most frequently reported route of exposure in the environmental contamination incidents reviewed. In all situations in which food supplies were contaminated, sequelae were rapid and severe. Symptoms were generally similar to those noted in occupational exposures (i.e., as summarized in Table 7-1), with skin and CNS effects predominating. Contamination of water supplies, in presumably much lower concentrations, did not always result in clear-cut responses; only in relatively high-dose situations were clear clinical illnesses apparent. In situations in which toxic chemicals could not be identified in food or water, unequivocal adverse health effects were rarely detected. Direct contact is important in the occupational environment where dermal and CNS effects have been reported, but seems to be a less likely route of exposure for other environmental contamination incidents and for populations around chemical disposal sites. When direct contact does occur, dermal and CNS effects are likely. Inhalation, unless at relatively high concentrations or as a result of fire, is likewise not as important a route of exposure, although hematopoietic, reproductive, and CNS effects have been reported in the occupational literature.

Attempts have been made to relate health effects associated with these environmental contamination incidents to possible effects of similar chemicals present at chemical disposal sites. However, these disposal site exposures tend to be low-dose, and direct measures of exposure in such situations are rarely available, with the exception of those chemicals that accumulate in body tissues (e.g., PBB's, PCB's, DDT, and lead). In the absence of direct methods of ascertaining exposure, it is extremely difficult to detect a dose-response relationship in exposed populations and therefore difficult to show an association between health status and chemical exposure. The relationships between environmental chemicals and adverse health effects will only be feasible when the population-at-risk is clearly defined, when the chemical exposure can be quantified, and when the disease endpoint is clearly delineated.

A summary of non-disposal site contamination incidents and their reported health effects follows:

Contamination of food supplies

Contamination of food supplies by organic pesticides and fungicides resulted in varying effects. Endrin (Coble et al., 1967; Davies and Lewis, 1956) and parathion (Diggory et al., 1977; Kanagaratnam et al., 1960) caused severe CNS effects, hexachlorobenzene (Cam and Nigogosyan, 1963; Schmid, 1960) caused a severe cutaneous porphyria, whereas low-dose DDT exposure (Kreiss et al., 1981b) caused no observed effect.

Contamination of food by organo-metallic compounds such as the episode of methylmercury poisoning in Iraq in 1971 resulted in severe dose-dependent CNS defects (Bakir et al., 1973). Bioconcentration of organic mercury in fish and shellfish in Minamata Bay, Japan resulted in a severe epidemic of neurologic disease in adults (Kurland et al., 1960), and severe neurologic sequelae in children exposed in utero (Harada, 1978).

Other episodes of chemical contamination of food supplies resulted in various sequelae, such as liver disturbances from 4,4'-diaminodiphenylmethane (Kopelman et al., 1966), CNS disturbances from ortho-cresyl-phosphates (Smith and Spaulding, 1959; Vora et al., 1962), and chloracne, reproductive disorders, and other disturbances from PCB's and polychlorinated dibenzofurans (PCDF) (Chen et al., 1980; Kuratsune et al., 1972, 1976).

Contamination of water supplies

One report of phenol in drinking water noted a syndrome consisting of diarrhea, dark urine, mouth sores, and burning in the mouth (Baker et al., 1978). No differences in clinical or laboratory findings were noted between exposed individuals and comparisons when retested six months after the episode. Acrylamide contamination of a well used by a Japanese family was reported to have caused severe CNS disturbances, although blood tests and liver enzyme levels were apparently normal (Igisu et al., 1975).

Peripheral nervous system and hematological effects were noted in families in rural areas of Scotland using drinking water contaminated with lead from lead-lined pipes in their homes (Beattie et al., 1972). These effects were similar to those noted in occupational exposures.

Certain areas of Taiwan have very high natural background levels of arsenic in groundwater which have been associated with high rates of skin cancer and "black foot disease," a disorder of the peripheral circulation. Similar studies in the United States have failed to replicate this finding although average concentrations were lower (mean of 16.5 ppb with range of 0-2150 ppb in rural areas; mean of 48 ppb with range of 0-860 ppb in urban areas) (Morton et al., 1976) than in Taiwan (most wells were in the range of 400-600 ppb with a total range of 10-1820 ppb) (Tseng, 1977).

Cadmium was released from a Japanese chemical factory into a river that was subsequently used both as a source of drinking water and for irrigation of crops (Friberg et al., 1974). Individuals who consumed this water developed a syndrome involving bone pain ("itai-itai" or "ouch-ouch"), proteinuria and glucosuria. It is unclear, in spite of a number of reports, to what extent this syndrome is attributable to cadmium alone; it is likely that an interaction between cadmium and nutritional status was involved.

Contamination due to unusual environmental circumstances

An explosion occurred in a plastics manufacturing plant in Seveso, Italy, where 2,3,7,8-tetrachlorodibenzo-p-dioxin (TCDD) was routinely produced as a by-product in the production of trichlorophenol (Dardanoni and Fara, 1982; De Carli et al., 1982; Sirchia et al., 1982). Following the explosion, TCDD was identified in several areas in the vicinity of the plant. The residential population began to experience various skin conditions, including chloracne, within a few days after the incident. TCDD concentrations varied in soil around the plant and no clear pattern or gradient emerged. A survey for dermatologic problems did not observe a correlation between frequency of problems and environmental concentrations except in the immediate vicinity of the plant. Cytogenetic and immunologic studies of subsets of the exposed population did not reveal alterations from baseline comparison levels. No respiratory symptoms were reported. As of three years after the incident, no sequelae other than chloracne and minor dermatological problems were apparent.

Another incident involving TCDD occurred in Missouri, where waste oil containing TCDD was sprayed on residential, recreational and work areas for dust control. Once the problem was discovered, a case-comparison study (Stehr et al., 1984) was initiated to examine health effects in a group of individuals at high risk of exposure as compared to those not at risk. No cases of chloracne or porphyria cutanea tarda were noted, nor were any discernible differences noted in immune, renal, or hepatic function between exposed and comparison subjects. The authors emphasized that this should be considered a pilot study and also noted that a method for measuring the body burden of TCDD would be extremely helpful in studies of this sort.

Children in Idaho (Landrigan et al., 1976) and Texas (Landrigan et al., 1975a; 1975b) acquired high blood lead levels due to exposure to lead emitted from stacks at a lead smelter. No respiratory symptoms or overt physiologic deficits were observed in these children although findings in some cases were suggestive of subclinical motor neuropathy.

Polyhalogenated biphenyls have been released into the environment by contamination of animal feed with PBB's (Bekesi et al., 1978; Landrigan et al., 1979), PCB-contaminated sewage sludge later used as fertilizer (Baker et al., 1980), and presumably as PCB by-products of pesticide manufacturing released into a stream (Kreiss et al., 1981a). In each of these instances, a direct measure of exposure, in the form of serum levels of the chemical, was available. It is not yet apparent whether these serum PBB or PCB levels are associated with adverse health effects.

Methyl isocyanate gas was released into the air from a Bhopal, India industrial source, resulting in the death of more than 2,000 people from bronchospasm and acute pulmonary edema due to chemical pneumonia (Anonymous, 1984). An estimated 20,000 to 100,000 more sustained injuries involving lung damage, skin burns and eye ulcerations. As this incident occurred only recently, no chronic effects can be expected to be observed in the near future. No epidemiological studies have yet been undertaken and all figures reported are estimates.

PROLONGED LOW-DOSE EXPOSURES OF LARGE SEGMENTS OF THE POPULATION

Epidemiologic Evidence of Adverse Health Effects of Chemical Contamination of Water Supplies

Organic chemicals of anthropogenic sources are widely found in drinking water supplies. The problem is not a new one. In 1956, Middleton and Rosen (1956) examined raw and finished water from five mid-western U.S. cities and reported the presence of aromatic compounds, insecticides, kerosene, phenols, polycyclic hydrocarbon compounds, and synthetic detergents. With increasing sophistication of methods for chemical analysis, it is now possible to identify many organic chemicals in drinking water and to measure their concentrations in nanograms per liter. Summarizing reports from different countries, Kraybill (1981) noted that of 2,221 organic chemicals found in water supplies on a worldwide basis, 765 were present in drinking water. Of these, 20 (2.6%) were recognized carcinogens, 23 (3.0%) were suspected carcinogens, 18 (2.4%) were carcinogenic promoters, and 56 (7.3%) were mutagens. Most of the chemicals reported in these surveys are volatile or low molecular weight organics which can easily be recovered from water (Laseter and Dowty, 1977); they are estimated to account for only 10 to 20 percent of total organic chemicals present in drinking water. Methods to analyze the nonvolatile organic fraction are still evolving but most of the nonvolatile and high molecular weight compounds remain unidentified.

Organic chemicals enter water supply systems from a number of major sources. The largest amounts arise from industrial discharges, municipal waste water discharges, agricultural runoff, and decomposition of natural organic matter (humus). Other more localized sources include, among others, landfill leachates, accidental chemical spills, and leaking underground storage tanks.

Disinfection of drinking water by chlorination is another mechanism by which chemicals may enter the water supply. Within the past 10 years, investigators have identified several organic by-products of the chlorination disinfection process. Among these compounds are chloroform, a known animal carcinogen; bromodichloromethane; and dibromochloromethane. Chloroform invariably occurs in water which has been chlorinated, while it is absent or present at much lower concentrations in unchlorinated groundwater or in raw surface water prior to chlorination (Williamson, 1981). However, nonchlorinated groundwater is by no means free of organic chemicals, as shown in Table 7-2. The significance of this finding is that a number of epidemiologic studies of water quality and cancer risk use populations consuming groundwater or non-chlorinated water as a "nonexposed" standard for comparison for calculating cancer risk associated with exposure to surface or chlorinated water supplies. These populations may not be appropriate comparisons on this basis.

Identification and quantification of the array of organic chemicals in drinking water is a complex and costly procedure and, even now, most of these compounds are rarely measured and none are routinely monitored. Hence, one of the major problems in assessing human health risks associated with chemical contamination of drinking water is the measurement or estimation of human exposure. Several pertinent questions need to be addressed in this regard. Can the epidemiologist utilize quantifiable indices of exposure such as direct chemical analyses of specific or total chemical concentrations, or indirect measures of exposure such as chlorinated surface water vs. unchlorinated ground water? If direct measures of analysis can be used, how variable are concentrations of these chemicals over time and place, and what is the relevant time interval (latent period) between exposure and any adverse health effect? Are there any biological indicators which the epidemiologist can utilize to represent human exposure to organic chemicals in water?

Epidemiologic investigations of organic chemicals in water have been virtually limited to considerations of cancer risk. Isolated reports can be found that attempt to investigate other health effects; for example, Clark et al., (1982) reported a study of a population whose wells were contaminated by leachates from a pesticide disposal site and who reportedly manifested serum enzyme activities suggestive of liver dysfunction. However, most systematic epidemiologic studies of the past decade have focused on cancer outcomes. These studies can be divided into two broad categories which are determined by the unit of observation: a) aggregate risk studies in which geographical areas (counties, cities, states) are the unit of analysis; and, b) individual risk studies.

Aggregate risk studies can be performed quite rapidly because public data sources generally provide the required information on cancer mortality, demographic characteristics, and water quality for geographical areas which are the unit of analysis. Fourteen aggregate risk studies were reported for different geographical areas of the U.S. between 1975 and 1982 (Bean et al., 1982; Buncher, 1975; Cantor et al., 1978; Carlo and Mettlin, 1980; DeRouen and Diem, 1977; Harris et al., 1977; Hogan et al., 1979; Kuzma et al., 1977; Meinhardt et al., 1975; Page et al., 1976; Reiches, 1977; Salg, 1977; Spivey and Sloss, 1977; Tuthill and Moore, 1980) and are summarized in Appendix 7-II. The most common cancer sites statistically associated with various measures of population exposure to organic chemicals in water were bladder, stomach, and rectum; other cancer sites that showed statistically significant relationships were pancreas, lung, colon, blood cells, and the total urinary tract and gastrointestinal tract. There was considerable inconsistency among cancer sites associated with water quality in these fourteen studies. The problems with drawing causal inferences from aggregate risk studies are many and are inherent to the nature of the aggregate unit of analysis: first, there is great variability of individual exposure within a geographical unit; second, no individual measurements of exposure are known; third, potential confounders are seldom known and cannot be linked to individuals; and fourth, individual exposures cannot be linked to those persons who developed cancer. Aggregate

TABLE 7–2. Some organic chemicals detected in groundwater wells used for drinking water.

Chemical	Maximum Concentration Reported (ppb)*	Location*	Evidence for Carcinogenicity**
Acetone	3,000	New Jersey	NR
Benzene	330	New Jersey	Group 1 (1982b)
Bromoform	20	Delaware	NR
Butyl benzyl phthalate	38	New York	Group 3 (1982a)
Carbon tetrachloride	400	New Jersey	Group 2B (1982b)
Chloroform	490	New Jersey	Group 2B (1982b)
Chloromethane	44	Massachusetts	NR
Cyclohexane	540	New York	NR
Dibromochloropropane (DBCP)	137	Arizona	Group 2B (1979b)
Dibromochloromethane	55	New York	NR
1,1-Dichloroethane	7	Maine	NR
1,2-Dichloroethane	250	New Jersey	Group 2B (1979b)
1,1-Dichloroethylene	280	New Jersey	NR
1,2-Dichloroethylene	323	Massachusetts	NR
Dichloromethane	3,000	New Jersey	Group 3 (1982b)
Di-n-butyl phthalate	470	New York	NR
1,4-Dioxane	2,100	Massachusetts	Group 2B (1982b)
Ethylbenzene	2,000	New Jersey	NR
Ethylene dibromide (EDB)	300	Hawaii	Group 2B (1982b)
bis(2-Ethylhexyl)phthalate	170	New York	NR
gamma-Hexachlorocyclohexane (Lindane)	22	California	Group 2B (1982b)
Isopropyl benzene	290	New York	NR
Parathion	4.6	California	NR
Tetrachloroethylene	1,500	New Jersey	Group 3 (1982b)
Toluene	6,400	New Jersey	NR
1,1,1-Trichloroethane	5,440	Maine	Group 3 (1979b)
1,1,2-Trichloroethane	20	New York	Group 3 (1979b)
Trichloroethylene	27,300	Pennsylvania	Group 3 (1982b)
Trifluorotrichloroethane	135	New York	NR
Vinyl chloride	50	New York	Group 1 (1982b)
Xylene	300	New Jersey	NR

*As reported in USEOP/CEQ, 1981.
**As reported in WHO/IARC (year listed in parentheses):
Group 1: Sufficient evidence of carcinogenicity in humans.
Group 2A: Sufficient evidence of carcinogenicity in animals, limited evidence in humans.
Group 2B: Sufficient evidence of carcinogenicity in animals, inadequate data in humans.
Group 3: Cannot be classified due to inadequate data.
NR: Not reviewed by WHO/AIRC.

risk studies may be useful for justifying more costly epidemiologic investigations of exposed individuals but they leave unresolved the question of causal relationships and provide no basis for quantitative risk assessment. A reasonable conclusion from the fourteen reported studies is that there are areas in the U.S. where cancer risk and various (usually indirect) measures of water quality are associated, and the nature of this association should be investigated by conducting individual risk studies.

Five individual risk studies, all of them case-comparison in design, have been reported in the literature (Alavanja et al., 1978; Brenniman et al., 1980; Gottlieb and Carr, 1982; Struba, 1979; Young et al., 1981) and are summarized in Appendix 7-III. Each of the five studies obtained cases and comparisons from state death registries and utilized data on exposure and potential confounders from information on the death certificates and secondary public sources. Since relatives of decedents were not interviewed, the investigators had no knowledge of smoking or dietary habits, residential mobility of individuals, personal consumption habits, or other factors which may have affected cancer risk. Crude indirect measures of exposure to organics in water were utilized, such as access to surface vs. groundwater, or chlorinated vs. nonchlorinated water at place of usual residence. Significant associations with water quality were found for: mortality from cancer of the bladder in two of the five studies; cancer of the colon in three of five studies; and cancer of the rectum in four of five studies. These three cancer sites were the only ones studied in all five case-comparison reports.

One cohort study relating water source (groundwater, chlorinated water in small towns, or chlorinated water in cities) to cancer incidence for liver, kidney, and bladder was reported from an ongoing investigation of residents of Washington County, Maryland (Wilkins and Comstock, 1981). No statistically significant associations with cancer incidence rates were observed, but the power of this study was low, given the relatively small population size (31,000 persons) for calculating incidence rates over a 12 year interval. Only 45 cases of liver cancer occurred in the period of follow-up.

Cantor (1982) presented a progress report on a large bladder cancer case-comparison study conducted by the National Cancer Institute (NCI). From the 10 cancer registries supported by NCI in each of the SEER (Surveillance, Epidemiology and End Results) program areas, 3,000 incident bladder cancer cases and 6,000 comparisons were interviewed with respect to lifetime residential history, smoking habits, occupation, coffee and artificial sweetener use, and several other factors. From 1,000 public water utilities in each of the 10 study areas, the investigators collected historical information on raw water sources, treatment practices, and geographical areas served. Approximately 50% of persons in one area (Iowa) served in 1975 by chlorinated surface or groundwater sources had a different water source in 1955, demonstrating that recent change in water source could lead to significant misclassification of exposure status in this area if exposures of 20 years ago were the most relevant to cancer risk. The overall results of this major study have not been reported to date.

At this time, the epidemiologic evidence on the issue of organic chemicals in water and cancer risk is no more than embryonic. Some associations found for aggregate units of analysis were also observed in a few case-comparison studies. Exposure, as measured by a crude and indirect measure of organic water quality, appears to be associated with a small but significantly increased risk of cancer of the colon, rectum, and/or bladder. The studies to date are neither consistent in their associations with the same set of target organ sites, nor did they convincingly control for potential confounding factors. Hence, one cannot conclude that a causal relationship is by any means established. However, the generally positive nature of the associations continues to justify further pursuit of the issue.

The questions to be addressed are: what further evidence should be sought, and where should new investigations go to shed light on the issue? Further aggregate risk studies are likely to be nonproductive. Clearly, we need some large case-comparison interview studies and cohort studies that might be fortuitously superimposed on other research objectives. Given the rare frequency of bladder and rectal cancer, a cohort study will need to involve many person-years of follow-up; however, the complexity of assessing exposure to organics in water may entirely preclude the conduct of any cohort study of cancer risk.

Assessment of relevant individual exposure to organics in water remains the primary barrier to progress. The usual indirect measures of exposure based on surface water vs. groundwater, or chlorinated vs. unchlorinated water sources are unsatisfactory, given accumulating evidence for an array of organics in both groundwater and unchlorinated drinking water. The resulting misclassification of exposure status, if nondifferential among cases and comparisons, will bias risk estimates towards the null hypothesis, and this bias may be severe. If indirect measures of exposure must be used for the present, a better index of exposure needs to be developed by consulting with knowledgeable water chemists and engineers. In this regard, research needs include establishing a system for grading surface water with respect to potential contamination, utilizing knowledge of upstream dischargers and water treatment methods; ascertaining whether well-depth and proximity of wells to agricultural runoff affect water quality; and establishing whether considerable temporal variation in the types and concentrations of organic chemicals present occurs over both short and long time spans in surface and groundwater sources.

Answers to these questions could lead to the development of more refined surrogate (indirect) measures of exposure to organic chemicals in water through a combination of interview studies and historical data available from each water utility. These surrogates might then reasonably be validated, at least for currently obtainable water samples, by comparison with direct chemical analyses.

Exposure estimates based on direct identification and quantification of organics in water seems an unfeasible approach to epidemiologic studies, given the large array of chemicals, the uncertainty of cancer risk related to nearly all of the chemicals, the multistage process of carcinogenesis, and the possibility of interaction between chemicals. Complexity and cost of chemical analyses would also seem to argue against direct measurements for individual risk studies.

It appears that the immediate need is for a collaborative research effort between epidemiologists, analytical chemists, and water quality engineers, to characterize human exposure to organics in water. This collaboration needs to be directed by the epidemiologist's research objectives, specifically relating cancer risk at least to an ordinal ranking of cancer-relevant exposures. If available, a biological indicator could best characterize this exposure, since it would provide a personal integrated exposure estimate. Further epidemiologic studies of cancer risk and water quality, in the absence of progress on exposure estimation, seem unlikely to advance our knowledge of the nature of this relationship.

SUMMARY

Chapter 7 considers health effects of environmental chemical exposures that are not directly related to chemical disposal sites. Exposures to well-defined, specific populations are examined in three ways. First, a review of the occupational health literature was undertaken in order to determine what adverse health effects have been reported to be associated with the list of selected chemicals found at chemical disposal sites developed in Chapter 2. In many cases, chemicals were implicated in causing disease, but lack of adequate documentation and quantitation of actual exposures hindered the assessment of cause and effect relationships.

Second, this chapter includes a review of a number of accidental chemical contamination incidents not related to disposal sites. The chemicals involved included polyhalogenated biphenyls, organic pesticides, trace metals and organo-metallic compounds, and other miscellaneous organic chemicals. Episodes described involved accidental release of chemicals into the environment causing contamination of food, drinking water, and other media, both in the United States and in other countries. Reported effects were very much dose-dependent, although direct measurements of exposure were seldom given. Predominant health effects were predictable on the basis of data from the occupational health literature. In all situations in which food supplies were contaminated, sequelae were rapid and severe. Contamination of water supplies, in presumably much lower concentrations, did not always result in clear-cut responses. In situations in which toxic chemicals could not be identified in food or water, unequivocal adverse health effects were rarely detected.

Third, the effects of prolonged, low-dose exposures to large segments of the population are examined. As an illustration, chemical contamination of water supplies is reviewed. Man-made organic chemicals have been found in drinking water for many years. The number and variety of chemicals found continue to increase as analytical capabilities improve. The identified chemicals comprise only 10 to 20 percent of the total organic matter present and, in general, are volatile or of low molecular weight. Many of them have carcinogenic or mutagenic properties. Chlorinated organic compounds have been found in untreated well water but are generally present at higher levels in chlorine-treated water than in untreated water.

Epidemiologic investigations of organic chemicals in water have focused on the risk of cancer. Aggregate risk studies for cancer are summarized. The most common target organ sites related to organic chemicals in water were: bladder, stomach, colon, and rectum. Numerous problems arise in drawing causal inferences from such studies. Perhaps most importantly, individual exposures cannot be linked to those individuals who developed cancer. However, aggregate risk studies may be useful for providing justification for further epidemiological investigations of exposed populations. Five case-comparison individual risk studies were reviewed and significant associations with water quality were found for: bladder cancer in two studies, colon cancer in three studies, and rectal cancer in four studies.

In a large case-comparison study of bladder cancer conducted by the National Cancer Institute, 50 percent of the persons in one study area had a change in source of water between the years 1955 and 1975. If a current or recent drinking water source is assumed to be the same as a drinking water source used 20 years ago, significant misclassification of exposure status may result. Such flaws in the data may create a false relationship or preclude finding a causal relationship between cancer and contaminants in drinking water.

Although measurements to date have been crude, exposure to organic chemical-contaminated water appears to be associated with a small but significantly increased risk of bladder, colon, and rectal cancer. Large case-comparison and cohort studies are needed because of the low frequency of these disease endpoints. Definitive questions about these and related issues are posed for cooperative effort and resolution by water chemists, engineers, and epidemiologists.

References

Abedin, Z.; Cook, R.C., Jr.; Milberg, R.M. (1980) Cardiac toxicity of perchloroethylene (a dry cleaning agent). South. Med. J. 73:1081-1083.

Adams, R.M. (1983) Occupational skin disease. New York: Grune and Stratton. 456p.

Aksoy, M.; Dincol, K.; Akgun, T.; Erdem, S.; Dincol, G. (1971) Haematological effects of chronic benzene poisoning in 217 workers. Br. J. Ind. Med. 28:296-302.

Aksoy, M.; Dincol, K.; Erdem, S.; Akgun, T.; Dincol, G. (1972) Details of blood changes in 32 patients with pancytopenia associated with long-term exposure to benzene. Br. J. Ind. Med. 29:56-64.

Alavanja, M.; Goldstein, I.; Susser, M. (1978) Case control study of gastrointestinal and urinary tract cancer mortality and drinking water chlorination. In: Jolley, R.L.; Gorchev, H.; Hamilton, D.H., Jr.; eds. Water chlorination: environmental impact and health effects. Vol. 2. Ann Arbor, MI: Ann Arbor Science Publishers, Inc. [Now Stoneham, MA: Butterworth Medical and Scientific Publishers]. p. 395-409.

Anonymous. (1984) India's chemical tragedy: death toll at Bhopal still rising. Chem. Eng. News 62(50):6-7.

Araki, S.; Honma, T. (1976) Relationships between lead absorption and peripheral nerve conduction velocities in lead workers. Scand. J. Work Environ. Health 2:225-231.

Axelson, O.; Dahlgren, E.; Jansson, C.D.; Rehnlund, S.O. (1978) Arsenic exposure and mortality: a case-referent study from a Swedish copper smelter. Br. J. Ind. Med. 35:8-15.

Baerg, R.D.; Kimberg, D.V. (1970) Centrilobular hepatic necrosis and acute renal failure in "solvent sniffers". Ann. Intern. Med. 73:713-720.

Baker, E.L., Jr.; Landrigan, P.J.; Barbour, A.G.; Cox, D.H.; Folland, D.S.; Ligo, R.N.; Throckmorton, J. (1979) Occupational lead poisoning in the United States: clinical and biochemical findings related to blood lead levels. Br. J. Ind. Med. 36:314-322.

Baker, E.L., Jr.; Landrigan, P.J.; Bertozzi, P.E.; Field, P.H.; Basteyns, B.J.; Skinner, H.G. (1978) Phenol poisoning due to contaminated drinking water. Arch. Environ. Health 33:89-94.

Baker, E.L., Jr.; Landrigan, P.J.; Glueck, C.J.; Zack, M.M., Jr.; Liddle, J.A.; Burse, V.W.; Housworth, W.J.; Needham, L.L. (1980) Metabolic consequences of exposure to polychlorinated biphenyls (PCB) in sewage sludge. Am. J. Epidemiol. 112:553-563.

Bakir, F.; Damluji, S.F.; Amin-Zaki, L.; Murtadha, M.; Khalidi, A.; Al-Rawi ,N.Y.; Tikriti, S.; Dhahir, H.I.; Clarkson, T.W.; Smith, J.C.; Doherty, R.A. (1973) Methylmercury poisoning in Iraq. Science 181:230-241.

Baloh, R.W.; Spivey, G.H.; Brown, C.P.; Morgan, D.; Campion, D.S.; Browdy, B.L.; Valentine, J.L.; Gonick, H.C.; Massey, F.J., Jr.; Culver, B.D. (1979) Subclinical effects of chronic increased lead absorption: a prospective study. II. Results of a baseline neurologic testing. J. Occup. Med. 21:490- 496.

Barr, M., Jr. (1973) Teratogenicity of cadmium chloride in two stocks of Wistar rats. Teratology 7:237-242.

Battigelli, M.C. (1960) Mercury toxicity from industrial exposure: a critical review of the literature--part 1. J. Occup. Med. 2:337-344.

Bean, J.A.; Isacson, P.; Hausler, W.J., Jr.; Kohler, J. (1982) Drinking water and cancer incidence in Iowa. I. Trends and incidence by source of drinking water and size of municipality. Am. J. Epidemiol. 116:912-923.

Beattie, A.D.; Dagg, J.H.; Goldberg, A.; Wang, I.; Ronald, J. (1972) Lead poisoning in rural Scotland. Br. Med. J. 2:488-491.

Beattie, A.D.; Moore, M.R.; Goldberg, A.; Finlayson, M.J.W.; Graham, J.F.; Mackie, E.M.; Main, J.C.; McLaren, D.A.; Murdoch, R.M.; Stewart, G.T. (1975) Role of chronic low-level lead exposure in the aetiology of mental retardation. Lancet 1:589-592.

Bekesi, J.G.; Holland, J.F.; Anderson, H.A.; Fischbein, A.S.; Rom, W.; Wolff, M.S.; Selikoff, I.J. (1978) Lymphocyte function of Michigan dairy farmers exposed to polybrominated biphenyls. Science 199:1207-1209.

Brenniman, G.R.; Vasilomanolakis-Lagos, J.; Amsel, J.; Namekata, T.; Wolff, A.H. (1980) Case-control study of cancer deaths in Illinois communities served by chlorinated or nonchlorinated water. In: Jolley, R.L.; Brungs, W.A.; Cumming, R.B.; eds. Water chlorination: environmental impact and health effects. Vol. 3. Ann Arbor, MI: Ann Arbor Science Publishers, Inc. [now Stoneham, MA: Butterworth Medical and Scientific Publishers]. p. 1043-1057.

Buchet, J.P.; Roels, H.; Bernard, A.; Lauwerys, R. (1980) Assessment of renal function of workers exposed to inorganic lead, cadmium or mercury vapor. J. Occup. Med. 22:741-750.

Buncher, C.R. (1975) Cincinnati drinking water: an epidemiologic study of cancer rates. Prepared for the City of Cincinnati Board of Health. Available from: University of Cincinnati Medical Center, Division of Epidemiology and Biostatistics, Cincinnati. 123p.

Cam, C.; Nigogosyan, G. (1963) Acquired toxic porphyria cutanea tarda due to hexachlorobenzene: report of 348 cases caused by this fungicide. J. Am. Med. Assoc. 183:88-91.

Cantor, K.P. (1982) Epidemiological evidence of carcinogenicity of chlorinated organics in drinking water. Environ. Health Perspect. 46:187-195.

Cantor, K.P.; Hoover, R.; Mason, T.J.; McCabe, L.J. (1978) Associations of cancer mortality with halomethanes in drinking water. J. Nat. Cancer Inst. 61:979-985.

Caplan, Y.H.; Backer, R.C.; Whitaker, J.Q. (1976) 1,1,1-Trichloroethane: report of a fatal intoxication. Clin. Toxicol. 9:69-74.

Carlo, G.L.; Mettlin, C.J. (1980) Cancer incidence and trihalomethane concentrations in a public drinking water system. Am. J. Public Health 70:523-525.

Carpenter, C.P.; Shaffer, C.B.; Weil, C.S.; Smyth, H.F., Jr. (1944) Studies on the inhalation of 1:3-butadiene; with a comparison of its narcotic effect with benzol, toluol, and styrene, and a note on the elimination of styrene by the human. J. Ind. Hyg. Toxicol. 26:69-78.

Chen, P.H.; Gaw, J.M.; Wong, C.K.; Chen, C.J. (1980) Levels and gas chromatographic patterns of polychlorinated biphenyls in the blood of patients after PCB poisoning in Taiwan. Bull. Environ. Contam. Toxicol. 25:325-329.

Chernoff, N. (1973) Teratogenic effects of cadmium in rats. Teratology 8: 29-32.

Clark, C.S.; Meyer, C.R.; Gartside, P.S.; Majeti, V.A.; Specker, B.; Balistreri, W.F.; Elia, V.J. (1982) Environmental health survey of drinking water contamination by leachate from a pesticide waste dump in Hardeman County, Tennessee. Arch. Environ. Health 37:9-18,125.

Clarkson, T.W.; Nordberg, G.F.; Sager, P.R.; editors. (1983) Reproductive and developmental toxicity of metals. New York: Plenum Press. 840p.

Coble, Y.; Hildebrandt, P.; Davis, J.; Raasch, F.; Curley, A. (1967) Acute endrin poisoning. J. Am. Med. Assoc. 202:489-493.

Dardanoni, L.; Fara, G. (1982) TCDD (Italy). In: Dardanoni, L.; Miller, R.W.; eds. Plans for clinical and epidemiologic follow-up after area-wide chemi cal contamination. Proceedings of an international workshop, March 17-19, 1980, Washington, DC. Washington, DC: National Academy Press. p. 3-33.

Davies, G.M.; Lewis, I. (1956) Outbreak of food-poisoning from bread made of chemically contaminated flour. Br. Med. J. 2:393-398.

Dawson, J.P.; Thayer, W.W.; Desforges, J.F. (1958) Acute hemolytic anemia in the newborn infant due to naphthalene poisoning: report of two cases, with investigations into the mechanism of the disease. Blood 13:1113-1125.

De Carli, L.; Mottura, A.; Nuzzo, F.; Zei, G.; Tenchini, M.L.; Fraccaro, M.; Nicoletti, B.; Simoni, G.; Mocarelli, P. (1982) Cytogenetic investigations of the Seveso population exposed to TCDD. In: Dardanoni, L; Miller, R.W.; eds. Plans for clinical and epidemiologic follow-up after area-wide chemical contamination. Proceedings of an international workshop, March 17-19, 1980, Washington, DC. Washington, DC: National Academy Press. p. 292-319.

DeRouen, T.A.; Diem, J.E. (1977) Relationships between cancer mortality in Louisiana drinking water source and other possible causative agents. In: Hiatt, H.H.; Watson, J.D.; Winsten, J.A.; eds. Origins of human cancer. Book A. Incidence of cancer in humans. Cold Spring Harbor, NY: Cold Spring Harbor Laboratory. p. 331-345.

Diggory, H.J.P.; Landrigan, P.J.; Latimer, K.P.; Ellington, A.C.; Kimbrough, R.D.; Liddle, J.A.; Cline, R.E.; Smrek, A.L. (1977) Fatal parathion poisoning caused by contamination of flour in international commerce. Am. J. Epidemiol. 106:145-153.

Edmonds, L.D.; Anderson, C.E.; Flynt, J.W., Jr.; James, L.M. (1978) Congenital central nervous system malformations and vinyl chloride monomer exposure: a community study. Teratology 17:137-142.

Edmonds, L.D.; Falk, H.; Nissim, J.E. (1975) Letter: Congenital malformations and vinyl chloride. Lancet 2:1098.

Fahim, M.S.; Fahim, Z.; Hall, D.G. (1976) Effects of subtoxic lead levels on pregnant women in the state of Missouri. Res. Comm. Chem. Path. Pharmacol. 13:309-331.

Fairbanks, V.F. (1967) Copper sulfate-induced hemolytic anemia: inhibition of glucose-6-phosphate dehydrogenase and other possible etiologic mechanisms. Arch. Intern. Med. 120:428-432.

Feldman, R.G. (1978) Urban lead mining: lead intoxication among deleaders. N. Engl. J. Med. 298:1143-1145.

Feldman, R.G.; Hayes, M.K.; Younes, R.; Aldrich, F.D. (1977) Lead neuropathy in adults and children. Arch. Neurol. 34:481-488.

Feldman, R.G.; Mayer, R.M.; Taub, A. (1970) Evidence for peripheral neurotoxic effect of trichloroethylene. Neurology 20:599-606.

Feldman, R.G.; Niles, C.A.; Kelly-Hayes, M.; Sax, D.S.; Dixon, W.J.; Thompson, D.J.; Landau, E. (1979) Peripheral neuropathy in arsenic smelter workers. Neurology 29:939-944.

Fink, J.N. (1983) Hypersensitivity pneumonitis. In: Rom, W.N., ed. Environmental and occupational medicine. Boston: Little, Brown and Company. p. 215-220.

Fischbein, A.; Wolff, M.S.; Lilis, R.; Thornton, J.; Selikoff, I.J. (1979) Clinical findings among PCB-exposed capacitor manufacturing workers. Ann. N.Y. Acad. Sci. 320:703-715.

Folland, D.S.; Schaffner, W.; Ginn, H.E.; Crofford, O.B.; McMurray, D.R. (1976) Carbon tetrachloride toxicity potentiated by isopropyl alcohol. J. Am. Med. Assoc. 236:1853-1856.

Fowler, B.A.; Weissberg, J.B. (1974) Arsine poisoning. N. Engl. J. Med. 291:1171-1174.

Friberg, L.; Hammarstrom, S.; Nystrom, A. (1953) Kidney injury after chronic exposure to inorganic mercury. Ind. Hyg. Occup. Med. 8:149-153.

Friberg, L.; Piscator, M.; Nordberg, G.F.; Kjellstrom, T.; editors. (1974) Health effects of cadmium in the general environment in Japan. In: Cadmium in the environment. 2nd ed. Cleveland: CRC Press. p. 137-195.

Funatsu, I.; Yamashita, F.; Ito, Y.; Tsugawa, S.; Funatsu, T.; Yoshikane, T.; Hayashi, M.; Kato, T.; Yakushiji, M.; Okamoto, G.; Yamasaki, S.; Arima, T.; Kuno, T.; Ide, H.; Ide, I. (1972) Polychlorbiphenyls (PCB) induced fetopathy. I. Clinical observation. Kurume Med. J. 19:43-50.

Gottlieb, M.S.; Carr, J.K. (1982) Case-control cancer mortality study and chlorination of drinking water in Louisiana. Environ. Health Perspect. 46:169-177.

Grabski, D.A. (1961) Toluene sniffing producing cerebellar degeneration. Am. J. Psychiat. 118:461-462.

Grandjean, P.; Arnvig, E.; Beckmann, J. (1978) Psychological dysfunctions in lead-exposed workers: relation to biological parameters of exposure. Scand. J. Work Environ. Health 4:295-303.

Granstein, R.D.; Sober, A.J. (1981) Drug- and heavy metal-induced hyperpigmentation. J. Am. Acad. Dermatol. 5:1-18.

Greenburg, L.; Mayers, M.R.; Heimann, H.; Moskowitz, S. (1942) Effects of exposure to toluene in industry. J. Am. Med. Assoc. 118:573-578.

Griggs, R.C. (1964) Lead poisoning: hematologic aspects. Prog. Hematol. 4:117- 137.

Gurney, R. (1943) Tetrachlorethane intoxication: early recognition of liver damage and means of prevention. Gastroenterology 1:1112-1126.

Gutch, C.F.; Tomhave, W.G.; Stevens, S.C. (1965) Acute renal failure due to inhalation of trichlorethylene. Ann. Intern. Med. 63:128-134.

Harada, M. (1978) Congenital Minamata disease: intrauterine methylmercury poisoning. Teratology 18:285-288.

Harris, D.K.; Adams, W.G.F. (1967) Acro-osteolysis occurring in men engaged in the polymerization of vinyl chloride. Br. Med. J. 3:712-714.

Harris, R.H.; Page, T.; Reiches, N.A. (1977) Carcinogenic hazards of organic chemicals in drinking water. In: Hiatt, H.H.; Watson, J.D.; Winsten, J.A.; eds. Origins of human cancer. Book A. Incidence of cancer in humans. Cold Spring Harbor, NY: Cold Spring Harbor Laboratory. p. 309-330.

Heyman, A.; Pfeiffer, J.B., Jr.; Willett, R.W. Taylor, H.M. (1956) Peripheral neuropathy caused by arsenical intoxication: a study of 41 cases with observations on the effects of BAL (2,3-dimercapto-propanol). N. Engl. J. Med. 254:401-409.

Hogan, M.D.; Chi, P.Y.; Hoel, D.G. (1979) Association between chloroform levels in finished drinking water supplies and various site-specific cancer mortality rates. J. Environ. Pathol. Toxicol. 2:873-887.

Homberger, E.; Reggiani, G.; Sambeth, J.; Wipf, H.K. (1979) Seveso accident: its nature, extent and consequences. Ann. Occup. Hyg. 22:327-367.

Husman, K. (1980) Symptoms of car painters with long-term exposure to a mixture of organic solvents. Scand. J. Work Environ. Health 6:19-32.

Husman, K.; Karli, P. (1980) Clinical neurological findings among car painters exposed to a mixture of organic solvents. Scand. J. Work Environ. Health 6:33-39.

Igisu, H.; Goto, I.; Kawamura, Y.; Kato, M.; Izumi, K.; Kuroiwa, Y. (1975) Acrylamide encephaloneuropathy due to well water pollution. J. Neurol. Neurosurg. Psychiatry 38:581-584.

Infante, P.F.; Wagoner, J.K.; McMichael, A.J.; Waxweiler, R.J.; Falk, H. (1976a) Genetic risks of vinyl chloride. Lancet 1:734-735.

Infante, P.F.; Wagoner, J.K.; Waxweiler, R.J. (1976b) Carcinogenic, mutagenic and teratogenic risks associated with vinyl chloride. Mutat. Res. 41:131-141.

Kanagaratnam, K.; Wong, H.B.; Tan, K.H. (1960) Parathion poisoning from contaminated barley. Lancet 1:538-542.

Kantor, A.F.; Curnen, M.G.M.; Meigs, J.W.; Flannery, J.T. (1979) Occupations of fathers of patients with Wilms's tumour. J. Epidemiol. Comm. Health 33:253-256.

Kazantzis, G. (1965) Chronic mercury poisoning: clinical aspects. Ann. Occup. Hyg. 8:65-71.

Kleinfeld, M.; Tabershaw, I.R. (1954) Trichloroethylene toxicity: report of five fatal cases. Arch. Ind. Hyg. Occup. Med. 10:134-141.

Knox, J.W.; Nelson, J.R. (1966) Permanent encephalopathy from toluene inhalation. N. Engl. J. Med. 275:1494-1496.

Kopelman, H.; Robertson, M.H.; Sanders, P.G.; Ash, I. (1966) The Epping jaundice. Br. Med. J. 1:514-516.

Kraybill, H.F. (1981) Carcinogenesis of synthetic organic chemicals in drinking water. J. Am. Water Works Assoc. 73:370-372.

Kreiss, K.; Zack, M.M.; Kimbrough, R.D.; Needham, L.L.; Smrek, A.L.; Jones, B.T. (1981a) Association of blood pressure and polychlorinated biphenyl levels. J. Am. Med. Assoc. 245:2505-2509.

Kreiss, K.; Zack, M.M.; Kimbrough, R.D.; Needham, L.L.; Smrek, A.L.; Jones, B.T. (1981b) Cross-sectional study of a community with exceptional exposure to DDT. J. Am. Med. Assoc. 245:1926-1930.

Kuratsune, M.; Masuda, Y.; Nagayama, J. (1976) Some of the recent findings concerning Yusho. In: U.S. Environmental Protection Agency, Office of Toxic Substances. National Conference on Polychlorinated Biphenyls (Nov. 19-21, 1975, Chicago, Illinois): conference proceedings. EPA-560/6-75-004. Washington, DC: The Agency. p. 14-29.

Kuratsune, M.; Yoshimura, T.; Matsuzaka, J.; Yamaguchi, A. (1972) Epidemiologic study on Yusho, a poisoning caused by ingestion of rice oil contaminated with a commercial brand of polychlorinated biphenyls. Environ. Health Perspect. 1:119-128.

Kurland, L.T.; Faro, S.N.; Siedler, H. (1960) Minamata disease: the outbreak of a neurologic disorder in Minamata, Japan, and its relationship to the ingestion of seafood contaminated by mercuric compounds. World Neurol. 1:370-395.

Kuzma, R.J.; Kuzma, C.M.; Buncher, C.R. (1977) Ohio drinking water source and cancer rates. Am. J. Public Health 67:725-729.

Lancranjan, I.; Popescu, H.I.; Gavanescu, O.; Klepsch, I.; Serbanescu, M. (1975) Reproductive ability of workmen occupationally exposed to lead. Arch. Environ. Health 30:396-401.

Landrigan, P.J.; Baker, E.L., Jr.; Feldman, R.G.; Cox, D.H.; Eden, K.V.; Orenstein, W.A.; Mather, J.A.; Yankel, A.J.; Von Lindern, I.H. (1976) Increased lead absorption with anemia and slowed nerve conduction in children near a lead smelter. J. Pediatr. 89:904-910.

Landrigan, P.J.; Gehlbach, S.H.; Rosenblum, B.F.; Shoults, J.M.; Candelaria, R.M.; Barthel, W.F.; Liddle, J.A.; Smrek, A.L.; Staehling, N.W.; Sanders, J.F. (1975a) Epidemic lead absorption near an ore smelter: the role of particulate lead. N. Engl. J. Med. 292:123-129.

Landrigan, P.J.; Whitworth, R.H.; Baloh, R.W.; Staehling, N.W.; Barthel, W.F.; Rosenblum, B.F. (1975b) Neuropsychological dysfunction in children with chronic low-level lead absorption. Lancet 1:708-712.

Landrigan, P.J.; Wilcox, K.R., Jr.; Silva, J., Jr.; Humphrey, H.E.B.; Kauffman, C.; Heath, C.W., Jr. (1979) Cohort study of Michigan residents exposed to polybrominated biphenyls: epidemiologic and immunologic findings. Ann. N.Y. Acad. Sci. 320:284-294.

Langolf, G.D.; Chaffin, D.B.; Henderson, R.; Whittle, H.P. (1978) Evaluation of workers exposed to elemental mercury using quantitative tests of tremor and neuromuscular functions. Am. Ind. Hyg. Assoc. J. 39:976-984.

Laseter, J.L.; Dowty, B.J. (1977) Association of biorefractories in drinking water and body burden in people. Ann. N.Y. Acad. Sci. 298:547-556.

Lilis, R.; Gavrilescu, N.; Nestorescu, B.; Dumitriu, C.; Roventa, A. (1968) Nephropathy in chronic lead poisoning. Br. J. Ind. Med. 25:196-202.

Lindstrom, K. (1973) Psychological performance of workers exposed to various solvents. Work Environ. Health 10:151-155.

Lloyd, J.W.; Moore, R.M., Jr.; Breslin, P. (1975) Background information on trichloroethylene. J. Occup. Med. 17:603-605.

Meinhardt, T.J.; Marienfeld, C.J.; Miller, R.S.; Wright, H.T. (1975) River water and cancer mortality. Prepared by Environmental Health Surveillance Center, University of Missouri. 25p. Available from: U.S. Environmental Protection Agency Library, San Francisco.

Middleton, F.M.; Rosen, A.A. (1956) Organic contaminants affecting the quality of water. Public Health Rep. 71:1125-1133.

Miller, J.M.; Chaffin, D.B.; Smith, R.G. (1975) Subclinical psychomotor and neuromuscular changes in workers exposed to inorganic mercury. Am. Ind. Hyg. Assn. J. 36:725-733.

Miller, R.W. (1971) Cola-colored babies: chlorobiphenyl poisoning in Japan. Teratology 4:211-212.

Moore, M.R.; Meredith, P.A.; Goldberg, A. (1977) A retrospective analysis of blood-lead in mentally retarded children. Lancet 1:717-719.

Morgan, D.P.; Lin, L.I. (1978) Blood organochlorine pesticide concentrations, clinical hematology and biochemistry in workers occupationally exposed to pesticides. Arch. Environ. Contam. Toxicol. 7:423-447.

Morgan, J.M.; Hartley, M.W.; Miller, R.E. (1966) Nephropathy in chronic lead poisoning. Arch. Intern. Med. 118:17-29.

Morton, W.; Starr, G.; Pohl, D.; Stoner, J.; Wagner, S.; Weswig, P. (1976) Skin cancer and water arsenic in Lane County, Oregon. Cancer 37:2523-2532.

Moskowitz, S.; Shapiro, H. (1952) Fatal exposure to methylene chloride vapor. Arch. Ind. Hyg. Occup. Med. 6:116-123.

Needleman, H.L.; Gunnoe, C.; Leviton, A.; Reed, R.; Peresie, H.; Maher, C.; Barrett, P. (1979) Deficits in psychologic and classroom performance of children with elevated dentine lead levels. N. Engl. J. Med. 300:689-695.

New, P.S.; Lubash, G.D.; Scherr, L.; Rubin, A.L. (1962) Acute renal failure associated with carbon tetrachloride intoxication. J. Am. Med. Assoc. 181:903-906.

Nordstrom, S.; Beckman, L.; Nordenson, I. (1978a) Occupational and environmental risks in and around a smelter in northern Sweden. I. Variations in birth weight. Hereditas 88:43-46.

Nordstrom, S.; Beckman, L.; Nordenson, I. (1978b) Occupational and environmental risks in and around a smelter in northern Sweden. III. Frequencies of spontaneous abortion. Hereditas 88:51-54.

Nordstrom, S.; Beckman, L.; Nordenson, I. (1979) Occupational and environmental risks in and around a smelter in northern Sweden. V. Spontaneous abortion among female employees and decreased birth weight in their offspring. Hereditas 90:291-296.

Page, T.; Harris, R.H.; Epstein, S.S. (1976) Drinking water and cancer mortality in Louisiana. Science 193:55-57.

Pagnotto, L.D.; Elkins, H.B.; Brugsch, H.G. (1979) Benzene exposure in the rubber coating industry: a follow-up. Am. Ind. Hyg. Assoc. J. 40:137-146.

Pazderova-Vejlupkova, J.; Lukas, E.; Nemcova, M.; Pickova, J.; Jirasek, L. (1981) Development and prognosis of chronic intoxication by tetrachlordibenzo-p-dioxin in men. Arch. Environ. Health 36:5-11.

Pinto, S.S. (1976) Arsine poisoning: evaluation of the acute phase. J. Occup. Med. 18:633-635.

Pittelkow, R. (1975) Occupational dermatoses. In: Zenz, C., ed. Occupational medicine: principles and practical applications. Chicago: Year Book Medical Publishers, Inc. p. 191-230.

Poindexter, C.A.; Greene, C.H. (1934) Toxic cirrhosis of the liver: report of a case due to long continued exposure to carbon tetrachloride. J. Am. Med. Assoc. 102:2015-2017.

Recknagel, R.O. (1967) Carbon tetrachloride hepatotoxicity. Pharmacol. Rev. 19:145-208.

Reiches, N.A. (1977) Epidemiologic investigation of the relationship between chemical contaminants in drinking water and cancer mortality. Columbus, OH: Ohio State University. 187p. Dissertation. Available from: University Microfilms International, Ann Arbor, MI.

Rosenberg, H.G. (1974) Systemic arterial disease and chronic arsenicism in infants. Arch. Pathol. 97:360-365.

Salaman, M.H.; King, A.J.; Williams, D.I.; Nicol, C.S. (1944) Prevention of jaundice resulting from antisyphilitic treatment. Lancet 2:7-8.

Salg, J. (1977) Cancer mortality rates and drinking water in 346 counties of the Ohio River Valley Basin. 136p. Available from: U.S. Environmental Protection Agency, Health Effects Research Laboratory, Cincinnati, OH.

Schmid, R. (1960) Cutaneous porphyria in Turkey. N. Engl. J. Med. 263: 397-398.

Schroeder, H.A. (1965) Cadmium as a factor in hypertension. J. Chronic Dis. 18:647-656.

Schwartz, L. (1936) Dermatitis from synthetic resins and waxes. Am. J. Public Health 26:586-592.

Schwetz, B.A.; Leong, B.K.J.; Gehring, P.J. (1975) Effect of maternally inhaled trichloroethylene, perchloroethylene, methyl chloroform, and methylene chloride on embryonal and fetal development in mice and rats. Toxicol. Appl. Pharmacol. 32:84-96.

Sirchia, G.G.; Group for Immunological Monitoring. (1982) Exposure to TCDD: immunologic effects. In: Dardanoni, L.; Miller, R.W.; eds. Plans for clinical and epidemiologic follow-up after area-wide chemical contamination. Proceedings of an international workshop, Mar. 17-19, 1980, Washington, DC. Washington, DC: National Academy Press. p. 234-266.

Smetana, H. (1939) Nephrosis due to carbon tetrachloride. Arch. Intern. Med. 63:760-777.

Smith, H.V.; Spalding, J.M.K. (1959) Outbreak of paralysis in Morocco due to ortho-cresyl phosphate poisoning. Lancet 2:1019-1021.

Smith, R.G.; Vorwald, A.J.; Patil, L.S.; Mooney, T.F., Jr. (1970) Effects of exposure to mercury in the manufacture of chlorine. Am. Ind. Hyg. Assoc. J. 31:687-700.

Spivey, G.H.; Sloss, E. (1977) Cancer and chlorinated drinking water. Part II. Prepared for U.S. Environmental Protection Agency, Health Effects Research Laboratory, Cincinnati. Available from: The Agency, Cincinnati.

Stahl, C.J.; Fatteh, A.V.; Dominguez, A.M. (1969) Trichloroethane poisoning: observations on the pathology and toxicology in six fatal cases. J. Forsensic Sci. 14:393-397.

Stehr, P.A.; Stein, G.F.; Webb, K.; Schramm, W.; Gedney, W.B.; Donnell, H.D.; Ayres, S.; Falk, H.; Sampson, E.J.; Smith, S.J.; Steinberg, K.K. ([1984]) Pilot epidemiologic study of health effects due to 2,3,7,8-tetrachlorodibenzo-p-dioxin contaminations in Missouri. 25p. Available from: Center for Environmental Health, Centers for Disease Control, U.S. Department of Health and Human Services, Atlanta.

Stewart, R.D.; Fisher, T.N.; Hosko, M.J.; Peterson, J.E.; Baretta, E.D.; Dodd, H.C. (1972) Experimental human exposure to methylene chloride. Arch. Environ. Health 25:342-348.

Strauss, N. (1944) Hepato-toxic effects following occupational exposure to halowax (chlorinated hydrocarbons). Rev. Gastroenterol. 11:381-396.

Struba, R.J. (1979) Cancer and drinking water quality. Chapel Hill, NC: University of North Carolina. 156p. Dissertation.

Taher, S.M.; Anderson, R.J.; McCartney, R.; Popovtzer, M.M.; Schrier, R.W. (1974) Renal tubular acidosis associated with toluene "sniffing". N. Engl. J. Med. 290:765-768.

Thomas, L.B.; Popper, H.; Berk, P.D.; Selikoff, I.; Falk, H. (1975) Vinyl-chloride-induced liver disease: from idiopathic portal hypertension (Banti's

syndrome) to angiosarcomas. N. Engl. J. Med. 292:17-22.
Travers, E.; Rendle-Short, J.; Harvey, C.C. (1956) Rotherham lead-poisoning outbreak. Lancet 2:113-116.
Tseng, W.P. (1977) Effects and dose-response relationships of skin cancer and blackfoot disease with arsenic. Environ. Health Perspect. 19:109-119.
Tuthill, R.W.; Moore, G.S. (1980) Drinking water chlorination: a practice unrelated to cancer mortality. J. Am. Water Works Assoc. 72:570-573.
U.S. Department of Health and Human Services, Centers for Disease Control. (1975) Vinyl chloride and congenital malformations: Ohio. Morb. Mort. Wkly. Rep. 24:245-246.
U.S. Department of Health, Education, and Welfare; National Institute for Ocupational Safety and Health. (1976) Recommended standard for occupational exposure to cadmium. DHEW(NIOSH) Pub. No. 76-192. Cincinnati: The Institute. 5p.
U.S. Department of Health, Education and Welfare; National Institute for Occupational Safety and Health. (1977) Criteria for a recommended standard for occupational exposure to polychlorinated biphenyls (PCBs). DHEW (NIOSH) Publication No. 77-225. 232p. Available from: The Institute, Cincinnati, OH.
USDHEW/NIOSH SEE U.S. Department of Health, Education, and Welfare; National Institute for Occupational Safety and Health
USDHHS/CDC SEE U.S. Department of Health and Human Services, Centers for Disease Control
USEOP/CEQ SEE U.S. Executive Office of the President, Council on Environmental Quality
U.S. Executive Office of the President, Council on Environmental Quality. (1981) Contamination of ground water by toxic organic chemicals. Washington, DC: The Council. 84p.
Valaes, T.; Doxiadis, S.A.; Fessas, P. (1963) Acute hemolysis due to naphthalene inhalation. J. Pediatrics 63:904-915.
von Krough, G.; Maibach, H.I. (1983) Contact urticaria. In: Adams, R.M., ed. Occupational skin disease. New York: Grune and Stratton. p. 58-69.
Vora, D.D.; Dastur, D.K.; Braganca, B.M.; Parihar, L.M.; Iyer, C.G.S.; Fondekar, R.B.; Prabhakaran, K. (1962) Toxic polyneuritis in Bombay due to ortho-cresyl-phosphate poisoning. J. Neurol. Neurosurg. Psychiatry 25:234-242.
Waters, E.M.; Gerstner, H.B.; Huff, J.E. (1977) Trichloroethylene. I. An overview. J. Toxicol. Environ. Health 2:671-707.
Webster, W.S. (1978) Cadmium-induced fetal growth retardation in the mouse. Arch. Environ. Health 33:36-42.
Wedeen, R.P.; Mallik, D.K.; Batuman, V. (1979) Detection and treatment of occupational lead nephropathy. Arch. Intern. Med. 139:53-57.
WHO/IARC SEE World Health Organization, International Agency for Research on Cancer
Wilkins, J.R., III; Comstock, G.W. (1981) Source of drinking water at home and site-specific cancer incidence in Washington County, Maryland. Am. J. Epidemiol. 114:178-190.
Williamson, S.J. (1981) Epidemiological studies on cancer and organic compounds in U.S. drinking waters. Sci. Total Environ. 18:187-203.
Wilson, J.G. (1977) Environmental chemicals. In: Wilson, J.G.; Fraser, F.C.; eds. Handbook of teratology. Vol. 1. General principles and etiology. New York: Plenum Press. p. 357-385.

Wilson, R.H.; McCormick, W.E.; Tatum, C.F.; Creech, J.L. (1967) Occupational acroosteolysis. J. Am. Med. Assoc. 201:577-581.

Winegar, D.A.; Levy, B.S.; Andrews, J.S., Jr.; Landrigan, P.J.; Scruton, W.H.; Krause, M.J. (1977) Chronic occupational exposure to lead: an evaluation of the health of smelter workers. J. Occup. Med. 19:603-606.

World Health Organization, International Agency for Research on Cancer. (1979a) IARC monographs on the evaluation of the carcinogenic risk of chemicals to humans. Vol. 19. Some monomers, plastics and synthetic elastomers, and acrolein. Lyon, France: The Agency. 513p.

World Health Organization, International Agency for Research on Cancer. (1979b) IARC monographs on the evaluation of the carcinogenic risk of chemicals to humans. Vol. 20. Some halogenated hydrocarbons. Lyon, France: The Agency. 609p.

World Health Organization, International Agency for Research on Cancer. (1979c) IARC monographs on the evaluation of the carcinogenic risk of chemicals to humans. Supplement 1 to Vol. 1-20. Chemicals and industrial processes associated with cancer in humans. Lyon, France: The Agency. 48p.

World Health Organization, International Agency for Research on Cancer. (1980) IARC monographs on the evaluation of the carcinogenic risk of chemicals to humans. Vol. 23. Some metals and metallic compounds. Lyon, France: The Agency. 438p.

World Health Organization, International Agency for Research on Cancer. (1982a) IARC monographs on the evaluation of the carcinogenic risk of chemicals to humans. Vol. 29. Some industrial chemicals and dyestuffs. Lyon, France: The Agency. 416p.

World Health Organization, International Agency for Research on Cancer. (1982b) IARC monographs on the evaluation of the carcinogenic risk of chemicals to man. Supplement 4 to Vol. 1-29. Chemicals, industrial processes and industries associated with cancer in humans. Lyon, France: The Agency. 292p.

Yoshimura, T. (1974) Epidemiological study on Yusho babies born to mothers who had consumed oil contaminated by PCB. [In Japanese]. Fukuoka Acta Med. 65:74-80.

Young, T.B.; Kanarek, M.S.; Tsiatis, A.A. (1981) Epidemiologic study of drinking water chlorination and Wisconsin female cancer mortality. J. Nat. Cancer Inst. 67:1191-1198.

Zenz, C., editor. (1980) Developments in occupational medicine. Chicago: Year Book Medical Publishers, Inc. 477p.

APPENDIX 7–I
Incidents of Accidental Environmental Chemical Exposure

A) ORGANIC CHEMICALS

Chemical, Route of Exposure, Circumstances	Reference	Study Design	Measure of Exposure	Study Findings
1) Phenol: Ingestion. Wisconsin, 1974: Railroad accident led to spill of phenol. Contaminated well water consumed by residents.	a) Baker et al., 1978	Case-comparison Exposed: N=39 Comparisons: N=119 Blinded study; clinical exam and lab tests performed without knowledge of exposure status.	Indirect: Concentration of phenol in well water and proximity of residence.	Acute symptoms of diarrhea, sores and dark urine, burning of the mouth occurred only in exposed. Clinical findings 6 months after the incident revealed significant differences between exposed and comparisons.
2) 4,4'-Diaminodiphenylmethane (4,4'-DAPM): Ingestion. U.K., 1965: Flour was contaminated by a spilled canister of 4,4'-DAPM during transport by truck.	a) Kopelman et al., 1966	Case series: N=84	Indirect: All patients had eaten bread from the bakery to which the contaminated flour had been delivered.	Abdominal pain, fever and generalized aches, enlarged liver, and jaundice with increased bilirubin. Symptoms ranged from mild to severe.
3) Methyl isocyanate (MIC): Inhalation. India, 1984: MIC, used as intermediate in pesticide production, accidently released from industrial source into air in highly populated area.	a) Anonymous, 1984	News reports. No epidemiological studies yet conducted due to recency of the incident.	Direct: ? Indirect: Proximity to cloud of MIC gas released.	More than 2,000 deaths due to acute pulmonary edema. An estimated 20,000 to 100,000 people sustained injuries involving lung damage, skin burns, and eye ulcerations. Incident too recent for possible chronic effects to be observed.

Appendix 7-I

Incidents of Accidental Environmental Chemical Exposure

Chemical, Route of Exposure, Circumstances	Reference	Study Design	Measure of Exposure	Study Findings
4) Ortho-cresyl phosphate [species unspecified*] (OCP): Ingestion. Bombay, 1960: Contaminated mustard oil used for cooking. Means of contamination was not established.	a) Vora et al., 1962	Case series: N=58 (32 examined in detail)	Indirect: Patients had purchased oil from several shops from which OCP-contaminated oil samples were later collected.	Acute peripheral polyneuritis, primarily affecting distal segments of limbs. Pain and paralysis, followed by "glove and stocking" sensory impairment. Characteristic high-stepping, lurching gait. Occasional bradycardia, fever and diarrhea.
5) Ortho-cresyl phosphate [species unspecified*] (OCP): Ingestion. Morocco, 1959: OCP-contaminated olive oil consumed by residents was supplied by a single wholesaler to shops in all affected areas.	a) Smith and Spaulding, 1959	Case series: N > 2000	Indirect: Self-reported history of consumption of dark" oil purchased from shops supplied by wholesaler who distributed contaminated oil.	Acute peripheral polyneuritis, primarily affecting distal segments of limbs. Pain and parasthesia followed by "glove and stocking" sensory impairment. Characteristic high-stepping, lurching gait. Occasional bradycardia, fever, and diarrhea.

* Chemical species unspecified, however, effects noted in study findings are consistent with effects known to be produced by Tri-ortho-cresyl phosphate (TOCP).

Appendix 7-I

Incidents of Accidental Environmental Chemical Exposure

Chemical, Route of Exposure, Circumstances	Reference	Study Design	Measure of Exposure	Study Findings
6) Acrylamide: Ingestion and direct contact. Japan, 1974: Acrylamide used in construction of a sewage system near patients' residence. Well water used for drinking and bathing was contaminated.	a) Igisu et al., 1975	Case series: N=5	Indirect: High concentrations of acrylamide found in well water.	CNS disturbances--subacute mental confusion and truncal ataxia, followed by polyneuropathy. More severe in 3 adults than in 2 children. Liver enzyme levels and various blood tests were normal.
7) 2,3,7,8-Tetrachlorodibenzo-p-dioxin (TCDD): Direct contact, inhalation, ?ingestion. Missouri, 1971: TCDD was a contaminant in waste oils sprayed on residential, recreational, and work areas for dust control.	a) Stehr et al., 1984	Case-comparison: Exposed: N=68 Comparisons: N=36 Blinded study; clinical exam and lab tests performed without knowledge of exposure status.	Indirect: Residence or employment in exposed areas, or high rate of contact with contaminated soil.	No significant differences between exposed and comparisons in hepatic, dermatologic, immunologic, and neurologic status. Authors noted that this should be considered a pilot study.

Appendix 7-I

Incidents of Accidental Environmental Chemical Exposure

Chemical, Route of Exposure, Circumstances	Reference	Study Design	Measure of Exposure	Study Findings
8) TCDD: Direct contact, inhalation, ingestion.	a) Dardanoni and Fara, 1982	Survey: Residents in the vicinity of the plant.	Indirect: Proximity of residence to the plant.	Higher frequency of dermatologic problems in residents nearest the plant, but no gradient beyond area of closest proximity.
Italy, 1976: An industrial explosion dispersed TCDD over a large area.		Case-comparison: Cases: N=163 Comparisons: N=51 Study attempted to correlate the presence or absence of chloracne with exposure.	Indirect: Residence in contaminated area during and after the explosion, animal deaths, eating own or neighbors' garden products, playing or working outdoors.	No differences between cases and comparisons in exposure history.
	b) Sirchia et al., 1982	Cohort: Exposed children: N=48 Unexposed comparisons: different subjects at each follow-up screening. Tests of immunologic function were performed blind (without knowledge of exposure status).	Indirect: Proximity of residence to plant.	Several test results (e.g. total serum complement hemolytic activity, and lymphocyte response to mitogens) showed higher values for exposed than for unexposed children. Results were not anticipated since TCDD experiments in animals suggest an immunosuppressant effect. Suggests that perhaps the effects noted are due to immunostimulation, or that misclassification of exposure status occurred.

Appendix 7-I

Incidents of Accidental Environmental Chemical Exposure

Chemical, Route of Exposure, Circumstances	Reference	Study Design	Measure of Exposure	Study Findings
	c) De Carli et al., 1982	Cohort: Examined women who had induced abortions. Exposed: N=13 Unexposed comparisons: N=12	Indirect: Proximity of residence to plant.	Frequency of chromosomal aberrations was significantly higher in fetal tissue from exposed than from unexposed. No difference observed for the other tissues.
		Examined maternal blood, placenta, umbilical cord and fetal tissue.		
		Cohort: Acute exposure: N=145 Chronic exposure: N=69 Unexposed comparisons: N=87	Indirect: Proximity of residence and duration of time spent in plant.	No statistically significant differences in frequency of chromosomal aberrations.
9) Polychlorinated biphenyls (PCB): Ingestion. Japan, 1968: Rice oil used for cooking was contaminated by PCB during manufacture.	a) Kuratsune et al., 1972	Case series: N=189 adults	Indirect: Self-reported use of rice oil, nearly all of which was produced or shipped by one company during a specific period of time.	Disease called "Yusho." Brown pigmentation of nails and skin, acneform skin eruptions. Increased eye discharge, swelling of upper eyelids. Feeling of fatigue and weakness. Reproductive--13 births to exposed mothers. 2/13 were stillborn; 10/13 showed unusually stained skin pigmentation.
More recent studies suggest formation of polychlorinated dibenzofurans (PCDF) during heating of contaminated oil may have had a role in causation of Yusho.	b) Kuratsune et al., 1976			

Appendix 7-I

Incidents of Accidental Environmental Chemical Exposure

Chemical, Route of Exposure, Circumstances	Reference	Study Design	Measure of Exposure	Study Findings
10) PCB: Ingestion. Taiwan, 1979-1980: Rice oil contaminated by PCB.	a) Chen et al., 1980	Case series: N > 1900	Direct: Serum PCB levels	Epidemic of an undescribed "peculiar skin disease." Reported to be similar to 1968 Japanese "Yusho" outbreak as described in #9 above.
11) PCB: Ingestion. Alabama, 1979: PCB found in environmental samples (sediments, water, fish) downstream from a chemical plant.	a) Kreiss et al., 1981a	Survey: A majority of residents in the community were examined. N=458	Direct: Serum PCB levels.	Serum PCB levels were positively correlated with serum cholesterol, and -GT, and borderline and definite hypertension. No correlation with SGPT and total bilirubin. No adverse reproductive events were associated with serum PCB levels.
12) PCB: Direct contact, ingestion. Indiana, 1960's-70's: PCB-contaminated sewage sludge used for fertilizer.	a) Baker et al., 1980	Survey: Four groups with varying exposures were evaluated. Exposed: N=126 Comparisons: N=22	Direct: Serum PCB levels.	No chloracne or symptoms of systemic toxicity. No correlations between serum PCB levels and tests of hematologic, hepat or renal function. Serum PCB correlated positively with plasm triglycerides suggesting that PCB may alte lipid metabolism.

Appendix 7-I

Incidents of Accidental Environmental Chemical Exposure

Chemical, Route of Exposure, Circumstances	Reference	Study Design	Measure of Exposure	Study Findings
13) PBB: Direct contact, ingestion. Michigan, 1973: Packaging mix-up resulted in PBB being mixed with cattle feed. Contamination of dairy animals resulted in widespread human exposure.	a) Landrigan et al., 1979	Cohort: Six cohorts with varying exposures identified: 3 groups at risk for high exposure: N=3820 3 groups with low-level exposure: N=725	Direct: Serum PBB levels	No association found between any symptom prevalence rates and serum PBB levels. No statistically significant differences in lymphocyte number or function. The cohort identified is to be followed prospectively for chronic disease.
	b) Bekesi et al., 1978	Case-comparison: Exposed: N=45 Comparisons: N=123	Direct: Serum PBB levels	Those with serum PBB levels had increased numbers of "null" cells (lymphocytes with no detectable surface markers), decreases in lymphocytes that form rosettes with sheep erythrocytes, and altered lymphocyte function. No consistent correlation between serum PBB concentration and altere lymphocytes, however, no unexposed subject showed such abnormalities.
B) ORGANIC PESTICIDES				
1) DDT: Ingestion. Alabama, 1978-1979: Downstream from a pesticide plant, high DDT metabolite residues detected in fish consumed by community residents.	a) Kreiss et al., 1981b	Survey: N=499	Direct: Serum DDT levels.	No acute health effects were demonstrated. Serum DDT levels positively associated with serum cholesterol, triglycerides, and -GT. No correlation with SGPT and total bilirubin levels. Study did not address possible long-term effects.

Appendix 7-I

Incidents of Accidental Environmental Chemical Exposure

Chemical, Route of Exposure, Circumstances	Reference	Study Design	Measure of Exposure	Study Findings
2) Endrin: Ingestion Egypt, 1966: Bread was made from treated seed intended exclusively for planting. Bread made from flour contaminated with endrin was eaten. Contamination assumed to occur during transport or storage.	a) Coble et al., 1967	Case series: N=4	Direct: Serum or urine endrin levels in 3 of 4 cases. Indirect: History of contaminated bread consumption in 4 of 4 cases.	3 of 4 persons who ate contaminated bread were affected. CNS--headache, dizziness, confusion, convulsions. Nausea and vomiting. Recovery was rapid and complete.
3) Endrin: Ingestion. U.K., 1956: Endrin had leaked onto the floor of a wagon and was absorbed by sacks of flour being transported at a later date. Flour subsequently baked into bread.	a) Davies and Lewis, 1956	Case series: N=59 Persons reported to local physicians for medical care.	Indirect: All cases purchased bread from the one bakery which had received contaminated flour.	In mild cases, dizziness, disorientation, weakness in legs, accompanied by nausea. In severe cases (30 subjects), convulsions of sudden onset. Severity of effects related directly to amount of contaminated bread consumed. Recovery was fairly rapid and apparently complete.

Appendix 7-I

Incidents of Accidental Environmental Chemical Exposure

Chemical, Route of Exposure, Circumstances	Reference	Study Design	Measure of Exposure	Study Findings
4) Hexachlorobenzene (perchlorobenzene): Ingestion. Turkey, 1955-1959: Wheat seed treated with the fungicide hexachlorobenzene was used for consumption rather than for planting, its intended purpose.	a) Cam and Nigogosyan, 1963	Case series: N=348 Cases in a region of Turkey were examined in the hospital. Total number of people affected has been estimated at N > 3000.	Indirect: History of consumption of treated seed intended for planting, or residence in area where treated seed was distributed.	Porphyria cutanea tarda due to interference with porphyrin metabolism in the liver. Symptoms manifested most severely in child primarily in males, and were characterized by: - Blistering and ulceration of skin - Hypertrichosis and hyperpigmentation, with depigmented scars. - Corneal opacities - Atrophy of hands and muscles - Hepatomegaly - Red urine with porphyrinuria - Weight loss and anorexia, but no abdominal pain. No neurological or mental disturbances observed. Symptoms exacerbated by exposure to sunlight. Acute skin manifestations disappear after discontinuation of intake, but relapses may occur during summer months.
5) Parathion: Ingestion Singapore, 1959: Barley became contaminated with parathion at some unestablished point during shipping from Europe.	a) Kanagaratnam et al., 1960	Case series: N=53 Persons reported to local hospitals for medical care.	Indirect: Cases purchased and consumed barley from several shops, all of whose barley came from one consignment from one importer.	38 of 53 exposed were affected. 9 children died. In severe cases, collapse and coma, accompanied by sweating, muscle fasciculation, excessive salivation, and dyspnea due to increased lung secretions. Constricted pupils. In mild cases, vomiting and diarrhea, headache and malaise. All who survived recovered completely.

Appendix 7-I

Incidents of Accidental Environmental Chemical Exposure

Chemical, Route of Exposure, Circumstances	Reference	Study Design	Measure of Exposure	Study Findings
6) Parathion: Ingestion. Jamaica, 1976: Flour contaminated by parathion at unestablished point during transport from Europe.	a) Diggory et al., 1977	Case series: N=79 Persons reported to local hospitals.	Direct: Paranitrophenol found in urine of 9 of 15 cases. Indirect: History of ingestion of baked goods from contaminated flour.	Acute organophosphate poisoning in which 17 of 79 cases died. In mild cases, nausea, vomiting, and pinpoint pupils. Severe cases resulted in convulsions, stupor or coma, accompanied by excessive salivation, diplopia, dyspnea, muscle fasciculations, and bradycardia. Depressed plasma or RBC cholinesterase levels observed.
C) METALS				
1) Arsenic: Ingestion. Oregon, 1958-71: Naturally occurring high levels of arsenic in well water.	a) Morton et al., 1976	Cross-sectional correlation: Census tract rates of skin cancer were correlated with arsenic levels in drinking water.	Indirect: Residence in tracts with high arsenic levels in drinking water.	No relationship between rates of skin cancer (basal cell and squamous cell carcinomas) and arsenic levels in drinking water, although arsenic levels were lower than in other studies where positive correlations have been found. Mean arsenic level in rural areas was 16.5 ppb (range 0-2150 ppb), and in urban areas was 4.8 ppb (range 0-860 ppb).
2) Arsenic: Ingestion. Taiwan, 1965: Naturally occuring high levels of arsenic in water from artesian wells, in use since 1900.	a) Tseng, 1977	Survey: N=40,421 Inhabitants of the area of endemic high arsenic levels.	Indirect: Arsenic concentrations in the well water of study subjects.	Manifestations of chronic arsenicism were skin cancer, hyperpigmentation, and keratosis. Also, blackfoot disease, a local term for a peripheral vascular disorder resulting in gangrene of the extremities. A clear dose-response gradient was observed between consumption of arsenical water and development of each of the above disorders. Arsenic levels in well water ranged from 10 to 1820 ppb.

Appendix 7-I

Incidents of Accidental Environmental Chemical Exposure

Chemical, Route of Exposure, Circumstances	Reference	Study Design	Measure of Exposure	Study Findings
3) Arsenic: Ingestion. Chile, 1968-71: Naturally occurring levels of arsenic in city water supply. Reduced by filtration plant in 1970.	a) Zaldivar, 1974	Case series: N=457 Cases presented to the regional hospital for evaluation.	Direct: Arsenic content of hair. Indirect: Residence in area served by water supply with high level of arsenic.	Manifestations of chronic arsenic poisoning were: cutaneous lesions (including hyperkeratosis and squamous-cell carcinoma), Raynaud's disease, bronchitis, weight loss and general debility. Thrombosis and myocardial infarctions were prominent events in children. Mean arsenic level in drinking water was 600 ppb (range 50-960 ppb).
4) Cadmium: Ingestion. Japan, 1950's-60's: Cadmium discharged into a river by a chemical plant. Downstream water used for drinking and irrigation of rice fields.	a) Friberg et al., 1974	Surveys and case-control studies.	Direct: Blood and urine levels of calcium and phosphate. Indirect: Residence in endemic area; estimated cadmium intake.	Major symptoms of "Itai-itai" (ouch-ouch) disease: - Skeletal deformity, leg and lumbar back pain, multiple fractures. - Gastroenteropathy. - Proteinuria, glucosuria, renal tubular dysfunction. - Osteomalacia with increased serum alkaline phosphatase and decreased serum inorganic phosphate. Deficient nutritional status in conjunction with cadmium poisoning may be a contributing etiological factor.

Appendix 7-I

Incidents of Accidental Environmental Chemical Exposure

Chemical, Route of Exposure, Circumstances	Reference	Study Design	Measure of Exposure	Study Findings
5) Lead: Ingestion. Scotland, 1971: Domestic water supply contaminated with lead from lead plumbing and storage tanks.	a) Beattie et al., 1972	Case series: N=20 Four families.	Direct: Blood lead levels, erythrocyte protoporphyrin, and urinary -ALA and coproporphyrin.	9 of 20 subjects had unequivocal evidence of excessive lead exposures, 5 of whom showed clinical symptoms of lead poisoning. Clinical symptoms included acute abdominal pain, impaired renal function, arthropathy and tremor. Hematological abnormalities manifested by anemia, reticulocytosis, and basophilic stippling. Biochemical evidence of exposure included high blood lead levels, and increased levels of urinary -ALA and coproporphyrin, and erythrocyte protoporphyrin.
6) Lead: Inhalation, ingestion. Idaho, 1974: Lead emitted from stacks at a large smelter, contaminating air and soil.	a) Landrigan et al., 1976	Case-comparison: Exposed: N=919 Comparisons: N=128 Children living around the smelter were examined.	Direct: Blood lead (PbB) and erythrocyte protoporphyrin (EP) levels. Indirect: Environmental sample analysis.	PbB found to increase with duration of residence near smelter and with estimated exposure to lead levels in air, soil and dust based on environmental sampling. PbB correlated negatively with hematocrit value and positively with EP levels. Hematologic consequences were anemia and increased EP. No frank neurological disease, however, dose-response relationship between PbB and motor nerve conduction velocity may be suggestive of subclinical motor neuropathy.

Appendix 7-I

Incidents of Accidental Environmental Chemical Exposure

Chemical, Route of Exposure, Circumstances	Reference	Study Design	Measure of Exposure	Study Findings
7) Lead: Inhalation, ingestion. Texas, 1971: Particulate lead emitted from smelter.	a) Landrigan et al., 1975a	Survey: Random sample of families residing in proximity to the smelter. Total persons: N=1369 Age 1-19: N=758	Direct: Blood lead levels.	Blood lead level was elevated (> 40 ug/100ml) in 53% of children, age 1-9 years old within 1.6 km of the smelter, and in 18% of those from 1.6 to 6.6 km. Beyond that distance and in older persons, levels were lower.
	b) Landrigan et al., 1975b	Case-comparison: Children aged 3-15 years. Exposed: N=46 Comparisons: N=78 Examinations conducted blind.	Direct: Blood lead levels.	Neuropsychological function was evaluated. Exposed children showed decrease in performance IQ and subclinical impairment of fine motor function (slowing in finger-wrist tapping test). No difference between exposed and comparisons in full-scale and verbal IQ scores, behavior, or hyperactivity ratings.

Appendix 7-I

Incidents of Accidental Environmental Chemical Exposure

Chemical, Route of Exposure, Circumstances	Reference	Study Design	Measure of Exposure	Study Findings
8) Mercury: Ingestion. Japan, 1950's: Methylmercury was discharged into a bay by a plastics factory, and was transmitted through ingestion of seafood.	a) Kurland et al., 1960	Survey: N=83 Examination of residents around the bay.	Direct: Mercury measured in autopsy tissues. Indirect: Proximity of residence to bay; source of food supply.	"Minamata disease": Cell degeneration of the brain, resulting in progressive and severe neurological symptoms. Numbness of the extremities followed by ataxia, abnormal involuntary movements, generalized muscle wasting, upper motor neuron involvement and severe emotional and intellectual impairment. Most cases resulted in permanent severe disability or death.
	b) Harada, 1978	Case series: N=40 Children born with congenital Minamata disease.	Indirect: Clinical diagnosis of characteristic symptoms. High levels of mercury in hair of mothers of affected children, 2 to 5 yrs. after birth.	Methylmercury transferred through placenta while in utero or through mother's milk. Mothers of affected babies not severely affected. No abnormalities observed in affected children until 6 months after birth. Extremely high rate of severe neurological and mental symptoms. Diffuse cerebral damage resembling cerebral palsy; however, some symptoms characteristic of congenital Minamata disease as compared to cerebral palsy due to to other factors: - a more marked degree of intelligence disturbance, - cerebellar symptoms and strabismus more common - undifferentiated right and left sided symptoms - high rate (60%) of microcephaly. No chromosomal abnormalities observed.

Appendix 7-I

Incidents of Accidental Environmental Chemical Exposure

Chemical, Route of Exposure, Circumstances	Reference	Study Design	Measure of Exposure	Study Findings
9) Mercury: Ingestion, inhalation, direct contact. Iraq, 1972: Methylmercury compounds used as fungicides on seed grain intended for planting only. Flour and bread were made from the treated grain.	a) Bakir et al., 1973	Case series: N=6530	Direct: Blood and hair methylmercury levels; some autopsy tissue samples. Indirect: History of having eaten homemade bread made from treated grain.	Severity of symptoms was dose-dependent. Symptoms included ataxia, visual disturbances, and slurred speech and hearing difficulties. 459 deaths due to CNS failure. Cardiovascular, gastrointestinal and urinary system involvement was rare.

APPENDIX 7–II
Aggregate Risk Assessment Studies of Drinking Water and Cancer

Reference	Study Data	Measure of Exposure	Controlled Confounding Variables	Study Findings*
Bean et al., 1982	Age-adjusted, sex-specific cancer incidence rates in selected Iowa communities, 1969-1978.	Source of public water supply designated as surface vs. groundwater.	Occupation, socioeconomic status, mobility.	Significant association in males and females between exposure to drinking water from surface water and cancer of the rectum and lung.
Buncher, 1975	Age-adjusted, sex- and race-specific cancer mortality rates in Cincinnati and Hamilton County, Ohio as compared to other Ohio and Kentucky counties, 1950-1969.	Ohio River serving or not serving as drinking water source.	Urbanicity, socioeconomic status, occupation, population density.	No significant association observed between exposure to Ohio River as drinking water source and high cancer mortality rate in Cincinnati, Ohio.
Cantor et al., 1978	Age-adjusted, sex-specific cancer mortality rates in whites in 76 U.S. counties, 1968-1971.	Trihalomethane (THM) concentrations of municipal water supply.	Urbanicity, occupation, education, population density, percent foreign-born.	Significant association between THM levels in drinking water and cancer in: White males: bladder (r = 0.38) White females: bladder (r = 0.45), lung (r = 0.22).
Carlo and Mettlin, 1980	Age-adjusted, sex- and race-specific cancer incidence rates by census tract in Erie County, New York, 1973-1976.	Direct measure of total trihalomethane (THM) concentration in water supply used in census tract.	Urbanicity, occupation, socioeconomic status, mobility, percent nonwhite.	Significant association in white males between THM concentrations in drinking water and cancer of the pancreas (r = 0.16).

* r = correlation coefficient

Appendix 7-II. Aggregate Risk Assessment Studies of Drinking Water and Cancer

Reference	Study Data	Measure of Exposure	Controlled Confounding Variables	Study Findings
DeRouen and Diem, 1977	Sex- and race-specific cancer mortality rates in 64 Louisian counties, 1950-1969.	Percent of county population using Mississippi River as source of drinking water.	Urbanicity, socioeconomic status, occupation, geographic location of county.	Significant association between percent of county using Mississippi River as drinking water source and cancer in: White males: total urinary tract, total gastrointestinal tract, total cancers White females: total gastrointestinal tract, total cancers Nonwhite males: total gastrointestinal tract Nonwhite females: total urinary tract, total gastro-intestinal tract, total cancers
Harris et al., 1977	Sex- and race-specific cancer mortality rates in 88 Ohio counties.	Percent of county using surface water as drinking water source.	Urbanicity, socioeconomic status, occupation, industrialization of county, population density.	Significant association between exposure to drinking water from surface water and cancer in: Males: stomach, bladder, total cancers Females: stomach
Hogan et al., 1979	Age-adjusted, sex- and race specific cancer mortality rates in selected U.S. counties, 1950-1969.	Chloroform concentration of county drinking water supply, as measured in 1975.	Urbanicity, occupation, education, socioeconomic status, population density, percent foreign-born, percent nonwhite.	Significant association between chloroform levels in drinking water and cancer in: White males: colon, rectum White females: bladder, colon, rectum

Appendix 7-II. Aggregate Risk Assessment Studies of Drinking Water and Cancer

Reference	Study Data	Measure of Exposure	Controlled Confounding Variables	Study Findings
Kuzma et al., 1977	Age-adjusted, sex- and race-specific cancer mortality rates in 88 Ohio counties, 1950-1969.	Source of drinking water used by a majority of county residents (as reported in 1963) designated as surface vs. ground-water.	Urbanicity, socioeconomic status, occupation, population density.	Significant association between exposure to drinking water from surface water and cancer in: White males: bladder, stomach, total malignancies White females: stomach
Meinhardt et al., 1975	Age-adjusted, age- and sex-specific cancer mortality rates in whites in 13 Missouri counties, 1960-1972.	Seventy percent of county population using the Missouri and/or the Mississippi River as source of drinking water.	Population density.	No significant association between drinking water from river water source and downstream gradient in cancer mortality rates.
Page et al., 1976	Age-adjusted, sex- and race-specific rates in 64 Louisiana counties, 1950-1969.	Percent of county population using Mississippi River as source of drinking water.	Urbanicity, occupation, socioeconomic status.	Significant association between percent of county using Mississippi River as drinking water source and cancer in: White males: total urinary tract, total gastorintestinal tract, total cancers White females: total gastrointestinal tract

Appendix 7-II. Aggregate Risk Assessment Studies of Drinking Water and Cancer

Reference	Study Data	Measure of Exposure	Controlled Confounding Variables	Study Findings
Reiches, 1977	Age-adjusted, age-, sex-, and site-specific cancer mortality rates in whites in 187 U.S. counties, 1966-1974.	Source of public water supply designated as surface vs. groundwater raw and finished concentrations of 6 organi compounds.	Urbanicity, socioeconomic status, education, industrialization.	No significant association between organic content of drinking water and nine site-specific cancers.
Salg, 1977	Age-adjusted, sex-, race-, and site-specific cancer mortality rates in 346 counties of seven contiguous states within the Ohio River Valley Basin, 1950-1969.	Percent of county population using surface water as source of drinking water; percent of population using prechlorinated drinking water.	Urbanicity, socioeconomic status, occupation, population density, education, percent nonwhite, percent foreign-born, mobility.	Significant association between percent surface water usage and/or percent prechlorination and cancer in: White males: bladder, rectum, lymphosarcoma, reticulosarcoma White females: rectum, multiple myeloma
Spivey and Sloss, 1977	Age-adjusted, sex-specific cancer mortality and incidence rates for whites in selected communities in Los Angeles County, California, 1950-1970.	Qualitative estimates of chloroform concentration of drinking water based on organic content and source of water.	Socioeconomic status, occupation, mobility.	No significant association between estimated chloroform concentration of drinking water and cancer mortality or incidence rates.
Tuthill and Moore, 1980	Age-adjusted, sex-specific mortality rates in all Massachusetts communities using surface water as drinking water source in 1949.	Average past chlorine dose, recent trihalomethane (THM) concentration, and recent chlorine dose.	Socioeconomic status, occupation, mobility, education, percent foreign-born, percent nonwhite.	Significant association between recent THM concentration in drinking water and recent chlorine dose, and cancer of the stomach in males and females, and cancer of the rectum in males.

APPENDIX 7–III
Individual (Case-Comparison) Risk Assessment Studies of Drinking Water and Cancer

Reference	Cases	Comparisons	Measure of Exposure	Controlled Confounding Variables	Study Findings*
Alavanja et al., 1978	3,446 total urinary tract and total gastrointestinal cancer deaths in 7 New York counties, 1968-1970.	3,444 noncancer deaths from same years matched on age, race, sex, foreign vs. U.S. born, county of residence.	Water source of residence at death designated as surface vs. ground and chlorinated vs. nonchlorinated.	Urbanicity, occupation.	Significant association between exposure to chlorinated drinking water and cancer in: Males: bladder (OR = 2.02), liver and kidney (OR = 2.76), esophagus (OR = 2.39), stomach (OR = 1.67), colon (OR = 1.99), rectum (OR = 2.33), pancreas (OR = 2.62), total cancer (OR = 2.09). Females: stomach (OR = 2.23), total cancer (OR = 1.44).
Brenniman et al., 1980	3,208 white, total urinary tract and total gastrointestinal cancer deaths in Illinois, 1973-1976.	43,666 white, nonmatched noncancer deaths in Illinois from same years.	Water source of residence designated as chlorinated vs. nonchlorinated groundwater.	Urbanicity, age, race, sex.	Significant association between exposure to chlorinated groundwater and cancer in: Females: rectum (OR = 1.35), total gastrointestinal cancer (OR = 1.15).
Gottlieb and Carr, 1982	11,349 deaths from 17 organ cancer sites in 20 southern Louisiana counties, 1960-1975.	11,349 randomly selected noncancer deaths matched on age, race, sex, year of death, county group.	Water source of residence at death designated as surface vs. ground and chlorinated vs. nonchlorinated.	Urbanicity, occupation, industrialization of county.	Significant association between exposure to chlorinated drinking water and cancer of the rectum (OR = 1.68) and breast (OR = 1.58).

*OR = Odds ratio

Appendix 7-III. Individual (Case-Comparison) Risk Assessment Studies of Drinking Water and Cancer

Reference	Cases	Comparisons	Measure of Exposure	Controlled Confounding Variables	Study Findings*
Struba, 1979	700-1500 cases per cancer site for bladder, colon, and rectal cancer deaths in North Carolina, 1975-1978.	Similar numbers of non-cancer deaths from same years matched on age, race, sex, geographical region of residence.	Water source of residence designated as surface vs. ground and chlorinated vs. nonchlorinated.	Urbanicity, occupation, socioeconomic status, place of birth.	Significant association between exposure to chlorinated water or surface water in rural (but not urban) areas and cancer of the bladder (OR = 1.54), rectum (OR = 1.53), and colon (OR = 1.30).
Young et al., 1981	8,029 white female cancer deaths in 28 Wisconsin counties, 1972-1977.	8,029 noncancer deaths matched on age, race, sex, year of death, county of residence.	Water source of residence at death designated as surface vs. ground and chlorinated vs. nonchlorinated.	Urbanicity, occupation, marital status.	Significant association between exposure to chlorinated drinking water and cancer of the colon (OR = 1.35).

*OR = Odds ratio

8

Disposal Site Health Studies

INTRODUCTION

The U.S. Environmental Protection Agency (EPA) (USEPA/OSMCD, 1980) has estimated that there are approximately 18,500 municipal solid waste disposal sites, 23,000 sites for disposal of sewage sludge, and over 100,000 industrial disposal sites. Until recently, there were 546 disposal sites in the United States listed on the EPA National Priority List (NPL) (USEPA/OSWER, 1983b; 1983c). Depending on the status of cleanup activities or the discovery of additional sites, these numbers are modified annually. In October 1984, EPA added 128 sites to the NPL and proposed another 244 for inclusion. This brings the total number of sites eligible for cleanup under the Superfund to 786 (USEPA, 1984a; 1984b).

The purpose of this chapter is to provide a review of available information on population studies of health effects associated with chemical disposal sites. By virtue of the fact that there are so many identified disposal sites, it is easy to presume that virtually every community has found or will, in the future, find it necessary to deal with issues related to disposal of chemical materials. Claims of adverse health have sometimes been based on media reporting of a few selected incidents.

Major scientific data bases, including Enviroline, Medline, Toxline, NTIS, and Embase, were searched by cross-referencing health consequences to chemical disposal and specific NPL sites and the list of representative chemicals described in Chapter 2. A review of the citations revealed that relatively few investigations of health effects in populations potentially exposed to chemicals from disposal sites were available in refereed scientific journals. The majority of literature citations were found to be news media reports or articles printed in the lay press.

Several damage and incident reports regarding health effects at individual sites of interest were requested and received from the EPA Office of Health and Environmental Assessment. In addition, the Centers for Disease Control (CDC) contributed information on current epidemiology, health risk, and health assessment activities related to human exposure at disposal sites.

CDC has commissioned a series of studies to assemble comprehensive bibliographies for selected NPL sites. A review of the final draft bibliography for the Love Canal site (CH2M Hill, 1984) cited more than 500 identifiable documents. Only 9 have been published in refereed journals; of these, 3 citations described human health effects, 2 described animal investigation with voles, 3 discussed analytical chemistry and 1 dealt with engineering. The remainder of the citations pertained to internal government documents, letters and reports; legal briefs, testimony and hearings; contractor reports and memos; and environmental monitoring reports and related correspondence.

A California-based group has recently conducted a nationwide mail questionnaire survey to determine ongoing epidemiological study activities at state health departments and university epidemiology departments. The survey requested information for all studies of potential health effects from chemical disposal in each state. Although not available to the Executive Scientific Panel at this time, the preliminary information gathered by this survey is presently under evaluation (personal communication with Dr. Raymond R. Neutra, California Department of Health Services, Berkeley, California). Additional information was discovered in a report prepared by a study group at the University of Pittsburgh (Marsh and Caplan, 1984). Their review of the literature described 16 studies of adverse health effects at disposal sites.

To supplement these data, a staff telephone inquiry was directed at several state and local departments of health and environmental protection. Several additional documents were obtained in this manner. This telephone survey revealed either a lack of present or proposed public health investigations in communities surrounding specific sites, or a reluctance to share preliminary reports prior to their completion. Impending enforcement action was yet another reason for the reluctance of several agencies to provide reports.

Of the sites listed in the NPL Hazardous Waste Site Summaries (USEPA/-OSWER, 1983b), approximately 30% were noted to have enforcement actions pending. It may be presumed that some of these actions are related to claims involving adverse health effects in addition to other claims for property damage. A review of the Grad report (USC/Senate, 1982a,b) and the Six Case Studies (USC/Senate, 1980b) indicated that several simultaneous actions may be in progress for any site. For example, in the case of Jackson Township, N.J., the Council on Environmental Quality (USEOP/CEQ, 1980) noted the following activities:

- State of New Jersey, New Jersey Department of Environmental Protection v. Jackson Township
- Warren J. Aldelund, Jr. et al. (96 residents) v. Jackson Township
- Carl L. Battaglia et al. (residents) v. SCM Corporation

Early attempts to obtain reports regarding potential health effects at this site may have been hampered by these enforcement actions. In another request, the Michigan Department of Health denied access to an epidemiology report on the Berlin and Farro site on the basis of a ruling by the State Attorney General's Office due to enforcement action.[1] Similar problems were

---

1. A copy of this report was later made available to project staff, but too late to be reviewed in the final report (Oudbier et al., 1981).

encountered in attempts to retrieve information from the State of New York about the Love Canal site.

Key documents were identified to provide information on public injury and damages reputed to be caused by environmental chemicals (USEPA/OSMCD, 1980; USC/Senate, 1980a,b; 1982a,b; USEOP/CEQ, 1979; 1980). A brief synoptic listing was then compiled to provide a cross-reference for the number of citations with the degree of reporting and ongoing or proposed investigations for each site noted. The key elements were then coupled with information from descriptions of NPL hazardous waste sites (USEPA/OSWER, 1983b,c; USEPA, 1984a,b).

The list (Appendix 8-I) contains information on over 60 chemical disposal sites that are considered to be a representative sampling of the types of studies and the status of available or ongoing investigations. The list is not considered to be all-inclusive, but serves as a working document to be updated as additional information becomes available. The sole purpose for the list is to direct attention to available information at disposal sites and to identify specific items for which additional information might be requested of appropriate study groups. The text is not meant to provide interpretation of the findings associated with the listing and does not include sufficient detail to permit such an evaluation. These matters are addressed later in this chapter or in the original reference source. The following observations were made by cross-checking the NPL list with the list of site activities and investigating agencies:

- Several sites believed to present a threat to public health in 1980 were no longer being carried on the 1984 NPL list.[2]
- Some incident reports focused on an industrial accident rather than a specific disposal site (e.g., reservoir damage in Kernersville, N.C., apparently the result of vandalism at the Destructo Chemway Corporation in 1977).[3]
- Federal interagency activity in the form of health assessment or certification of a public health threat was noted at disposal sites not presently included on the NPL.[4]
- Sites were referred to by a variety of names which may include any or all of the following: company responsible for the chemicals, city, county, aquifer, or any other of a variety of geographic designations or indicators of property ownership.[5]

---

2. While this is more than likely a reflection of Superfund cleanup activities, it does not exempt the site from consideration in this report, as epidemiological investigations for adverse health effects may have taken

3. These reports were removed from consideration as they did not fall into the scope of work of this study.

4. EPA usually updates the NPL during the third quarter of each year.

5. Revised additions of the NPL frequently make note of synonyms where changes have been indicated.

At present there is no known registry for investigations at chemical disposal sites. Although such a registry for the long-term followup of populations exposed to environmental chemicals has been proposed (Heath, 1981; USDHHS/CDC/CEH, 1984), the proposal has not been implemented. Because most investigations have not been reported in the scientific literature, the number of proposed, ongoing, or completed studies cannot be determined, nor can their study designs be evaluated.

METHODS AND FINDINGS OF PUBLIC HEALTH INVESTIGATIONS

Appendix 8-II summarizes available reports on public health investigations of human exposure at chemical disposal sites. Included in this appendix are 21 chemical disposal sites and a description of findings from 29 studies. Emphasis is placed on the chemicals involved, exposure criteria, health endpoints, study design, populations studied, and purported findings attributed to chemical exposure at the site.

Chemicals Involved

Appendix 8-II indicates that for the 21 sites summarized, polyhalogenated biphenyls were listed 5 times, pesticides 6 times, solvents 16 times, and metals 7 times. Of the studies at these sites, 25 (86%) were concerned with multiple chemical exposures. Perhaps the best known example of potential multi-chemical exposures occurred at the Love Canal site where a mixture of 82 chemicals has been identified.

Methods to Establish Exposure

A measurement of body burden was used as an index of exposure at 7 (33%) sites of the 21 listed in Appendix 8-II . Specifically, 2 of the 7 utilized blood level measurements of either PCB's or chlorinated organic pesticides; volatile organics were determined by urine screening at 3 sites, while arsenic exposure was determined by hair analysis at another site. A new technique for the chemical analysis of exhaled breath was also attempted at one site.

A targeting technique was frequently used as an alternative for estimating relative exposure. In this manner, exposure inferences could be made between study populations on the basis of concentric, targeted exposure zones radiating from either the leachate field or the site epicenter. Study design frequently took into consideration distance of residence from the disposal site, occurrence of work and/or recreational activities associated with site exposure, and measurement of disposal site chemicals in the air, food or water. At 7 sites, quantitative analyses of air samples were made to determine the concentration of detectable odors traceable to the nearby chemical disposal site. Chemical exposure was attributed to ingestion of a contaminated water supply at 11 of the 21 disposal sites (52%).

Information Sources

A general medical questionnaire was used for gathering information on health effects at disposal sites in 9 of 29 studies listed. This was accompanied by physical examination and/or ancillary clinical testing in several studies. Questionnaires were presented either by personal interviews, via mail, or over the telephone. The majority of questionnaire surveys relied on participant recall. Hospital or physician medical incident reports were evaluated in 30% of the studies, and nationwide, state, county, or community cancer mortality records were reviewed in 27%. Similar information for reproductive outcomes and birth defects was reported in 19% of the studies.

Indicators of Health Effects

Study endpoints for specific indicators of health effects included results of physical examination (8 studies), cancer incidence (8 studies), reproductive outcomes (13 studies), neurological changes (3 studies), and cytogenetic abnormalities (2 studies). Occurrence rates of specific illnesses were sought in several studies. For example, investigations involving leukemia were described at 3 sites. Medical care utilization was evaluated in 33% of the studies, while quality of life, parentally reported social and intellectual skills of children, and owner reported pet health were assessed less frequently. Subjective complaints such as headache, tiredness, eye irritation, itching, or difficult breathing were evaluated to some extent in 11 studies.

Study Design

The epidemiologic methods used to investigate health effects varied widely among the 29 studies reviewed. The methods most frequently utilized were the case-comparison study (5 studies) and the cross-sectional study (7 studies). Cohort methods (4 studies) and case incident reports (3 studies) were used less frequently. Quick survey techniques were used in 3 of the studies. Six additional reports described various pilot studies conducted by volunteers, while 3 investigations were considered to be prospective studies with cohort designs.

Population Size

The power of statistical tests used to detect population differences in serious health effects, such as increased mortality, cancer, or reproductive abnormalities, was extremely limited because of the size of the populations studied. With few exceptions, the number of study subjects involved has been quite small. This problem has been noted in several reports as a possible explanation for negative results. Forty percent of the site investigations reported fewer than 100 participants. Seventy-five percent of the studies involved fewer than 1,000 participants, and only 10% included more than 1,000 participants. The remaining 15% did not report sample size. The largest study for which data are available included only 3,639 persons.

A recent Love Canal study reported by the U.S. Congress (USC/OTA, 1983), endeavored to interview as many residents of the area as possible, but the exact number of subjects studied was not cited. The total number of residents at risk based on census tract data was 4,897, and it was estimated that only 300 of those lived in close proximity to the site. In 2 other Love Canal reports, fewer than 50 exposed persons were involved in the recently reported cytogenetic analyses (Heath et al., 1984), while 220 births were analyzed in a second study (USC/OTA, 1983).

Putative Investigation Findings

To date, none of the investigations listed in Appendix 8-II has provided sufficient evidence to support the hypothesis that a causal link exists between exposure to chemicals at a disposal site and latent or delayed adverse health effects in the general populace, although several positive associations have been shown. Two Love Canal studies have suggested a decrease in birth weight, while one study raised the question of an association with spontaneous abortions. A slower growth rate for children has also been noted in connection with this site (USC/OTA, 1983). Among those investigations with blood level measurements of circulating PCB's and/or pesticides, statistically significant positive correlations have been found between PCB's and circulating triglycerides, PCB's and cholesterol, DDT and cholesterol, and both DDT and PCB's with gamma-glutamyl transpeptidase (Kreiss et al., 1981a,b). Kreiss et al., (1981a) have also shown a positive correlation between PCB's and hypertension. Another study reported cases of transitory liver damage associated with exposure near a site containing carbon tetrachloride, hexachlorocyclopentadiene, and hexachlorobicycloheptadiene (Clark et al., 1982; Meyer, 1983).

Numerous reports have noted an increased frequency of acute health complaints including eye, skin, respiratory, and/or joint irritation as well as menstrual complaints. At least two investigations attempted to link the negative impact of such symptoms with the offensive odors originating from the disposal site; these studies also investigated the relationship of noxious odors to the residents' quality of life and sense of well-being. In one of these investigations, it was noted that residents indicated improvement when they left the site (Satin et al., 1983a,b). Yet another investigation made note of the significant role of emotional stress as a contributor to the increased incidence of adverse health complaints (New Jersey, 1983). Interpretations of study data based solely on self-reported, subjective complaints should be considered tentative at best and not conclusive evidence of causal relationships.

OBSERVATIONS ON THE CONDUCT OF EPIDEMIOLOGIC STUDIES AT CHEMICAL DISPOSAL SITES

Chapter 5 provided an awareness of the difficulties faced by public health investigators in the interpretion of epidemiological studies. There is a disparity between what public health investigators believe to be ideal study conditions and what is feasible in real life situations at a disposal site. An overview of the currently available reports of public health investigations provides an opportunity to comment on the design of studies in terms of their time course, location, factors prompting study, and response from investigating agencies.

Time Course of Public Health Studies

Interest in potential adverse health effects resulting from chemical exposure at disposal sites is relatively recent; the earliest studies did not appear until the 1970's. Table 8-1 focuses on the sequence of key events that have occurred in public health studies at 21 sites. The table provides information

TABLE 8–1. Sequence of key events at 21 disposal sites.

Period of Active Waste Disposal	Location / Site Name*		Onset of Initial Health Studies at Site	Abatement Method Used to Discontinue Exposure	Time from Site Closing to Study	Time from Discontinued Exposure to Study
1942-1962	(CA)	McColl	1981	Excavation (proposed method)	19 yrs after	Proposed
1940's-1977	(MA)	New Bedford	1981	Closed to shell fishing (1979); excavation (1982)	4 yrs after	2 yrs after
1971-1977	(MA)	Silresim	1983	Excavation; encapsulation (1978-1981)	6 yrs after	2-5 yrs after
1853-1970	(MA)	Wells G&H "Woburn"	1981	Wells closed (1979)	11 yrs after	2 yrs after
197?-1977	(ME)	McKin	1978	Alternate water source (1977); ongoing excavation	1 yr after	1 yr after
1954-1978	(MI)	Montague	Proposed 1980	Wells closed (1978)	2 yrs after	Proposed 2 yrs after
1930's-1940's	(MN)	Perham	1972	Well closed (1972); encapsulated (1982)	32 yrs after	1 mo after
1970-1974	(NJ)	GEMS	1982	Excavation (1983)	8 yrs after	1 yr before
1972-1980	(NJ)	Jackson Township	1980	Advisory not to use well water (1978)	? mos	2 yrs after
1965-1970	(NJ)	Krysowaty Farm	1981	Excavation (1981)	11 yrs after	During
Not known	(NJ)	Pomona Oaks	1982	Residents advised to seek another water source	?	During
1972-1980	(NJ)	Price Landfill	1982	88% of wells closed (1980); site containment (1982)	2 yrs after	2 yrs after
1971 (Mar.-Dec.)	(NJ)	Reich's Farm "Dover Township"	1974	Wells closed: Zone I (1974); Zone II (1976)	3 yrs after	During
?-1980	(NJ)	Rutherford Area	1980	?	? mos	Before
1966-1980	(NY)	Brookfield Landfill	1982	Excavation (1982)	2 yrs after	During
1953-1975	(NY)	Hooker "Hyde Park"	1979	Encapsulation (1978)	4 yrs after	1 yr after
1920's-1953	(NY)	Love Canal	1978	Local residents evacuated; re-encapsulation (1977-present)	25 yrs after	During
1975-1977	(PA)	ABM Wade	1979	Fire (Feb. 1978); excavation	2 yrs after	8 mos after fire
1948-1981	(PA)	Drake Chemical	1982	Excavation (1982)	1 yr after	During
1962-1978	(PA)	Stanley Kessler	1979	Wells closed; excavation (1979)	1 yr after	During
1964-1972	(TN)	Velsicol Chemical Corp. "Hardeman County"	1978	Wells closed (1978)	6 yrs after	3 mos-1 yr after

*Listed in alphabetical order by state abbreviation and site name on the period of active waste disposal, the onset of the initial health study, and the time and method of exposure reduction or termination. On the basis of this information, observations have been made concerning the temporal relationships between site closings or exposure abatement and the initiation of studies.

Among the 21 sites, studies were initiated or proposed to begin more than 10 years after the start of active waste disposal at approximately 72% of the sites, while 44% were initiated more than 25 years after the start of disposal activity. The majority (55%) of health investigations at disposal sites were proposed or begun within 4 years after the site had been closed. Of particular interest is the temporal relationship between the initiation of remedial action at a site and the beginning of a health investigation (i.e., time to response). The average time to response for 8 studies initiated in the 1970's was approximately 6 months or less, while the time to response for 13 studies initiated in the 1980's was 3 times longer (approximately 18 months). This difference is probably a reflection of the increasing number of agencies that have become involved in coordinating site activities and conducting investigations.

It is interesting to note that approximately 55% of the public health investigations described in Table 8-1 were initiated after exposure was discontinued or cleanup activity had begun. Restriction of the use of groundwater as a drinking source was the most common method for resolving exposure (52%). Evacuation of residents occurred at one site, and some form of excavation or encapsulation occurred at almost all sites.

Location of Health Studies

Table 8-2 summarizes the data from Appendix 8-I and provides information on the number of completed or proposed epidemiology/health studies and reports for both NPL and non-NPL disposal sites. The table incorporates additional information from the NPL (USEPA/OSWER, 1983a; USEPA, 1984c) and an indication of the degree of state enforcement actions in participation with the Superfund program under CERCLA (USEPA/OSWER, 1984b). The largest number of population-based studies (10) have occurred in New Jersey. At least one investigation has been completed or proposed in each of only 24 states. Studies have occurred most frequently in states that are both heavily populated and have large industrialized regions.

Factors Prompting Studies and Investigating Agencies

Information provided in Appendix 8-II lists the initiating study group for 29 health investigations at 21 disposal sites. With few exceptions, the health agency responsible for the investigation was rarely described as the study originator. Complaints by concerned residents were cited as the primary impetus for conducting the study at 76% of the sites. Citizen complaints related to tainted water were listed at 6 sites, while unpleasant odors in the air were listed at 10 sites. Three investigations were begun as a result of local physician concerns, and two were initiated following environmental monitoring by governmental agencies.

Once the disposal site problems had been identified, most investigations (86%) were undertaken by government agencies. State and local health agencies participated in 20 studies, while federal agencies, such as CDC, EPA, and

TABLE 8–2. Health study activities in relation to the NPL and CERCLA enforcement actions.

State	NPL Sites(1)		NPL Sites Completed or Ongoing Studies(2)	NPL Sites Proposed Studies(2)	Proposed and Completed Non-NPL Site Studies(2)	Sites Subject to State Enforcement Actions(3)
	1983	1984				
AL	7	9	1	–	–	1
AZ	5	6	1	–	–	5
CA	19	60	4	–	–	399
CO	9	14	1	–	1	1
ID	4	4	–	1	–	0
IL	11	22	1	–	–	56
IN	17	21	2	–	–	9
KS	4	7	1	–	–	12
LA	4	6	1	–	1	*
MA	16	21	3	1	–	30
ME	5	6	1	1	–	*
MI	44	59	2	–	1	96
MN	23	34	1	–	–	67
MO	7	15	1	1	–	*
MT	5	8	–	1	–	0
NC	3	7	1	–	–	*
NJ	85	95	10	–	1	692
NY	29	58	4	–	3	77
OH	22	28	3		–	50
PA	39	49	5	1	2	50
RI	6	7	–	–	–	*
TN	6	8	1	–	–	27
TX	11	25	–	–	1	40
VA	4	10	1	–	1	13
24	385	579	44	6	11	1,625

(1) Based on information from the National Priority List (NPL) (USEPA/OSWER, 1983a; USEPA, 1984c).
(2) As listed in Appendix 8-I.
(3) Based on ASTSWMO Survey (USEPA/OSWER, 1984b) for pending or resolved civil enforcement actions from 1981 to November 1983. It is not known whether the enforcement sites listed are on the National Priority List. A site may be subject to more than one action.
* Did not respond to ASTSWMO Survey.

HADWC-I

NIOSH, contributed to 12 of the studies listed in Appendix 8-II. At least 5 listed sites were the subject of Congressional reports (USC/SENATE, 1980a; 1980b).

Investigations by non-government study groups accounted for 14% of the studies reviewed. These studies were generally contracted out to university and private study groups by government agencies faced with staffing or time limitations, or by concerned citizen groups who were disenchanted by the response from government agencies. The few studies that were available for consideration in this report may not be representative of the degree of ongoing activity in this sector. In an atmosphere where both individual health and personal property are frequently involved, there is a danger that objective scientific reporting may be relegated to a position of secondary importance in the face of impending enforcement actions.

The most comprehensive source of data regarding federal and state enforcement actions is a survey report prepared by the Association of State and Territorial Solid Waste Management Officials (ASTSWMO) and entitled State Cleanup Programs for Hazardous Substance Sites and Spills (ASTSWMO, 1983). As demonstrated in Table 8-2, the extent of enforcement action frequently exceeds the number of health effect studies and the number of sites listed for a state on the NPL.

Observations on Study Milieu

A review of the milieu surrounding study activities at disposal sites has revealed that the responding agencies frequently do not assume a primary, active role in the selection of sites for study. This may be either a function of the complex sociopolitical environment or perhaps a reflection of the scope of the problem at hand.

Data provided by the EPA for observed chemical releases to environmental media at 546 NPL sites indicate that releases to air have occurred at only 20% of the sites. In contrast, groundwater releases were recorded at 75%, and releases to surface water were recorded for 56% of the sites (USEPA/OSWER, 1984a). On the basis of these data one might expect a similar proportion of air to water complaints to be listed in Appendix II under column "Reason For Study." However, it appears that investigators receive public complaints of unpleasant odors more frequently than complaints of contaminated drinking water. Odors tend to affect smaller populations in close proximity to disposal sites since vapors are readily diluted, whereas contaminated water is generally distributed to larger populations. Initiating a full investigation based solely on odor complaints appears to be a response to an over-anxious public rather than an exercise of responsible scientific judgement. Investigations based on poorly selected populations with an inadequate number of participants are frequently of limited scientific value and a misappropriation of valuable scientific resources.

It is difficult to explain the geographic distribution of investigations on the basis of objective, scientific priorities. Investigations may be compromised by artificial or politically motivated delineations. Frequently, health effect studies have been conducted at a particular site without full consideration of the contributions of other potential sources of contamination in the area. For example, in a densely populated and heavily industrialized region served by a single aquifer, there are usually multiple sources of chemical pollution of the groundwater. At one time, the Cohansey Aquifer supplied water to a major portion of Ocean County, New Jersey. It has been determined that at least 10 NPL sites in adjoining communities contribute to the pollution problem in this aquifer (USEPA/OSWER, 1983b; 1983c). In such a case, a broader perspective might be gained by studying effects over the entire aquifer rather than by considering individual sites. Similar scenarios could be described for studies of aquifers supplying Long Island, New York or the San Gabriel Valley, California. It would appear that vast areas of major pollution will require a more unified research effort.

There is no scientific explanation for the delay in the onset of health studies at many of the 21 sites reviewed in this report. Of particular interest is the pesticide problem at the Velsicol Chemical Site in Toone, Tennessee. Although there was sufficient knowledge to be concerned about public exposure to certain pesticides in the 1960's, no investigations by public health officials of sites where exposure has occurred have been completed. The Velsicol Chemical Site was classified as a possible hazard in 1964. Results of the first studies of health effects by private sector organizations were not available until 1982.

One explanation for the lack of relevant data on acute health effects resulting from chemical disposal site pollution may be the efficiency, in certain situations, with which government agencies take steps to prevent continued human exposure. A review of the voluminous information on well closings and the banning of groundwater use in communities where chemical contamination has been documented by environmental monitoring indicates that the risk of exposure is considered to be unacceptable at almost any level. Wells are closed by community leaders because there is no ethical alternative. Consequently, the exposure levels are generally kept below the threshold of observable response, and the temporal relationship for evaluating acute effects is usually lost prior to study initiation. These factors compound the difficulties of interpreting an already complex series of events.

In both the conduct of a study and the reporting of study findings, it is of paramount importance that every effort be made to maintain responsible communication between scientific professionals, government officials, and the affected community. The perception of risk of adverse health effects is a complex matter that in itself has both health and social impacts, particularly when the relevant issues of chemical exposure are not brought into perspective for the community at risk. In recent years both investigating scientists and government officials have become increasingly responsive to the public's need to understand environmental exposure to chemicals. The resultant improvements in public policy can be seen by comparing the methods used to communicate study findings during the 1970's at Love Canal (Levine, 1983) with the recent study of benzene exposure in the Pomona Oaks community in New Jersey (New Jersey, 1984).

The analytical objectivity of reports from investigating scientists and the official government response described in the Levine report resulted in a breakdown of communication and seldom offered consideration for concerns of community residents at Love Canal (Levine, 1983). In contrast, the New Jersey investigating team at Pomona Oaks (New Jersey, 1984) took an active role in addressing community concerns. They established an effective network of communication that provided opportunities for the exchange of information and addressed questions from the community. To avoid miscommunication, test results were reported by telephone or letter to the study residents as soon as they were available. The Pomona Oaks report uses clear language to express the issues studied and can be readily understood by all those concerned. The introduction to the report places the issue of exposure to environmental chemicals into perspective for the community and is consistent with the study findings and recommendations. This study demonstrates that through improved communication, responsible reporting of study findings, and an increased awareness of matters involving health concerns in affected communities exposed to environmental chemicals, a relationship of trust can be maintained between scientist and citizen.

EVALUATION OF DISPOSAL SITES STUDIES

Information on each site was considered on an individual basis as there were numerous dissimilarities in both the study design and the approach used by the investigators. For purposes of evaluation, a narrative (Appendix 8-III) was prepared to summarize the studies at each of the 21 sites discussed in this chapter.

One purpose of this chapter is to direct attention to the difficulties encountered by investigators in interpreting the relationship between health effects and exposure to chemicals from disposal sites. Unlike studies of communicable diseases, for which the epidemiologist usually has definitive diagnostic criteria, complete and accurate exposure data, and clinical tests with sufficient specificity to establish the etiology, investigations of chemically induced disease or disorders seldom incorporate these characteristics. Nevertheless, clarification of the conditions of exposure and precision in the measurement of the health effect under investigation are of no less importance for the confirmation of study findings. Epidemiological findings can be interpreted more readily if the following criteria and concerns are satisfied:

- Identification of the chemicals at the site.
- Characterization of the degree of exposure as accurately as possible.
- Careful measurements of health and illness status, using methods that are applied consistently in a standardized manner.
- Assessment of the relationship between the degree of exposure and the magnitude of response.
- Selection of appropriate study populations and suitable reference groups.
- Attention to sources of confounding and bias.

Approaches to the consideration of these concerns have been the subject of recent reports prepared by CDC. The manual entitled S.P.A.C.E. for Health (USDHHS/CDC, 1984) provides an overview of the steps involved and the feasibility of conducting an investigation involving exposure to environmental chemicals. In another report (USDHHS/CDC/CEH, 1983), guidelines are presented for the uniform methods used to collect health effect information that is needed to assess exposure to chemicals. A report (USDHHS/CDC/CEH, 1984) on the utility, criteria, and methods for the use of a non-occupational environmental disease registry in the conduct of epidemiologic studies has been prepared by CDC for the Agency for Toxic Substances and Disease Registry (ASTDR). This report provides additional insight for investigators who become involved in gathering health information for long-term registry studies of chemical exposure of populations at disposal sites. These manuals should serve as useful guides to aid health investigators in guarding against omitting important considerations in future studies.

A review of Appendix 8-III has shown that the methods used to conduct studies and to collect health information did not always follow the guidelines recommended by CDC. Difficulties were most frequently observed in addressing concerns for measuring relevant exposure, identifying a relationship between the health effect studied and a specific chemical exposure, and establishing sufficient study sample size in the absence of study confounding and bias.

Almost all study groups were able to identify the chemicals present in the site at the time of the investigation. Available technology in analytical chemistry can detect the presence of even minute quantities of pollutant chemicals in most environmental media. However, determining the actual amount of human exposure is exceedingly more complex than the detection of chemicals in air, soil or water.

Estimates of exposure were frequently extrapolated from indirect measurements (e.g., location of residence in proximity to the disposal site, duration of water consumption and water usage patterns, variations in wind direction). Data of these types are collected from questionnaire surveys and are subject to bias when study participants are asked to recall past experience or when a researcher must base estimates on assumptions and circumstantial information rather than on hard data. Methods of an invasive nature (e.g., blood and tissue sampling) offer convincing information but are costly and seldom available to investigators to apply to large populations. Those studies where exposure had been substantiated by objective measurements demonstrating a body burden for a specific chemical have considerable credibility and provide data in support of the extent of the relevant biological exposure. Studies such as those at New Bedford, Massachusetts (Appendix 8-III, number II); Hyde Park, New York (Appendix 8-III, number XVI); and Perham, Minnesota (Appendix 8-III, number VII) have demonstrated the use of analytical techniques to detect chemicals that are persistent in the body after exposure.

Study groups that relied solely on self-reported claims for multiple, nonspecific or subjective health endpoints rarely presented convincing evidence in support of their findings. Data for such health endpoints are generally collected from health questionnaire surveys that are subject to participant recall bias. In contrast, studies where data were assembled to determine if the diagnostic criteria for a particular disease or chemical entity could be substantiated were able to follow a logical course that was relatively free of participant bias. The study at the Velsicol Chemical Site (Appendix 8-III, number XXI) demonstrated the use of established clinical chemistry techniques to investigate serum enzyme changes following exposure to carbon tetrachloride. The use of cancer and reproductive outcome registries -- for example, Woburn, Massachusetts (Appendix 8-III, number IVa) and Love Canal, New York (Appendix 8-III, number XVIIc) -- have provided useful data over long time periods. Such examples are in striking contrast to most of the studies reviewed here.

The relationship between the degree of exposure and the magnitude of response is difficult to assess, as sites often involve multiple chemicals with exposure levels below the threshold of response. Study findings have shown inherent strengths, whenever information from animal experimentation or industrial epidemiological experience can be offered in support of effects of exposure to chemicals at disposal sites. Several studies related medical findings for a specific chemical exposure to data from toxicological research. Exemplary is the study at Perham, Minnesota (Appendix 8-III, number VII), where data were assembled to establish diagnostic criteria in support of arsenical poisoning. A dose-response relationship was evident by the severity of the effects and was further substantiated when the presence of arsenic was detected in the hair of those individuals with the most severe exposure.

Many studies were conducted in emotionally charged situations, under constraints of time, and with limited funds. Furthermore, many exposed populations near chemical disposal sites were of insufficient number to enable definitive interpretation of the findings. A major problem encountered by many investigators was the selection of an appropriate comparison population. This problem was usually resolved by using whatever population the investigating group found available at the time of the study. Those studies that were most successful in addressing this concern relied heavily on information from historical residence and health registries.

Designing a study which minimizes confounding and bias is a difficult challenge for any investigator. With so many factors to consider as a result of varying lifestyles and personal interests, even the most carefully designed studies can be flawed by uncompensated sources of confounding and bias. As demonstrated so vividly in the study at Pomona Oaks, New Jersey (Appendix 8-III, number XI), both lifestyle and the home environment were found to contribute significantly to the exposure in question. Wherever study findings are based on self-reported claims and where enforcement actions for property and medical injuries are imminent, investigators must look beyond their findings to consider the special interest group. In many of the studies reviewed, authors commented openly on the interpretation of study findings in relation to concerns expressed by exposed residents. The study at Price Landfill, New Jersey (Appendix 8-III, number XII) offers insight into the role of stress and the increased number of symptoms reported in populations concerned by exposure. Findings from this study suggest that increased concern or stress may be an important factor in the etiology of health complaints among individuals with low level exposures.

An ideal investigation would include a series of chemical disposal sites containing chemicals with similar potential biological effects, and would study (by methods other than self-reporting) the incidence of disorders that have been associated with those chemicals. Results from large studies of this nature could be pooled to provide study populations of sufficient sample size. Current activities by CDC investigators regarding exposures to PCB's, lead, and asbestos are an indication of this trend.

Even though the criteria and concerns (see itemized list above) are frequently not met, conclusions must still be reached, even if tentative. The inability to demonstrate a causal relationship between exposure and health effects at chemical disposal sites does not mean that effects do not exist nor that they will not be found in the future. Conversely, an accumulation of positive associations does not necessarily establish that an observed disease or disorder has been caused by exposure to chemicals from disposal sites. Regardless of the strength of either assumption, prudent public health policy requires that exposure be eliminated or at least reduced until the scientific issues are resolved.

It has been shown that without a valid index of exposure and precise measurements of health endpoints, it becomes inherently more difficult to establish claims of adverse health effects. This is a reflection of both the difficulty associated with applying scientific methods to answering apparently simple questions concerning health as they are raised by communities that live close to disposal sites, and the fact that our present experience in conducting health effect studies at disposal sites is limited. There is a need to recognize that the research effort necessary to address such questions has not been fully implemented at this time.

In a recent report prepared by the Comptroller General regarding health-related responsibilities of the Department of Health and Human Services for implementation of the Superfund, it was noted that these activities were adversely affected as a result of limited staffing in conjunction with a lack of sufficient funds (USC/GAO, 1984). The implementation of the Agency for Toxic Substances and Disease Registry (ATSDR) has been hindered by the lack of clarity as to the purpose and intent of this national exposure and disease registry. The full implementation of such an agency would provide a valuable resource and focal point for health issues relating to environmental exposure by chemicals. In addition, the agency could assist in determining the extent to which full-scale health or epidemiology studies should be undertaken. As many reports are released without peer review, ATSDR could provide a mechanism to include scientifically objective peer review of these studies.

SUMMARY

A review of the literature for studies of health effects in populations near chemical disposal sites has revealed that relatively few scientific reports are available in refereed journals. The bulk of available information is derived from incident reports in the lay press, or documents prepared for either enforcement actions or legislative proceedings. This chapter discusses some of the more significant issues raised during an analysis of several of these reports and describes the difficulty in applying classical epidemiologic principles to affected communities.

Studies of acute health effects offer the most promise for establishing causal relationships between observed conditions and chemical exposures. Acute effects are usually observed in close temporal relationship to exposure, and actual body burdens can sometimes be documented. Detection of a latent health effect (e.g., cancer) is more complex because the etiology is less specific, and a close temporal relationship to an identifiable exposure rarely exists. Studies based on delayed endpoints are often adversely affected by inadequate population size, background incidence of disease, subject recall bias, and the absence of accurate exposure data.

Further, studies may also have limitations imposed by the quality of toxicologic information on the chemicals in the site. In the absence of these data, it is difficult for investigators to select relevant clinical measurements to assess health status. Frequently, the clinical tests used in these circumstances lack specificity, making it more difficult to differentiate between effects from a disposal site chemical exposure and the background incidence of similar diseases in the exposed population.

Residents in proximity to a disposal site are prone to link any excess of unusual or apparently relevant medical conditions in the community to the chemicals in the site. Thus, concern by local citizens has been the most common determinant for the selection of a site for investigation. These communities are often small, and evaluation of the biological significance of moderately elevated frequencies of disease is very difficult.

References

Association of State and Territorial Solid Waste Management Officials. (1983) State cleanup programs for hazardous substance sites and spills: a report to the U.S. Environmental Protection Agency. Washington, DC: The Association. 87p.

ASTSWMO SEE Association of State and Territorial Solid Waste Management Officials

CH2M Hill. (1984) Love Canal document citations. Prepared for Love Canal health experts. 172p. Available from: CH2M Hill, Reston, VA.

Clark, C.S.; Meyer, C.R.; Gartside, P.S.; Majeti, V.A.; Specker, B.; Balistreri, W.F.; Elia, V.J. (1982) Environmental health survey of drinking water contamination by leachate from a pesticide waste dump in Hardeman County, Tennessee. Arch. Environ. Health 37:9-18,125.

Heath, C.W., Jr., chairman. (1981) Report of HHS/CCERP Subcommittee on Use of Long-term Registries for Health Followup of Populations Exposed to Toxic Hazards. 15p. Available from: Centers for Disease Control, U.S. Department of Health and Human Services, Atlanta, GA.

Heath, C.W., Jr.; Nadel, M.R.; Zack, M.M.; Chen, A.T.L.; Bender, M.A.; Preston, R.J. (1984) Cytogenetic findings in persons living near the Love Canal. J. Am. Med. Assoc. 251:1437-1440.

Kreiss, K.; Zack, M.M.; Kimbrough, R.D.; Needham, L.L.; Smrek, A.L.; Jones, B.T. (1981a) Association of blood pressure and polychlorinated biphenyl levels. J. Am. Med. Assoc. 245:2505-2509.

Kreiss, K.; Zack, M.M.; Kimbrough, R.D.; Needham, L.L.; Smrek, A.L.; Jones, B.T. (1981b) Cross-sectional study of a community with exceptional exposure to DDT. J. Am. Med. Assoc. 245:1926-1930.

Levine, A. (1983) Psychosocial impact of toxic chemical waste dumps. Environ. Health Perspect. 48:15-17.

Marsh, G.M.; Caplan, R.J. (1984) Evaluating health effects of exposure at hazardous waste sites: a review of the state-of-the-art with recommendations for future research. Working paper No. 84-01. September 1984 reprint. 136p. Available from: Center for Environmental Epidemiology, Graduate School of Public Health, University of Pittsburgh, Pittsburgh, PA.

Meyer, C.R. (1983) Liver dysfunction in residents exposed to leachate from a toxic waste dump. Environ. Health Perspect. 48:9-13.

New Jersey. Department of Health, Environmental Health Hazard Evaluation Program. (1983) Health survey of the population living near the Price landfill, Egg Harbor Township, Atlantic County. Prepared in cooperation with Atlantic County Health Department. 29p. Available from: The Program, [Trenton, NJ].

New Jersey. Department of Health, Environmental Health Program. (1984) Pomona Oaks exposure assessment: volatile organics in wellwater and indoor air. 80p. Available from: [The Program, Trenton, NJ].

Oudbier, A.J.; Eyster, J.T.; Lock, J.H. (1981) Berlin-Farro respiratory study: December 1981. 28p. Available from: [Michigan Department of Public

Health, Lansing, MI].
Satin, K.; Deane, M.; Leonard, A.; Neutra, R.; Gravitz, N.; Harnly, M.; Green, R. (1983a) The McColl Site Health Survey: executive summary. Prepared by the California Department of Health Services, Epidemiological Studies Section. 7p. Available from: The Department, Berkeley, CA.
Satin, K.; Deane, M.; Leonard, A.; Neutra, R.; Gravitz, N.; Harnly, M.; Green, R. (1983b) The McColl Site Health Survey: an epidemiological and toxicological assessment of the McColl site in Fullerton, California. Prepared by the California Department of Health Services, Epidemiological Studies Section. 179p. Available from: The Department, Berkeley, CA.
USC/GAO SEE U.S. Congress, General Accounting Office
U.S. Congress, General Accounting Office. (1984) HHS' implementation of Superfund health-related responsibilities. GAO/HRD-84-62. Washington, DC: The Office. 40p. Available from: The Office, Gaithersburg, MD.
U.S. Congress, Office of Technology Assessment. (1983) Habitability of the Love Canal area: an analysis of the technical basis for the decision on the habitability of the emergency declaration area--a technical memorandum. OTA-TM-M-13. Washington, DC: The Office. 58p.
U.S. Congress, Senate. Environment and Public Works Committee. (1980a) Health effects of toxic pollution: a report from the Surgeon General and a brief review of selected environmental contamination incidents with a potential for health effects. Prepared by the Surgeon General, Department of Health and Human Services, and the Congressional Research Service, Library of Congress. Committee print. 96th Cong., 2nd sess. Serial No. 96-15. Washington, DC: The Committee 312p. Available from: U.S. Government Printing Office, Washington, DC.
U.S. Congress, Senate. Environment and Public Works Committee. (1980b) Six case studies of compensation for toxic substances pollution: Alabama, California, Michigan, Missouri, New Jersey, and Texas. Committee print. 96th Cong., 2nd sess. Serial No. 96-13. Washington, DC: The Committee. 521p. Available from: U.S. Government Printing Office, Washington, DC.
U.S. Congress, Senate. Environment and Public Works Committee. (1982a) Injuries and damages from hazardous wastes--analysis and improvement of legal remedies. A report to Congress in compliance with section 301(e) of the Comprehensive Environmental Response, Compensation, and Liability Act of 1980 by the "Superfund Section 301(e) Study Group". Part 1. Report and comments. Prepared by F.P. Grad, P.A. Porter, and the staff of the Legislative Drafting Research Fund of Columbia University. Committee print. 97th Cong., 3rd sess. Serial no. 97-12. Washington, DC: The Committee 344p. Available from: National Technical Information Service, Springfield, VA.
U.S. Congress, Senate. Environment and Public Works Committee. (1982b) Injuries and damages from hazardous wastes--analysis and improvement of legal remedies. A report to Congress in compliance with section 301(e) of the Comprehensive Environmental Response, Compensation, and Liability Act of 1980 by the "Superfund Section 301(e) Study Group". Part 2. Appendices. Prepared by F.P. Grad, P.A. Porter, and the staff of the Legislative Drafting Research Fund of Columbia University. Committee print. 97th Congress, 3rd session. Serial no. 97-12. Washington, DC: The Committee 501p. Available from: National Technical Information Service, Springfield, VA.
USC/OTA SEE U.S. Congress, Office of Technology Assessment

USC/SENATE SEE U.S. Congress, Senate

U.S. Department of Health and Human Services, Centers for Disease Control. (1984) System for prevention, assessment, and control of exposures and health effects from hazardous sites (S.P.A.C.E. for Health). Prepared by the Center for Environmental Health and the Center for Professional Development and Training. 187p. Available from: Chronic Diseases Division, Center for Environmental Health, Centers for Disease Control, Atlanta, GA.

U.S. Department of Health and Human Services, Centers for Disease Control, Center for Environmental Health. (1983) Uniform methods to collect information on mortality, serious diseases and human exposure to toxic substances: recommendations from an epidemiologic perspective. 289p. Available from: Center for Environmental Health, Centers for Disease Control, Atlanta, GA.

U.S. Department of Health and Human Services, Centers for Disease Control, Center for Environmental Health. (1984) Criteria and methods for establishing and maintaining exposure and outcome registries for environmental public health problems. Prepared for the U.S. Public Health Service, Agency for Toxic Substances and Disease Registry. 123p. Available from: [The Agency, Atlanta, GA].

USDHHS/CDC SEE U.S. Department of Health and Human Services, Centers for Disease Control

USDHHS/CDC/CEH SEE U.S. Department of Health and Human Services, Centers for Disease Control, Center for Environmental Health

U.S. Environmental Protection Agency. (1984a) Hazardous waste sites: descriptions of 244 sites on proposed update #2 to National Priorities List, October 1984. 2nd printing. HW-8.4. Washington, DC: The Agency. 269p.

U.S. Environmental Protection Agency. (1984b) Hazardous waste sites: descriptions of sites on current National Priorities List, October 1984. HW-8.5. Washington, DC: The Agency. 569p.

U.S. Environmental Protection Agency. (1984c) National Priorities List: 786 current and proposed sites in order of ranking and by state, October 1984. Dec. rev. ed. HW-7.2. Washington, DC: The Agency. 84p.

U.S. Environmental Protection Agency, Office of Solid Waste and Emergency Response. (1983a) Hazardous waste sites: National Priorities List. HW-7.1. Washington, DC: The Office. 60p.

U.S. Environmental Protection Agency, Office of Solid Waste and Emergency Response. (1983b) Hazardous waste site descriptions: National Priorities List--final rule. HW-8.1. Washington, DC: The Office 422p.

U.S. Environmental Protection Agency, Office of Solid Waste and Emergency Response. (1983c) Hazardous waste site descriptions: National Priorities List--proposed update. HW-8.2. Washington, DC: The Office. 134p.

U.S. Environmental Protection Agency, Office of Solid Waste and Emergency Response. (1984a) Extent of the hazardous release problem and future funding needs: CERCLA Section 301(a)(1)(C) study--final report. 61p. Available from: The Office, Washington, DC.

U.S. Environmental Protection Agency, Office of Solid Waste and Emergency Response. (1984b) State participation in the Superfund Program: CERCLA Section 301(a)(1)(E) study--final report. 83p. Available from: The Office, Washington, DC.

U.S. Environmental Protection Agency, Oil and Special Materials Control Division. (1980) Damages and threats caused by hazardous material sites.

EPA 430/9-80/004. Washington, DC: The Agency. 258p.
USEOP/CEQ SEE U.S. Executive Office of the President, Council on Environmental Quality
USEPA SEE U.S. Environmental Protection Agency
USEPA/OSMCD SEE U.S. Environmental Protection Agency, Oil and Special Materials Control Division
USEPA/OSWER SEE U.S. Environmental Protection Agency, Office of Solid Waste and Emergency Response
U.S. Executive Office of the President, Council on Environmental Quality. (1979) Environmental quality 1979: tenth annual report of the Council on Environmental Quality. Washington, DC: U.S. Government Printing Office. 816p. Available from: U.S. Government Printing Office, Washington, DC.
U.S. Executive Office of the President, Council on Environmental Quality. (1980) Environmental quality 1980: eleventh annual report of the Council on Environmental Quality. Washington, DC: The Council. 497p. Available from: U.S. Government Printing Office, Washington, DC.

APPENDIX 8–I

ABBREVIATIONS

CDC	=	Centers for Disease Control
6CS	=	Six Case Studies (82) **
CEQ	=	Council on Environmental Quality (7,8,9,10)
CR	=	Congressional Report (20)
Ct.	=	County
DAT	=	Damages and Threats (81)
DOH	=	Department of Health
EPA	=	Environmental Protection Agency
GRAD	=	Grad Report (12,13)
IGS	=	Independent Group Study
LHD	=	Local Health Department Report
NIOSH	=	National Institute for Occupational Safety and Health
NPL	=	National Priority List (74,75,77,78)
SGR	=	Surgeon General Report (83)
SHA	=	State Health Agency Report
STS	=	Staff telephone survey to State and County EPA or Department of Health
TPS	=	Top Priority Site for State
" "	=	Dump site synonym or nickname

* The information described in this appendix was accumulated from multiple sources to provide an identification listing of source documents in alphabetical order by state and site. It is not intended to indicate that a causal hypothesis was supported by the reference cited. The list does not provide an interpretation of the findings associated with the citation, nor does it include sufficient detail to permit such an evaluation, as such matters are addressed elsewhere in this chapter or in the original reference source.

** See References for Appendix 8-I (p. 276).

APPENDIX 8-I

Cross-referenced Listing of Disposal Site Activities and Investigating Agencies

STATE/ SITE NAME	SOURCE	DESCRIPTION	REF. #
ALABAMA/			
McIntosh Plant	NPL:	(Olin & Ciba-Geigy on site landfills) agricultural/pesticide chemicals, synthetic resins, mercury. 2,200 nearby residents receive drinking water 3 mi. from plant, wetland within 1 mi.	(77,78)
	GRAD:	Latent medical injuries: possibly teratogenic; from eating fish containing mercury. Injury: waters closed to fishing; aquifer contaminated.	(12,13)
ARIZONA/			
Mountain View Est.	NPL:	Asbestos/soil and air.	(77,78)
"Globe" **TPS**	IGS:	Trailer park community exposed. Study describes events that took place and evaluation of the residents, permanent relocation of residents; and decontamination plans. Potential for lung cancer discussed.	(22)
CALIFORNIA/			
Celtor Chem Works	NPL:	Metals/Cd/Hg/Cu/Pb/As/Zn.	(77,78)
"Hoopa Valley"	CDC:	This series of documents describes health risk assessment and consultation child health assessment; evaluates potential food chain problem; reviews soil sampling plan; requests public health advisory.	(15,29,57)
Lathrop "Occidental Chemical" "Hooker Chemical"	NPL:	Listed in 1980 but not carried on 1983 list. Dibromochloropropane (DBCP) contamination of drinking water well from on-site waste disposal.	(77,78)

	DAT:	Multiple toxic chemicals and pesticides along with known animal carcinogens; DBCP (known animal and potential human carcinogen, may also cause male sterility); Lindane (toxic pesticide - known animal carcinogen). Diethylformamide (DEF) (potential damage to central nervous system). Radioactive waste. Determined a threat to public health.	(81)
McColl "Fullerton"	NPL:	Oil refinery acid sludge/mixed complex organic and sulfur-containing compounds.	(77,78)
	SHA:	State Health study prompted by resident irritation; "no serious medical problem found."	(1,37, 60,89)
Stringfellow	NPL:	Spent acids/organics/heavy metals with contamination of surface and groundwater.	(77,78)
	DAT:	Waste leaching to the Santa Anna River.	(68,81)
COLORADO/			
Jefferson Ct. Landfill	LHD:	Incident report describing improper disposal of rat bait--sodium monofluoracetate in a sanitary landfill.	(24)
Rocky Mtn. Arsenal	NPL:	Used to manufacture and demilitarize mustard gas and chemical munitions; also pesticides and herbicides. Chemicals migrate from site via ground water.	(74,75)
	GRAD:	Toxic substances in wells 8 mi. away. Pollution of wells and crop damage.	(12,13)
IDAHO/			
Kellogg	CDC:	Kellogg, revisited, proposed study. Additional correspondence.	(21,33)
ILLINOIS/			
Byron Salvage	NPL:	Cyanide plating wastes; soil, surface and ground water contamination.	(77,78)
	DAT:	Serious threat to two aquifers.	(81)
INDIANA/			
MIDCO I	NPL:	Midwest Solvent Recovery Co. Multiple chemicals: solvents/acids/As/	(77,78)

		Pb/paint sludge/cyanides; in surface, ground and soil contamination.	
	DAT:	Nearest residence within 1 mi.	(81)
Seymour Recycling **TPS**	NPL:	Solvents/cyanides/acids/ hexachlorocyclopentadiene (C-56) (pesticide byproducts) mixed chemicals; soil and groundwater contaminated.	(77,78)
	CR:	Describes cleanup remedies and problems at the site.	(20)
KANSAS/			
Doepke-Holliday	NPL:	Mixed chemicals/solvents/metals Potential drinking water contamination.	(77,78)
	CDC:	Correspondence, investigation reports; and health assessment.	(26,28, 30,40)
LOUISIANA/			
Calcasieu Parish	STS:	Private citizen group sponsored Univ. of Texas, School of Public Health study. Study may implicate effects from two nearby waste sites.	
Old Inger Oil Refinery (Darrow) **TPS**	NPL:	Reclaimed oil waste overflowed from ponds and swamp; groundwater & soil contaminated by organic chemicals.	(77,78)
	GRAD:	Hexachlorobenzene (HCB) in soil, vegetation, and cattle. Injuries: high plasma levels in residents.	(12,13)
	DAT:	HCB bioacumulates in adipose tissue and can result in liver enzyme damage.	(81)
MAINE/			
McKin Dump Site "Gray"	NPL:	Wells contaminated by Tamano tanker fuel waste and industrial process wastes.	(77,78)
	GRAD:	Trichloroethylene (TCE) found in aquifer. Latent medical injuries: cancer, and mutagenicity. Injury: wells supply 750 residents.	(12,13)
	SGR:	750 families affected; 16 wells capped. Acute health effects: skin rashes, loss of equilibrium, bladder and liver disorders. Chronic health effects: none reported.	(83)

	SHA:	DOH proposed an epidemiology study in 1980 - no information available at this time.	
	CEQ:	(1979) same as above.	(8)
	DAT:	No mention of resident injuries.	(81)
MASSACHUSETTS/			
New Bedford **TPS**	NPL:	PCB--air/water/food chain contamination.	(77,78)
	CDC:	Proposal for PCB health survey by Massachusetts Health Research Inst., Inc. and State Dept. of Public Health. Study protocol.	(39)
	SHA:	Survey of area fish eaters to detect PCB body burdens.	(86,38)
Silresim "Lowell"	NPL:	Reclamation wastes--oil/solvents/ heavy metals/polychlorinated biphenyls (PCB's)/pesticides--in ground/surface water; drinking source for Lawrence. Air odor aggravated by hot weather.	(77,78)
	IGS:	Boston University School of Public Health, in conjunction with the State Dept. of Environmental Quality Engineering, has completed an area-wide health study.	(54)
	GRAD:	Rotting drums 1/4 mi. from homes; soil, air & water pollution. Latent medical injuries: possible cancer, mutagenicity, teratogenicity. Injury: water, state paid for cleanup.	(12,13)
	CEQ:	(1979) info as above.	(8)
Woburn	NPL:	Chlorinated solvents/lead/arsenic. Public wells closed.	(77,78)
	SHA:	Cancer incidence and environmental hazards study.	(55)
	CDC:	Proposed cancer study protocol.	(70)
	IGS:	Harvard University Woburn Health Study.	(31) (32)

	DAT:	Contaminants include As/Pb/Zn. Increased childhood leukemia noted.	(81)
MICHIGAN/			
Berlin & Farro	NPL:	Pesticide sludge; air contamination. public health threat. Governor declared emergency in 1980.	(77,78)
	STS:	DOH has an epidemiology study. Site in litigation.	
	SHA:	DOH report on respiratory study.	(53)
Gratiot Ct. Landfill "St Louis" **TPS**	NPL:	Polybrominated biphenyls (PBB's) contaminate aquifer supplying water to the region.	(77,78)
	GRAD:	Chlorinated & brominated organic chemicals: DDT, benzene. Latent medical injuries: possible cancer, mutagenic, teratogenic. Site cleanup workers claim numerous health problems.	(12,13)
	CEQ:	(1980) same as above.	(9)
Montague "White Lake" "Muskegon"	NPL:	Not on list.	
	GRAD:	Hooker Chemical dump site. Mirex, chloroform, carbon tetrachloride, chlorinated hydrocarbons, C-56, C-58, and other organic material found leaking to surface and ground water; plume extends to a recreational lake. Latent medical injuries: cancer, mutagenicity, teratogenicity. Injury: polluted water.	(12,13)
	SGR:	Addition chemicals noted: TCE, As, tetrachlorodibenzo-p-dioxin (TCDD). 12 domestic wells have been closed. 12 families affected. Acute health effects: headaches & nausea reported. Chronic health effects: none reported.	(83)
	6CS:	Same as above.	(82)
	CEQ:	(1979) same as above.	(8)
	DAT:	Same as above.	(81)

MINNESOTA/			
Perham	NPL:	Groundwater contamination by arsenic pesticide.	(77,78)
	EPA:	Hazardous waste damage report.	(79)
	CDC:	Study of arsenic exposure conducted with Minnesota Dept. of Health.	
	GRAD:	Arsenic migrated to well water; several people hospitalized.	
	SGR:	13 people affected. Acute health effects: gastrointestinal distress, 1 lost use of legs for 6 mo. Chronic health effects: none reported.	
	DAT:	Same as above.	(81)
MISSOURI/			
Dioxin	NPL:	Several sites listed including: Minker/Stout/Romainer Creek/ Shenandoah Stables/Times Beach/ Quail Run Mobil Manor. All relate to dioxin exposure.	(77,78)
	DAT:	6 yr. old girl suffered inflammatory reaction of kidney and bladder bleeding. 10 adults with symptoms; diarrhea, headache, nausea, polyarthralgias and persistent skin lesions. Numerous farm animals died.	
	GRAD:	Same as DAT.	(12,13)
	6CS:	Same as above.	(82)
	CDC:	Pilot epidemiologic study	(87,90,91)
MONTANA/			
E. Helena Site	NPL:	Pb/Cd/As/air/soil/surface and ground water. CDC and State Health Dept. to screen child health. Montana State Univ. to study effects on vegetation.	(77,78)
	CDC:	East Helena, Montana Child Health Study Protocol. CDC plan to do health screen for Pb effects.	

NEW JERSEY/

Diamond Alkali Co. "Newark" Dioxin Sites"	NPL:	Manufactured numerous chemicals, including 2,4,5-trichlorophenoxyacetic acid (2,4,5-T). Multiple industrial sites with potential resident exposures include: Blue Spruce Int.; Brady Iron and Metals; Chemical Insecticide Co.; Diamond Shamrock; Gividan Chemical Co.	(77,78)
	EPA:	13 State EPA pollution reports.	(44)
Elizabeth "Relocated Bayway"	STS:	Study not available at this time; the study includes behaviorial testing and site interviews.	
	EPA:	Analyses of exhaled breath of 355 urban residents for volatile organic compounds in Elizabeth and Bayonne, NJ.	
Gloucester Environmental Management Services Landfill "GEMS"	NPL:	Organic solvents/industrial chemicals in pits contaminating ground and surface water and air.	(77,78)
	SHA:	DOH report; toluene/other organics. Residents complain of eye and respiratory irritation; study includes respiratory testing and community questionnaire.	(45,46, 48,49)
Jackson Township Landfill "Legler"	NPL:	Mixed waste including liquid hazardous waste; Cohansey Aquifer contaminated with organic compounds; 40 wells contaminated.	(77,78)
	STS:	Findings from community citizen questionaire were reviewed by state DOH. Ocean County health questionnaire survey conducted in 1980. The matter is now being handled by State officials.	
	SHA:	Report on evaluation of citizen health survey and health effects from contaminated water supply.	
	SHA:	Report on Ocean County Health Census for Jackson Township well users; includes indoor air sampling.	(6)
	GRAD:	Legler community in the Pine Barrens.	(12,13)

		Acute injuries: skin ailments. Latent medical injuries: possible nervous system, blood, bone marrow, respiratory, digestive tract damage; cancer, mutagenicity, teratogenicity. Residents complain of kidney ailments, cancer, miscarriages, skin rashes. Injury to aquifer, property value. Numerous legal actions by residents.	
	CEQ:	(1980) same info as above.	(9)
	DAT:	Contamination includes: chloroform/ methylene chloride/benzene/TCE/ acetone/ethylbenzene. No other dependable source of water.	(81)
Kin-Buc	NPL:	Chlorinated organic compounds; PCB's; other liquid chemical wastes contaminating nearby river and Cohansey Aquifer.	(77,78)
	EPA:	Hazardous waste disposal damage report. Bulldozer operator killed in explosion at landfill.	(80)
Krysowaty Farm	NPL:	Ground and surface water contaminated with volatile organics. Chemicals at the site: paint and dye waste; waste solvent/ sludge. Residential well 900 feet from the site.	(77,78)
	SHA:	DOH report on health survey of well users.	(51)
Pomona Oaks	NPL:	Well contamination by volatile organic chemicals: benzene, 1,2-dichloroethane, 1,1,2-trichloroethane, methylene chloride. Residents advised to seek alternate water source.	(75)
	SHA:	DOH report on health survey of well users includes indoor air sampling.	(50)
Price Landfill "Atlantic City"	NPL:	Benzene/chloroform/TCE/metals/organic chemicals; leachate contaminated groundwater to 35 residential local wells. Well field to Atlantic City is 1 mi. downgrade of site.	(77,78)
	SHA:	Report on health effects survey.	(47)
	GRAD:	Listed as Atlantic City. 7 of 12	(12,13)

		municipal wells affected in 1980. Latent medical injuries: possible cancer, mutagenicity, teratogenicity, kidney and liver damage, nervous disorders. To date: 1 kidney damage, 1 cancer death, 1 chemically induced epilepsy may be attributed to drinking water. Injury to water resource; property value decreased.	
Reichs Farm & Dover Municipal Well #4	NPL:	Toluene, styrene, carbon tetrachloride; wells shut down due to halogenated organic solvents from unknown source.	(77,78)
"Cohansey Aquifer"	STS:	Ocean County DOH conducted an illness survey; found no health effects.	
	EPA:	Report presents an analysis of the land disposal damage incident. Includes a summary of the findings from the Ocean County illness survey.	(11)
	GRAD:	Toxic chemicals leached into groundwater from Union Carbide. Latent medical injuries: possible cancer, mutagenicity, teratogenicity. 148 private wells contaminated, decline in property value, aquifer contaminated. Out of court settlement.	(12,13)
	6CS:	Same as above.	(82)
	DAT:	Same as above.	(81)
Rutherford	NPL:	Listed as 1) Universal Oil Products Chemical Division, (East Rutherford); 2) Scientific Chemical Processing Inc. (Carlstadt) as Newark Dioxin (Gividan site), or Ventron/Velsicol (Woodridge).	(77,78)
	SHA:	Investigation of leukemia cluster by NJ DOH.	(3,16)
S. Brunswick Landfill	NPL:	Ground and surface water contaminated by mixed chemicals (vinyl chloride/toluene/ benzene/heavy metals. Private wells affected 3 mi. from site.	(77,78)
	GRAD:	Latent medical injuries: possible cancer, teratogenicity, mutagenicity.	(12,13)

		Injuries to water resource, property value.	
	DAT:	Same as above.	(81)
NEW YORK/			
Brookfield	LHD:	Report on odors from landfill which had accepted liquid chemical wastes.	(61)
	IGS:	Report by New York Public Interest Research Group reviewing health implications at 6 municipal landfills.	(58)
Hooker "Hyde Park"	NPL:	(C-56)/trichlorophenols/chlorinated benzofluorides. Ground and surface water contaminated. Dioxin found in creek running through 500-home residential community.	(77,78)
	NIOSH:	Numerous complaints of odors, skin irritation and metal corrosion. Used modified NHANES study questionnaire and medical exam to study 50 residents and numerous employees from nearby industry.	(59) (63)
Love Canal	NPL:	Numerous migrating toxic chemicals.	(77,78)
	SHA:	Numerous reports on cancer incidence, reproductive outcomes, and cytogenetics.	(18,19,23 67,84,85)
	CDC:	Bibliography on Love Canal (update).	(73)
	GRAD:	Acute medical injuries: birth defects. Latent medical injuries: possible cancer, mutagenicity, teratogenicity and fetotoxicity. Injuries: 100 families evacuated, groundwater resource, decrease in property value.	(12,13)
	SGR:	Mixture of 82 chemicals present on site; 1100 families affected. Acute health effects: headaches, respiratory discomfort, skin ailments, fatigue, irritability, ear infections, hair loss, nervous disorders/depression. Chronic health effects: birth defects, retardation, cancer, miscarriages, deafness, cytogenetic aberrations.	(83)

	6CS:	Same as above.	(82)
	CEQ:	(1979) same as above.	(8)
Moureau Lake GE Glens Fall "Caputo Landfill"	NPL:	Wells to Fort Edwards contaminated with PCB's.	(77,78)
	CDC:	Correspondence regarding health assessment within affected communities. Additional documents identified are: 1) door to door survey by local RN (Tudor Survey). 2) NY State cancer cluster study for this area.	(27)
	SHA:	Report on influence of PCB's on birthweight and gestation for occupational exposure near the GE site.	(65,66)
Nassau Ct., Long Island	NPL:	Numerous sites on Long Island with contamination of groundwater source by metals, including Cd. Listed as: Old Bethpage Landfill, Port Washington Landfill, Syosset Landfill, etc.	(74,75)
	LHD:	Reports on groundwater Cd contamination.	(5,34,35)
	IGS:	Report by New York Public Interest Research Group, Inc. called Toxics on Tap describes problems with Long Island groundwater.	
	GRAD:	All 3 aquifers to the area are contaminated from industrial wastes, municipal treatment plants and highway runoff. Latent medical injuries: possible cancer, and mutagenicity.	(12,13)
Tifft Farm	CDC:	Correspondence of health consultation at the site. Apparently several waste drums had been found at this nature preserve and some concern was noted regarding exposure to visitors.	(14,64)
Sayville	EPA:	Leachate damage assessment report on groundwater contamination in drinking wells.	(62)

NORTH CAROLINA/

Waste Industries	NPL:	Listed in 1980. This site is not on the 1984 NPL.	(77,78)
	DAT:	Flemington Landfill in New Hanover Ct. TCE/benzene/vinyl chloride/dichloromethane/Pb/Zn/phenol, etc. leached to groundwater; found in wells of several homes.	(81)

OHIO/

Chem Dyne	NPL:	Mixed pesticides/PCB/solvents/metals in air/soil/water.	(77,78)
	DAT:	Phenol/benzene/TCE/cyanides/As; instances of periodic sickness of nearby workers and people using recreation facilities.	(81)
Laskin/ Poplar Oil	NPL:	Metals/PCB/organics runoff into drinking water.	(77,78)
	CDC:	Correspondence for site evaluation; health consultations; biological sampling; and health assessment.	(25)
Summit National Liquid Serv.	NPL:	Solvents/paint sludge/waste oil/C-56. Runoff threatens Berlin Reservoir; groundwater and soil contaminated.	(77,78)
	DAT:	Acetone/methyl ethyl ketone (MEK)/ toluene/latex/oils/Mirex/C-56.	(81)

PENNSYLVANIA/

ABM Wade "Chester"	NPL:	Benzene and naphthalene derivatives, chromium, copper, lead, nickel, zinc, diphenylamine, phenanthrene, rubber reclamation waste containing toxic and carcinogenic compounds.	(77,78)
	CDC:	CuCN; phenol; benzene. Dump fire in 1979 resulted in hospitalization of 47 firemen. Health effects: skin rashes and eye inflammation. National Health Service Corps exam found no evidence of acute health effects in 86 residents/14 children.	(69)
	DAT:	Same as above. Acute effects of chemicals: eye and	(81)

		respiratory irritation. Long term effects: CNS depression, hepatic, renal and bone marrow disorder. Known, suspect carcinogens.	
Drake Chemical "Kilsdonk Chem. Co." "Lock Haven"	NPL:	Cyanides, beta-naphthylene, o-toluidine, C-56, Fenac, dichlorobenzene and mixed chemical/organic leachate through park. Air/soil/water contamination.	(77,78)
	NIOSH:	Review of exposed workers; review toxicology impact assessment and site sampling.	(71)
	EPA:	Toxicological impact assessment of Drake Chemical Inc. (NUS Corporation for EPA).	(52)
	SHA:	Health Surveillance Protocol.	(56)
	CDC:	Clinton County health consultation for cancer mortality and congenital malformations.	(76,88,93)
Lehigh Electric "Old Forge"	NPL:	High levels of PCB contaminated soil. Site is in floodplain; probable ground and surface water contamination.	(77,78)
	STS:	Report of study on PCB blood levels sampling. Not available.	
Lindane Dump Allegheny Ct.	NPL:	Lindane pesticide and industrial waste leaching into Allegheny River near Clearview Water Supply Co.	(77,78)
	STS:	No adverse health effects noted in the community. Odors emanating from the leachate stream and ballfield have been identified as organic vapors. Noted by state EPA Superfund office.	
Montgomery City	GRAD:	Leachate from landfill: vinyl chloride/ TCE/benzene. Latent medical injuries: possible cancer, mutagenicity, teratogenicity. Contamination of drinking water.	(12,13)
Neville	NPL:	No longer on list.	(77,78)
Island Pk.	DAT:	Park built on a previous dump site.	(81)

		Benzene/phenols/cyanide/Hg/coal tar residues/parathion. Workers at park reported problems: eye irritation, blood in urine, etc. Allegheny Ct. Health Dept. study cited 26% respondents with health problems while working at the park.	
Palmerton Zn Pile	NPL:	Zn/Cd in drinking wells; soil and air pollution.	(77,78)
	STS:	Reports of vegetation and livestock damages. Univ. of Pittsburgh to work with CDC in developing a study protocol for Cd exposure in the area.	
Stanley Kessler "King of Prussia"	NPL:	Solvent wastes, TCE, disposed of in floor drain contaminated groundwater feeding upper Merion Reservoir which supplies 170,000 people.	(77,78)
	SHA:	Report of a DOH investigation of liver cancer mortality in Montgomery County. No acute illness reported. 6% of residents had TCE in urine.	(36,95)
RHODE ISLAND/			
Picillo Farm	NPL:	Solvents/oils/pesticides/PCB's/paint sludges/explosives. Air; soil; ground and surface water contaminated. Nearby residents served by private wells/ reservoir 1.5 mi. from site.	(77,78)
	STS:	Risk assessment and feasibility study. No epidemiology study/no effects noted.	
Stamina Mills	NPL:	TCE solvents, oils and contaminants. organic solvents found in 41 of 53 wells. Alternate water source in use.	(77,78)
	STS:	Risk assessment and feasibility study. No epidemiology study/no effects noted.	
TENNESSEE/			
N. Hollywood Dump **TPS**	NPL:	Pesticides/metals in soil/surface water, potential for groundwater.	(77,78)
	CDC:	Protocol for the study of possible human exposure to the health effects of toxic chemicals from the dump.	(41)

Velsicol Chem. Corp. "Toone" "Hardeman County"	NPL:	Organic contaminants in local wells.	(77,78)
	IGS:	University of Cincinnati reports on physical examination and clinical chemistry findings in residents. Chemicals found in well water: carbon tetrachloride, chlorobenzene, chloroform, tetrachloroethylene, C-56	(4,17,42)
	GRAD:	Endrin, dieldrin, aldrin, heptachlor and other pesticide residues pollute family wells. Acute medical injuries: 1 death, liver and urinary tract ailments, dizziness, nausea, skin rashes, numbness. Latent medical injuries: possible cancer, teratogenicity. Injury: aquifer contaminated, property value.	(12,13)
	SGR:	35 families affected. Acute health effects: one death, numbness, anxiety, paralysis and depression. Chronic health effects: unknown.	(83)
	CEQ:	40 families affected.	(8)
	DAT:	Same as above.	(81)
TEXAS/			
Dallas	CDC:	Childhood Pb exposure around secondary smelters; proposed study design and questionnaire.	(72)
French LTD	NPL:	Heavy metals/phenols/PCB/acids/solvents. Groundwater and San Jacinto River contaminated.	(77,78)
	STS:	No epidemiologic study.	
VIRGINIA/			
Eastern Shore Arsenic	LHD:	Report of fatal arsenic poisoning from contaminated well water.	(2)
Saltville Waste Disposal Ponds	NPL:	Hg contaminated waste contaminates 50 mi. of Holston River which is closed to recreational use. Olin Corp.	(77,78)
	DAT:	Evidence that contamination extends to TVA Cherokee Reservoir 100 miles from site. No reported Hg related illnesses.	(81)

References for Appendix 8-I

1. Chemical Manufacturer's Association. (1983) Technical and epidemiological studies of the McColl site (Fullerton, CA). 2p. Prepared by the Data Development Work Group. Available from: The Association, Washington, DC.
2. Armstrong, C.W.; Stroube, R.B.; Rubio, T.; Siudyla, E.A.; Miller, G.B. (1984) Outbreak of fatal arsenic poisoning caused by contaminated drinking water. Arch. Environ. Health 39:276-279.
3. Burke, T.A.; Gray, S.; Krawiec, C.M.; Katz, R.J.; Preuss, P.W.; Paulson, G. (1980) Environmental investigation of clusters of leukemia and Hodgkin's Disease in Rutherford, New Jersey. J. Med. Soc. N.J. 77:259-264.
4. Clark, C.S.; Meyer, C.R.; Gartside, P.S.; Majeti, V.A.; Specker, B.; Balistreri, W.F.; Elia, V.J. (1982) Environmental health survey of drinking water contamination by leachate from a pesticide waste dump in Hardeman County, Tennessee. Arch. Environ. Health 37:9-18, 125.
5. Colton, D.; Braids, O.C.; MacCallum, D.R.; Miller, D.W.; Sgambat, J.P. (1979) Supplemental report to the Nassau-Suffolk Regional Planning Board on the current status of ground-water investigations for organic chemicals in the Nassau-Suffolk area. 85p. Available from: Geraghty & Miller, Inc., Port Washington, NY.
6. Conomos, M.G. (1983) A health census of a community with groundwater contamination: Jackson Township, 1980. Trenton, NJ: New Jersey State Department of Health. 113p.
7. U.S. Executive Office of the President, Council on Environmental Quality. (1975) Environmental quality 1975: sixth annual report of the Council on Environmental Quality. Washington, DC: U.S. Government Printing Office. 763p. Available from: U.S. Government Printing Office, Washington, DC.
8. U.S. Executive Office of the President, Council on Environmental Quality. (1979) Environmental quality 1979: tenth annual report of the Council on Environmental Quality. Washington, DC: U.S. Government Printing Office. 816p. Available from: U.S. Government Printing Office, Washington, DC.
9. U.S. Executive Office of the President, Council on Environmental Quality. (1980) Environmental quality 1980: eleventh annual report of the Council on Environmental Quality. Washington, DC: The Council. 497p. Available from: U.S. Government Printing Office, Washington, DC.
10. U.S. Executive Office of the President, Council on Environmental Quality. (1982) Environmental quality 1982: thirteenth annual report of the Council on Environmental Quality. Washington, DC: The Council. 330p. Available from: U.S. Government Printing Office, Washington, DC.
11. Ghassemi, M. (1976) Analysis of a land disposal damage incident involving hazardous waste materials, Dover Township, New Jersey: final report. Prepared for the U.S. Environmental Protection Agency, Office of Solid Waste Management Programs by TRW Systems Group, Inc., Redondo Beach, CA. Available from: National Technical Information Service, Springfield, VA.
12. U.S. Congress, Senate. Environment and Public Works Committee. (1982) Injuries and damages from hazardous wastes--analysis and improvement of legal remedies. A report to Congress in compliance with section 301(e) of the Comprehensive Environmental Response, Compensation, and Liability Act of 1980 by the "Superfund Section 301(e) Study Group". Part 1. Report

and comments. Prepared by F.P. Grad, P.A. Porter, and the staff of the Legislative Drafting Research Fund of Columbia University. Committee print. 97th Cong., 3rd sess. Serial no. 97-12. Washington, DC: The Committee 344p. Available from: National Technical Information Service, Springfield, VA.

13. U.S. Congress, Senate. Environment and Public Works Committee. (1982) Injuries and damages from hazardous wastes--analysis and improvement of legal remedies. A report to Congress in compliance with section 301(e) of the Comprehensive Environmental Response, Compensation, and Liability Act of 1980 by the "Superfund Section 301(e) Study Group". Part 2. Appendices. Prepared by F.P. Grad, P.A. Porter, and the staff of the Legislative Drafting Research Fund of Columbia University. Committee print. 97th Cong., 3rd sess. Serial no. 97-12. Washington, DC: The Committee 501p. Available from: National Technical Information Service, Springfield, VA.
14. Hainline, A., Jr. (1983) Tifft Farm Nature Preserve, Buffalo County, New York. [Memorandum, dated September 12, to Deputy Director, Chronic Disease Division, Center for Environmental Health, approved by Assistant Director for Health Services, Centers for Disease Control]. 3p. Available from: [Center for Environmental Health, Centers for Disease Control, U.S. Department of Health and Human Services, Atlanta, GA].
15. Hainline, A., Jr. (1983) Hoopa Valley Reservation soil sampling plan. [Memorandum dated September 27, from Chief of Clinical Chemistry Standardization Section, Center for Environmental Health to Superfund Implementation Group, approved by Assistant Director for Health Services]. 3p. Available from: [Center for Environmental Health, Centers for Disease Control, U.S. Department of Health and Human Services, Atlanta, GA].
16. Halperin, W.; Altman, R.; Stemhagen, A.; Iaci, A.W.; Caldwell, G.; Mason, T.; Bill, J.; Abe, T.; Clark, J.F. (1980) Epidemiologic investigation of clusters of leukemia and Hodgkin's disease in Rutherford, New Jersey. J. Med. Soc. N. J. 77:267-273.
17. Harris, R.H.; Rodricks, J.V.; Highland, J.H.; Papadopulos, S.S. ([1984]) Adverse health effects at a Tennessee hazardous waste disposal site. 22p. Available from: Center for Energy and Environmental Studies, Princeton University, Princeton, NJ.
18. Heath, C.W., Jr.; Nadel, M.R.; Zack, M.M.; Chen, A.T.L.; Bender, M.A.; Preston, R.J. (1984) Cytogenetic findings in persons living near the Love Canal. J. Am. Med. Assoc. 251:1437-1440.
19. Heath, C.W., Jr.; Nadel, M.R.; Zack, M.M.; Stutzman, C.D.; Chen, A.T.L.; Kimbrough, R.D.; VanderMeer, D.C.; Figler, J.; Bender, M.A.; Kale, R.P.; Makar, M.S.; Pyatt, B.E.; Preston, R.J.; Au, W.W.; Gooch, P.C.; Ho, T.; Luippold, H. (1983) Study of cytogenetic patterns in persons living near the Love Canal. 45p. Available from: U.S. Department of Health and Human Services, Atlanta, GA.
20. Hirschhorn, J. S. ([1985]) Statement of senior associate of U.S. Congress, Office of Technology Assessment. In: U.S. Congress, House of Representatives. Public Works and Transportation Committee. Hazardous waste contamination of water resources. Hearing before the Subcommittee on Investigations and Oversight. 98th Cong., 2nd sess. In press. Delivered on March 13, 1984. 10p. Available from: Committee on Public Works and Transportation, U.S. House of Representatives.
21. Idaho. Department of Health and Welfare. (1983) Kellogg revisited--1983:

proposed protocol. [36p]. Available from: [Center for Environmental Health, Centers for Disease Control, U.S. Department of Health and Human Services, Atlanta, GA].
22. Ing, R.; Falk, H.; Lemen, R.A.; Carson, G.; Kelter, A.; Sarn, J.; Gray, M. (1981) Community asbestos exposure in Globe, Arizona. J. Pediatr. 99:409-411.
23. Janerich, D.T.; Burnett, W.S.; Feck, G.; Hoff, M.; Nasca, P.; Polednak, A.P.; Greenwald, P.; Vianna, N. (1981) Cancer incidence in the Love Canal area. Science 212:1404-1407.
24. Johnson, C.J. (1977) Toxic soluble waste disposal in a sanitary landfill site draining to an urban water supply. Am. J. Public Health 67:468-469.
25. Jones, G. (1982) Memorandum on Laskin/Poplar Oil Company, dated December 20, to Peter McCumiskey. 2p. Available from: [Center for Environmental Health, Centers for Disease Control, U.S. Department of Health and Human Services, Atlanta, GA].
26. Jones, G. A. (1983) Doepke-Holliday Site, Johnson County, Kansas. [Memorandum, dated February 2, from Chief, Superfund Implementation Group, to E.J. Skowronski, Public Health Advisor, EPA Region VII]. 1p. Available from: [Center for Environmental Health, Centers for Disease Control, U.S. Department of Health and Human Services, Atlanta, GA].
27. Jones, G. (1983) G.E. Moreau hazardous waste site in South Glen Falls, New York. [Memorandum 057C, dated June 13, from Chief, Superfund Implementation Group, to D. Knorowski, public health advisor, EPA Region II]. 12p. Available from: [Center for Environmental Health, Centers for Disease Control, U.S. Department of Health and Human Services, Atlanta, GA].
28. Jones, G. A. (1983) Doepke-Holliday site, Johnson County, Kansas. [Letter, dated October 19, from Chief, Superfund Implementation Group, to E.J. Skowronski, Public Health Advisor, EPA Region VII]. 1p. Available from: [Center for Environmental Health, Centers for Disease Control, U.S. Department of Health and Human Services, Atlanta, GA].
29. Jones, G.A. (1983) Hazardous waste tests, Project 9-4/5, Hoopa Valley Indian Reservation. [Memorandum 1102c, dated December 6, to J. Mulder, Public Health Advisor, EPA Region IX]. 1p. Available from: [Center for Environmental Health, Centers for Disease Control, U.S. Department of Health and Human Services, Atlanta, GA].
30. Kay, R.L., Jr. (1983) Reassessment of Superfund Site Review: Doepke-Holliday Site, Johnson County, Kansas. [Memorandum 2427s, dated October 4, to F.S. Lisella, Center for Environmental Health, approved by C. Tate, Jr., E. Kilbourne, and H. Falk]. 3p. Available from: [Center for Environmental Health, Centers for Disease Control, U.S. Department of Health and Human Services, Atlanta, GA].
31. Lagakos, S.W.; Wessen, B.J.; Zelen, M. (1984) Analysis of contaminated well water and health effects in Woburn, Massachusetts: Technical report No. 3. Prepared by the Harvard School of Public Health, Department of Biostatistics for the Siam Institute for Mathematics and Society. 64p. Available from: The Department, Harvard School of Public Health, Boston, MA.
32. Lagakos, S.W.; Wessen, B.J.; Zelen, M. (1984) Woburn Health Study: an analysis of reproductive and childhood disorders and their relation to environmental contamination. 153p. Available from: Department of Biostatistics, Harvard School of Public Health, Boston, MA.

33. Landrigan, P.J.; Baker, E.L., Jr.; Feldman, R.G.; Cox, D.H.; Eden, K.V.; Orenstein, W.A.; Mather, J.A.; Yankel, A.J.; Von Lindern, I.H. (1976) Increased lead absorption with anemia and slowed nerve conduction in children near a lead smelter. J. Pediatr. 89:904-910.
34. Lieber, M.; Perlmutter, N.M.; Frauenthal, H.L. (1964) Cadmium and hexavalent chromium in Nassau County ground water. J. Am. Water Works Assoc. 56:739-747.
35. Lieber, M; Welsch, W.F. (1954) Contamination of ground water by cadmium. J. Am. Water Works Assoc. 46:541-547.
36. Long, F.A.; Schweitzer, G.E., editors. (1982) Risk assessment at hazardous waste sites. Based on a symposium sponsored by the ACS Committee on Environmental Improvement at the 183rd Meeting of the American Chemical Society, March-April, Las Vegas, Nevada. ACS Symposium Series 204. Washington, DC.: American Chemical Society. 128p.
37. Lowrance, W.W., editor. (1981) Assessment of health effects at chemical disposal sites: proceedings of a symposium held in New York City on June 1-2, by the Life Sciences and Public Policy Program of The Rockefeller University. New York City: The University. 166p.
38. Massachusetts. Department of Public Health. (1982) The New Bedford PCP study-preliminary findings. 6p. Available from: The Department, Boston, MA.
39. Massachusetts. Department of Public Health, Division of Environmental Health Assessment; and Massachusetts Health Research Institute, Inc. (1983) Proposal for a greater New Bedford, MA PCB health survey. Draft report. 106p. Available from: [The Institute or The Department, Boston, MA].
40. McKinley, D.K. (1983) Doepke-Holliday site. Memorandum, dated August 30, to E. Skowronski, EPA Public Health Advisor, approved by K. Biggs, Superfund Chief]. 2p. Available from: [Center for Environmental Health, Centers for Disease Control, U.S. Department of Health and Human Services, Atlanta, GA].
41. Memphis-Shelby County Health Department. (1983) Protocol for the study of possible human exposure to and health effects of toxic chemicals from the Hollywood dump, Memphis, Tennessee. Draft Report. Prepared in cooperation with the Tennessee Department of Public Health and the U.S. Centers for Disease Control, Center for Environmental Health. 81p. Available from: Center for Environmental Health, Centers for Disease Control, U.S. Department of Health and Human Services, Atlanta, GA.
42. Meyer, C.R. (1983) Liver dysfunction in residents exposed to leachate from a toxic waste dump. Environ. Health Perspect. 48:9-13.
43. Montana. Department of Health and Environmental Sciences, Air Quality Bureau. (1983) Kellogg revisited--1983: proposed protocol. 165p. Available from: [Center for Environmental Health, Centers for Disease Control, U.S. Department of Health and Human Services, Atlanta, GA].
44. New Jersey. Department of Environmental Protection. (1984) Newark dioxin sites data summary. 20p. Available from: The Department.
45. New Jersey. Department of Health, Environmental Health Hazard Evaluation Program. (1982) Briar Lake health survey--appendum to "Health survey of the population living near the Gloucester Environmental Management Services (GEMS) landfill." In: Summary of Health Studies Conducted at GEMS landfill in Camden County by the New Jersey Department of Health. p. 24-40. Available from: The Program, [Trenton,

NJ].
46. New Jersey. Department of Health, Environmental Health Hazard Evaluation Program. (1982) Pulmonary function testing of a population living near the Gloucester environmental management services (GEMS). Prepared in cooperation with the Camden County Health Department. In: Summary of Health Studies Conducted at GEMS landfill in Camden County by the New Jersey Department of Health. p. 41-46. Available from: The Program, [Trenton, NJ].
47. New Jersey. Department of Health, Environmental Health Hazard Evaluation Program. (1983) Health survey of the population living near the Price landfill, Egg Harbor Township, Atlantic County. Prepared in cooperation with Atlantic County Health Department. 29p. Available from: The Program, [Trenton, NJ].
48. New Jersey. Department of Health, Environmental Health Hazard Evaluation Program. (1983) Health survey of the population living near Gloucester Environmental Management Services (GEMS) landfill. Prepared in cooperation with the Camden County Health Department. In: Summary of health studies conducted at GEMS landfill in Camden County by the New Jersey Department of Health. p. 4-23. Available from: The Program, [Trenton, NJ].
49. New Jersey. Department of Health; Environmental Health Hazard Evaluation Program. (1983) Environmental air sampling both within and outside of homes located near the Gloucester Environmental Management Services (GEMS) landfill, conducted during winter 1981-82 and summer 1982. In: Summary of health studies conducted at GEMS landfill in Camden County by the New Jersey Department of Health. p. 47-69. Available from: The Program, [Trenton, NJ].
50. New Jersey. Department of Health, Environmental Health Program. (1984) Pomona Oaks exposure assessment: volatile organics in wellwater and indoor air. 80p. Available from: [The Program, Trenton, NJ].
51. New Jersey. Department of Health, Environmental Health Program. (1984) Health survey of the population living near the Krysowaty Farm: abandoned hazardous waste disposal site. 34p. Available from: [The Program, Trenton, NJ].
52. NUS Corporation. (1983) Toxicological impact assessment of Drake Chemical, Inc. Prepared for the U.S. Environmental Protection Agency, Hazardous Site Control Division. 69p. Available from: [NUS Corporation, Gaithersburg, MD].
53. Oudbier, A.J.; Eyster, J.T.; Lock, J.H. (1981) Berlin-Farro respiratory study: December 1981. 28p. Available from: [Michigan Department of Public Health, Lansing, MI].
54. Ozonoff, D.; Colten, M.E.; Cupples, A.; Heeren, T.; Schatzkin, A.; Mangione,T.; Dresner, M.; Colton, T. (1983) Silresim area health study: report of findings. Prepared for the Massachusetts Department of Public Health and Department of Environmental Quality Engineering by Boston University, School of Public Health, and the University of Massachusetts (Boston), Center for Survey Research. 114p. Available from: [School of Public Health, Boston University, Boston, MA].
55. Parker, G.S.; Rosen, S.L. (1981) Woburn: cancer incidence and environmental hazards: 1969-1978. Prepared by Massachusetts Department of Public Health. 53p. Available from: The Department, Boston, MA.
56. Pennsylvania. Department of Health. (1983) Health surveillance protocol:

planned activities related to the Drake Chemical Superfund site, Lock Haven, Clinton County, Pennsylvania. 10p. Available from: The Department of Health, Harrisburg, PA.
57. Rau, E.H. (1983) [Evaluation of potential chemical toxicity hazards at Hoopa Valley, California Superfund sites. Letter, with attached summary, dated Dec. 20, from U.S. Department of Health and Human Services, Indian Health Service (Redding, CA), to E. Livermore, Executive Director of Hoopa Valley Indian Housing Authority, and J. Ransom, U.S. Department of Housing and Urban Development, Office of Indian Programs (San Francisco)]. 5p. Available from: [Center for Environmental Health, Centers for Disease Control, U.S. Department of Health and Human Services, Atlanta, GA].
58. Reinganum, C.B.; Hang, W.L.T. (1983) Toxic wastelands: environmental and public health implications of six New York City municipal landfills. New York: New York Public Interest Research Group, Inc. 98p.
59. Rothenberg, R. (1981) Morbidity study at a chemical dump--New York. Morb. Mortal. Wkly. Rep. 30:293-294.
60. Satin, K.; Deane, M.; Leonard, A.; Neutra, R.; Gravitz, N.; Harnly, M.; Green, R. (1983) The McColl Site Health Survey: executive summary. Prepared by the California Department of Health Services, Epidemiological Studies Section. 7p. Available from: The Department, Berkeley, CA.
61. Schultz, S. (1983) Report on the Brookfield Health Survey. 61p. Available from: Bureau of Preventable Diseases, New York City Health Department, NY.
62. Shuster, K.A. (1976) Leachate damage assessment. Case study of the Sayville solid waste disposal site in Islip (Long Island), New York. EPA/530/SW-509. Cincinnati, OH: U.S. Environmental Protection Agency, Office of Solid Waste Management Programs. 18p.
63. Singal, M.; Kominsky, J.R.; Schulte, P.A.; Landrigan, P.J. (1982) Health hazard evaluation report. 79-022-789. Washington, DC: U.S. Department of Health and Human Services, National Institute for Occupational Safety and Health. 86p. Available from: National Technical Information Service, Springfield, VA.
64. Stehr, P. (1983) Superfund site review for the Tifft Farm Nature Preserve, Erie County, New York. [Memorandum 2268S, dated September 14, from EIS Officer, Center for Environmental Health, approved by F. Liseela, E. Kilbourne, and H. Falk]. 3p. Available from: [Center for Environmental Health, Centers for Disease Control, U.S. Department of Health and Human Services, Atlanta, GA].
65. Taylor, P.R.; Lawrence, C.E.; Hwang, H.; Paulson, A.S. (1984) Polychlorinated biphenyls influence on birthweight and gestation. 12p. Available from: C.E. Lawrence, Center for Laboratories and Research, New York Department of Health, Albany, N.Y.
66. Taylor, P.R.; Stelma J.; Lawrence, C.E. ([1984]) Relation of polychlorinated biphenyls to birthweight and gestational age in the offspring of occupationally exposed mothers. 31p. Available from: principal author, National Cancer Institute, U.S. Department of Health and Human Services, Bethesda, MD.
67. Thomas, L., chairman. (1980) Report of the Governor's Panel to Review Scientific Studies and the Development of Public Policy on Problems Resulting from Hazardous Wastes. 61p. Available from: New York State Legislature, Albany, NY.

68. Trezek, G.J. (1984) Engineering case study of the Stringfellow Superfund site. 151p. Available from: U.S. Congress, Office of Technology Assessment, Washington, DC.
69. Falk, H.; Heath, C.W., Jr.; Stein, G.F.; Caldwell, G.G.; Drotman, D.P. (1980) Multiple toxic chemicals in an illegal dump, Chester, Pennsylvania. [Memoranda EPI-80-12-1 and EPI-80-12-2, dated October 30, 1979, and November 12, 1980, respectively, from Chronic Diseases Division, Center for Environmental Health, to the director, U.S. Centers for Disease Control, Atlanta, GA]. 8p. Available from: [Center for Environmental Health, Centers for Disease Control, U.S. Department for Health and Human Services, Atlanta, GA].
70. Caldwell, G.G.; Heath, C.W., Jr.; Cutler, J.L.; French, J. (1981) Cancer in Woburn, Massachusetts. [Memoranda EPI-80-37-1 and EPI-80-37-2, dated January 8, 1980, and September 16, 1981, respectively, from Chronic Diseases Division, Center from Environmental Health to the director, U.S. Centers for Disease Control, Atlanta, GA]. 19p. Available from: [Center for Environmental Health, Centers for Disease Control, U.S. Department of Health and Human Services, Atlanta, GA].
71. Moody, P.L. (1981) Drake Chemical, Inc., Lockhaven, Pennsylvania: interim Report (HETA 81-248). 8p. Available from: National Institute for Occupational Safety and Health, U.S. Department of Health and Human Services, Cincinnati, OH.
72. U.S. Department of Health and Human Services, Centers for Disease Control, Center for Environmental Health. (1982) Childhood lead exposure around secondary smelters, Dallas, Texas. Draft report. 17p. Available from: [The Center, Atlanta, GA].
73. CH2M Hill. (1984) Love Canal document citations. Prepared for Love Canal health experts. 172p. Available from: CH2M Hill, Reston, VA.
74. U.S. Environmental Protection Agency. (1984) Hazardous waste sites: descriptions of sites on current National Priorities List, October 1984. HW-8.5. Washington, DC: The Agency. 569p.
75. U.S. Environmental Protection Agency. (1984) Hazardous waste sites: descriptions of 244 sites on proposed update #2 to National Priorities List, October 1984. 2nd printing. HW-8.4. Washington, DC: The Agency. 269p.
76. Budnick, L.D. (1983) Drake Superfund site, Lock Haven, Clinton County, Pennsylvania: analysis of county-wide cancer mortality and congenital malformation incidence rates. [Memorandum, dated July 19, and report from EIS Officer, Chronic Diseases Division, Center for Environmental Health, Centers for Disease Control, to J. Logue, Director of Division of Environmental Health, Pennsylvania Department of Health, Harrisburg, PA]. 13p. Available from: U.S. Environmental Protection Agency, Philadelphia, PA.
77. U.S. Environmental Protection Agency, Office of Solid Waste and Emergency Response. (1983) Hazardous waste site descriptions: National Priorities List: final rule. HW-8.1. Washington, DC: The Office 422p.
78. U.S. Environmental Protection Agency, Office of Solid Waste and Emergency Response. (1983) Hazardous waste site descriptions: National Priorities List--proposed update. HW-8.2. Washington, DC: The Office. 134p.
79. U.S. Environmental Protection Agency, Office of Solid Waste Management Programs. (1975) Arsenic poisoning in Minnesota. In: Hazardous waste disposal damage reports. EPA/530/SW-151. Washington, DC: The Agency.

p. 1-2. Available from: National Technical Information Service, Springfield, VA.
80. U.S. Environmental Protection Agency, Office of Solid Waste Management Programs. (1975) Fatality at a New Jersey industrial landfill. In: Hazardous waste disposal damage reports. EPA/530/SW-151. Washington, DC: The Agency. p. 6-8. Available from: National Technical Information Service, Springfield, VA.
81. U.S. Environmental Protection Agency, Oil and Special Materials Control Division. (1980) Damages and threats caused by hazardous material sites. EPA 430/9-80/004. Washington, DC: The Agency. [258]p.
82. U.S. Congress, Senate. Environment and Public Works Committee. (1980) Six case studies of compensation for toxic substances pollution: Alabama, California, Michigan, Missouri, New Jersey, and Texas. Committee print. 96th Congress, 2nd session. Serial No. 96-13. Washington, DC: The Committee. 521p. Available from: U.S. Government Printing Office, Washington, DC.
83. U.S. Congress, Senate. Environment and Public Works Committee. (1980) Health effects of toxic pollution: a report from the Surgeon General and a brief review of selected environmental contamination incidents with a potential for health effects. Prepared by the Surgeon General, Department of Health and Human Services, and the Congressional Research Service, Library of Congress. Committee print. 96th Congress, 2nd session. Serial No. 96-15. Washington, DC: The Committee 312p. Available from: U.S. Government Printing Office, Washington, DC.
84. Vianna, N.J.; Polan, A.K. (1984) Incidence of low birth weight among Love Canal residents. Science 226:1217-1219.
85. Vianna, N.J. (1980) Adverse pregnancy outcomes--potential endpoints of human toxicity in the Love Canal: preliminary results. In: Porter, I.H.; Hook, E.B.; eds. Human embryonic and fetal death. New York: Academic Press, Inc. p. 165-168.
86. Weaver, G. (1984) PCB contamination in and around New Bedford, Mass. Environ. Sci. Technol. 18:22A-27A.
87. Webb, K.; Ayres, S.; Slavin, R.; Knutsen, A.; Roodman, S.; Gedney, W.B.; Schramm, W.; Hotchkiss, R.L.; Miller, R.; Donnell, H.D. (1984) Results of a pilot study of health effects due to 2,3,7,8-tetrachlorodibenzodioxin contamination-Missouri. Morb. Mortal. Wkly. Rep. 33:54-61.
88. Budnick, L.D.; Sokal, D.C.; Falk, H.; Logue, J.N.; Fox, J.M. (1984) Cancer and birth defects near the Drake Superfund site, Pennsylvania. Arch. Environ. Health 39:409-413.
89. Satin, K.; Deane, M.; Leonard, A.; Neutra, R.; Gravitz, N.; Harnly, M.; Green, R. (1983) The McColl Site Health Survey: an epidemiological and toxicological assessment of the McColl site in Fullerton, California. Prepared by the California Department of Health Services, Epidemiological Studies Section. 179p. Available from: The Department, Berkeley, CA.
90. Stehr, P.A.; Stein, G.F.; Webb, K.; Schramm, W.; Gedney, W.B.; Donnell, H.D.; Ayres, S.; Falk, H.; Sampson, E.J.; Smith, S.J.; Steinberg, K.K. ([1984]) A pilot epidemiologic study of health effects due to 2,3,7,8-tetrachlorodibenzo-p-dioxin contaminations in Missouri. 25p. Available from: Center for Environmental Health, Centers for Disease Control, U.S. Department of Health and Human Services, Atlanta, GA.
91. Anonymous. (1984) Health-risk estimates for 2,3,7,8-tetrachlorodibenzodioxin in soil. [Partial reprint of Vol. 33, No. 3 of Morb. Mortal. Wkly.

Rep.]. J. Am. Med. Assoc. 251:878, 883.
92. Feinglass, E.J. (1973) Arsenic intoxication from well water in the United States. New Engl. J. Med. 288:828-830.
93. Budnick, L.D.; Sokal, D.C.; Falk, H.; Logue, J.N.; Fox, J.M. (1984) Cancer and birth defects near the Drake Superfund site, Pennsylvania. Arch. Environ. Health 39:409-413.
94. Wallace, L.; Pellizzari, E.; Hartwell, T.; Zelon, H.; Sparacino, C.; Whitmore, R. ([1984]) Analyses of exhaled breath of 355 urban residents for volatile organic compounds. 6p. Available from: U.S. Environmental Protection Agency, Washington, DC.
95. Pearson, L., Lindemuth, E., Witte, E. J. (1981) Trichloroethylene exposure: Pennsylvania. Morb. Mortal. Wkly. Rep. 30:226-233.

APPENDIX 8–II

SUMMARY OF METHODS AND FINDINGS OF SELECTED INVESTIGATIONS FOR HEALTH CONSEQUENCES RESULTING FROM CHEMICAL EXPOSURE OF POPULATIONS AT CHEMICAL DISPOSAL SITES

NPL SITE/ STATE*	REASON FOR STUDY	STUDY GROUP (REFERENCE)	PREDOMINANT CHEMICALS	ROUTE OF EXPOSURE AND CRITERIA USED	MAJOR HEALTH ENDPOINTS	STUDY DESIGN	SAMPLE SIZE	FINDINGS REPORTED
I McColl, CA	Odors and health complaints by residents	SHA (41,42) and local industry (1,21)	Oil sludge, sulfur compounds	Inhalation; targeted odor zones	General health; consultation of physicians and use of medical care facilities; cancer; reproductive outcomes; pet health	Cross section survey of 2 neighborhoods (mailed questionnaire)	1024 adults, 448 children	No excess cancers, no impairment of intellectual skills in children, and no "serious medical problems;" increased physician visits by children, and excess menstrual complaints by women
II New Bedford, MA	Public concern; contamination of a food chain; EPA monitoring	Proposed CDC (23) Phase I	PCB's	Food ingestion; circulating blood levels tested	Blood pressure; blood PCB levels	Prospective survey	800 adults	Not available
		Proposed CDC (23) Phase II	PCB's	Food ingestion; circulating blood levels tested	Liver enzyme induction; altered lipid metabolism; depressed immune function; neurotoxicity	Case-comparison; occupational and environmental exposures compared to control populations	150 adults	Not available

SHA - State Health Agency
CDC - Centers for Disease Control
EPA - Environmental Protection Agency

* Sites listed in alphabetical order by state abbreviation and site name

APPENDIX 8-II

SUMMARY OF METHODS AND FINDINGS OF SELECTED INVESTIGATIONS FOR HEALTH CONSEQUENCES RESULTING FROM CHEMICAL EXPOSURE OF POPULATIONS AT CHEMICAL DISPOSAL SITES

NPL SITE/ STATE	REASON FOR STUDY	STUDY GROUP (REFERENCE)	PREDOMINANT CHEMICALS	ROUTE OF EXPOSURE AND CRITERIA USED	MAJOR HEALTH ENDPOINTS	STUDY DESIGN	SAMPLE SIZE	FINDINGS REPORTED
II New Bedford, MA	Public concerns; contamination of a food chain; EPA monitoring	SHA (22) IGS (55) CDC (23)	PCB's	Food ingestion; circulating blood levels tested	Body burden PCB	2 Pilot study surveys	21 adults in one study and 51 adults in other study	More subjects than expected with circulating PCB levels in excess of 30 ppb
III Silresim, MA	Odors and health complaints by residents	IGS and SHA (34)	Oil, solvents, metals, PCB, pesticides	Inhalation; targeted exposure zones	General health; consultation of physicians and use of medical care facilities; adults/children; cancer incidence; mortality records; reproductive outcomes	Cross sectional survey: telephone and in-person interviews	1970 adults, 1100 children	No significant differences between exposed and control residents for cancer, reproduction, or mortality; greater frequency of respiratory symptoms, headache, fatigue, and heart problems
IV Woburn, MA	Leukemia cluster noted by local physician and SHA; resident complaints	SHA (35) CDC (47)	Pb, As, TCE	Ingestion of tainted well water; geographically targeted interviews	Childhood leukemia; renal, liver, and bladder cancer	Case-comparison: 12 childhood leukemias; 20 kidney and 5 liver cancers	Census tracts	Cluster; excess of male kidney cancer and leukemia; no association to chemical exposure
		IGS (17, 18,19)	Pb, As, TCE	Ingestion of tainted well water; geographically targeted interviews	Childhood leukemia; reproductive outcomes; childhood disease	Blinded phone health survey questionnaire	3,257 homes	Exposure associated with childhood leukemia, perinatal mortality, birth defects, childhood disorders

SHA - State Health Agency
CDC - Centers for Disease Control
IGS - Independent Group Study

APPENDIX 8-II

SUMMARY OF METHODS AND FINDINGS OF SELECTED INVESTIGATIONS FOR HEALTH CONSEQUENCES RESULTING FROM CHEMICAL EXPOSURE OF POPULATIONS AT CHEMICAL DISPOSAL SITES

NPL SITE/ STATE	REASON FOR STUDY	STUDY GROUP (REFERENCE)	PREDOMINANT CHEMICALS	ROUTE OF EXPOSURE AND CRITERIA USED	MAJOR HEALTH ENDPOINTS	STUDY DESIGN	SAMPLE SIZE	FINDINGS REPORTED
V McKin, Me	Physician report; well testing	CR (51) SHA	TCE	Ingestion of tainted well water	Bladder and liver disorder; skin rash; loss of equilibrium	Case incident report	750 families	One young girl with bladder and liver disorders
		CDC				Study Proposal for Surveillance of Exposed Persons.		
VI Montague, MI	Odors and health complaints by residents	CR (51,52)	Chloroform, CCl4, TCE, mirex, TCDD, As, hexachloro-cyclopentadiene	Ingestion of tainted well water and game fish	General health	Case incident report	12 homes	Residents report headache, nausea. No known latent effects
		SHA	Chloroform, CCl4, TCE, mirex, TCDD, As, hexachloro-cyclopentadiene	Ingestion of tainted well water and game fish	State Department of Health proposed an epidemiology study (1980)- No further information available at this time.			

SHA - State Health Agency
CDC - Centers for Disease Control
CR - Congressional Report

APPENDIX 8-II

SUMMARY OF METHODS AND FINDINGS OF SELECTED INVESTIGATIONS FOR HEALTH CONSEQUENCES RESULTING FROM CHEMICAL EXPOSURE OF POPULATIONS AT CHEMICAL DISPOSAL SITES

NPL SITE/ STATE	REASON FOR STUDY	STUDY GROUP (REFERENCE)	PREDOMINANT CHEMICALS	ROUTE OF EXPOSURE AND CRITERIA USED	MAJOR HEALTH ENDPOINTS	STUDY DESIGN	SAMPLE SIZE	FINDINGS REPORTED
VII Perham, MN	Hospital report of disease cluster	SHA (7,9) CR (6,51) EPA (50)	Arsenic trioxide	Ingestion of tainted well water; As levels in hair; estimation of water consumption	Neurologic and GI tract function	Case-comparison	13 homes	Gastrointestinal distress and neuropathy were consistent with exposure to arsenic.
VIII "GEMS" Landfill, Gloucester Environmental Management Services, NJ	Odors and health complaints by residents	SHA (27,28, 29,30)	Toluene, organic solvents	Inhalation; targeted exposure	General health; consultation of physicians and use of medical care facilities; respiratory function; reproductive outcomes	4 Cross-sectional cohort surveys; medical questionnaire	566 exposed residents; 163 controls	No excess medical or reproductive problems; excess respiratory symptoms, eye irritation, and odor complaints among the exposed group; no significant differences in pulmonary function
IX Jackson Township Landfill, NJ	Health complaints by residents	CR (6,11)	Benzene, chloroform, TCE, acetone, methylene chloride	Ingestion of tainted well water	General health; cancer; skin; reproductive outcomes; nervous system; blood; respiratory; GI tract	Case incident report	96 Residents	Skin ailments; complaints of kidney ailments, cancer, and miscarriages
		SHA (5)	Benzene, chloroform, TCE, acetone, methylene chloride	Ingestion of tainted well water; measurements of volatile organics in water and indoor air	Skin; reproductive outcomes	Cross-sectional; health census	560 exposed residents	Increased incidence of skin complaints

SHA - State Health Agency
CR - Congressional Report

APPENDIX 8-II

SUMMARY OF METHODS AND FINDINGS OF SELECTED INVESTIGATIONS FOR HEALTH CONSEQUENCES RESULTING FROM CHEMICAL EXPOSURE OF POPULATIONS AT CHEMICAL DISPOSAL SITES

NPL SITE/ STATE*	REASON FOR STUDY	STUDY GROUP (REFERENCE)	PREDOMINANT CHEMICALS	ROUTE OF EXPOSURE AND CRITERIA USED	MAJOR HEALTH ENDPOINTS	STUDY DESIGN	SAMPLE SIZE	FINDINGS REPORTED
X Krysowaty Farm, NJ	Resident complaints of health problems; EPA monitoring	SHA (26)	Benzidine, benzene, hexachloro-benzene, heavy metals	Ingestion of tainted well water; duration of use of well water	12 Subjective symptoms; pregnancy; consultation of physicians and use of medical care facilities	Cross-sectional survey	12 exposed households; 30 control households	Increased tiredness for females; questionable increase for miscarriages; no significant differences between exposed and unexposed for a variety of subjective complaints
XI Pomona Oaks, NJ	Resident complaints of foul well water	SHA (25)	Benzene, volatile organics	Inhalation; indoor air measurements; urinary phenol measurements	Potential exposure to a known carcinogen; GLOBAL 82 risk assessment model	Prospective cross-sectional survey	153 residents	GLOBAL 82 model estimated an increased cancer risk from inhalation of benzene at levels in excess of acceptable public health standards
XII Price Landfill, NJ	Contami-nated groundwater	SHA (31)	Benzene, chloroform, TCE, vinyl chloride, toluene, Pb, Zn, Cd	Ingestion of tainted well water from proven plume of contaminated aquifer	General health; consultation of physicians and use of medical care facilities; reproductive outcomes	Cross-sectional cohort	121 exposed residents; 123 controls	Excess muscle pain, rash and skin irritation, joint pain, pregnancy problems in exposed group; no long-term effects; (symptoms possibly stress-related)

SHA - State Health Agency

HADWC-J*

APPENDIX 8-II

SUMMARY OF METHODS AND FINDINGS OF SELECTED INVESTIGATIONS FOR HEALTH CONSEQUENCES RESULTING FROM CHEMICAL EXPOSURE OF POPULATIONS AT CHEMICAL DISPOSAL SITES

NPL SITE/ STATE*	REASON FOR STUDY	STUDY GROUP (REFERENCE)	PREDOMINANT CHEMICALS	ROUTE OF EXPOSURE AND CRITERIA USED	MAJOR HEALTH ENDPOINTS	STUDY DESIGN	SAMPLE SIZE	FINDINGS REPORTED
XIII Reich's Farm, Dover Township, NJ	Resident complaints of foul well water	EPA (10) CR (51) LHA	Acetone, toluene, styrene, CCl4, methyl ethyl ketone	Ingestion of tainted well water	Medical complaints; illness	Illness survey	23 families (48 persons)	No correlation between use of contaminated well water and reported illness
XIV "Rutherford," Universal Oil Products and Scientific Chemical Processing, Inc., NJ	Resident concerns for cluster of leukemia and Hodgkin's disease	SHA (3) EPA (12)	Benzene, TCE	Inhalation; residents lived within the school district	Incidence of leukemia and Hodgkin's disease	Case-comparison, interview questionnaire	13 cases leukemia; 9 cases Hodgkin's disease	Increased incidence of leukemia in age group 5-19 and Hodgkin's disease in age group 20-39; no significant exposure source consistently associated with all cases
XV # Brookfield Landfill, NY	Odor complaints by residents	LHA (43)	Non-specific odor	Inhalation; targeted exposure zones	Subjective symptoms; utilization of medical care facilities	Cross-sectional telephone and interview survey	555 households (454 exposed, 101 control)	Statistically significant increase in resident complaints and physician visits, especially for children
		IGS (38)			Verified physician diagnosis	Physician questionnaire	144 physicians	

LHA - Local Health Agency
SHA - State Health Agency
IGS - Independent Group Study
CR - Congressional Report

Not listed on NPL.

APPENDIX 8-II

SUMMARY OF METHODS AND FINDINGS OF SELECTED INVESTIGATIONS FOR HEALTH CONSEQUENCES RESULTING FROM CHEMICAL EXPOSURE OF POPULATIONS AT CHEMICAL DISPOSAL SITES

NPL SITE/ STATE*	REASON FOR STUDY	STUDY GROUP (REFERENCE)	PREDOMINANT CHEMICALS	ROUTE OF EXPOSURE AND CRITERIA USED	MAJOR HEALTH ENDPOINTS	STUDY DESIGN	SAMPLE SIZE	FINDINGS REPORTED
XVI "Hooker Hyde Park," NY	Citizen complaints of odors and skin irritation	CDC (40) NIOSH (44)	TCDD, many chlorinated hydrocarbons	Residence or employment near landfill; pesticide blood levels	Physical examination of skin, mucous membranes, and thyroid; blood pressure; laboratory tests of urine and blood	Cross-sectional community study, interview questionnaire, and NHANES data comparisons	428 persons (residents and workers near landfill)	No indicators of ill health were attributable to the waste site.
XVII Love Canal, NY	Odors and health complaints by residents	SHA (16)	82 Organic chemicals and metals	Inhalation; ingestion; contact in targeted exposure zones; breath analysis	Cancer incidence	Retrospective cohort by census tract	225 households with 700 people in a census tract containing 4,897 people	Respiratory cancer rate higher than NY state, but consistent with Niagara area; no excess liver cancer, lymphoma, or leukemia
		SHA (54)	82 Organic chemicals and metals	Inhalation; ingestion; contact in targeted exposure zones	Reproductive outcomes	Cohort study by questionnaire	460 live births	Excess of spontaneous abortions and low birth weight infants in wet areas
		SHA (53)	82 Organic chemicals and metals	Inhalation; ingestion; contact in targeted exposure zones	Birth weight; reproductive outcomes	Retrospective follow-up	1201 females, 617 births	Low birth weight observed in swale area residents during period of active dumping only

SHA - State Health Agency
CDC - Centersfor Disease Control
NIOSH - National Institute for Occupational Safety and Health

APPENDIX 8-II

SUMMARY OF METHODS AND FINDINGS OF SELECTED INVESTIGATIONS FOR HEALTH CONSEQUENCES RESULTING FROM CHEMICAL EXPOSURE OF POPULATIONS AT CHEMICAL DISPOSAL SITES

NPL SITE/ STATE*	REASON FOR STUDY	STUDY GROUP (REFERENCE)	PREDOMINANT CHEMICALS	ROUTE OF EXPOSURE AND CRITERIA USED	MAJOR HEALTH ENDPOINTS	STUDY DESIGN	SAMPLE SIZE	FINDINGS REPORTED
XVII Love Canal, NY		EPA (37)	82 Organic chemicals and metals	Residence near site	Chromosome breakage	Pilot study	36 resident volunteers in comparison with historic controls	No overall increase in aberrations; supernumery acentric fragments observed
		CDC (2, 14,15)	82 Organic chemicals and metals	Detection of high chemical levels in basement air	Cytogenetic abnormalities; reproductive outcomes	Blinded cohort study with matched comparisons	46 exposed residents	No significant difference in chromosomal aberrations or sister chromatid exchange (SCE)
		SHA (32) State Gov. (45)	82 Organic chemicals and metals			Review of existing studies for Governor		No acute intoxication by chemicals at site; long-term hazards inconclusive
XVIII ABM Wade, PA	Complaints of acute symptoms following fire at site.	CDC (46) SHA	Benzene and naphthalene deriatives, Cr, Pb, Ni, Zn, Cu	Inhalation; direct contact; targeted exposure zones	General health; consultation of physicians and use of medical care facilities; respiratory symptoms; skin rashes; child health profile with clinical laboratory tests	a)Non-random door-to-door survey; b)local physician interviews; c)firemen questionnaire	86 adults; 28 children (14 children with medical examination); 35 firemen	No history of adverse health effects in residents or from physical examination of 14 children; firemens' complaints of headache, cough and skin rashes could not be conclusively related to this fire.

SHA - State Health Agency
CDC - Centers for Disease Control

APPENDIX 8-II

SUMMARY OF METHODS AND FINDINGS OF SELECTED INVESTIGATIONS FOR HEALTH CONSEQUENCES RESULTING FROM CHEMICAL EXPOSURE OF POPULATIONS AT CHEMICAL DISPOSAL SITES

NPL SITE/ STATE*	REASON FOR STUDY	STUDY GROUP (REFERENCE)	PREDOMINANT CHEMICALS	ROUTE OF EXPOSURE AND CRITERIA USED	MAJOR HEALTH ENDPOINTS	STUDY DESIGN	SAMPLE SIZE	FINDINGS REPORTED
XIX Drake Chemical, PA	Worker complaints to NIOSH; community concern for bladder cancer; EPA monitoring	EPA (33) CDC (49,56) SHA (36) NIOSH (48)	Beta-naphthylamine, cyanides, benzidene, benzene and its derivatives, Fenac	Inhalation; direct contact	Cancer mortality, reproductive outcomes	County cancer mortality and reproductive outcome comparisons; ongoing population surveillance	Based on census data and cancer registries	No indication of a general environmental exposure to a bladder carcinogen; no excess of congenital malformations
XX Stanley Kessler, "King of Prussia," PA	Resident complaints	SHA (20)	TCE	Ingestion of tainted well water; body burden determined by urine test	Liver cancer	Case-comparison	117 Residents	No acute illness reported
XXI Velsicol Chemical Corp., "Hardeman County," "Toone," TN	Odors and health complaints by residents	IGS (4,13, 24) CR (51) SHA (8,39)	CCl4, chlorobenzene, chloroform, hexachlorobicyloheptadiene, tetrachloroethylene	Ingestion of tainted well water; measurement of carbon tetrachloride in well water; indoor air sampling	Health questionnaire; clinical examination; clinical laboratory tests	Case-comparison	82 exposed residents	Transitory liver injury was documented (based on) liver function tests

SHA - State Health Agency
CDC - Centers for Disease Control
NIOSH - National Institute for Occupational Safety and Health
IGS - Independent Group Study
CR - Congressional Report

References for Appendix 8-II

1. Chemical Manufacturers Association. (1983) Technical and epidemiological studies of the McColl site (Fullerton, CA). 2p. Prepared by the Data Development Work Group. Available from: The Association, Washington, DC.
2. Bender, M.; Preston, R. (1983) Cytogenetic patterns in Love Canal, NY, residents. [Partial reprint of Vol. 32, No. 20 of Morb. Mortal. Wkly. Rep.]. J. Am. Med. Assoc. 249:3159-3160.
3. Burke, T.A.; Gray, S.; Krawiec, C.M.; Katz, R.J.; Preuss, P.W.; Paulson, G. (1980) Environmental investigation of clusters of leukemia and Hodgkin's Disease in Rutherford, New Jersey. J. Med. Soc. N.J. 77:259-264.
4. Clark, C.S.; Meyer, C.R.; Gartside, P.S.; Majeti, V.A.; Specker, B.; Balistreri, W.F.; Elia, V.J. (1982) Environmental health survey of drinking water contamination by leachate from a pesticide waste dump in Hardeman County, Tennessee. Arch. Environ. Health 37:9-18, 125.
5. Conomos, M.G. (1983) A health census of a community with groundwater contamination: Jackson Township, 1980. Trenton, NJ: New Jersey State Department of Health. 113p.
6. U.S. Executive Office of the President, Council on Environmental Quality. (1980) Environmental quality 1980: eleventh annual report of the Council on Environmental Quality. Washington, DC: The Council. 497p. Available from: U.S. Government Printing Office, Washington, DC.
7. Crossland, J. (1977) Wastes endure. Environment 19(5):6-13.
8. Tennessee. Department of Public Health, Pollution Control Board. (1971) Transcript of hearing on the Velsicol Chemical Company waste disposal site in Hardeman County, March 4, Jackson, Tennessee. 36p. Available from: The Control Board, Nashville, TN.
9. Feinglass, E.J. (1973) Arsenic intoxication from well water in the United States. New Engl. J. Med. 288:828-830.
10. Ghassemi, M. (1976) Analysis of a land disposal damage incident involving hazardous waste materials, Dover Township, New Jersey: final report. Prepared for the U.S. Environmental Protection Agency, Office of Solid Waste Management Programs by TRW Systems Group, Inc., Redondo Beach, CA. Available from: National Technical Information Service, Springfield, VA.
11. U.S. Congress, Senate. Environment and Public Works Committee. (1982) Injuries and damages from hazardous wastes--analysis and improvement of legal remedies. A report to Congress in compliance with section 301(e) of the Comprehensive Environmental Response, Compensation, and Liability Act of 1980 by the "Superfund Section 301(e) Study Group". Part 1. Report and comments. Prepared by F.P. Grad, P.A. Porter, and the staff of the Legislative Drafting Research Fund of Columbia University. Committee print. 97th Congress, 3rd sess. Serial no. 97-12. Washington, DC: The Committee. 344p. Available from: National Technical Information Service, Springfield, VA.
12. Halperin, W.; Altman, R.; Stemhagen, A.; Iaci, A.W.; Caldwell, G.; Mason, T.; Bill, J.; Abe, T.; Clark, J.F. (1980) Epidemiologic investigation of clusters of leukemia and Hodgkin's disease in Rutherford, New Jersey. J. Med. Soc. N. J. 77:267-273.
13. Harris, R.H.; Rodricks, J.V.; Highland, J.H.; Papadopulos, S.S. ([1984])

Adverse health effects at a Tennessee hazardous waste disposal site. 22p. Available from: Center for Energy and Environmental Studies, Princeton University, Princeton, NJ.

14. Heath, C.W., Jr.; Nadel, M.R.; Zack, M.M.; Stutzman, C.D.; Chen, A.T.L.; Kimbrough, R.D.; VanderMeer, D.C.; Figler, J.; Bender, M.A.; Kale, R.P.; Makar, M.S.; Pyatt, B.E.; Preston, R.J.; Au, W.W.; Gooch, P.C.; Ho, T.; Luippold, H. (1983) Study of cytogenetic patterns in persons living near the Love Canal. 45p. Available from: U.S. Department of Health and Human Services, Atlanta, GA.
15. Heath, C.W., Jr.; Nadel, M.R.; Zack, M.M.; Chen, A.T.L.; Bender, M.A.; Preston, R.J. (1984) Cytogenetic findings in persons living near the Love Canal. J. Am. Med. Assoc. 251:1437-1440.
16. Janerich, D.T.; Burnett, W.S.; Feck, G.; Hoff, M.; Nasca, P.; Polednak, A.P.; Greenwald, P.; Vianna, N. (1981) Cancer incidence in the Love Canal area. Science 212:1404-1407.
17. Lagakos, S.W.; Wessen, B.; Zelen, M. (1984) Synopsis: the Woburn health study. An analysis of reproductive and childhood disorders and their relation to environmental contamination. 21p. Available from: Department of Biostatistics, Harvard School of Public Health, Boston, MA.
18. Lagakos, S.W.; Wessen, B.J.; Zelen, M. (1984) Woburn Health Study: an analysis of reproductive and childhood disorders and their relation to environmental contamination. 153p. Available from: Department of Biostatistics, Harvard School of Public Health, Boston.
19. Lagakos, S.W.; Wessen, B.J.; Zelen, M. (1984) Analysis of contaminated well water and health effects in Woburn, Massachusetts: Technical report No. 3. Prepared by the Harvard School of Public Health, Department of Biostatistics for the Siam Institute for Mathematics and Society. 64p. Available from: The Department, Harvard School of Public Health, Boston.
20. Long, F.A.; Schweitzer, G.E., editors. (1982) Risk assessment at hazardous waste sites. Based on a symposium sponsored by the ACS Committee on Environmental Improvement at the 183rd Meeting of the American Chemical Society, March-April, Las Vegas, Nevada. ACS Symposium Series 204. Washington, DC.: American Chemical Society. 128p.
21. Lowrance, W.W., editor. (1981) Assessment of health effects at chemical disposal sites: proceedings of a symposium held in New York City on June 1-2, by the Life Sciences and Public Policy Program of The Rockefeller University. New York City: The University. 166p.
22. Massachusetts. Department of Public Health. (1982) The New Bedford PCP study-preliminary findings. 6p. Available from: The Department, Boston, MA.
23. Massachusetts. Department of Public Health, Division of Environmental Health Assessment and Massachusetts Health Research Institute, Inc. (1983) Proposal for a greater New Bedford, MA PCB health survey. Draft report. 106p. Available from: [The Institute or The Department, Boston, MA].
24. Meyer, C.R. (1983) Liver dysfunction in residents exposed to leachate from a toxic waste dump. Environ. Health Perspect. 48:9-13.
25. New Jersey. Department of Health, Environmental Health Program. (1984) Pomona Oaks exposure assessment: volatile organics in wellwater and indoor air. 80p. Available from: [The Program, Trenton, NJ].
26. New Jersey. Department of Health, Environmental Health Program. (1984) Health survey of the population living near the Krysowaty Farm: aban-

doned hazardous waste disposal site. 34p. Available from: [The Program, Trenton, NJ].
27. New Jersey. Department of Health, Environmental Health Hazard Evaluation Program. (1983) Environmental air sampling both within and outside of homes located near the Gloucester Environmental Management Services (GEMS) landfill, conducted during winter 1981-82 and summer 1982. In: Summary of health studies conducted at GEMS landfill in Camden County by the New Jersey Department of Health. p. 47-69. Available from: The Program, [Trenton, NJ].
28. New Jersey. Department of Health, Environmental Health Hazard Evaluation Program. (1982) Pulmonary function testing of a population living near the Gloucester environmental management services (GEMS). Prepared in cooperation with the Camden County Health Department. In: Summary of Health Studies Conducted at GEMS landfill in Camden County by the New Jersey Department of Health. p. 41-46. Available from: The Program, [Trenton, NJ].
29. New Jersey. Department of Health, Environmental Health Hazard Evaluation Program. (1982) Briar Lake health survey-- appendum to "Health survey of the population living near the Gloucester Environmental Management Services (GEMS) landfill." In: Summary of Health Studies Conducted at GEMS landfill in Camden County by the New Jersey Department of Health. p. 24-40. Available from: The Program, [Trenton, NJ].
30. New Jersey. Department of Health, Environmental Health Hazard Evaluation Program. (1983) Health survey of the population living near Gloucester Environmental Management Services (GEMS) landfill. Prepared in cooperation with the Camden County Health Department. In: Summary of health studies conducted at GEMS landfill in Camden County by the New Jersey Department of Health. p. 4-23. Available from: The Program, [Trenton, NJ].
31. New Jersey. Department of Health, Environmental Health Hazard Evaluation Program. (1983) Health survey of the population living near the Price landfill, Egg Harbor Township, Atlantic County. Prepared in cooperation with Atlantic County Health Department. 29p. Available from: The Program, [Trenton, NJ].
32. New York. Department of Health. (1981) Love Canal: a special report to the Governor and legislature. Prepared with the assistance of the New York Departments of Transportation and Environmental Conservation. Albany, NY: New York Department of Health 69p.
33. NUS Corporation. (1983) Toxicological impact assessment of Drake Chemical, Inc. Prepared for the U.S. Environmental Protection Agency, Hazardous Site Control Division. 69p. Available from: [NUS Corporation, Gaithersburg, MD].
34. Ozonoff, D.; Colten, M.E.; Cupples, A.; Heeren, T.; Schatzkin, A.; Mangione, T.; Dresner, M.; Colton, T. (1983) Silresim area health study: report of findings. Prepared for the Massachusetts Department of Public Health and Department of Environmental Quality Engineering by Boston University, School of Public Health and the University of Massachusetts (Boston), Center for Survey Research. 114p. Available from: [Public Health, Boston University, Boston, MA].
35. Parker, G.S.; Rosen, S.L. (1981) Woburn: cancer incidence and environmental hazards: 1969-1978. Prepared by Massachusetts Department of

Public Health. 53p. Available from: The Department, Boston, MA.
36. Pennsylvania. Department of Health. (1983) Health surveillance protocol: planned activities related to the Drake Chemical Superfund site, Lock Haven, Clinton County, Pennsylvania. 10p. Available from: The Department, Harrisburg, PA.
37. Picciano, D. (1980) Love Canal chromosome study [letter to the editor]. Science 209:754, 751-756.
38. Reinganum, C.B.; Hang W.L.T. (1983) Toxic wastelands: environmental and public health implications of six New York City municipal landfills. New York: New York Public Interest Research Group, Inc. 98p.
39. Rima, D.R.; Brown, E.; Goerlitz, D.F.; Law, L.M. (1967) Potential contamination of the hydrologic environment from the pesticide waste dump in Hardeman County, Tennessee. Administrative report to the Federal Water Pollution Control Administration. 41 p. Available from: Water Resources Division, Geological Survey, U.S. Department of the Interior, [Washington, DC].
40. Rothenberg, R. (1981) Morbidity study at a chemical dump--New York. Morb. Mortal. Wkly. Rep. 30:293-294.
41. Satin, K.; Deane, M.; Leonard, A.; Neutra, R.; Gravitz, N.; Harnly, M.; Green, R. (1983) The McColl Site Health Survey: an epidemiological and toxicological assessment of the McColl site in Fullerton, California. Prepared by the California Department of Health Services, Epidemiology Study Section. 179p. Available from : The Department, Berkeley, CA.
42. Satin, K.; Deane, M.; Leonard, A.; Neutra, R.; Gravitz, N.; Harnly, M.; Green, R. (1983) The McColl Site Health Survey: executive summary. Prepared by the California Department of Health Services, Epidemiological Studies Section. 7p. Available from: The Department, Berkeley, CA.
43. Schultz, S. (1983) Report on the Brookfield Health Survey. 61p. Available from: Bureau of Preventable Diseases, New York City Health Department.
44. Singal, M.; Kominsky, J.R.; Schulte, P.A.; Landrigan, P.J. (1982) Health hazard evaluation report. 79-022-789. Washington, DC: U.S. Department of Health and Human Services, National Institute for Occupational Safety and Health. 86p. Available from: National Technical Information Service, Springfield, VA.
45. Thomas, L., chairman. (1980) Report of the Governor's Panel to Review Scientific Studies and the Development of Public Policy on Problems Resulting from Hazardous Wastes. 61p. Available from: New York State Legislature, Albany, NY.
46. Falk, H.; Heath, C.W., Jr.; Stein, G.F.; Caldwell, G.G.; Drotman, D.P. (1980) Multiple toxic chemicals in an illegal dump, Chester, Pennsylvania. [Memoranda EPI-80-12-1 and EPI-80-12-2, dated October 30, 1979, and November 12, 1980, respectively, from Chronic Diseases Division, Center for Environmental Health, to the director, U.S. Centers for Disease Control, Atlanta, GA]. 8p. Available from: [Center for Environmental Health, Centers for Disease Control, U.S. Department for Health and Human Services, Atlanta, GA].
47. Caldwell, G.G.; Heath, C.W., Jr.; Cutler, J.L.; French, J. (1981) Cancer in Woburn, Massachusetts. [Memoranda EPI-80-37-1 and EPI-80-37-2, dated January 8, 1980 and September 16, 1981. respectively, from Chronic Diseases Division, Center fron Environmental Health to the director, U.S. Centers for Disease Control, Atlanta, GA. 19p. Available from: [Center for Environmental Health, Centers for Disease Control, U.S. Department

of Health and Human Services, Atlanta, GA].
48. Moody, P.L. (1981) Drake Chemical, Inc., Lockhaven, Pennsylvania: interim Report (HETA 81-248). 8p. Available from: National Institute for Occupational Safety and Health, U.S. Department of Health and Human Services, Cincinnati, OH.
49. Budnick, L.D. (1983) Drake Superfund site, Lock Haven, Clinton County, Pennsylvania: analysis of county-wide cancer mortality and congenital malformation incidence rates. [Memorandum, dated July 19, and report from EIS Officer, Chronic Diseases Division, U.S. Center for Environmental Health, to J. Logue, Director of Division of Environmental Health, Pennsylvania Department of Health, Harrisburg, PA]. 13p. Available from: U.S. Environmental Protection Agency, Philadelphia, PA.
50. U.S. Environmental Protection Agency, Office of Solid Waste Management Programs. (1975) Arsenic poisoning in Minnesota. In: Hazardous waste disposal damage reports. EPA/530/SW-151. Washington, DC: The Agency. p.1-2. Available from: National Technical Information Service, Springfield, VA.
51. U.S. Congress, Senate. Environment and Public Works Committee. (1980) Health effects of toxic pollution: a report from the Surgeon General and a brief review of selected environmental contamination incidents with a potential for health effects. Prepared by the Surgeon General, Department of Health and Human Services, and the Congressional Research Service, Library of Congress. Committee print. 96th Congress, 2nd sess. Serial No. 96-15 .Washington, DC: The Committee. 312p. Available from: U.S. Government Printing Office, Washington, DC.
52. U.S. Congress, Senate. Environment and Public Works Committee. (1980) Six case studies of compensation for toxic substances pollution: Alabama, California, Michigan, Missouri, New Jersey, and Texas. Committee print. 96th Congress, 2nd sess. Serial No. 96-13. Washington, DC: The Committee. 521p. Available from: U.S. Government Printing Office, Washington, DC.
53. Vianna, N.J.; Polan, A.K. (1984) Incidence of low birth weight among Love Canal residents. Science 226:1217-1219.
54. Vianna, N.J. (1980) Adverse pregnancy outcomes--potential endpoints of human toxicity in the Love Canal: preliminary results. In: Porter, I.H.; Hook, E.B.; eds. Human embryonic and fetal death. New York: Academic Press, Inc. p. 165-168.
55. Weaver, G. (1984) PCB contamination in and around New Bedford, Mass. Environ. Sci. Technol. 18:22A-27A.

APPENDIX 8–III

I. McColl Site, Fullerton, California
Study by: Satin et al. (1983a;b)

This study was conducted by a mailed questionnaire survey comparing a population near oil fields and a chemical disposal site with nearby populations adjacent to oil fields alone. Information was collected for a broad range of potential health outcomes both acute and latent in nature. No direct measurements were made for exposure to specific chemicals and the study did not identify a specific health outcome that might be expected from the suspect chemicals at the disposal site.

The reliance on self-reported complaints and the fact that the response rate was higher from the suspect (82%) than the two comparison areas (69%), presented a bias within this study. As a consequence the authors described its results as modest. The appreciably greater frequency of complaints of odor and their particular pattern among McColl residents were consistent with the view that the odors were related to the site and not merely due to the oilfields, which also were present near the comparison areas. Furthermore, the high prevalence of complaints of respiratory symptoms, eye irritation, and nausea among McColl residents indicated that the reported symptoms were not coming from only a few vocal individuals. The findings did not suggest that there were any differences between the areas with respect to the prevalence of cancer, adverse reproductive effects or morbidity. However, the methods and the response rates for this study were not sufficient to support definitive interpretation. The results of the study did not establish that the chemical disposal site was responsible for the greater prevalence of troublesome symptoms. The authors acknowledge that there may be subject bias due to the air pollution for those individuals residing nearby, particularly in the block immediately adjacent to the site, and this could well have influenced the perception of overall health.

II. New Bedford, Massachusetts
Report by: Weaver (1984)

Information from: Massachusetts Health Research Institute Inc.; Massachusetts Department of Public Health, Division of Environmental Health Assessment (Massachusetts, 1983)
Study by: Massachusetts Department of Public Health (Massachusetts, 1982)

The documents reviewed describe a series of pilot studies, conducted by the Masssachusetts Department of Health and the Centers for Disease Control, to establish body burden of PCB's from blood samples of volunteers residing in the New Bedford region. Included in the study population were occupationally exposed individuals and consumers of large quantities of fish. The first survey, involving 21 participants, was conducted in 1981. In this survey, 9 individuals (43%) were found to have circulating blood levels of PCB's greater than or equal to 30 ppb. Similar findings were reported for a second survey conducted one year later in a group of 51 volunteers. In both surveys, several individuals were found to have PCB levels greater than or equal to 100 ppb, a level frequently associated with occupational exposure or cases of overt poisoning.

To date, no occurrence of illness related to PCB exposure has been noted by the local medical community, and the questions of cancer and hypertensive mortality are being addressed in a prospective epidemiology study proposed by the Centers for Disease Control and the Massachusetts Department of Public Health.

These surveys document a body burden for PCB's in those individuals tested; however, as indicated in the author's discussion, interpretation of the findings are limited due to the presence of bias. The study population was not randomly selected, offered no consideration of occupational exposure, and did not include a regional comparison group. In addition, the study population was not of a sufficient size to afford relevant comparisons with the general population in the New Bedford region. As the proposed studies of health effects are prospective, no conclusion can be reached at this time.

III. Silresim, Massachusetts
Study by: Ozonoff et al. (1983)

A good response was obtained from both the exposed and comparison populations but no serious disorder was detected by this telephone and in-person interview study, though it was necessarily limited by relying on self-reported measures of ill health. The exposed population reported more respiratory symptoms than the authors considered to be explainable solely in terms of their significantly heavier smoking habits. Moreover, the authors argued that the similar frequency of respiratory symptoms in exposed subjects who regarded their local air as "bad" and in those who did not suggested that the effect could not be simply attributed to biased recall. The other part of the study concerned a comparison of stated causes of death among Lowell residents and could only hope to detect a very marked effect, given that there were only 191 and 300 deaths in the exposed and comparison populations, respectively.

IV. Woburn, Massachusetts
Study A by: Centers for Disease Control
(Caldwell et al., 1981)
and
Massachusetts Department of Public Health
(Parker and Rosen, 1981)

Using standard methods (mortality statistics, tumor registries, and questionnaire surveys), these studies showed that the incidence rates of renal cancer and childhood leukemia in Woburn in the period 1969-1978 were higher than those found by the Third National Cancer Survey (TNCS) in 1969-1971. In addition, the incidences of cancers of the bladder and liver were similar in the two surveys. Because renal cancer has been increasing (Finger-Kantor et al., 1976), it is likely that comparison with data from the standard population over a more recent time period would have resulted in a slightly lower observed-to-expected ratio than the 1.5:1.0 ratio found. The fact that comparisons were only investigated for childhood leukemia (with negative results), greatly limited the evaluation of the findings about individuals with other malignancies.

The conclusions of the authors are justified in that the available data fail to establish any association between environmental hazards and increases in childhood leukemia and renal cancer. This investigation established that the overall incidence of childhood leukemia was elevated in Woburn. However, the hypothesis that such an increased incidence could be associated with the contamination of drinking water was neither supported nor refuted by the study findings.

Study B by: Lagakos et al. (1984)

This extensive study of childhood leukemia in relation to contaminated water (from 2 wells and in 2 neighborhoods) is limited by the fact that it was prompted not only by the excess itself (no case was found in addition to those known to the residents' group) but also because a cluster of 6 cases, sufficient in itself to "explain" the excess, was known from the outset to be close to the contaminated wells. This study must therefore primarily be a hypothesis-generating exercise.

The authors themselves referred to the difficulties of interpreting data about an excess of some disease in a community that was selected for study because of the extreme incidence. However, they did not consider that such arguments could apply to the excess found subsequently in the period 1980-1983, when 4 cases of leukemia were recorded compared to 1.36 expected. Nevertheless, since 3 of these 4 cases had no exposure to the well water in question (and the fourth only to a minimal extent), the excess in the later period does not provide support for the hypothesis of a relationship between the earlier excess and contaminated water. In addition, the detailed case-comparison study was further limited by involving all cases of childhood leukemia but only a non-random group of comparisons (see below). The authors' conclusions that children having greater access to water from the wells in question are at higher risk of childhood leukemia cannot, therefore, be regarded as being supported by their findings.

The methods by which other possible health effects of the contaminated water were pursued are open to other criticisms. These health effects were not independently ascertained but were identified in the course of the same phone calls in which details of exposure to contaminated water were obtained. There is, however, no means of knowing if willingness of eligible residents to be interviewed by phone was influenced either by the occurrence of any of the relevant disorders, or by their proximity to the relevant areas. The opportunity for bias is evident from the low proportion of households contacted for which successful interviews with eligible residents were obtained -- namely, eligible interviews with only 52% of the 6219 housholds reached by phone, which accounted for 32% of the 10,310 listed phones (excluding obvious business phones). Similar refusal rates in different parts of the town do not directly address this point, since the authors do not refer to households with eligible women or children. This, together with the low proportion of total residents studied, prevents evaluation of the findings of spontaneous abortion, low birth-weight, perinatal deaths, certain congenital malformations, and childhood disorders. However, it may be noted that the numbers of cases involved and the size of the relative risks are never very striking. In addition, although the excesses observed receive much of the attention, little emphasis is given to the deficiencies of the study's implementation.

The definition of one group of birth defects (which showed an excess) deserves comment, namely "environmental" birth defects. This grouping of disorders included neural tube defects, mental retardation, cerebral palsy, cleft palate, Down's syndrome, and other chromosomal abnormalities, and was justified on the grounds that all are due to environmental factors. In fact, specific environmental causes have not been established for the majority of these disorders, and, indeed, neural tube defects represent almost the only example to show etiological evidence of a marked effect of external (though unidentified) factors.

The authors conclusion that their findings "strongly suggest that water from the 2 wells is linked to a variety of adverse health effects" cannot be supported because the evidence they present is inherently weak.

V. McKin Site, Gray, Maine
Information from: U.S. Senate (USC/Senate, 1980a)

No study was described in the papers provided in relation to this chemical disposal site. The information was derived from a Congressional report describing an incident of well water contamination. Sixteen wells were found to be contaminated with trichloroethylene and 1,1,1-trichloroethane and were ordered closed after a young girl was found to have bladder and liver disorders purported to be associated with exposure to the water. Unsubstantiated information, based on a small incidence rate, is subject to errors in interpretation by all sectors of the community and government. In addition, it decreases the opportunity to conduct a valid study in the future as a consequence of the element of implied subject bias.

VI. Montague, Michigan
Information from: U.S. Senate (USC/Senate, 1980a; 1980b)

The documents (Congressional incident reports) mainly described details of the chemical disposal site and the concern it aroused. Twelve households are known to have been exposed to contaminated drinking water though an unknown number of individuals have eaten contaminated fish. An epidemiological study by the Michigan Department of Public Health was proposed in 1980 but the study design is not known. The exposed population has reported acute symptoms of headache and nausea but no chronic effects have been found to date. The available information relies on self-reported effects. It is not possible to comment further on the basis of the information available.

VII. Perham, Minnesota
Study by: Feinglass (1973)

Report by: U.S. EPA (USEPA/OSWMP, 1975)

Among the studies included in this review, this was the only one that involved an established entity -- namely, arsenical poisoning. The exposed population was clearly restricted to those individuals using a specific water source. In this case, both the chemical in question and the route of exposure could be determined, and in addition, the level of exposure could be estimated. With quantitative knowledge of the degree of well contamination, the exposure was estimated from individual drinking habits and confirmed by the discovery of toxic levels of arsenic in the hair of the most severely affected individuals. The health effect outcome could be projected from established toxicologic data bases for the compound, and verified by hospital records. Because the acute adverse health outcome occurred in close temporal relationship to the exposure, a causal relationship was easily identified. The conditions surrounding this study helped to minimize the extent of confounding and bias. The conclusions described by this study could be confirmed as the study effectively considered the criteria necessary to establish a causal relationship between chemical exposure to arsenic and an adverse health effect.

VIII. GEMS Landfill, Gloucester Township, Camden County, New Jersey
Studies by: New Jersey Department of Health,
Environmental Health Hazard Evaluation Program.

Study A: A health survey of the Fox Chase population.
New Jersey Department of Health (New Jersey, 1983b)

A high proportion (86%) of 122 households living close to the landfill responded to a questionnaire survey about odors and adverse health effects, with a lower response (69%) from a comparison group of 61 unexposed households. Ninety-one percent of the exposed population had been bothered by odors, whereas only 13% of the comparison group expressed concern. Certain exposed subgroups, defined in terms of smoking habits and sex, reported more respiratory symptoms and eye irritation than did the comparable unexposed group. This study relied upon self-reported odors and symptoms, and as such is subject to participant bias. The authors reasonably concluded that it was difficult to associate the differences solely with air pollution because of the unexplained subgroup confounding differences (which may well have been due to chance).

Study B: A Survey of the Briar Lake population.
New Jersey Department of Health (New Jersey, 1982a)

This small questionnaire survey found that all 46 residents living in the 14 exposed homes were bothered by odors in contrast to only 13% (20 individuals) among the 163 residents of respondents in the comparison area, the latter representing 69% of those households approached. As in the case of the previous survey, this study relied on self-reporting and is subject to participant bias. More respiratory symptoms were reported by the exposed group than by the comparison group but there was no appreciable difference in serious medical problems.

The authors were cautious in interpreting the significance of these findings because of the possible effects of the heightened concern among the residents as a consequence of the odor problem and the publicity about chemicals at the site.

Study C: Pulmonary function testing.
New Jersey Department of Health (New Jersey, 1982b; 1983a)

In this small study of 69 exposed and 24 unexposed individuals, no significant differences in the proportions of individuals with abnormal pulmonary functions were detected between the two groups. The authors noted that the 19 exposed adult females tested had a significantly lower forced vital capacity and forced expiratory volume than the 9 women in the comparison group. The small number of participants prevents the drawing of any firm conclusions. The authors noted that symptomatic individuals tended to have better pulmonary function results than non-symptomatic individuals, the opposite of what would be expected.

IX. Jackson Township, New Jersey
Study by: Conomos (1983)

This study was originated following complaints by residents about water quality in the Legler Community of Jackson Township, New Jersey, during 1978, and the discovery by the New Jersey Department of Environmental Protection of a variety of volatile organic chemicals contaminating the Cohansey aquifer. The source of the contamination was found to be a nearby landfill known to have received waste chemicals. Evidence of concern for potential health effects became apparent during January 1980 when a local citizens committee, comprising 94 households involved in a class action lawsuit, completed a self-administered mail questionnaire health survey. Staff of the State health department analyzed the responses and found a possible association of skin and eye irritation (New Jersey, 1980). Subsequently, the State Health Department sponsored a health census that included an expanded questionnaire with particular emphasis focused on skin disease and water use patterns. In addition to water sampling, indoor air testing was conducted.

The health census was administered by a team of trained health nurses to a total of 560 individuals representing 150 households in the affected community. This represented a 92.6% response rate as compared to an 87% response rate from the citizens' survey. Duration of well water use and the depth of wells were used to represent extent of exposure by ingestion, and frequency of bathing and washing was the measure of exposure by skin contact. The findings were irregular and inconsistent, with a suggestion that increased skin contact with the contaminated water was related to an increased frequency of redness, itching, and other skin problems which were not of great clinical importance. An increasing trend for reports of hospitalization and kidney illness diagnosed to be of infectious origin was noted with increasing years of well water usage at all well depths and in a subgroup for shallow well users. The percentage of participants reporting problems of ill health appeared greater among shallow well users; however, the authors were justified in expressing reservations about this observation as there was an insufficient number of deep well users available in the study area to validate this finding. No excessive frequency of reported adverse reproductive outcomes was observed among Jackson study residents.

The study encountered difficulties as it was conducted in a milieu where many of the participants were involved in a class-action suit, had been subject to previous surveys, and were aware of the potential health problems under evaluation. A discussion of recall bias was presented by the authors. This study suffered from many of the problems of similar studies: the number of study subjects was small, the measures of exposure were indirect and inexact, the study subjects were all exposed to some degree, and the basis for calculating expected values for the effects was not sound. The results show no important health effects related to exposure to the pollution.

X. Krysowaty Farm, New Jersey
Study by: New Jersey Department of Health (New Jersey, 1984a)

This chemical disposal site was closed in February 1982 following the discovery that ground and surface waters were being degraded by hazardous chemicals and suspect carcinogens, including benzidine, benzene, and hexachlorobenzene. Heavy metals (not identified in the report) had been documented by the New Jersey Department of Environmental Protection to be present in nearby residential wells. Analytical testing revealed that some of the compounds were detected at levels approximately three orders of magnitude higher than EPA recommended safe drinking water standards.

The New Jersey Department of Health, in June 1982, responded to complaints of health problems and, at the request of the local health department, conducted a questionnaire survey of 12 households near the site and a comparison group of 30 households several miles from the site. The objective of this cross-sectional health survey was to determine whether symptoms of illness were more prevalent among residents near the Krysowaty Farm site.

Information describing the method by which study households were contacted, the degree of training for the individuals who administered the questionnaires, and the selection of comparison households is not readily available. While the comparison group received water from a municipal water supply not known to be contaminated at the time of the study, there is no mention as to whether this water source was supplied from ground or surface water nor the degree of pre-treatment, if any. The survey relied on self-reported information to identify the relative risk for a wide variety of 12 subjective symptoms, for pregnancy, and for use of related medical care.

As noted by the authors, interpretation of the study findings was made difficult by numerous problems. There was a notably low response rate (50%) among the comparison households, and disparity existed between the populations for sex, age, tobacco use, and nonsite-related chemical exposure. A greater number of comparison households (33%) were bothered by bad water taste as compared with 13% of the study population. The number of households studied were too few to permit conclusions and the increased concern or stress of living near the site may have led to the increased reporting of symptoms among people living near the site. A statistically significant risk was determined for tiredness in females, and a questionable increase was observed for miscarriages in the study group. No appreciable differences were found for reported medical conditions. The authors feel that the levels of exposure were very low and the likelihood of causal relationship to the problem and complaints were unlikely.

XI. Pomona Oaks, New Jersey
Study by: New Jersey Department of Health (New Jersey, 1984b)

This investigation was prompted by resident complaints of bad taste and odor in drinking water in the New Jersey community of Pomona Oaks. Environmental water sampling found levels of volatile organics exceeding the recommended New Jersey guidelines of 100 ppb. Particular concern was expressed when levels of benzene (a known carcinogen) were detected, and residents were advised to discontinue use of well water for all potable purposes. The New Jersey Department of Environmental Protection then became involved in a series of studies to assess the potential effects of non-potable use of the contaminated water in a cross-sectional survey of approximately 153 area residents. The studies focused on assessing the types and levels of exposure in this community as determined by indoor air and water sampling. An additional goal was to determine whether urinary phenol, a metabolite of benzene, could provide a useful measurement as an index of exposure to benzene. A followup study assessed the amount of benzene inhaled when showering with the contaminated water. The environmental measurements for benzene were then applied to a risk assessment model (GLOBAL82) to serve as a basis for developing public health policy recommendations.

The environmental sampling demonstrated that more chemicals were found in the indoor air than in the water. In the full-scale study, 70% of the homes had total volatile organic levels in air over 99 ppb, while only 46% had water levels that high. The authors noted that homes frequently contained multiple sources of volatile organics in addition to the contaminated water. Additional sources of volatile organics included attached garages, cigarette smokers, use of kerosene space heaters, wood stoves, fireplaces, paint, and cleaning solvents. This report clearly demonstrates the need for considering secondary sources of volatile organics in the home as they may have a profound health impact and affect the quality of indoor air.

Although residents had been advised not to use the water for potable purposes, a relatively large proportion of the population commented that it tasted bad. In the full-scale study, 14% of the participants noted skin rashes, whereas none of the participants in the pilot study reported having skin problems.

The authors concluded that "if environmental exposures could be related to urinary phenol they are likely to be associated with a long-term, higher level of benzene exposure than observed in Pomona Oaks." Only 11% of the study participants in the full scale study showed an increased phenol level over 20 mg/L in contrast to 24% observed in the pilot study. A cross-tabulation of benzene exposure and elevated levels of urinary phenol did not show a statistically significant association. As noted by the authors, the populations studied (11 housholds in the pilot study, 37 households in the full-scale study, and 14 households in the followup study) were of insufficient size to pursue statistical analysis of the data to include cross-tabulation to control for confounders.

When applied to the GLOBAL82 risk assessment model, the level of risk associated with the quantity of benzene found in indoor air samples exceeded public health standards and indicated an increased estimated risk for cancer. A finding of concern not addressed in the study is an evaluation of the considerable proportion of individuals surveyed who apparently continued to consume the contaminated water, despite previous warnings. The cancer risk that might be associated with continued drinking of the contaminated water is unknown.

This study is exemplary in that it sought to ascertain potential cancer risk in a prospective mode without soliciting self-reported ailments from the affected community. As an adequate time has not elapsed between exposure and potential latent effects such as cancer, such outcomes cannot be ascertained at present. Fortunately, no significant illnesses were reported, and the area residents have been advised to take specific steps to protect their health until the current water supply can be replaced with a clean source.

XII. Price Landfill, Atlantic County, New Jersey
Study by: New Jersey Department of Health (New Jersey, 1983c)

Prepared by: Environmental Health Hazard Evaluation Program, New Jersey Department of Health in cooperation with Atlantic County Health Department.

In 1980, private wells within a one mile radius of the Price Landfill were found to contain concentrations of total volatile organics exceeding 100 ppb. This was a small study involving 38 housholds (68% of the group exposed to contaminated ground water) and 40 comparison households (76% of the comparison group). The majority of exposed respondents had been connected to a new water supply some months prior to this survey. Although there was no obvious difference in the proportion of the cases and comparisons who were troubled by any specific medical disorder, the exposed population reported a greater frequency of certain symptoms including muscle pain, skin irritation, and abdominal pain. The investigators considered that the number and frequency of symptoms were greater than would be expected given the known toxicity of the comparatively low concentrations of chemicals found in the water. The possibility of a relationship between the effects and the increased concern among the exposed population was discussed.

As the authors indicated, the study design, together with the local concern about the disposal site, made recall bias a real possibility, but the contribution that such concern might have made to the observed difference in symptoms could not be evaluated.

XIII. Reich's Farm - Dover Township, New Jersey
Report by: Ghassemi (1976)

This report, compiled for the EPA Office of Hazardous Waste Management, was prepared by the TRW Systems Group, and describes an analysis of a land disposal incident in which industrial wastes including phenol and volatile organics, toluene and styrene, leached from a disposal site to contaminate 148 wells in the Pleasant Plains section of Dover Township, New Jersey. The report's primary focus is on the economic, political, and regulatory aspects of the problem. It is included in this appendix because a summary of the Pleasant Plains illness survey was included to describe the health implications of the incident. The data collected in connection with this study were obtained through field interviews and telephone inquiries.

The Pleasant Plains illness survey was conducted by the Disease Control Section of Ocean County. Interviews were conducted at random for users of 23 area wells (23 households consisting of 48 individuals) to determine the possible correlation between consumption of tainted well water and the incidence of illness and medical complaints. The study included 10 contaminated wells, 4 non-contaminated wells, and 9 wells for which analytical testing was not available. Reported illnesses involving the kidney, stomach, liver, and gallbladder were distributed among those using all 3 categories of wells. As reported in the information reviewed, no correlation could be established between the use of contaminated well water and reported illness. The author concludes that although some of the chemicals found at the site could have posed a possible health hazard if ingested, there were no substantiated and medically documented cases of illness.

It is not surprising that no conclusion could be reached from such a small population sampling. None of these inquiries met even the most minimal criteria for a definitive study. Apparently no acute health problems occurred that were of sufficient severity to alarm the public. However, the wastes were deposited in 1971 and this analysis conducted in 1975; therefore, any latent effects would not have been reported. The information provided is anecdotal, and it is not possible to comment further on the health survey summarized in this report.

XIV. Rutherford, New Jersey
Study by: Halperin et al. (1980)
and
Burke et al. (1980)

This study showed that the leukemia incidence for the age category of less than 20 years was higher in Rutherford than that recorded by the Third National Cancer Survey (TNCS) or the Upstate New York Cancer Registry (6 cases observed compared to 0.6 expected). Otherwise the study could only be considered hypothesis-generating, because it was prompted by the cluster in Pierrepoint School and also because questions were added to the survey as interviewees supplied leads. Furthermore, its power would have been limited by the selection of comparisons from 6th graders enrolled at the Pierrepoint School in 1977-78 because they could not be age-matched with the cases (though better matching may have been obtained with another set of comparisons primarily chosen for the study of Hodgkin's disease).

Similar limitations are also relevant to the study of Hodgkin's disease since there was evidence of an excess before the study and also because the study did not attempt to test hypotheses which had been defined in advance. Such limitations subject this investigation to study bias.

The authors concluded, with justification, that available technology was unable to determine whether the excess disease incidence represented a chance event or a cause unique to Rutherford.

The examination of the environmental features at Rutherford revealed no unusual source of either ionizing or non-ionizing radiation in the area, nor anything unusual about its water supply. This preliminary examination was supplemented by sampling air, water, and soil for the estimation of certain chemicals and radiation. Particular attention was given to benzene and to ionizing radiation, but neither for these nor for any of the other hazards investigated was there evidence of unusual exposures as compared with other communities in heavily industrialized areas.

The methods used to pursue the aims of this study were reasonable and the conclusions justified.

XV. Brookfield Landfill, Staten Island, New York
Study by: Schultz (1983)

Foul and irritating odors were released when sewer line construction passed thru this municipal landfill, which had accepted unknown quantities of waste oil and liquid chemical residues. This telephone and personal interview survey, conducted by the New York City Department of Health during a six month period in 1982, was prompted by citizen complaints of nausea, vomiting, and eye and throat irritation. The initial phase of the investigation was designed to determine if people living in close proximity to the landfill experienced more health complaints and reported more medical visits than people living further away. The second phase of the investigation was designed to validate these findings for study participants and determine the medical diagnoses made by the treating physicians.

Of the 555 households selected for study from the 1980 census enumeration blocks, only 258 (51%) were successfully interviewed, and 297 replacement households had to be incorporated into the study. A greater response rate was observed for those residents living closer to the landfill, and medical care for this group was greater (45%) than that found in the comparison area (25%). Claims of a variety of subjective complaints including chest, breathing, eye, nasopharynx, nerve (headache), and emotional problems showed significant differences between the target and comparison populations. Several items on the survey, including pregnancy outcome and occupation, could not be evaluated as they were not uniformly recorded, and cancer frequency was so low that a meaningful analysis was not possible. No attempt was made to determine if individuals had made multiple visits to a physican for exposure-related complaints during the study period.

Of the 419 individuals (63%) who permitted a release of medical records, a response of 70% was provided by residents nearest the landfill while only 48% of the comparison group gave permission. Verification of physician visits was documented for 218 participants (66%), while 22% of the physician questionnaires were not completed. An increase in physician visits among participants who resided progressively closer to the landfill was validated by physician records, and of those complaints registered, 61% were related to respiratory problems.

Environmental sampling studies by both the U.S. and New York City Environmental Protection Agencies, and the New York City Departments of Sanitation and Health were unable to identify any specific chemical substance to which the community was exposed. The only exposure that could be documented was that of odor, which was described as a subjective response to low concentrations of complex mixtures of volatile chemicals perceived by residents living near the site. The resident health survey noted a distribution gradient for odor perception among study participants, with 63% resident complaints observed in the zone nearest to the landfill and 28% noted in the comparison group. No attempt was made to distinguish daily length of exposure for study participants.

This investigation lacked clarity as to a chemical exposure and a definitive disease or syndrome present in the community. As a result, the authors were able to detect bias within the first phase of the study due to reliance on self-reported complaints, disparity in mean age and family size between study and comparison areas, and the fact that a greater number of replacement participants were required for the comparison area than for the study area. The authors acknowledged that the subject bias due to the odor could have influenced the perception of overall health for individuals residing near the landfill. Bias was also demonstrated in the second phase of the study because permission to validate medical claims with the treating physicians was inversely proportional to the distance of the home from the landfill. Participant recall bias was evident because both over- as well as under-reporting of complaints, symptoms, and medical visits were documented in the records of responding physicians.

The authors' finding that "the reported illness complaints do not appear to have been influenced by the smoking habits of those living in the same household," may require additional consideration, as smoking is recognized as a primary cause of respiratory ailments. The study was able to detect only small differences in chest, breathing, eye, ear, or nose problems reported between smoking and non-smoking households within both the study and the comparison areas.

XVI. Hyde Park Landfill, Niagara Falls, New York

Study by: Singal et al. (1982)

Report by: Rothenberg (1981)

This study of 428 persons living or working near the Hyde Park Landfill was conducted by the National Institute for Occupational Safety and Health (NIOSH). Although the major portion of the study population consisted of individuals employed in local industry adjacent to the site, approximately 50 local residents were included as a subgroup. The investigation was conducted following complaints of odors and skin irritation, which were attributed to chemical vapors emanating from over 80,000 tons of chemicals that had been deposited at the site since 1953. Environmental sampling revealed concentrations of chlorinated hydrocarbons, including lindane, mirex, and dioxins in concentrations of parts per billion (ppb) in all locations tested.

The investigation consisted of an interviewer-administered questionnaire modeled after that used by the Health and Nutrition Examination Survey (HANES), and a limited physical examination that focused on skin, mucous membranes, thyroid, and blood pressure. In addition, urine analysis and blood testing for liver enzymes, hematologic parameters, and creatinine were performed. Body burden for lindane and mirex was determined for 20% of the study participants. The study utilized a matched comparison population from the HANES sample with consideration for age, sex, income, and marital status.

No increases in the incidence of spontaneous abortion or chloracne, or of abnormalities in liver injury enzyme activity, hemoglobin, red cell count, or serum creatinine were observed. There was an increased odds ratio for hiatal hernia and other minor upper gastrointestinal signs, the significance of which is not known.

The cross-sectional approach taken by this study group is most useful for the detection of existing disease, especially when being targeted at a unique health outcome related to a specific chemical exposure rather than to intermittent, low-level exposure from multiple chemicals. Interpretation of the study findings are limited by the low participation rate (59%). The authors could not assess latent effects, such as cancer, because adequate time has not elapsed between exposure and outcome. As no definite associations could be determined, the authors could draw no definitive link between the site and adverse health effects.

XVII. Love Canal, Niagara Falls, New York
Study A: Vianna (1980)
and
Study B: New York Department of Health (New York, 1981)

Following widespread concern about the possible adverse health effects caused by the chemical disposal site at Love Canal, this study was initiated to investigate the incidences of spontaneous miscarriage and of low birth weight (defined as 5.5 lb or less). The reasons offered by the authors for choosing these reproductive health endpoints -- "to balance rapid assessment with sensitivity" (Study A), and that "the sensitivity of the human reproductive process to toxic chemical exposure is well established" (Study B) -- are not persuasive. Evaluation of this field interview study is limited by the lack of details about the numbers of eligible individuals, of those approached, and the acceptance rate. With the exception of birth weights, which were checked by referring to birth certificates, the study was also limited because it relied on recall. Attempts to corroborate statements made by residents with reference to medical records were made wherever possible, although these results were not presented in this report.

No significant excess of congenital defects was found in any part of the study area, although a possible exception was made with regard to "wetland homes" and then only when compared to "dry properties." Reference to the questionable nature of this observation is made in both reports.

In the study area as a whole, no significant excesses of spontaneous abortions were observed. Data from homes on the two streets directly adjacent to the Love Canal showed that there was a slight excess of spontaneous abortions (12 observed compared to 8.2 expected) along one of the streets and no excess on the other. In the "historical water area," a significant excess was observed when compared with "dry properties" or with data published in a Canadian survey (Warburton & Fraser, 1963). There are, however, significant problems in using the Canadian study as a standard of comparison because it excluded certain groups of women: those with only spontaneous abortions (and no live-born child), those with successful or attempted induced abortion, and those with any illegitimate births.

A higher proportion of low birth weights was recorded in the "historical wet area" than in "dry properties." It is noteworthy that only one underweight infant was recorded on the two streets directly adjacent to the canal. The definition of low birth weight was generous and the one paper quoted in support of this measurement did not use this particular cutoff level, presenting data only in terms of mean birth weight.

Doubts exist in the use of these endpoints as reliable measures, and as described by the authors of one report, "they must be considered tentative" (Study A). The main limitation is that much of the data are based on recall. As the study was conducted against a background of intense local concern, it would hardly be surprising that residents living close to Love Canal recalled more miscarriages 10 or more years previously (the period in which the reported excess was most marked) than did those living further away.

Study C: Vianna and Polan (1984)

In a retrospective followup study, Vianna and Polan compared the incidence of low birth weight among infants born in the Love Canal area from 1940-1953 (the active dumping period) and 1954-1978 (the period after dumping). The probands were infants whose parents lived in houses adjacent to, but not abutting, the canal, and to which drainage of leachate from the Canal was facilitated by swales (natural depressions), some of which crossed the Canal. Adult residents were interviewed using a prepared questionnaire to provide information about past medical, therapeutic, social, occupational, and pregnancy histories. Interviews were completed for a study population consisting of 1201 females, representing a 92.7% response rate. Birth certificate information was verified for all infants; those that weighed 2500 g or less were considered low birth weight children. Additional consideration was made for sex ratios, smoking, education, disease patterns, and temporal trends. The authors concluded that individuals born to mothers living in the swale area during the period of active dumping were at greater risk of low birth weight than were those born to mothers living in homes abutting the canal.

As with other studies at disposal sites, in the absence of historical environmental data, this study relies on weak evidence of exposure. Exposure was inferred on the basis of geographical or topographical location of population subsets, which may have received maximum exposure to chemicals, and it was impossible to relate exposure to any toxic agent. Another major difficulty of study design was that the investigators were not certain all infants born in the area during the study period were included in the study. The authors justly note the potential for recall bias in response to the interview questionnaires, and the failure to rule out alcoholism in mothers. A peculiar temporal relationship could be noted in that there was no excess risk of low birth weight for babies born after the period of active dumping. It cannot be ascertained whether this could be due to the intensity of exposure during the dumping period, but in neither period, that is, 1940-1953 and 1954-1978, was there an excess of low birth weight children in the homes directly adjacent to the Love Canal. The study provides a well-supported hypothesis that a real excess of low birth weights occurred in the swale areas during the period of active dumping.

Study D: Janerich et al. (1981)

This study concluded from an examination of data in the New York Cancer Registry that there was no evidence of higher cancer rates in the area close to Love Canal than in upstate New York, or in the case of respiratory cancer, than in the entire city of Niagara Falls. This simple study had clear-cut objectives which were successfully achieved.

Study E: Heath et al. (1983; 1984)

No significant differences in the frequency of chromosomal aberrations or sister chromatid exchange were found between residents living close to the Love Canal and comparisons living in other parts of Niagara Falls. The authors pointed out the lack of direct evidence for the relationship, if any, between the measures that were studied and disorders such as cancer and birth defects. In both Love Canal and the comparison groups, cigarette smoking was associated with an increase in sister chromatid exchange frequency, pointing to the importance of taking such factors into account in studies of this type.

XVIII. ABM Wade, Chester, Pennsylvania
Study by: Centers for Disease Control
(Falk et al., 1980)

Concerns of local residents and historical, anecdotal information about symptoms of cough and rash among firemen exposed to fumes during a fire at this site, led to a voluntary questionnaire survey which was distributed among the 45 firemen involved. Local physicians were interviewed about illnesses that might be related to the fire and to the disposal site in general. In addition, a non-random, door-to-door survey was carried out among residents in the immediate vicinity of the disposal site to determine whether or not there were symptoms of adverse health effects, and to decide if a more formal study was warranted.

No firm conclusions were drawn from the survey of fire-fighters, because it was carried out two years after the fire in question. In the absence of an appropriate temporal relationship, it was not possible to distinguish any effects that could be related to this particular fire as compared with all other fires. Such information must be considered to be subject to recall bias. The survey among local residents showed no evidence of acute health effects that might be related to the fire or to the disposal site. For reasons referred to by the authors, this study lacked sufficient power to be of value. The main conclusions concerned recommendations about cleaning up the site and restricting access, all of which seemed sensible and reasonable.

XIX. Drake Site, Lockhaven, Pennsylvania
Studies by: Budnick (1983); Budnick et al. (1984)

Report by: National Institute for Occupational Safety and Health (Moody, 1981)

This study consisted of an examination of age-adjusted mortality rates from bladder cancer in Clinton County, Pennsylvania. The impetus for the study stemmed from longstanding concern about the relationship between naphthylamine and bladder cancer at the Drake site and in 3 adjacent counties lying within 20 miles of Lockhaven. Whereas the rates of bladder cancer in females living in the study area during the target time period dropped in all but one county and overall for the whole state, the rates for males rose in all four counties but remained steady for the state. No information was provided in the report about the numbers of people in the four counties who either worked at the Drake site or were likely to have been exposed to its chemical constituents.

The report offers no evidence to link the mortality changes among males with living in the vicinity of the Drake site. Given that the use of naphthylamine at the site did not begin until 1962, this fact would appear to weigh against a conclusion that changes in mortality rates were related to the Drake site, as neither the temporal relationship nor the the extent of the health effect seems consistent with that interpretation. The author referred to this point, stating "it is quite possible that the data do not reflect the fate of the Lockhaven townspeople who lived near and/or worked in the site".

No appreciable clustering of congenital malformations was noted in the review of a random sample amounting to 50% of the birth certificates in Clinton in the period 1973-1978. However, birth certificates are not a sensitive measure of the incidence of malformations, and since the projected total number of malformations was only 34, this part of the study was unlikely to produce reliable results due to the insufficient numbers involved.

XX. Stanley Kessler, King of Prussia, Pennsylvania
Information from: Long and Schweitzer (1982)

This review document describes the inadvertent disposal of trichloroethylene (TCE) which leached into nearby lagoons and well water supplies in Montgomery County, Pennsylvania. Although hundreds of people were apparently exposed to TCE in their drinking water, no signs of acute illness were reported. In this study, the chemical entity was identified and its presence monitored by chemical analysis. As indicated by the Pennsylvania Department of Health, samples of 100 wells yielded measurable concentrations of TCE in 70 cases, some of which exceeded 1,000 ppb. These findings brought about a study of exposure and liver cancer in Bucks and Montgomery County.

Apparently, the body burden of exposed residents was evaluated by means of urine screening for detectable trichloro-metabolites. Seven (6%) of the 117 residents screened showed detectable levels of trichloroethanol and trichloroacetic acid in their urine. The following facts could not be determined from the information provided: a) the methods used for selecting the study population, b) the temporal relationship between exposure and urine screening, and c) consideration of confounding factors.

Liver cancer deaths in the suspect area were compared with similar deaths in unexposed populations within the same county. The information provided does not indicate whether this is an ongoing prospective study or simply a documentation of exposure. It is not possible to comment further as the information is incomplete and additional data are not available.

XXI. Velsicol Chemical Company, Hardeman County, Tennessee
Study by: Clark et al. (1982)

Report by: Meyer (1983)

Prompted by complaints of contamination in well water, this study had the notable strength that it did not rely solely on self-reported symptoms but used more objective measures of both exposure and adverse effects. Almost all the individuals who lived in households using well water with concentrations of carbon tetrachloride exceeding 150 ug/L were identified. Specimens of blood and urine were obtained from these residents, and samples of air and water were collected from their homes. Similar sampling was conducted for a comparison group of area residents with no well contamination.

No evidence of carbon tetrachloride nor of other selected chlorinated organic compounds was found in any of the urine specimens tested. Higher serum activity levels of alkaline phosphatase and serum glutamic oxaloacetic transaminase were found in the exposed group, although the concentrations of total bilirubin and the activity of serum glutamic pyruvic transaminase were found to be lower than the comparisons. However, it was noted that the values observed for the liver function tests in both the exposed and the comparison groups were within what is usually considered to be the normal range of values for these tests. Exposure to contaminated water was reduced from essentially only bathroom use to no household use whatsoever. On repeat testing 3 months later, a significant reduction was observed for individual liver function tests, and no significant differences between the groups were detected. The fact that six of seven individuals with "borderline liver enlargement" were in the exposed group is difficult to evaluate because of the nature of this physical sign.

The authors were restrained in their conclusions that the findings of transient liver injury were related to exposure to contaminated drinking water, because the mean value of all biochemical levels were within the normal range.

References for Appendix 8-III

Budnick, L.D. (1983) Drake Superfund site, Lock Haven, Clinton County, Pennsylvania: analysis of county-wide cancer mortality and congenital malformation incidence rates. [Memorandum, dated July 19, and report from EIS Officer, Chronic Diseases Division, U.S. Center for Environmental Health, to J. Logue, Director of Division of Environmental Health, Pennsylvania Department of Health, Harrisburg, PA]. 13p. Available from: U.S. Environmental Protection Agency, Philadelphia, PA.

Budnick, L.D.; Sokal, D.C.; Falk, H.; Logue, J.N.; Fox, J.M. (1984) Cancer and birth defects near the Drake Superfund site, Pennsylvania. Arch. Environ. Health 39:409-413.

Burke, T.A.; Gray, S.; Krawiec, C.M.; Katz, R.J.; Preuss, P.W.; Paulson, G. (1980) Environmental investigation of clusters of leukemia and Hodgkin's Disease in Rutherford, New Jersey. J. Med. Soc. N.J. 77:259-264.

Caldwell, G.G.; Heath, C.W., Jr.; Cutler, J.L.; French, J. (1981) Cancer in Woburn, Massachusetts. [Memoranda EPI-80-37-1 and EPI-80-37-2, dated January 8, 1980, and September 16, 1981, respectively, from Chronic Diseases Division, Center for Environmental Health to the director, U.S. Centers for Disease Control, Atlanta, GA. 19p. Available from: [Center for Environmental Health, Centers for Disease Control, U.S. Department of Health and Human Services, Atlanta GA].

Clark, C.S.; Meyer, C.R.; Gartside, P.S.; Majeti, V.A.; Specker, B.; Balistreri, W.F.; Elia, V.J. (1982) Environmental health survey of drinking water contamination by leachate from a pesticide waste dump in Hardeman County, Tennessee. Arch. Environ. Health 37:9-18,125.

Conomos, M.G. (1983) A health census of a community with groundwater contamination: Jackson Township, 1980. Trenton, NJ: New Jersey State Department of Health. 113p.

Falk, H.; Heath, C.W., Jr.; Stein, G.F.; Caldwell, G.G.; Drotman, D.P. (1980) Multiple toxic chemicals in an illegal dump, Chester, Pennsylvania. [Memoranda EPI-80-12-1 and EPI-80-12-2, dated October 30, 1979, and November 12, 1980, respectively, from Chronic Diseases Division, Center for Environmental Health, to the director, U.S. Centers for Disease Control, Atlanta, GA]. 8p. Available from: [Center for Environmental Health, Centers for Disease Control, U.S. Department for Health and Human Services, Atlanta].

Feinglass, E.J. (1973) Arsenic intoxication from well water in the United States. New Engl. J. Med. 288:828-830.

Finger-Kantor, A.L.; Meigs, J.W.; Heston, J.F.; Flannery, J.T. (1976) Epidemiology of renal cell carcinoma in Connecticut, 1935-1973. J. Nat. Cancer Inst. 57:495-500.

Ghassemi, M. (1976) Analysis of a land disposal damage incident involving hazardous waste materials, Dover Township, New Jersey: final report. Prepared for the U.S. Environmental Protection Agency, Office of Solid Waste Management Programs by TRW Systems Group, Inc., Redondo Beach, CA. Available from: National Technical Information Service, Springfield, VA.

Halperin, W.; Altman, R.; Stemhagen, A.; Iaci, A.W.; Caldwell, G.; Mason, T.; Bill, J.; Abe, T.; Clark, J.F. (1980) Epidemiologic investigation of clusters of leukemia and Hodgkin's disease in Rutherford, New Jersey. J. Med. Soc. N. J. 77:267-273.

Heath, C.W., Jr.; Nadel, M.R.; Zack, M.M.; Chen, A.T.L.; Bender, M.A.; Preston, R.J. (1984) Cytogenetic findings in persons living near the Love Canal. J. Am. Med. Assoc. 251:1437-1440.

Heath, C.W., Jr.; Nadel, M.R.; Zack, M.M.; Stutzman, C.D.; Chen, A.T.L.; Kimbrough, R.D.; VanderMeer, D.C.; Figler, J.; Bender, M.A.; Kale, R.P.; Makar, M.S.; Pyatt, B.E.; Preston, R.J.; Au, W.W.; Gooch, P.C.; Ho, T.; Luippold, H. (1983) Study of cytogenetic patterns in persons living near the Love Canal. 45p. Available from: U.S. Department of Health and Human Services, Atlanta, GA.

Janerich, D.T.; Burnett, W.S.; Feck, G.; Hoff, M.; Nasca, P.; Polednak, A.P.; Greenwald, P.; Vianna, N. (1981) Cancer incidence in the Love Canal area. Science 212:1404-1407.

Lagakos, S.W.; Wessen, B.J.; Zelen, M. (1984) Woburn Health Study: an analysis of reproductive and childhood disorders and their relation to environmental contamination. 153p. Available from: Department of Biostatistics, Harvard School of Public Health, Boston, MA.

Long, F.A.; Schweitzer, G.E., editors. (1982) Risk assessment at hazardous waste sites. Based on a symposium sponsored by the ACS Committee on Environmental Improvement at the 183rd Meeting of the American Chemical Society, March-April, Las Vegas, Nevada. ACS Symposium Series 204. Washington, DC.: American Chemical Society. 128p.

Massachusetts. Department of Public Health. (1982) The New Bedford PCP study-preliminary findings. 6p. Available from: The Department, Boston, MA.

Massachusetts. Department of Public Health, Division of Environmental Health Assessment, and Massachusetts Health Research Institute, Inc. (1983) Proposal for a greater New Bedford, MA PCB health survey. Draft report. 106p. Available from: [The Institute or The Department, Boston, MA].

Meyer, C.R. (1983) Liver dysfunction in residents exposed to leachate from a toxic waste dump. Environ. Health Perspect. 48:9-13.

Moody, P.L. (1981) Drake Chemical, Inc., Lockhaven, Pennsylvania: interim Report (HETA 81-248). 8p. Available from: National Institute for Occupational Safety and Health, U.S. Department of Health and Human Services, Cincinnati, OH.

New Jersey. Department of Health, Division of Epidemiology and Disease Control. (1980) Groundwater contamination and possible health effects in Jackson Township, New Jersey. 33p. Available from: The Department, Trenton, NJ.

New Jersey. Department of Health, Environmental Health Hazard Evaluation

Program. (1982a) Briar Lake health survey--appendum to "Health survey of the population living near the Gloucester Environmental Management Services (GEMS) landfill." In: Summary of Health Studies Conducted at GEMS landfill in Camden County by the New Jersey Department of Health. p. 24-40. Available from: The Program, [Trenton, NJ].

New Jersey. Department of Health, Environmental Health Hazard Evaluation Program. (1982b) Pulmonary function testing of a population living near the Gloucester environmental management services (GEMS). Prepared in co-operation with the Camden County Health Department. In: Summary of Health Studies Conducted at GEMS landfill in Camden County by the New Jersey Department of Health. p. 41-46. Available from: The Program, [Trenton, NJ].

New Jersey. Department of Health, Environmental Health Hazard Evaluation Program. (1983a) Environmental air sampling both within and outside of homes located near the Gloucester Environmental Management Services (GEMS) landfill, conducted during winter 1981-82 and summer 1982. In: Summary of health studies conducted at GEMS landfill in Camden County by the New Jersey Department of Health. p. 47-69. Available from: The Program, [Trenton, NJ].

New Jersey. Department of Health, Environmental Health Hazard Evaluation Program. (1983b) Health survey of the population living near Gloucester Environmental Management Services (GEMS) landfill. Prepared in cooperation with the Camden County Health Department. In: Summary of health studies conducted at GEMS landfill in Camden County by the New Jersey Department of Health. p. 4-23. Available from: The Program, [Trenton, NJ].

New Jersey. Department of Health, Environmental Health Hazard Evaluation Program. (1983c) Health survey of the population living near the Price landfill, Egg Harbor Township, Atlantic County. Prepared in cooperation with Atlantic County Health Department. 29p. Available from: The Program, [Trenton, NJ].

New Jersey. Department of Health, Environmental Health Program. (1984a) Health survey of the population living near the Krysowaty Farm: abandoned hazardous waste disposal site. 34p. Available from: [The Program, Trenton, NJ].

New Jersey. Department of Health, Environmental Health Program. (1984b) Pomona Oaks exposure assessment: volatile organics in wellwater and indoor air. 80p. Available from: [The Program, Trenton, NJ].

New York. Department of Health. (1981) Love canal: a special report to the Governor and legislature. Prepared with the assistance of the New York Departments of Transportation and Environmental Conservation. Albany, NY: New York Department of Health 69p.

Ozonoff, D.; Colten, M.E.; Cupples, A.; Heeren, T.; Schatzkin, A.; Mangione, T.; Dresner, M.; Colton, T. (1983) Silresim area health study: report of findings. Prepared for the Massachusetts Department of Public Health and Department of Environmental Quality Engineering by Boston University, School of Public Health, and the University of Massachusetts (Boston), Center for Survey Research. 114p. Available from: [School of Public Health, Boston University, Boston, MA].

Parker, G.S.; Rosen, S.L. (1981) Woburn: cancer incidence and environmental hazards: 1969-1978. Prepared by Massachusetts Department of Public Health. 53p. Available from: The Department, Boston, MA.

Rothenberg, R. (1981) Morbidity study at a chemical dump--New York. Morb. Mortal. Wkly. Rep. 30:293-294.

Satin, K.; Deane, M.; Leonard, A.; Neutra, R.; Gravitz, N.; Harnly, M.; Green, R. (1983a) The McColl Site Health Survey: executive summary. Prepared by the California Department of Health Services, Epidemiological Studies Section. 7p. Available from: The Department, Berkeley, CA.

Satin, K.; Deane, M.; Leonard, A.; Neutra, R.; Gravitz, N.; Harnly, M.; Green, R. (1983b) The McColl Site Health Survey: an epidemiological and toxicological assessment of the McColl site in Fullerton, California. Prepared by the California Department of Health Services, Epidemiological Studies Section. 179p. Available from: The Department, Berkeley, CA.

Schultz, S. (1983) Report on the Brookfield Health Survey. 61p. Available from: Bureau of Preventable Diseases, New York City Health Department.

Singal, M.; Kominsky, J.R.; Schulte, P.A.; Landrigan, P.J. (1982) Health hazard evaluation report. 79-022-789. Washington, DC: U.S. Department of Health and Human Services, National Institue for Occupational Safety and Health. 86p. Available from: National Technical Information Service, Springfield, VA.

U.S. Congress, Senate. Environment and Public Works Committee. (1980a) Health effects of toxic pollution: a report from the Surgeon General and a brief review of selected environmental contamination incidents with a potential for health effects. Prepared by the Surgeon General, Department of Health and Human Services, and the Congressional Research Service, Library of Congress. Committee print. 96th Cong., 2nd sess. Serial No. 96-15. Washington, DC: The Committee 312p. Available from: U.S. Government Printing Office, Washington, DC.

U.S. Congress, Senate. Environment and Public Works Committee. (1980b) Six case studies of compensation for toxic substances pollution: Alabama, California, Michigan, Missouri, New Jersey, and Texas. Committee print. 96th Cong., 2nd sess. Serial No. 96-13. Washington, DC: The Committee. 521p. Available from: U.S. Government Printing Office, Washington, DC.

USC/SENATE SEE U.S. Congress, Senate

U.S. Environmental Protection Agency, Office of Solid Waste Management Programs. (1975) Arsenic poisoning in Minnesota. In: Hazardous waste disposal damage reports. EPA/530/SW-151. Washington, DC: The Agency. p. 1-2. Available from: National Technical Information Service, Springfield, VA.

USEPA/OSWMP SEE U.S. Environmental Protection Agency, Office of Solid Waste Management Programs

Vianna, N.J. (1980) Adverse pregnancy outcomes--potential endpoints of human toxicity in the Love Canal: preliminary results. In: Porter, I.H.; Hook, E.B.; eds. Human embryonic and fetal death. New York: Academic Press, Inc. p. 165-168.

Vianna, N.J.; Polan, A.K. (1984) Incidence of low birth weight among Love Canal residents. Science 226:1217-1219.

Warburton, D., Fraser, F. C. (1963) Spontaneous abortion in man: data from reproductive histories collected in a medical genetics unit. Am. J. Human Genet. 16:1-25.

Weaver, G. (1984) PCB contamination in and around New Bedford, Mass. Environ. Sci. Technol. 18:22A-27A.

9

International Aspects of Chemical Waste Disposal

Public officials in a number of countries are aware of the potential for adverse health effects which may arise from chemical waste disposal as a consequence of industrialized development (Suess and Huismans, 1983). The World Health Organization (WHO) Regional Office for Europe (EURO) in Copenhagen, Denmark, and the International Register of Potentially Toxic Chemicals (IRPTC) of the United Nations Environment Programme (UNEP) in Geneva, Switzerland, provided information regarding identified and potential health problems associated with point source solid waste deposits in Europe and elsewhere. Both offices willingly collaborated and provided published and unpublished reports of their programs. In most countries, however, the effort and concern has been primarily concentrated on improved hazardous waste management and thus less data were available concerning health effects. It became clear that more material on health effects was available through the departments of health and environment in individual countries and local authorities within several countries. These sources were not easily accessed and the degree of coverage in this report varies among countries depending on the amount of material readily available.

IRPTC/UNEP was most helpful in providing information from a number of their regional offices and programs, as was EURO. In addition, TOXLINE, MEDLINE, and CANCERLINE databases were examined for reports and articles, but little additional pertinent material was identified. Reports from western Europe and especially Japan provided some information on documented episodes of adverse health effects or local pollution due to chemical disposal sites. A number of individuals and authorities were also requested to supply additional information regarding their problems. While some did so, the absence of a reply cannot be considered conclusive evidence that such episodes had not occurred. The following review briefly summarizes the type of data presently available regarding health effects. In most cases, information regarding health effects was abstracted from reports dealing with risk management of past exposures or deduced from reports on the health effects of general water or air pollution. Since experience elsewhere is different from that in the United States, some detailed comments regarding the experiences of other countries are worthwhile in view of the implications to present and future public health strategies regarding risk assessment and risk management of chemical disposal sites.

In view of the difficulties of obtaining specific information on chemical disposal sites, it has proven more practical in preparing this report to consider the question of health effects in relation to environmental pollution as a whole, with special emphasis on water and situations where pollution is predominantly from point source disposal sites. In many countries, the prevailing opinion was expressed that water pollution is the major problem associated with chemical disposal sites.

INTERNATIONAL COLLABORATION ON CONTROLLING HAZARDOUS WASTE

A number of international bodies have been involved in the development and collaboration in efforts to control hazardous chemical waste.

The International Programme on Chemical Safety

In the late 1970's, a number of foreign governments decided that increased determined collaborative efforts should be made regarding the control of toxic chemicals. Such efforts were of special concern to the developing nations which lacked resources to acquire the expertise needed to solve the complex health and environmental problems caused by the use of chemicals. In May 1977, the Thirtieth World Health Assembly, adopted a resolution which resulted in the establishment of the International Programme on Chemical Safety (IPCS) (WHO/IPCS, 1983). The resolution addressed the growing use of chemicals due to rapid industrialization and new technologies together with the resultant environmental pollution, as well as the increasing needs for recognition of chemical safety in the health policies of all countries. The Assembly proposed a resolution to the Director-General to study and report on the problem of concern for acute and chronic health effects on both present and future generations, as it was "disturbed by the increasing number of accidental releases of chemicals into the environment, resulting in adverse effects on health of epidemic proportions" (WHO/IPCS, 1983).

In January 1979, the Director-General proposed to the Executive Board of WHO the creation of an International Programme on Chemical Safety among whose objectives would be to:

- "carry out and disseminate evaluations of the risk to human health from exposure to chemicals, based on existing information and data;

- encourage the use and improvement, and in some cases the validation of methods for laboratory testing and epidemiological studies that are suitable for health risk evaluations; and propose appropriate methods for assessing health risks, hazards, benefits and exposure" WHO/IPCS, 1983).

Major aspects of the Programme were to include: evaluations of the adverse effects that chemicals may produce on human health; guiding principles for exposure limits to exogenous and toxic chemical substances in food, air, water, and soil; guidelines on appropriate methods for exposure measurement and assessment, toxicity testing, epidemiological and clinical studies, and risk evaluation and hazard assessment; information on dealing with chemical accidents; and case studies of past chemical accidents.

In August 1979, a meeting was held in Copenhagen on European cooperation on the control of chemicals in the environment (WHO/EURO, 1980a). Although most countries had legislation in this area, there was a need to ensure greater consistency in both scope and content. Further, because most epidemiological studies had concentrated on occupational rather than on general population hazards, more consistent epidemiological study methods were required.

Representatives from the Economic Commission for Europe (ECE), the International Labour Organisation (ILO) and the Organization for Economic Cooperation and Development (OECD) provided information on the activities of their organizations. All parties emphasized the importance of controling environmental chemicals and their potential adverse health effects as the major reason for establishing the Programme. In April 1980, a Memorandum of Understanding was signed by the executive heads of UNEP, ILO, and WHO, pledging to promote the development of the IPCS and to encourage collaboration at both national and international levels (WHO/IPCS, 1983).

EUROPEAN CONCEPTS IN FORMULATING HEALTH POLICIES AND STRATEGIES

The European components of health policy formulation have been the subject of several recent reviews (Suess and Huismans, 1983; Waste discharge, 1982; WHO/EURO, 1981, 1983). The reviews largely cover hazardous waste management with few specific comments on defined health effects, although the possibility of such and the need for further investigations were of considerable concern.

Since the 1970's, there has been increasing European public awareness and concern about the growth of the chemical industry and the potential risks that chemical disposal may pose to both human health and the quality of the environment. While hazardous waste management is still evolving, many problems lack clear-cut solutions. For example, there is a lack of international agreement on a universally acceptable and comprehensive definition of what constitutes hazardous waste. Such a classification is difficult because of the complex mixtures involved and the variable concentrations of individual waste components. Out of mutual interest and concern, UNEP and EURO have collaborated on an authoritative text on policy guidelines and a code of practice for hazardous waste management (Suess and Huismans, 1983).

The text emphasizes the problems developing countries may face as they industrialize or import hazardous waste from developed countries. It also reaches a number of conclusions and recommendations for the development of guidelines and for appropriate mechanisms to control hazardous waste management. The following statements illustrate the nature of the background concepts that governed formulation of European strategies and policies with respect to hazardous waste management (Suess and Huismans, 1983).

Hazardous waste management policy should be consistent with policies to control the use of hazardous chemicals, to conserve resources, to preserve health, and to protect the environment. Most hazardous waste comes from industries that are among the most important to the growth and maintenance of a modern industrial society. Improper or inadequate methods for the disposal of hazardous waste have excessively polluted air, land, ground and surface waters. Recognition that such pollution can lead to death, injury or serious impairment of health is well established from episodes such as the methylmercury poisoning at Minamata Bay, Japan, during the 1950's (Harada, 1978; Kurland et al., 1960). The great extent to which land has been polluted creates public concern which, in turn, leads to extensive government action at considerable cost. Although the UNEP/EURO policy guidelines state that "a very large amount of hazardous waste has been disposed of to air, surface waters, and land without apparent harmful effects, [it was felt that] nevertheless, a government's clear duty to safeguard public health demands that it has a policy for the management of hazardous waste."

Definition of Hazardous Waste in the European Context

European scientists and legislators have the same problem as in this country, that is, a satisfactory definition of hazardous waste. The definition varies among countries according to their respective social, political, and economic conditions; terms are modified to reflect environmental conditions prevalent in a particular country. Because most legal definitions of "hazardous waste" are not valid outside their country of origin, the UNEP/EURO guidelines suggest pragmatic working definitions, focusing more on the hazard characteristics of a waste than on its form or composition (Suess and Huismans, 1983). The use of exclusive or inclusive lists was proposed as two possible approaches to the problem of adequately defining what constitutes a hazardous waste.

The exclusive list presented by UNEP/EURO (Suess and Huismans, 1983) defines a list of known wastes that present no significant short-term handling or long-term environmental hazards. Health hazards are not included in this definition but may need to be considered. Hazardous wastes, therefore, are defined by exclusion; that is, they are those wastes not listed. This approach has been used successfully in some circumstances in the United Kingdom. While one advantage of the exclusive list is that it is relatively simple to ensure that the listed materials are not hazardous, those materials excluded from the list may well be only marginally hazardous. In addition, interpretations of what constitute the above listed hazards may differ when only qualitative, subjective criteria are used. This method of definition may not provide waste producers, waste disposers, and regulatory authorities the criteria they need to define wastes of uncertain health hazards.

Inclusive lists of hazardous waste are more widely employed for regulatory purposes, either with or without accompanying criteria. This approach is currently employed in Belgium, Denmark, France, the Federal Republic of (West) Germany, the Netherlands, Sweden, the United Kingdom, and the United States (Suess and Huismans, 1983). The lists describe hazardous compounds which result as wastes from certain industries, wastes containing specific components, or specific waste streams identified by the processes from which they originate. For example, in the Netherlands, the Law on Chemical Wastes of 1976 uses an inclusive list of hazardous components of waste and provides concentration limits to restrict the wastes acceptable for landfill disposal. There is generally substantial international agreement on those components of a waste that cause it to be classified as hazardous. These components were previously listed in Chapter 2 and are reprinted here as Table 9-1 (CEC, 1978). International differences tend to arise when quantitative criteria are assigned (Suess and Huismans, 1983).

TABLE 9–1. List of toxic or dangerous substances and materials (adapted from the Council of the European Communities, 1978).

Acids and/or basic substances used in the surface treatment and finishing of metals
Antimony and compounds
Arsenic and compounds
Asbestos
Beryllium and compounds
Biocides and phytopharmaceutical substances
Cadmium and compounds
Chemical laboratory materials, not identifiable and/or new, with unknown effects on the environment
Chlorinated solvents
Chromium (VI) compounds
Cyanide compounds
Ethers
Isocyanates
Lead and compounds
Mercury and compounds
Metal carbonyls
Organic solvents
Organohalogenated compounds, excluding inert polymeric materials and other substances referred to in this list or covered by other CEC Directives concerning the disposal of toxic or dangerous waste
Peroxides, chlorates, perchlorates and azides
Pharmaceutical compounds
Phenolic compounds
Polycyclic aromatic hydrocarbons (carcinogenic)
Selenium and compounds
Soluble copper compounds
Tarry materials from refining and tar residues from distilling
Tellurium and compounds
Thallium and compounds

WHO/EURO officials indicate that the definition of hazardous waste may be of academic interest but has greater practical importance. In the European context, environmental pollution by nitrates and sulfates, compounds not currently classified as hazardous wastes, is now regarded as presenting problems of great potential importance in terms of health and ecology (Duffus and Waddington, 1983). UNEP has also explored the prioritizing of chemicals or substances that are harmful at the global level, but only some of these studies are pertinent to the problem of defined chemical "hazardous waste" in disposal sites (UNEP/IRPTC, 1984).

The issue of toxic waste arising from domestic sources has also been addressed. A recent report from the Federal Republic of Germany describes the formation of dioxins in smoke from incinerators dealing exclusively with domestic waste (Anonymous, 1984b). Finnecy (1983) and other European reports have tended to emphasize the need for a total approach to environmental pollution in which chemical disposal sites comprise only one component.

In evaluating the European experience in hazardous waste control, a number of other reports have been reviewed. The British House of Lords, Select Committee on Science and Technology (Great Britain, 1981a) reported on the large differences among European countries in methods of controlling and disposing of chemical waste. At one extreme, Denmark has a "Kommunekemi," a centralized disposal company to which all hazardous wastes in Denmark must be consigned. It provides treatment of oily wastes, chemical treatment for inorganic wastes, incineration, an effluent water system, and landfill for non-toxic filter cakes and other residues. At the other extreme, the Netherlands' Chemical Waste Act prohibits, with few exceptions, the disposal of chemical waste on land. A few facilities for incineration, detoxification, and distillation are available but export of chemical wastes is normal procedure. In the Federal Republic of Germany, which has a much larger chemical industry than Denmark or the Netherlands, waste disposal is controlled regionally, with a high degree of public sector involvement. A notable facility exists at Schwabach in Bavaria, where 12 towns and 7 country districts have combined to set up a chemical waste disposal unit serving an area of 21,000 square kilometers, with 3.7 million inhabitants and 4,000 industrial companies. Waste producers in this region must send their hazardous wastes to the Schwabach facility where disposal methods include landfill, wastewater and oily waste treatment, chemical and physical treatment, incineration, and tire mastication. The unit handles all waste except cyanide salts (which are deposited in salt mines), and recyclable oils and solvents.

In a survey of European chemical disposal practices, Mills (1983) concluded that there are no western European nations where industries are solely responsible for hazardous waste treatment and disposal. In the United Kingdom and in Ireland, where public sector involvement is generally limited to a controlling and enforcing role, such industries have limited fiscal responsibilities. No such investment is made in Denmark, France, the Federal Republic of Germany, the Netherlands, or in the Scandinavian countries, where massive public sector investment in hazardous waste management is the rule. The situation is less clear in Belgium, Italy, and Luxembourg.

Mills (1983) estimated that the hazardous waste produced in western Europe was 40 million metric tons per year in comparison with a 1980 reported treatment capacity of only 6 million metric tons per year. He concluded that landfill is the predominant disposal route for hazardous waste in western Europe. Although frequently not officially acknowledged, landfill is also the predominant disposal route in most individual western European nations. While known in the Federal Republic of Germany, Ireland, Italy, Switzerland, and the United Kingdom, it may also be inferred for Austria, Belgium, France, the Netherlands, Norway, Sweden, and Luxembourg. No evidence was presented from Portugal and Spain. Denmark appears to be the only western European country in which disposal by landfill does not predominate.

STUDIES IN INDIVIDUAL COUNTRIES

Canada

Canada's Environmental Protection Program Directorate has examined problems associated with chemical disposal; however, a limited number of publications were available for review. A monograph by Burton et al. (1982) reviewed the overall risk of environmental hazards and described a number of case studies. These studies largely described environmental pollution due to occupational and agricultural emissions and effluents, accidents and spills, radiation, and diffuse pollution. Another study (Solsberg et al., 1983) discussed the problem of establishing a new chemical disposal site in light of the present environmental climate; however, no mention was made of any specific incidents relating to adverse health effects due to specific disposal sites. A manual on land application of treated municipal wastewater and sludge (Canada, 1984) reported that significant gaps still exist in our knowledge of the fate of organic contaminants in waste water and sludge as applied to land. With no existing evidence of adverse human health effects, continued vigilance was recommended.

No report was available on systematic Canadian epidemiological studies of the health effects of non-occupational chemical exposures. Most Canadian reviews make specific reference to U.S. experiences in this field.

Egypt

A review of the chemical waste situation in Alexandria indicates that there is a major problem of industrial pollution in Egypt (Hamza, 1983). The increase in pollution is reflected by decreased fish catches from the Nile, from lakes, and from the Mediterranean and Red Seas, and by degradation of water resources, deterioration of recreational areas, and severe operational problems in sewage treatment networks with resultant "serious effects on public health."

The degree of contamination by various compounds is documented by Hamza (1983). Levels quoted for many chemical compounds in the water supply are much higher than prudence would dictate. In addition, a considerable accumulation of mercury and DDT in aquatic organisms within Alexandria has occurred. Available information, however, suggests that the major adverse health effects from pollution relate primarily to infectious diseases due to sewage contamination of potable water supplies.

Federal Republic of Germany (West Germany)

Management of industrial hazardous waste in the Federal Republic of Germany has been reviewed by Defregger (1983). Over the last 15 years, it has been recognized that, compared with domestic wastes, hazardous wastes constitute relatively small quantities which must be disposed of in special facilities without detriment to health or the environment. Each of the ten Landers (or states) is responsible for these activities in its own region. Defregger (1983) reports that 3-4 million metric tons of "special waste" are generated per year in the Federal Republic of Germany. About 15% is disposed of by incineration, 35% by chemo-physical treatment, and 50% in secure landfill sites. There is no mention of significant water pollution or adverse health effects in this report and no additional information on these points was available. A report on the German Chemicals Act of 1980 (Schmidt-Bleek and Lange, 1983) discussed the general aspects of the Act, which aims to protect man and the environment from adverse effects by requiring chemical manufacturers to conduct specific tests to determine risk; no mention was made of specific identified health effects.

Brown et al. (1983) reported that major current environmental concerns relate primarily to contamination by compounds of cadmium and mercury, and polychlorinated biphenyls. Cadmium contamination of soil resulted from land disposal of sewage, sludge, and river sediments and from deposition of airborne combustion products. Brown et al. (1983) stated that "cadmium is now found in virtually all foods in high concentrations. Several thousand Germans have kidney disease directly attributable to cadmium intake." No other references were found to confirm these statements; such incidents seem rare in Europe despite general concern about cadmium contamination.

An emerging environmental principle in the Federal Republic of Germany is that of pollution prevention (Brown et al., 1983); that is, prevention of pollution before environmental contamination and the resultant damage can occur. Environmental policies based on this principle focus on the development of forms of production which make economical use of raw materials and energy and which cause the least possible harm to the environment.

The Federal Republic of Germany in a number of ways has some of the most advanced technology for disposal of hazardous wastes. One example is the facility at the Gesellschaft, owned by the Bavarian state (Gesellschaft zur Beseitigung von Sondermull in Bayern, 1983). Unsatisfactory past practices of uncontrolled chemical waste disposal near Hamburg have been reported (Spilker, 1984) but no episode of adverse health effects was described.

France

France's current government policy regarding hazardous waste concentrates on (a) reduction of the amount of waste produced at the source, (b) an increase in waste recovery, and (c) more satisfactory methods of disposal (Le Roy, 1983). Approximately 18 million metric tons of "special waste" (likely to contain toxic substances presenting risks to the environment) are produced yearly, primarily generated by industry. This includes 2 million metric tons containing toxic substances in concentrations such that particular care must be given to its disposal. In addition to waste-processing facilities operated by waste producers themselves, more centralized collective waste disposal systems have also been developed. It was reported that past disposal practice was not always satisfactory and that some disposal sites may have serious effects on the environment. No definitive health effects were reported in relation to such disposal sites.

India

Episodes of adverse health effects to chemicals in India have been reported primarily in relation to food contamination, for example, ortho-cresyl phosphate poisoning in Bombay in the early 1960's (Vora et al., 1962). Sundaresan et al. (1983a), outlining the extent of the hazardous waste problem in India, noted that the increase over the last three decades of industrial production of gaseous, liquid, and solid waste that is traditionally discharged to land, air, and water (including the sea) has caused considerable environmental degradation. Production of organic chemicals and pesticides has expanded enormously. Legislation was enacted under the Water (Prevention and Control of Pollution) Act in 1974 and Air Act in 1981, however, not all industries as yet comply with the regulatory requirements (Sundaresan et al., 1983b).

The first recorded case of water pollution due to hazardous industrial waste was in Bihar in 1968; oil discharge resulted in a severe river fire. Another incident involved a fish kill near Goa caused by ammonia and arsenic discharge (Sundaresan et al., 1983a). Animal deaths due to consumption of water receiving chemical waste discharge were reported, however, specific information was not available. Episodes of crop pesticide poisoning due to accidents or occupational hazards have been documented, as has the recent accident in Bhopal in which methyl isocyanate gas was released from an industrial source into the air. In this latter incident, more than 2,000 people died of acute pulmonary edema due to chemical pneumonia, and an estimated 20,000 to 100,000 more sustained injuries involving lung damage, skin burns, and eye ulcerations (Anonymous, 1984a).

Conventional solid wastes generated by the phosphatic fertilizer industry and thermal power plants may cause considerable contamination of both surface and groundwater. Other industries generate chemicals of concern including pesticides, dyes and pigments, fertilizers, chromates, organic chemicals, acrylonitrile, and caustic chlorine; it is clear that contamination may be a considerable problem. A number of episodes have resulted in well pollution. There is some indication that the situation has been recognized and attempts are being made to control certain of these wastes in large industries (Sundaresan et al., 1983a). At present, primary industrial concern involves wastewater treatment in order to meet the requirements of the Water Pollution Control Boards. However, the sludge produced in waste-water treatment is rarely processed but instead allowed to accumulate (Sundaresan et al., 1983a). The need for more effective methods of disposal is well recognized.

Pollution of the environment due to chemical disposal sites is considerable and may present potential health hazards. However, as yet, there are no available data indicating that such chemical wastes have produced significant adverse health effects. Most reported adverse health episodes refer to occupational exposures or accidents.

Italy

There have been several episodes of general environmental pollution in Italy due to industrial chemicals, for example, 2,3,7,8-TCDD contamination in Seveso (Regione Lombardia, 1983). A second incident relating to groundwater pollution from an effluent lagoon containing halogenated aromatic compounds was reviewed intensively by an international Working Group (WHO/EURO, 1980b). The problem, first recognized in 1976, arose in Vicenza Province and a considerable amount of information was presented to the Working Group regarding the chemicals identified and the degree of pollution. The primary products of the industrial source were perfluorinated derivatives of intermediate chemicals used for pharmaceuticals and pesticides. Annual production of these intermediates had risen from 200 metric tons per year in 1973 to 2500 metric tons per year in 1980 (WHO/EURO, 1980b).

In 1977, water from some wells had been polluted by a number of fluorinated and chlorinated aromatic compounds. Analyses were carried out on samples from hundreds of wells in order to determine the distribution and concentration of pollutants in the water. The pollutant present in highest concentration, from 5 to 800 ug/L, was 4-chloro-3-nitro-*a*,*a*,*a*-trifluorotoluene. In wells from other provinces, the concentration of this chemical was found to be much lower, typically in the range of 0.5 to 90 ug/L.

Epidemiological studies were made on random samples of the population living in the area of pollution and on a comparison sample of people living outside the area. The former population was classified in terms of the water supply available to the family. The eating habits of the people were also studied to assess the amount of water consumed and the quantity used for cooking. At the same time, an epidemiological cross-sectional morbidity analysis was carried out, but no significant differences were found between the two groups. For 2 years, the factory workers were given hematological and clinical examinations at 6-month intervals but, again, no significant effects were observed. During this period, they had been exposed to atmospheric halotoluene concentrations of less than 1 mg/m^3 for unspecified time periods.

It was found that only a limited portion of the population had previously used wellwater for drinking purposes. Also, because the odor and color of water are unpleasant when 4-chloro-3-nitro- *a*,*a*,*a*-trifluorotoluene is present in concentrations of 50 ug/L or less, the Working Group concluded that the risk to the population from drinking contaminated water was very low. The data from the epidemiological cross-sectional analysis, however, will be useful in the event of a possible future follow-up study.

This recognized episode of pollution of a groundwater drinking water supply from chemical disposal site leachate is of interest, and is described in unusual detail in view of the rarity of such reports from Europe. However, it appears unlikely that an adverse health effect would have been detected under the relatively low exposure levels. In the case of Seveso, where there was an escape of dioxin into the surrounding area, much more detailed reports are available (Regione Lombardia, 1983). These are not pertinent to the present discussion of sites but provide useful background information in determining the potential health effects to humans of such exposures.

Japan

Japan has experienced heavy pollution of water and air along with serious diseases caused by exposures to toxic chemicals (Ichikawa, 1983). This resulted in the enactment of the Basic Law for Environmental Pollution Control in 1967. The early 1970's were a time of drastic changes in Japan's environmental pollution policies, provoked by increasing domestic as well as international concern on environmental issues. The Environment Agency was established in 1971 for monitoring environmental pollution and for the conservation of nature. The Chemical Substances Control Law was then enacted in 1973 and came into force in 1974. The purpose of the law was to prevent pollution of the environment by chemical substances which persist or have other such properties and which may possibly be harmful to human health (Japan International Cooperation Agency, 1982a).

Since 1974, the Environment Agency has been surveying the general presence and safety of chemicals in the environment, including both water and sediments. This ongoing study is being carried out in three stages. First, 50 substances were identified with a potential for high residues in atmosphere, water, and sediment. Second, a general survey of these 50 substances has been followed by more detailed studies on about 5 substances a year. Third, tests for effects on the ecosystem are carried out on those chemicals found to show a wide range of residual values. Wild life monitoring is also undertaken. The following chemicals have been proscribed: PCB's, hexachlorobenzene, polychlorinated naphthalene, aldrin, dieldrin, endrin, and DDT (Japan International Cooperation Agency, 1982a). The most important aspect of soil pollution related to problems caused by heavy metals, notably cadmium, copper, and arsenic and their compounds. Agricultural or livestock products harmful to human health are produced as a result of the presence of these chemicals in soil (Japan International Cooperation Agency, 1982b).

A law concerning special measures for the relief of victims of pollution-related diseases was promulgated in 1969 to provide nationwide uniform relief measures (Japan International Cooperation Agency, 1982c). It was later replaced by the Pollution-Related Health Damages Compensation Law in 1974. The major background to this legislation was given in the above reference.

The following is an example of a pollution-related disease. An outbreak of Minamata disease occurred in the southern parts of Kyushu island in 1956, but it was not until 1964 that the cause could be linked to effluents from a mercury-cell caustic-chlorine plant (Ichikawa, 1983). Another outbreak of Minamata disease occurred in 1965 in the Agano river basin, and following that, the Government restricted the production and use of mercury and mercury compounds, and production methods of caustic-chlorine were changed. The use of agricultural chemicals, such as insecticides and herbicides, synthesized from mercury compounds was also prohibited.

Minamata disease is now regarded as definitely proven to be caused by chemical pollution but at the time there was considerable uncertainty as to cause and effect relationships. Later, for legalities of victim compensation purposes, it was decided that in the absence of other significant environmental causes, an epidemiological study that shows the existence of a probable cause and effect relationship between a certain contaminant level and an outbreak of an epidemic disease may be accepted as evidence of sufficient cause (Japan International Cooperation Agency, 1982c). This policy of compensation to victims of pollution-related disease appears unique to Japan.

The following diseases in Japan are considered pollution-related:

Air pollution-related diseases:

- chronic bronchitis and its sequelae
- bronchial asthma and its sequelae
- asthmatic bronchitis and its sequelae
- pulmonary emphysema and its sequelae

Water pollution-related diseases:

- Minamata disease: resulted from industrial methylmercury poisoning due to ingestion of contaminated seafood

- Itai-Itai disease: resulted from industrial cadmium poisoning due to discharge of effluent into river water used for drinking and irrigation

- Chronic arsenic intoxication

The absence of recent reports of adverse health effects following chemical disposal suggests that further adverse health effects have not been identified in humans.

Environmental policy in Japan is an outgrowth of a health crisis situation which came about in the 1960's as a result of rapid industrialization in an area of high population density. This policy was developed rapidly with the highest priority given to the abatement of air pollution and to the control of toxic substances such as mercury, cadmium, and PCB's. Consequently, other aspects of pollution control, such as water pollution control and overall environmental protection, have received less attention. This approach has brought about a Japanese policy focused on health objectives.

The focus on a rapid reduction in health impairment required forceful measures to be taken without waiting for complete scientific data and often with little initial knowledge of economic costs and consequences. Industry responded with a high degree of acceptance of the relatively strict standards. Government sponsored loans and tax incentives have encouraged the development of new control technologies to meet the standards. Different environmental protection measures have been adopted in varying degrees. The implementation of emission standards and requirements has relied more on voluntary acceptance and persuasion than on coercion. Other control measures such as agreements on application of best available techniques have been used only to a limited extent. Sophisticated monitoring networks have been developed in all major areas. Besides their main function of continuous surveillance, they perform two other roles. One role is that of informing the public of the nature and extent of ambient concentrations of pollutants. The other is that of serving as an alerting system (Brown et al., 1983).

The evidence indicates that the national effort at present is concentrated on prevention of new sources of waste, the cleanup of older sources being dependent on need and urgency. Thus, it would appear that the Japanese authorities regard the immediate problem as largely under control with the emphasis in the future on better chemical disposal technology to avoid future potential risk. Pollution prevention is seen as a far cheaper approach than remedial action (Ichikawa, 1983).

The Netherlands

Problems associated with chemical disposal sites have been more extensively studied in the Netherlands than in most other countries in Europe. A detailed survey of the hazardous waste management problem was prepared by Erasmus (1983). The Netherlands is one of the most densely populated countries in the world. Its surface water is supplied primarily by the Rhine river, about 70 billion cubic meters per year. By the time it reaches the Netherlands, the Rhine is so polluted that purification for drinking purposes is extremely difficult. Waste contaminants include chloride salts, phenols, phosphates, nitrates, many types of chlorinated organic substances, and heavy metals such as mercury, cadmium, lead, copper and zinc. Specific problems relating to fluvial pollution sources are further described elsewhere (WHO/-EURO, 1982). With industrialization the country has become increasingly concerned about pollution, and there is evidence that there is a correlation between air pollution and respiratory diseases. Substances prohibited from being discharged into water or air still find a way into the ground in some form or another and pose a threat to groundwater and health. It is clear that the problem concerns chemical waste disposal both within and outside the country. In 1980, public attention focused on a site in Lekkerkerk, where 250 houses had been built on a belt of chemical waste collected over 100 years (van Duijvenbooden, 1983). Ultimately, about 150,000 tons of the polluted soil were dug out and removed for disposal elsewhere. Four thousand additional chemical disposal sites were then reported throughout the country. Since then the figure has increased to 5,000. It is estimated that 750 of these sites are situated in residential or water collecting areas while 350 others require remedial action with some urgency.

Sangster et al. (1984) studied inhabitants of an area where vegetables grown in cadmium-contaminated soil had been consumed. They concluded that the increase in cadmium excretion in urine and slight differences in renal function did not present potential health hazards. Based on a survey of 500 sites, the types and sources of soil contaminants in the Netherlands were classified. Approximately 21% of soil pollution came from chemical disposal sites (van Duijvenbooden, 1983). The Soil Purification Act went into operation in 1983 and provided guidelines for research and clean-up prioritization.

A survey of groundwater pollution during 1976-1978 (van Duijvenbooden, 1983) reported widespread contamination by halogenated organic compounds in the groundwater of the Netherlands. In most groundwater pumping stations, halogenated organic compounds were found in concentrations of greater than or equal to 0.01 ug/L. About 30% of all pumping stations showed levels of greater than or equal to 0.1 ug/L. It was concluded that in 90% of all water samples analyzed, chlorinated organic compounds were detected; in 16% of these samples, concentrations were detected at 1.0 ug/L or more.

An extensive series of studies was conducted on the presence of mutagenic activity in drinking water in 18 cities in the Netherlands (Kool, 1984; Kool et al., 1982, 1984). Of 20 drinking water types, 14 showed mutagenic activity in the Ames test. Twelve of the 15 cities which received their water from surface water or a mixture of surface and groundwater showed mutagenic activity, whereas only 2 of the 5 cities which used groundwater as a water source showed mutagenic activity. It was stated that chlorine treatment applied during drinking water preparation correlated well with direct mutagenic activity. A number of other references to the problem of mutagens, drinking water, and water pollution are discussed in other publications. In terms of water pollution, the relative significance of leakage from chemical disposal sites in comparison to the effects of chlorination on the burden of mutagens in drinking water is not clearly defined.

It would appear that in the Netherlands, where the potential has existed for a direct risk to human health, emergency interventions have been made according to local conditions. Although specific and definitive action has been taken regarding certain episodes of pollution, there does not appear to have been a significant episode of severe adverse health effects as a result of groundwater contamination despite the awareness of the authorities to such danger. The direction of future activities appears to be addressed to a prioritized cleanup of past problems and the avoidance or reduction of potential hazardous situations in the future.

Poland

Kempa (1983) reviewed low-waste producing technologies in the management of hazardous waste and the importance of protecting the environment in Poland. It was recommended that industrial wastes be divided into three classes: inactive, moderately troublesome, and highly hazardous wastes. Methods of control by waste type were discussed. Hazardous waste with the potential for polluting the environment is widespread in Poland and is reported to have resulted in serious environmental problems. It is clear that there is a great deal still to be done in the control of such waste. Despite widespread environmental pollution, no specific mention was made of defined health effects in relation to these deposits.

Sweden

A report on a chemical disposal site in Teckomatorp, Sweden, described the effects of illegal disposal of chlorinated phenols and phenoxy acids, eventually leading to the closing and razing of the factory and extensive clean-up activities (Ferm, 1982). Although water contamination was clearly demonstrated, there was no discussion of adverse health effects found in association with this episode. Despite excellent cancer registries throughout Sweden, no report associating cancer clusters with chemical disposal sites has yet been identified.

United Kingdom

After centuries of industrialization in the United Kingdom, there has been extensive environmental degradation. In 1947, Smith et al. (1982) estimated that 1/3 of 1% of the total land area of England had been made so derelict or contaminated by past industrial use that it was incapable of beneficial use without treatment. Planning legislation was enacted to address this problem; however, uncontrolled land dumping of chemical waste continued until enactment of the Deposit of Poisonous Waste Act in 1972 which established mechanisms to control toxic waste disposal (Smith et al., 1982). All landfill sites are now regulated under the Control of Pollution Act of 1974 (Burgess, 1983).

Research on hazardous waste in the United Kingdom is a major responsibility of the Harwell Laboratory Waste Research Unit (Harwell Laboratory, 1982). A 1974 Harwell Laboratory survey of hazardous waste disposal in Cheshire was used to estimate that in the United Kingdom, approximately 94% of industrial waste was disposed of on land, 4% by sea, 2% by incineration, and perhaps 1% by chemical treatment or other methods. The most toxic wastes were believed to be disposed of by incineration or chemical treatment. Stevens and Wilson (1982) estimated that 5.5 million metric tons of notifiable or hazardous waste are produced annually in the United Kingdom and an equivalent amount is probably disposed of on-site.

In a survey carried out in 1973-74 (Gray et al., 1974), some 2500 disposal sites of domestic or industrial waste were identified as having been operational for at least 27 months. Study of data on local hydrogeological characteristics suggested that only 51 of these had the potential to cause pollution of major or minor aquifers. Harwell Laboratory Waste Research Unit reports indicate that despite the fact that toxic chemicals are migrating from disposal sites via air, water, and soil, no serious human health effects have been established thus far.

In 1975, the Royal Society of London established a Study Group to review knowledge of the toxic effects on man and other organisms from long-term exposure to substances in the environment at concentrations not found to cause adverse health effects with short-term exposures. The Study Group concluded (Royal Society of London, 1978) that, with the exception of urban air pollutants implicated in the etiology of chronic bronchitis, there was no evidence to suggest that industrial pollutants had any major effect on the physical health and mortality of the general population. However, localized incidents of adverse health effects had been reported occasionally as a result of accidental exposure to high levels of industrial pollutants or contaminated food. The Study Group emphasized the necessity for creating an improved system of monitoring the results of human exposure to chemicals.

The most extensive source of information on health effects related to chemical exposure in the United Kingdom is the three volume report of a Select Committee on Science and Technology from the 1980-81 session of the British House of Lords (Great Britain, 1981a, b, c). The committee was established to consider the problems relating to a large landfill called Pitsea but reviewed the entire hazardous waste disposal experience within the United Kingdom and to a lesser extent in continental Europe.

Volume III of the Report (Great Britain, 1981c) reviews the written evidence from local authorities as well as organizations such as the Royal Society of London and the Institute of Public Health Engineers. Although concern about adverse health effects was the driving force behind the study, such effects had apparently not been documented in the general population. Instead, the House of Lords study and a report by Keen and Mumford (1975) emphasize the potential hazards to operators and drivers dealing with hazardous waste. Although the principal hazard is mechanical injury, deaths have occurred as a result of inhalation of fumes. Other exposures of chemical disposal workers have been known to occur and these reports suggest that such incidents may be the most likely source of future health problems related to chemical waste disposal.

Developing Nations

It is difficult to obtain information regarding chemical disposal sites and associated health effects within the developing third world. Previous reports on health problems in such countries have, for practical purposes, concentrated on communicable diseases and the need to provide clean water. However, there is a general impression that in these countries most chemical pollution, especially of water, occurs from uncontrolled discharges from industrial plants and agricultural practices. From a public health viewpoint, no distinction is made between diffuse and point source pollutions such as chemical disposal sites. It must be assumed that hazardous industrial wastes are being disposed of within the community and receiving relatively little attention. This situation is under review by IPCS currently; however, action is largely one of information gathering and encouraging the implementation of preventive legislation. Due to the relatively poor public health statistics in many developing countries, it may prove difficult to relate a particular episode of illness to chemical exposure unless high-dose exposures with acute effects can be identified, as in food contamination.

A report from Sri Lanka noted that significant levels of lead poisoning exist among families engaged in gold and silver recovery from jewellers' waste (Ramakrishna et al., 1982). This indicates the potential dangers associated with cottage industries in such countries and may, in fact, prove to be of relatively much greater immediate importance than the problems caused by chemical disposal sites from large industries. With the exception of Japan, no Asian country is known to have experienced any large-scale health hazard due to chemical waste disposal (Ichikawa, 1983).

Informed comments and views of international scientists suggest that in the Central Pacific and African regions, local practices of chemical waste disposal are often unsatisfactory, both in relation to industrial sites, and small businesses or workshops that dispose of their untreated waste locally or through municipal disposal facilities. Apart from the Republic of South Africa, there appears to be a minimum availability of modern chemical disposal practices. At present, it is difficult to carry out any epidemiological studies regarding chronic diseases in third world countries because of the logistic and statistical problems involved. On the other hand, some illnesses and poisonings of an acute nature associated with toxic chemical deposits are reported. Most definite reports relate to an occupational setting or accidental poisoning where the chemical of concern has entered the food supply.

At present, more detailed analyses of the health situations in Africa, Asia, and Latin America are being carried out by UNEP and the IPCS. However, no definitive reports are currently available. Health problems due to chemical waste disposal in developing countries may be more critical than believed, especially when one considers that such substances as asbestos have been disposed of without due care and control.

The practical difficulties of chemical disposal and potential adverse health effects in East Africa and the South Pacific have been reviewed briefly (Dahl and Baumgart, 1983; UNEP, 1982). Human exposures may be occurring at higher levels than in the industrial countries. Many small accidents have occurred but since there is little or no monitoring of toxic chemicals, any adverse health effects have gone largely unrecognized. The absence of reports on adverse health effects in many developing countries is relatively meaningless due to the absence of adequate data. At the present time, identification of specific site problems would appear to be only a small part of the problem of widespread uncontrolled chemical disposal practices in which the requirements of industrial and agricultural development are a predominant force in formulating governmental policies (Dahl and Baumgart, 1983; UNEP, 1982).

DISCUSSION

The potential problem of chemical disposal sites varies according to the degree of development in a country. Japan and several countries in Europe have recognized the significance of chemical waste since the 1950's. Since the mid-1960's, they have had increasingly stringent legislation regarding the generation and disposal of chemical waste. In these countries there are a variety of methods for dealing with chemical waste disposal. All possess good public health services and many have excellent health statistics.

While reports on adverse health effects of occupational exposures are quite extensive in the literature of these countries, conclusive reports on health effects as a result of ambient pollution, either of air or water, are limited. While the extensive reports describe the problem of hazardous waste management in great depth and emphasize national needs on the basis of protection of public health, documentation of definitive health damage is almost entirely lacking. Most reports of health effects from chemical exposure describe the results of transportation accidents or are related to occupational settings. In such cases, acute illness is usually observed as a result of high-dose exposures and the difficulty in identifying chronic health effects due to low-dose exposures is much greater as described in earlier chapters of this report. There is evidence in the United Kingdom that deaths among chemical disposal workers have occurred in relation to their work, but on examination, most such deaths were due to accidents unrelated to the chemical nature of the waste alone. Although employees engaged in chemical disposal are more likely to be exposed, when acute accidental deaths were excluded, such employees did not show an unusually high frequency of disease. Nevertheless, there is a general consensus that the health of such employees should be monitored and surveyed (Royal Society of London, 1978).

In most cases where health hazards were described, reference was made to the United States' experience at Love Canal in Niagara Falls, New York. Health problems conclusively causally related to chemical disposal sites were not known to either WHO/EURO or to the International Programme of Chemical Safety. The same was true for the United Kingdom's Department of the Environment (Royal Society of London, 1978) and confirmed by the Minister of Health. Nevertheless, authorities appear uncomfortable with any existing disposal practices which had caused or could lead to an increased contamination of the general environment; all emphasized the continuing need to improve chemical disposal in the future.

The difficulties of seeking evidence of isolated incidents of health effects (e.g., leukemia clusters) within large population groups and the logistics of such epidemiological studies as outlined in Chapter 5 were also emphasized by these spokesmen. They expressed interest in studies to determine whether health effects could be demonstrated among specific individuals who were exposed to high levels of suspected chemicals. EURO emphasized the need for such studies and systematic epidemiologic studies of general populations to insure both credibility and to allay concern, but identified no new technical approaches. The organization is exploring the possibility of identifying susceptible populations through biological markers. In Europe and Japan, the problem of health complaints relative to the proximity to a point source chemical disposal site apparently has not been as frequent as in the United States. However, the situation appears to be changing. Episodes of self-reported symptoms recently occurred following contamination of the River Dee in the United Kingdom by phenol and following reports of cattle death in the area of a chemical disposal site near Donnybrook in Scotland. According to the United Kingdom Department of the Environment, no evidence of significant exposures to any chemicals likely to be involved has been demonstrated in either case. Another episode of this type occurred near Hamburg where large chemical deposits have been identified (Spilker, 1984). The methodology for evaluating such episodes is still far from satisfactory (Royal Society of London, 1978).

In evaluating the relative scarcity of chemical-related reports of adverse health effects in industrialized countries, it is difficult to determine whether the lack of reports indicates a lack of health effects or that specific systematic studies have not been made to identify such effects. However, authorities appear to be fully aware of the issues and the need to identify such episodes. The apparent rarity of specific health effects conclusively related to chemical disposal site practices has not led to complacency by the authorities, but rather to a requirement that every attempt be made to insure improved waste disposal practices in the future as part of an overall environmental program. This view has had considerable impact in determining future public health policies at the national level. More intensified attempts are needed to obtain data from local authorities on levels of exposures that have occurred and possible associated health effects.

In the developing countries, high-dose exposures to toxic chemical waste almost certainly have occurred and do occur frequently but such exposures are poorly documented. Subtle chronic health effects are unlikely to be detected within the background of variation of disease patterns and the absence of good public health statistics.

A 1984 report on the State of the Environment (UNEP, 1984) noted how it is difficult to obtain scientifically sound information on how many incidents have been caused by chemical waste disposal or how serious the problem is worldwide. The true extent of the damage may not be accurately reflected by available reports. Nevertheless, even countries that have not uncovered serious problems do appreciate that past practices have been unsatisfactory and need to be improved. Laws controlling chemical disposal are now in effect in most developed countries. As these waste control practices have tightened, chemical industries have had to assume greater fiscal responsibility or find alternative methods for waste disposal. Some have exported wastes to countries where the laws are less strict, or less strictly enforced. These countries could well end up with the same sort of problems that brought the strict legislation of the originating countries in the first place. Developing countries would be particularly vulnerable to such pollution exports.

The need to clarify the definition of hazardous waste as it differs from country to country is widely acknowledged. Furthermore, in many parts of Europe, the problem of hazardous waste disposal is perceived not in terms of localized waste sites but as part of the overall problem of total waste disposal; that is, it takes into consideration waste from a multitude of sources including oil spills and agricultural practices. This has led to environmental waste disposal being placed in the context of other sources of chemical pollution in terms of health significance and not limited to its hazardous waste status. This overall approach is important in estimating the future potential problems since methodologies of chemical disposal are changing in different countries; the nature of waste could change according to the methodologies to be utilized. In the United Kingdom, disposal in landfills seems acceptable in certain cases, whereas in other countries an alternative to landfills is sought. In all countries, the problem of secondary waste generators, including domestic sources and small industries, was emphasized as a major source of uncontrolled pollutants.

Many studies emphasized the general problem of water pollution and the difficulty of separately determining the contribution of a chemical disposal site. This is especially important in relation to heavy metals. In the United Kingdom, farmers are thought to be the cause of 20% of the cases of pollution and the number of incidents is increasing. While not directly related to point source chemical disposal sites, such incidents are significant in the European context as indicative of the extent and the various types of chemical waste problems that may occur. This explains the awareness among European authorities that environmental problems should be addressed holistically rather than in the context of a compartmentalized approach.

A study comparing U.S. policies for regulating toxic chemicals with the policies of Britain, France and the Federal Republic of Germany highlighted some of the basic differences among these nations (Brickman et al., 1982). The conflict among the three branches of government in the U.S. creates barriers to implementation of policy and delegation of authority. In addition, the adversarial framework of regulation in the U.S. polarizes expert scientific opinion and illuminates the quasi-scientific nature of many of the chemical regulatory decisions. In contrast, the study found that European government branches tend to be more cooperative with each other, reducing the need for legislated controls on administrative discretion. Closed consultation permits mediation and negotiation among differing interests without polarizing conflicting opinion. This process simultaneously addresses both political and scientific issues and generally results in stable compromises that are seldom challenged in the courts.

SUMMARY

Information on chemical disposal sites and associated health effects was collected from a number of sources in Europe, Japan, and certain developing countries. While considerable information of a general nature was available on chemical disposal practices, actual data on disposal practices were scant, and specific studies of associated adverse health effects were not found.

Information provided by developed countries suggests that environmental pollution is regarded as a total problem of which chemical disposal sites are only a part. Thus, emphasis is on determining the effect of different disposal practices on the larger problem, for example, water pollution. In the developing world there is agreement that pollution from chemical waste disposal in uncontrolled sites seems more important than that associated with localized chemical disposal sites. The most serious episodes of adverse health effects due to chemical disposal have occurred in Japan and were related to industrial effluents. In Europe, no serious episodes of ill health have been reported which can conclusively be related to point source chemical disposal sites. Significant public health effects that have occurred in relation to chemical waste disposal have largely been due to accidental chemical release during transportation or directly from industrial facilities. A number of sources emphasized that uncontrolled and unsatisfactory disposal practices were more likely to be associated with small business and domestic operations than with large primary generators. Hence, the view in Europe, Japan, and developing countries is that the problems should be approached in an overall sense and the contribution of different sources evaluated and prioritized.

In most developing countries few definitive studies of health effects due to chemical disposal have been reported, but in many cases known disposal sites were poorly controlled and already overflowing. Furthermore, surveillance mechanisms to identify environmental contamination episodes are often unavailable. In many such countries, chemicals also reach the environment directly through industrial effluents.

Despite the absence of conclusive reports on defined health effects, there is no complacency among public health authorities, and there is general agreement that priorities should be given to systematic control of all types of chemical disposal sites and the development of new technologies for disposal. Differences in opinion as to the best practices exist; in certain of the more industrialized countries, there seems to be a consensus that the problem is under control, but in developing countries, the situation appears to be unsatisfactory. There is general agreement that chemical disposal sites must be monitored to avoid accidental contamination of the surrounding environment. It should be noted that episodes of pollution may remain unreported at the local level. Several episodes of contamination of water sources could have had serious consequences. In all countries, both developed and developing, considerable pertinent data may exist at the local authority level and a systematic attempt should be made to collect, collate, and evaluate such material.

The prevailing approach in foreign countries appears to focus on cleanup and the establishment of evaluation priorities according to the potential for health effects. Public policy should be dependent, not so much on the demonstration of health effects, but on control of chemicals at what would be deemed prudent levels. Nevertheless, the distinction between policy or control, and undesirably high exposure to chemicals and health effects is acknowledged universally. Development of policies on chemical regulation tends to be much less adversarial in European governments than in the United States due to differences in the framework of the governments and in the systems of checks and balances on the legislative process.

References

Anonymous. (1984a) India's chemical tragedy: death toll at Bhopal still rising. Chem. Eng. News 62(50):6-7.

Anonymous. (1984b) World poison warning. The Observer. July 8.

Brickman, R.; Jasanoff, S.; Ilgen T. (1982) Chemical regulation and cancer: a cross-national study of policy and politics. Executive summary. 6p. Available from: Program on Science, Technology, Society, Cornell University, Ithaca, NY.

Brown, R.D.; Ouellette, R.P.; Cheremisinoff, P.N. (1983) National environmental policies and research programs: a comparison among selected countries. Lancaster, PA: Technomic Publishing Co., Inc. 160p.

Burgess, C.D. (1983) Health and safety aspects of handling hazardous wastes in Great Britain. [From: Industrial hazardous waste management]. [UNEP] Ind. Environ. (special issue)4:7-10.

Burton, I.; Fowle, C.D.; McCullough, R.S. (1982) Living with risk: environmental risk management in Canada. Environmental Monograph No. 3. Toronto: University of Toronto, Institute for Environmental Studies 247p.

Canada. Environment Canada Environmental Protection Service. (1984) Manual for land application of treated municipal wastewater and sludge. EPS 6-EP-84-1. Ottawa: The Service. 216p.

CEC SEE Council of the European Communities

Council of the European Communities. (1978) Council directive of 20 March

1978 on toxic and dangerous waste (78/319/EEC). Off. J. Eur. Community 84:43-48.

Dahl, A.L.; Baumgart, I.L. (1983) State of the environment in the South Pacific. UNEP Regional Seas Reports and Studies No. 31. Prepared in cooperation with the South Pacific Commission, South Pacific Bureau for Economic Cooperation, and the Economic and Social Commission for Asia and the Pacific. Geneva: United Nations Environment Programme. 29p.

Defregger, F. (1983) Status and trends on the management of industrial hazardous waste in the Federal Republic of Germany. [From: Industrial hazardous waste management]. [UNEP] Ind. Environ. (special issue)4:15-21.

Duffus, J.H.; Waddington, J.I.; editors. (1983) Environmental toxicology: Proceedings of a course, [held] August 16-20, 1982, Edinburgh. European Cooperation on Environmental Health Aspects of the Control of Chemicals: interim document No. 13. Copenhagen: World Health Organization, Regional Office for Europe. 325p.

Erasmus, H. (1983) Industrial hazardous waste management in the Netherlands. [From: Industrial hazardous waste management]. [UNEP] Ind. Environ. (special issue)4:25-27.

Ferm, R. (1982) Teckomatorp, Sweden: waste problems from the perspective of a manufacturing industry. In: Highland, J., ed. Hazardous waste disposal: [Proceedings of a conference sponsored by the Society for Occupational and Environmental Health, held December 1980 in Washington, DC]. Ann Arbor, MI: Ann Arbor Science/The Butterworth Group [now Stoneham, MA: Butterworth Medical and Scientific Publishers]. p. 31-34.

Finnecy, E.E. (1983) Aspects of control of hazardous waste disposal in the European community. [From: Different waste disposal practices in Europe; papers presented at a meeting of SCI Water and Environment Group, held January 12, in London]. Chem. Ind. (June):416-420.

Gesellschaft zur Beseitigung von Sondermull in Bayern. (1983) Disposal of special refuse in Bavaria. 17p. Available from: International Register of Potentially Toxic Chemicals, United Nations Environment Programme, Geneva.

Gray, D.A.; Mather, J.D.; Harrison, I.B. (1974) Review of groundwater pollution from waste disposal sites in England and Wales, with provisional guidelines for future site selection. Q. J. Eng. Geol. 7:181-196.

Great Britain. Parliament, House of Lords. Science and Technology Select Committee. (1981a) Hazardous waste disposal. 1980-81 session, 1st report. Vol. I. Report. London: Her Majesty's Stationery Office. 85p.

Great Britain. Parliament, House of Lords. Science and Technology Select Committee. (1981b) Hazardous waste disposal. 1980-81 session, 1st report. Vol. II. Oral evidence. London: Her Majesty's Stationery Office. 502p.

Great Britain. Parliament, House of Lords. Science and Technology Select Committee. (1981c) Hazardous waste disposal. 1980-81 session, 1st report. Vol. III. Written evidence. London: Her Majesty's Stationery Office. 137p.

Hamza, A. (1983) Management of industrial hazardous wastes in Egypt. [From: Industrial hazardous waste management]. [UNEP] Ind. Environ. (special issue)4:28-32.

Harada, M. (1978) Congenital Minamata disease: intrauterine methylmercury poisoning. Teratology 18:285-288.

Harwell Laboratory. (1982) [Individual abstracts]. Waste Manage. Inf. Bull. 19.

Ichikawa, A.K. (1983) Environmental pollution control for hazardous wastes in

Japan. [From: Industrial hazardous waste management]. [UNEP] Ind. Environ. (special issue)4:33-38.
Japan International Cooperation Agency. (1982a) Assessment of existing chemicals in the environment. 11p. Available from: The Agency, [Tokyo].
Japan International Cooperation Agency. (1982b) Outline of government administration for preventing soil pollution. 18p. Available from: The Agency, [Tokyo].
Japan International Cooperation Agency. (1982c) Relief measures for the victims of pollution-related diseases. 13p. Available from: The Agency, [Tokyo].
Keen, R.C.; Mumford, C.J. (1975) Preliminary study of hazards to toxic waste disposal operators on ten landfill sites in Britain. Ann. Occup. Hyg. 18:213-228.
Kempa, E.S. (1983) Role of low-waste technologies in the management of hazardous waste in Poland. [From: Industrial hazardous waste management]. [UNEP] Ind. Environ. (special issue)4:39-45.
Kool, H.J. (1984) Organic mutagens and drinking water in the Netherlands: a study on mutagenicity of organic constituents in drinking water in the Netherlands and their possible carcinogenic effects. Leidschendam, the Netherlands: Rijksinstituut voor Drinkwatervoorziening, [National Institute For Water Supply]. 116p.
Kool, H.J.; van Kreijl, C.F.; de Greef, E.; van Kranen, H.J. (1982) Presence, introduction and removal of mutagenic activity during the preparation of drinking water in the Netherlands. Environ. Health Perspect. 46:207-214.
Kool, H.J.; van Kriejl, C.F.; van Oers, H. (1984) Mutagenic activity in drinking water in the Netherlands: a survey and a correlation study. Toxicol. Environ. Chem. 7:111-129.
Kurland, L.T.; Faro, S.N.; Siedler, H. (1960) Minamata disease. World Neurol. 1:370-395.
Le Roy, E. (1983) Processing hazardous waste in France. [From: Industrial hazardous waste management]. [UNEP] Ind. Environ. (special issue)4:46-51.
Mills, D.A. (1983) National hazardous waste management practices in western Europe. Chem. Indust. (June):421-426.
Ramakrishna, R.S.; Ponnampalam, M.; Brooks, R.R.; Ryan, D.E. (1982) Blood lead levels in Sri Lankan families recovering gold and silver from jewellers' waste. Arch. Environ. Health 37:118-120.
Regione Lombardia, Ufficio Speciale di Seveso. (1983) CEFIC Health Protection Committee meeting, scientific documentation kit. 283p. Available from: Regione Lombardia, Ufficio Speciale de Seveso, Servizio Documentazione Scientifica, Seveso, Italy.
Royal Society of London. (1978) Long-term toxic effects: final report of a Royal Society study group. London: The Society. 14p. Available from: National Technical Information Service, Springfield, VA.
Sangster, B.; de Groot, G.; Loeber, J.G.; Derks, H.J.G.M.; Krajnc, E.I.; Savelkoul, T.J.F. (1984) Urinary excretion of cadmium, protein, beta-2-microglobulin and glucose in individuals living in a cadmium-polluted area. Hum. Toxicol. 3:7-21.
Schmidt-Bleek, F.; Lange, A.W. (1983) Research aspects in connection with the German Chemicals Act. Regul. Toxicol. Pharmacol. 3:311-319.
Smith, E.T.; Chem, C.; Bromley, J. (1982) British experience--clean up: technical aspects. In: Highland, J., ed. Hazardous waste disposal, assessing the problem. Ann Arbor, MI: Ann Arbor Science/The Butterworth

Group [now Stoneham, MA: Butterworth Medical and Scientific Publishers]. p. 188-197.

Solsberg, L.B.; Parent, R.; Buist, I. (1983) A survey of chemical spill countermeasures: preliminary findings. In: Canada. Environment Canada. Proceedings of the Technical Seminar on Chemical Spills, October, Toronto. Toronto: Environment Canada. p. 131-135.

Spilker, R. (1984) Poisonous waste threatens Hamburg. Ambio. 13:124-125.

Stevens, C.; Wilson, D.C. (1982) Toxicology of wastes. Paper presented at the Institution of Public Health Engineers symposium on hazardous wastes, September 13-14, Nottingham, [England]. 31p. Available from: Harwell Laboratory, [Didcot, England].

Suess, M.J.; Huismans, J.W.; editors. (1983) Management of hazardous waste: policy guidelines and code of practice. WHO Regional Publications, European Series No. 14. Copenhagen: World Health Organization, Regional Office for Europe. 101p.

Sundaresan, B.B.; Subrahmanyam, P.V.R.; Bhide, A.D. (1983a) Hazardous waste scene in India. Presented at the Third International Symposium on Industrial and Hazardous Solid Wastes, March, Philadelphia. 22p. Available from: National Environmental Engineering Research Institute, Nagpur, India.

Sundaresan, B.B.; Subrahmanyam, P.V.R.; Bhide, A.D. (1983b) Overview of toxic and hazardous waste in India. [From: Industrial hazardous waste management]. [UNEP] Ind. Environ. (Special Issue)4:70-73.

UNEP SEE United Nations Environment Programme

UNEP/IRPTC SEE United Nations Environment Programme, International Register of Potentially Toxic Chemicals

United Nations Environment Programme. (1982) Public health problems in the coastal zone of the East African region: UNEP Regional Seas Reports and Studies No. 9. Prepared in cooperation with the World Health Organization. Geneva: The Programme. 29p.

United Nations Environment Programme. (1984) State of the environment 1983: selected topics. Nairobi, Kenya: The Programme. 38p.

United Nations Environment Programme, International Register of Potentially Toxic Chemicals. (1984) Second meeting of experts on listing of environmentally dangerous chemical substances and processes of global significance, November 21-25, 1983, Geneva. 109p. Available from: The Programme, Geneva.

Van Duijvenbooden, W. (1983) Emergency interventions. 20p. Available from: National Institute for Water Supply, Voorburg, the Netherlands.

Vora, D.D.; Dastur, D.K.; Braganca, B.M.; Parihar, L.M.; Iyer, C.G.S.; Fondekar, R.B.; Prabhakaran, K. (1962) Toxic polyneuritis in Bombay due to ortho-cresyl-phosphate poisoning. J. Neurol. Neurosurg. Psychiatry 25:234-242.

WASTE DISCHARGE SEE Waste discharge into the marine environment: principles and guidelines for the Mediterranean Action Plan

Waste discharge into the marine environment: principles and guidelines for the Mediterranean Action Plan. (1982) Prepared in collaboration with the Institute of Sanitary Engineering, Polytechnic of Milan, Italy, under the joint sponsorship of the World Health Organization, Regional Office for Europe, and the United Nations Environment Programme. Oxford, England: Pergamon Press. 422p.

WHO/EURO SEE World Health Organization, Regional Office for Europe

WHO/IPCS SEE World Health Organization, International Programme on Chemical Safety

World Health Organization, International Programme on Chemical Safety. (1983) International Programme on Chemical Safety (IPCS). Geneva: The Programme. 17p.

World Health Organization, Regional Office for Europe. (1980a) European Cooperation on Environmental Health Aspects of the Control of Chemicals: proposed UNEP/WHO project. Report on a consultation, August 20-22, 1979, Copenhagen. Copenhagen: The Regional Office. 35p.

World Health Organization, Regional Office for Europe. (1980b) Toxicological appraisal of halogenated aromatic compounds following groundwater pollution: report on a WHO Working Group. Copenhagen: The Regional Office. 58p.

World Health Organization, Regional Office for Europe. (1981) Rapid assessment of the situation of drinking-water and sanitation in Europe (resolution WHA 30.33): consolidated report. W3/62/3. 24p. Available from: The Regional Office, Copenhagen.

World Health Organization, Regional Office for Europe. (1982) Micropollutants in river sediments: report on a WHO Working Group, August 5-8, 1980, Trier, [West Germany]. EURO reports and studies No. 61. Copenhagen: The Regional Office. 85p.

World Health Organization, Regional Office for Europe. (1983) Methods for monitoring selected pollutants in sewage effluents and coastal recreational waters. Report of a joint WHO/UNEP Meeting, 24-26 November 1982, Rome. Copenhagen: The Organization. 109p.

10

Summary, Conclusions, and Recommendations

The Executive Scientific Panel has conducted a comprehensive search for information on the health effects of the uncontrolled disposal of waste chemicals and has solicited from the public additional relevant contributions to this subject. These efforts indicate that there are few published scientific reports of health effects clearly attributable to chemicals from uncontrolled disposal sites. Only one report offered convincing evidence that serious health effects had resulted from exposure to a specific chemical. The remaining reports did not provide adequate scientific evidence of statistically significant and biologically plausible cause and effect relationships between specific chemicals and human disease.

These results are not surprising, given the many technical difficulties inherent in establishing cause and effect relationships in such situations. For example, a major study of Love Canal area inhabitants reviewed in Chapter 8 reported an increased incidence of reduced birth weight for children born in certain limited topographical sections of the area during the period of active dumping from 1940 to 1953; and a more recent study reported increased prevalence of seven physician-diagnosed disorders in children as reported by their parents or guardians.[1] In the former, retrospective study, there was no opportunity to measure or estimate exposure, leaving only the physical location of the residence as the principal risk factor in the analysis of birth weights above and below 2500 grams. In the latter study, an inverse relationship of frequency of reporting and distance from the canal was suggested for 6 of 7 disorders (including learning problems, skin rashes, eye irritation, abdominal pain, hyperactivity and seizures), but concentrations of five index chemicals in the basement air of 200 residences did not exceed

---

1. Paigen, B.; Goldman, L.R.; Highland, J.H.; Magnant, M.M.; Steegman, A.T., Jr. (1985) Prevalence of health problems in children living near Love Canal. Haz. Was. Haz. Mat. 2: 23-43. (Not available for evaluation during this study).

1/100 of a continuous exposure safety level extrapolated from occupational standards. Further, none of the reported maladies were verified by examination of medical records.

Although these and several similar studies do not confirm a causal association, they cannot be interpreted to indicate that no risk exists but, rather, that no serious health effects clearly attributable to a specific chemical or group of chemicals have been identified to date by the study methods employed.

The panel found that there are a number of reports on chemical disposal sites in the files of federal, state, and local authorities that have not been and may never be published. There are many reports from private contractors and consultants to various government agencies which are incomplete, or are extremely difficult to obtain because they are involved in litigation.

Frequently, epidemiologic studies are undertaken in reaction to pressures from the community near a chemical disposal site, and local residents and the news media may already be convinced that an excess of illness is caused by chemicals from the site. Scientists undertaking the epidemiologic study generally find that the contents of the site have not been adequately identified and that quantitative measurements of the extent of exposure of individuals are seldom available. Indeed, the effects of environmental exposures in air, water, and food in the home and the workplace confound the interpretation of findings in studies of chemical disposal sites.

Experience to date is limited, but the data suggest that human exposures, in general, have not been of sufficient duration and concentration to have resulted in observable long-term health effects correlated with dumpsite exposures. Based on current knowledge, chronic health effects from long-term, low-level exposures would be expected to affect only a small percentage of those exposed. To date, the sizes of the exposed populations studied have seldom been large enough to correlate a slight increase in adverse health effects with specific chemical exposures. In the context of chronic or long-latency diseases such as cancer, sufficient time may not have elapsed for detection of any excess incidence of disease in populations around these sites.

The following lines of inquiry were pursued during the study: the number and distribution of chemical disposal sites; the toxic chemicals found most frequently in substantial quantities and having chemical and/or physical properties favoring migration from the site; the physical nature of sites, and routes by which toxic substances might migrate from sites and result in human exposure; the health effects these chemicals produce in humans or in experimental animals; the methods required to identify chemically induced health effects in individuals; and patterns of health effects in populations. The Panel has concluded that many uncontrolled chemical disposal sites contain substantial quantities of toxic substances, some of which have migrated from the sites and resulted in human exposure, and which have the potential to cause serious, irreversible, or even fatal illness if exposure is substantial or prolonged.

CONCLUSIONS AND RECOMMENDATIONS

1. Conclusion: Critical reviews of 29 investigations of the health of populations in the vicinity of chemical disposal sites have confirmed that, with one exception, evidence for causal association with occurrence of disease is weak. These investigations can be termed suggestive at best. The absence of

demonstrable effects on human health should not be taken as proof that no such effects exist, nor do positive associations necessarily prove that observed disease was caused by chemicals escaping from a site.

Recommendation:

Additional prospective and retrospective health studies using sensitive methods should be scheduled according to the precedence indicated for those chemical disposal sites on EPA's list of priority sites. The populations that have been exposed should be monitored for at least 20 years to detect any excess incidence of cancer or other chronic disease.

2. Conclusion: Retrospective epidemiologic health studies to date have relied heavily on the survey questionnaire, which is useful. However, prospective epidemiological studies involving biological indices of exposure, though more difficult to perform, can be of greater value in identifying causal relationships between exposures and associated diseases.

Recommendations:

a. Standards for protocols to be used in surveys of health effects should be developed, and such protocols should be used to the maximum extent feasible.

b. Interdisciplinary teams of scientists should be used in future studies of health effects associated with chemical disposal sites. Disciplines relevant to these studies include, but are not limited to, epidemiology, public health medicine, toxicology, environmental science and engineering, and biostatistics.

3. Conclusion: Investigations to date have been complicated by the inherent scientific difficulties of performing these studies, as well as by the climate of public concern in which the studies were performed.

Recommendations:

a. Because the nature and extent of exposure of persons near chemical disposal sites has been the primary unknown in studies to date, major efforts in prospective investigations should focus on characterizing chemical exposure by air, water, soil or food.

b. Retrospective studies often rely on indirect measures of exposure. These investigations should continue to use such measures, with full recognition of the scientific limitations of investigations performed without direct estimates of population exposure.

c. If subtle biological changes in body tissues are used to characterize health effects in populations in epidemiological studies, investigators should clearly interpret the significance of these measures to effects on human health. Failure of previous investigators to address this point adequately has led to major public misunderstanding of results from studies of chemical disposal sites.

d. Written results of investigations should be made public promptly and should be openly available to the scientific community, governmental agencies, and local communities.

4. Conclusion: The public has been an important source in alerting the scientific community and governmental agencies to the problem of disposal sites.

Recommendations:

a. An alert and informed citizenry should continue to be involved in alerting scientists and government of issues relating to disposal sites.

b. Principal investigators should keep local communities informed of the progress and conclusions of their work.

5. Conclusion: Community reactions to perceived health effects require serious consideration and evaluation. Coordination of investigations of potential health effects relating to chemical disposal sites will improve the quality of the individual studies and contribute to the scientific quality of the methods used. The issue of health effects in populations at chemical disposal sites is so emotionally charged, and the economic and social consequences are so great, that special efforts are needed to assure that investigations are of high quality and the results are subjected to appropriate professional scrutiny.

Recommendation:

The Centers for Disease Control should establish a mechanism for external and independent peer review of epidemiologic research proposals and should be designated as a clearinghouse for studies of health effects at chemical disposal sites, providing advice and assistance in design, the application of standard methods, monitoring research activities, and disseminating research results.

6. Conclusion: The current uncertainties associated with relating adverse human health effects to chemicals from either active or inactive disposal sites must not result in postponement of justifiable preventive or remedial action to reduce the potential health impact of these sites.

Recommendations:

a. Because of the numerous uncontrolled candidate sites identified for remedial action and the critical need for setting priorities, each site should be placed in perspective from the viewpoint of its potential effects on human health by the responsible public health agencies of the Federal Government.

b. In view of the uncertainties in relating adverse health effects to chemical disposal sites, it is appropriate to base remedial actions for these sites on estimated human exposures (based on human and environmental monitoring) until adequate information on the duration of exposure, toxicity of the chemicals, and the dose-response effects of the compounds found at individual sites is acquired.

7. Conclusion: Monitoring of emissions of chemical substances from disposal sites is necessary if potentially hazardous exposure of the surrounding community is to be avoided or minimized.

Recommendations:

a. Monitoring of leakage from contained chemical disposal sites should be performed routinely. Surrogate exposure indices should be explored, as well as the potential for use of ecological sentinels.

b. Available models for predicting the fate of chemical substances escaping from disposal sites should be evaluated and improved in order to provide better tools to model exposure.

8. Conclusion: The risk of health hazards can be reduced by appropriate treatment of toxic wastes prior to disposal. A number of such systems of treatment have been developed.

Recommendations:

a. The existing systems for treating chemical wastes prior to disposal should be utilized now for all active sites.

b. Additional research and development is needed to improve methods of reducing or treating chemical wastes to minimize the risk to humans and the environment.

9. Conclusion: Although solid evidence for health effects caused by chemicals from disposal sites is generally lacking, it is clear that a potential hazard is presented by many sites. Their contents, construction, and location favor pollution of air, water, and soil, which may ultimately result in human exposure.

Recommendations:

a. The EPA's efforts to identify and rank priority sites should be intensified.

b. The current practice of developing appropriate risk management measures as soon as serious risk is detected, without waiting for adverse health effects to appear, should be supported and adequately funded.

10. Conclusion: There is a great disparity between the public perception of the human health effects potentially attributable to chemicals in disposal sites and the current scientific evidence on this subject.

Recommendation:

The nature of risks from chemical disposal sites as compared to risks from other environmental agents should be conveyed to the public in a major educational effort. Brochures, tapes, films, slides, and discussions by capable scientists will help to increase understanding of this subject.

Appendix A

Chemical Synopses

Benzene

Information abstracted from TDB II, 1984 (No. 35) unless otherwise specified.

I. Chemical Identification

A. CAS Registry No.: 71-43-2

B. Synonyms: Cyclohexatriene

C. Molecular formula: C_6H_6

D. Structural formula:

II. Physical and Chemical Properties

A. Molecular weight: 78.11

B. Physical state (color, form): Clear, colorless liquid

C. Odor: Aromatic

D. Melting point: 5.5 deg C

E. Boiling point: 80.1 deg C

F. Solubilities:

Water: 1780 mg/L at 25 deg C

Organic solvents: Soluble in most organic solvents

G. Octanol/water partition coefficient (log P): 2.28

H. Density: 0.879 g/ml at 5 deg C

I. Dissociation constant: --

J. Viscosity: 0.652 cp at 20 deg C (CRC, 1983)

K. Surface tension (liquid interfacial tension with air): 28.9 mN/m (dynes/cm) at 20 deg C

L. Vapor pressure: 100 mm Hg at 26.1 deg C

M. Vapor density: 2.7 (air = 1)

N. Relative evaporation rate: Half-life values of 3-5 hours

O. Spectral properties: Maximum absorption at 243 nm, 249 nm, 256 nm, and 261 nm (in alcohol) (CRC, 1978)

III. Manufacturing/Use Information

United States production of benzene in 1975 was 3.42 billion kg (7.5 billion lb). Major uses of benzene are as a chemical intermediate in the manufacture of styrene and cyclohexane (NRC, 1980) and in the manufacture of detergents, pesticides, and solvents. It was formerly used to control screwworms on animals and as an ingredient in some early grain fumigants, but pesticidal uses have been banned.

IV. Environmental Fate/Exposure Potential

A. Potential for release from chemical disposal sites

Spills or releases of benzene to soil will evaporate very rapidly from the surface due to benzene's high vapor pressure; vapors may travel considerable distances. It has a low to moderate potential for adsorption to soil and is reported to leach into groundwater. Benzene is fairly soluble in water and may be removed from the atmosphere in rain.

B. Biological uptake/accumulation

Benzene has a low potential for bioconcentration since it is volatile. Benzene that enters surface waters will evaporate fairly readily or be removed by photodegradation and biodegradation.

C. Persistence

Benzene will be lost from soil due to evaporation and, to some extent, biodegradation, although it reportedly biodegrades slowly in soil. Benzene is fairly mobile in soil and would be expected to percolate into groundwater where it may be degraded by microorganisms. In aqueous solutions, evaporation occurs rapidly and is the primary loss mechanism in the winter, with a reported half-life of 13 days. In the spring and summer, the reported evaporation half-lives were 23 and 3.1 days, respectively, with biodegradation playing a major role. Photodegradation would contribute to benzene's removal under conditions unfavorable for microbial degradation (i.e., cold water, low nutrients). Benzene released into the atmosphere will photochemically degrade, with a half-life of about 6 days (much less in smog conditions). Benzene does not absorb light at wavelengths of 290 nm or longer; thus, it will not degrade by direct photolysis, but predominantly by reactions with hydroxyl radicals, resulting in products including phenol, nitrophenol, nitrobenzene, and formic acid.

D. Measured concentrations

Groundwater: The EPA water quality criteria document reported benzene concentrations of "greater than 100 ppb" (100 ug/L) (WHO/IARC, 1982a).

Surface water: Sampling surveys have detected benzene in the range of 1-7 ug/L in heavily industrialized areas and have found an average of 5.4 ug/L in 700 random U.S. sites.

Drinking water: An analysis from 5 U.S. cities revealed 0-0.3 ppb benzene. An analysis of contaminated drinking-water wells in New York, New Jersey, and Connecticut showed 30-300 ug/L.

Air: Average atmospheric concentrations in various U.S. areas are reported as 0.0014 ppm in rural/remote areas and 0.0028 ppm in suburban/urban areas, whereas the concentration near chemical factories using benzene ranged from 0.0006-0.034 ppm.

Soil: In soil near factories where benzene was used or produced, the concentration range was from less than 0.002 to 0.191 mg/kg.

Food: Benzene has been reported in several foods as follows: eggs, 0.50-1.90 mg/kg; heat-treated or canned beef, 0.002 mg/kg.

E. Human environmental exposure

Human exposure to any benzene released from chemical disposal sites would most likely occur via ingestion of drinking water from groundwater sources contaminated with benzene. Since benzene is very volatile, heavier than air, and its vapors reportedly travel a considerable distance, inhalation is a possible route of exposure.

Average daily intake: The estimated daily human exposure to benzene is 0.2 ug from water, and 183 ug from air.

Body burden: Benzene has been detected in 8 of 8 samples of mothers' milk from 4 U.S. urban areas. It was detected in the breath of persons without known exposure to benzene in concentrations of 47-64 ppm.

V. Health Effects

A. Human health effects

The major acute effect in humans is on the central nervous system, with symptoms occasionally progressing to pulmonary distress and cardiovascular failure. Other effects from acute exposure may include damage to the kidneys, lungs, and spleen. Delayed effects may appear and persist long after exposure. Symptoms from mild exposure include dizziness, euphoria, and headache.

Chronic exposure may result in moderate blood abnormalities (bone marrow depression resulting in leukopenia, anemia, and thrombocytopenia), but these effects may be reversible in the early stages. Exposure to high doses (greater than 100 ppm) for long periods of time may lead to irreversible aplasia. Although the results are not as clear with lower exposure levels (less than 100 ppm), case reports and epidemiological studies suggest a relationship between exposure to benzene and the development of leukemia (acute nonlymphocytic). An evaluation by IARC (WHO/IARC, 1982a) reported that there was sufficient evidence that benzene is carcinogenic to man. Although benzene crosses the human placenta, no definite results on developmental effects in humans were reported.

Absorption, Distribution, Metabolism, Excretion:

Benzene is eliminated rapidly by exhalation and by excretion in the urine. Studies on atmospheric exposure showed that of approximately 47% taken up, about 30% was retained, and about 17% was eliminated as unchanged benzene in expired air (WHO/IARC, 1982a). Metabolism occurs primarily in the liver, where benzene is converted to benzene oxide. This epoxide is a highly unstable, reactive intermediate and rearranges to form phenol, the major metabolite of benzene found in urine. Other metabolic products may include catechol, hydroquinone, and conjugated phenolic compounds.

B. Non-human mammalian health effects

Acute oral LD50 values have been reported to be 4,894 mg/kg for rats and 4,700 mg/kg for mice. The most commonly observed effect of chronic benzene toxicity in laboratory animals is a decrease in circulating lymphocytes (leukopenia). Bilateral cataracts have also been reported in rats subjected to prolonged benzene exposures. No evidence of carcinogenicity was seen in mice that received benzene by skin application. Other experiments in animals were considered by IARC (WHO/IARC, 1982b) to be inadequate to evaluate the carcinogenicity of benzene. Although chronic inhalation studies in rats have provided some evidence of embryolethality, reduced fetal weight, and various developmental abnormalities, there is no conclusive evidence of teratogenicity of benzene in animals. The metabolic products of benzene in animals are similar to those in humans.

C. Non-mammalian effects

The EPA (USEPA/OWRS, 1980) reviewed results of aquatic toxicology studies with benzene. The 96-hour LC50 values for six freshwater fish species ranged from 5,300 ug/L for the rainbow trout (Salmo gairdneri) to 386,000 ug/L for the mosquitofish (Gambusia affinis). For two freshwater invertebrate species, the cladoceran water fleas (Daphnia magna and pulex), the 48-hour LC50 values ranged from 203,000 ug/L to 620,000 ug/L.

D. Short-term assays

Benzene was not mutagenic in bacteria, yeast, *Drosophila melanogaster*, or mouse lymphoma cells in culture (WHO/IARC, 1982b). Benzene induced chromosomal aberrations in animals but did not induce dominant lethal mutations in mice. Benzene produced decreased DNA synthesis in animal bone marrow cells in vitro, and induced chromosomal aberrations in cultured leukocytes obtained from occupationally exposed humans.

VI. Exposure Standards and Recommendations

A. Air

1. Immediately Dangerous to Life or Health (IDLH): 2,000 ppm (Sittig, 1981)

2. OSHA standards: (1978 values cited in USDHHS/CDC, 1983)

   TLV-TWA (8 hr): 10 ppm
   Ceiling: 25 ppm
   Peak (10 min): 50 ppm

3. NIOSH recommendations:

   Ceiling (1 hr): 1 ppm (1976 value cited in USDHHS/CDC, 1983)

4. ACGIH (1980):

   TLV-TWA (8 hr): 10 ppm (30 mg/m^3)
   STEL (15 min): 25 ppm (75 mg/m^3)

5. Other:

   Under section 112 of the Clean Air Act, emissions from process vents of new ethylbenzene or styrene plants is limited to 5 ppm by volume.

B. Water

The following ambient water quality criteria are cited by EPA (USEPA/OWRS, 1980):

To protect human health from potential carcinogenic effects: Preferably zero. An increase in the human lifetime cancer risk by $1x10^{-5}$ due to ingestion of contaminated water and aquatic organisms is estimated to occur at a benzene concentration of 6.6 ug/L.

To protect freshwater aquatic life:	5.3 mg/L (acute toxicity) No available data on chronic toxicity

To protect marine aquatic life:	5.1 mg/L (acute toxicity) No available data on chronic toxicity

7-day SNARL = 12.6 mg/L (NRC, 1980). This calculation is based on animal data and ignores the suspected carcinogenicity of benzene in humans.

References for Benzene Synopsis

ACGIH SEE American Conference of Governmental Industrial Hygienists, Inc.

American Conference of Governmental Industrial Hygienists, Inc. (1980) Documentation of the threshold limit values. 4th ed. Cincinnati: The Conference. 486p.

CRC SEE CRC handbook of chemistry and physics

CRC handbook of chemistry and physics. (1978) 59th ed. Weast, R.C.; Astle, M.J.; eds. West Palm Beach, FL: CRC Press, Inc. 2488p.

CRC handbook of chemistry and physics. (1983) 64th ed. Weast, R.C.; Astle, M.J.; Berger, W.H.; eds. Boca Raton, FL: CRC Press, Inc. 2302p.

National Research Council. (1980) Drinking water and health. Vol. 3. Washington, DC: National Academy Press. 415p.

NRC SEE National Research Council

Sittig, M. (1981) Handbook of toxic and hazardous chemicals. Park Ridge, NJ: Noyes Publications. 729p.

TDB II SEE Toxicology data bank II

Toxicology data bank II. (1984) [Enhanced version of data base]. Benzene, TDB no. 35. (Last edited April 17, 1984). Bethesda, MD: U.S. Department of Health and Human Services, National Library of Medicine, Specialized Information Services. Magnetic-tape reels.

U.S. Department of Health and Human Services, Centers for Disease Control. (1983) NIOSH recommendations for occupational health standards. Morb. Mortal. Wkly. Rep. 32:(supplement)3S-22S.

USDHHS/CDC SEE U.S. Department of Health and Human Services, Centers for Disease Control

U.S. Environmental Protection Agency, Office of Water Regulations and Standards. (1980) Ambient water quality criteria for benzene. EPA-440/5-80-018. Washington, DC: The Office. 118p. Available from: National Technical Information Service, Springfield, VA.

USEPA/OWRS SEE U.S. Environmental Protection Agency, Office of Water Regulations and Standards

WHO/IARC SEE World Health Organization, International Agency for Research on Cancer

World Health Organization, International Agency for Research on Cancer. (1982a) IARC monographs on the evaluation of the carcinogenic risk of chemicals to humans. Vol. 29. Some industrial chemicals and dyestuffs. Lyon, France: The Agency. 416p.

World Health Organization, International Agency for Research on Cancer. (1982b) IARC monographs on the evaluation of the carcinogenic risk of chemicals to man. Supplement 4 to Vol. 1-29. Chemicals, industrial processes and industries associated with cancer in humans. Lyon, France: The Agency. 292p.

Carbon Tetrachloride

Information abstracted from TDB II, 1984 (No. 53) unless otherwise specified

I. Chemical Identification

A. CAS Registry No.: 56-23-5

B. Synonyms:

Methane tetrachloride
Perchloromethane
Tetrachloromethane

C. Molecular formula: CCl_4

D. Structural formula:

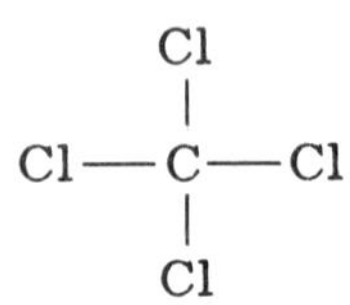

II. Physical and Chemical Properties

A. Molecular weight: 153.84

B. Physical state (color, form): Colorless, heavy liquid

C. Odor: Sweetish, aromatic, moderately strong

D. Melting point: -23 deg C

E. Boiling point: 76.5 deg C

F. Solubilities:

Water: 800 mg/L at 20 deg C (Verschueren, 1983); 1 ml/2,000 ml water

Organic solvents: Soluble in acetone; miscible with alcohol, benzene, chloroform, ether, carbon disulfide, petroleum ether, oils

G. Octanol/water partition coefficient (log P): 2.64

H. Density: 1.594 g/ml at 20 deg C

I. Dissociation constant: --

J. Viscosity: 0.969 cp at 20 deg C (CRC, 1983)

K. Surface tension:

Liquid interfacial tension with air: 45.0 mN/m (dynes/cm)

Liquid interfacial tension in contact with vapor: 26.95 dynes/cm at 20 deg C (CRC, 1983)

L. Vapor pressure: 91.3 mm Hg at 20 deg C

M. Vapor density: 5.32 (air = 1)

N. Relative evaporation rate: Reported half-life values range from minutes to days

O. Spectral properties: Maximum absorption at less than 200 nm (vapor)

III. Manufacturing/Use Information

Production of carbon tetrachloride in the United States in 1973 exceeded one billion lb (NRC, 1977). After peaking in 1974, production declined at a rate of 7.9% per year. In 1981, 726 million lb were produced. In both 1982 and 1983, 586 million lb were used. The major uses of carbon tetrachloride are as an intermediate in synthesizing fluorocarbons. Other uses are as a grain fumigant, fire extinguisher, anthelmintic, solvent for oils and fats, and a cleaning agent.

IV. Environmental Fate/Exposure Potential

A. Potential for release from chemical disposal sites

Spills or releases of carbon tetrachloride to soil should evaporate rapidly due to its high vapor pressure. Its low soil adsorption coefficient suggests that it would only be slightly adsorbed to sediments and soils and that it would readily migrate from chemical disposal sites into groundwater. The evaporation of carbon tetrachloride from water is a significant removal process; the reported evaporation half-life values in laboratory studies ranged from minutes to hours.

B. Biological uptake/accumulation

Carbon tetrachloride has a low bioconcentration potential, based on the common logarithm of the bioconcentration factors in trout and bluegill sunfish which are 1.24 and 1.48, respectively.

C. Persistence

Carbon tetrachloride will be lost rapidly from soil due to evaporation. Leaching in soil results in groundwater contamination. Data are unavailable on biodegradation in soil. Though data are limited, biodegradation has been noted in aqueous screening tests; under anaerobic conditions, degradation occurs in 16 days. Wilson and McNabb (1983) reported that biotransformation was possible in anoxic groundwater although there was practically no likelihood of degradation in oxygenated groundwater. In laboratory studies simulating methanogenic conditions, carbon tetrachloride was almost completely oxidized to carbon dioxide (Bouwer and McCarty, 1983a); under denitrification conditions, carbon from carbon tetrachloride was both assimilated into cell material and mineralized to carbon dioxide (Bouwer and McCarty, 1983b). In aqueous solutions, carbon tetrachloride is removed quickly by evaporation as evidenced by a half-life of minutes to hours. Based on field monitoring data, the estimated half-life in rivers is 3-30 days, and in lakes and groundwater is 3-300 days. Carbon tetrachloride is very stable in the troposphere with residence times of 30-50 years; its main loss mechanism is diffusion upwards to the stratosphere, where direct photolysis (higher energy irradiation: 195-254 nm) causes degradation. Carbon tetrachloride does not react significantly with any active species in the atmosphere (e.g., half-life for reaction with hydroxyl radicals is reported to be greater than 330 years).

D. Measured concentrations

Groundwater: Carbon tetrachloride concentrations in raw water for 27 U.S. cities were reported to be 3-20 ug/L. Clark et al. (1982) reported that in private drinking-water wells near a chemical disposal site in the Toone-Teague area of Hardeman County, Tennessee, the carbon tetrachloride concentrations ranged from 61 to 18,700 ug/L, with a median concentration of 1,500 ug/L; of 15 samples analyzed, all 15 showed detectable carbon tetrachloride concentrations. A concentration of 29,450 ug/L was later found in one of these wells.

Surface water: Average concentration was 0-9 ug/L, but it was 1-3 ug/L in 14 heavily industrialized rivers, and 9-47 ug/L in the Great Lakes. In marine waters, the concentration was 0.00012-0.00085 ug/L at the surface, and 0.00015 ug/L at a depth of 300 m.

Drinking water: In 181 U.S. cities whose drinking-water source was surface water, the concentration of carbon tetrachloride was 0.1-30.0 ug/L; in 39 U.S. cities whose drinking-water source was groundwater, it was 0.2-13.0 ug/L. Nine homes near Love Canal in Niagara Falls, New York, had a concentration range of 0-190 ug/L. In 80 U.S. cities, the concentrations were 0-4 ug/L, but 135-400 ug/L were measured in New Jersey and New York drinking-water wells.

Air: Atmospheric concentrations of carbon tetrachloride in urban/-suburban areas were 0-42.4 ppb and in rural/remote areas, 0.082-0.24 ppb, whereas the range in industrialized areas was 0-70 ppb. The amount of carbon tetrachloride in the atmosphere has been increasing because of its high emission rate and slow degradation. Clark et al. (1982) reported an air concentration of 600 ppb (3,600 ug/m^3) in a bathroom while a shower delivering contaminated water was being used.

Food: Carbon tetrachloride has been detected in the following foods: seafood, 2-209 ppb; fats, fruit, vegetables, meat, tea, and bread, 3-18 ppb, with oils and fats at the high end of the range; flour, 0.0002-2.61 ppm.

E. Human environmental exposure

Human exposure to carbon tetrachloride from chemical disposal sites could occur by dermal contact with wastes at the disposal site, by swimming in lakes and bathing or showering in water contaminated with carbon tetrachloride, by ingestion of drinking water from contaminated groundwater wells or surface waters, or by inhalation in the vicinity of disposal sites.

Average daily intake: The estimated intake is reported by TDB II (1984) to be 12-511 ug/day from drinking water, and 0.2-60.0 ug/day from air. Food may be contaminated by carbon tetrachloride when it is used as a grain fumigant, or contaminated by those herbicides, insecticides, and fungicides that contain carbon tetrachloride.

Occupational exposure: Inhalation of high concentrations of carbon tetrachloride is largely restricted to occupational settings. It was estimated at one time that 160,000 workers were potentially exposed.

Body burden: Only limited information was available. Carbon tetrachloride was detected in 5 of 6 samples of mothers' milk at four sites in Pennsylvania, New Jersey, and Louisiana.

V. Health Effects

A. Human health effects

Acute poisonings are usually due to inhalation of carbon tetrachloride in poorly ventilated rooms. Carbon tetrachloride acts as a central nervous system depressant; its acute toxicity effects are best characterized as hepatic nodular hyperplasia and cirrhosis, and renal dysfunction (NRC, 1977). Presence of hepatic and/or renal failure will vary among individuals (NRC, 1977); ingestion tends to cause hepatic lesions, whereas inhalation tends to cause renal damage (Kirsch-Volders, 1984). Other observed effects may include dizziness, headache, vomiting, and diarrhea (Kirsch-Volders, 1984). Dermal contact with carbon tetrachloride produces distinct pain with erythema, hyperemia, and wheal formation followed by vesication. The human oral lethal dose is estimated to be 50-500 mg/kg (between 1 tsp and 1 oz for a 70 kg or 150 lb person), although a fatal dose for children has been reported as low as 3 ml (NRC, 1977).

Chronic exposure to 45-100 ppm carbon tetrachloride results in gastrointestinal symptoms, such as nausea and vomiting; after exposures of 100-300 ppm, additional symptoms include apathy or mental confusion and weight loss (NRC, 1977). According to IARC (WHO/IARC, 1982), there is inadequate evidence for carcinogenicity due to carbon tetrachloride. However, one mortality study of laundry and dry cleaning workers who had been exposed to a variety of substances including carbon tetrachloride revealed excess deaths due to lung cancers (10 expected, 17 observed), cervical cancers (4.8 expected, 10 observed), liver cancers (1.7 expected, 4 observed), and leukemia (2.2 expected, 5 observed).

Absorption, Distribution, Metabolism, Excretion:

Carbon tetrachloride is rapidly absorbed through the skin, gastrointestinal tract, and the respiratory tract. It may be distributed to all organs and tissues, with the greatest amounts partitioning into body lipids. Carbon tetrachloride is metabolized by the microsomal mixed-function oxidase system and is excreted primarily by the lungs, 85% as the parent compound (NRC, 1977).

B. Non-human mammalian health effects

Oral LD50 values are 2,800 mg/kg for rats; 12,800 mg/kg for mice; and 6,380 mg/kg for rabbits. The acute effect of carbon tetrachloride inhalation on animals has been shown in various studies to result in central nervous system depression, with variations in dose, exposure, and species as follows: 6,350-7,500 ppm for 2 hours for mice; 14,300 ppm for 4 hours for cats; 21,000 ppm for 9-13 minutes for dogs. The percutaneous LD50 value for rats is 5,070 mg/kg. Symptoms of poisoning in animals after oral administration are loss of appetite, dullness, staggering gait, evidence of gastrointestinal disturbances with passage of blood-stained feces, constipation followed by diarrhea, collapse, and death.

Chronic studies of oral, subcutaneous, and rectal administration of carbon tetrachloride produced kidney damage and liver injuries leading to hepatic carcinomas in rats, mice, and hamsters, as well as mammary adenocarcinomas in subcutaneously injected rats. Other chronic effects include hepatic cirrhosis and necrosis, renal damage, changes in blood enzymes, and increased serum bilirubin (NRC, 1977). Experimental data indicate that carbon tetrachloride is not teratogenic to rats exposed orally, subcutaneously, or via inhalation, but it has been shown to reduce fetal weight after maternally toxic inhalation exposures (NRC, 1983). IARC (WHO/IARC, 1979) reported retarded development of fetuses and increased fetal mortality.

Absorption, Distribution, Metabolism, Excretion:

Carbon tetrachloride is rapidly absorbed by all routes. It is excreted primarily by the lungs as both the unchanged compound and as metabolites. It is metabolized by the microsomal mixed-function oxidase system (Casarett, 1980; NRC, 1977). Experimental evidence suggests that metabolism is a prerequisite to its hepatotoxicity (Casarett, 1980); it is hypothesized that a reactive trichloromethyl radical is generated, which induces lipid peroxidation, which in turn leads to hepatotoxicity.

Other reported metabolites of carbon tetrachloride are chloroform, phosgene, and carbon dioxide.

Susceptibility to carbon tetrachloride poisoning is enhanced by simultaneous use of DDT, hexachlorobenzene, polybrominated biphenyls, polychlorinated biphenyls, isopropanol, ethanol, and barbiturates. Poor nutritional status and perhaps calcium deficiency also enhance carbon tetrachloride toxicity.

C. Non-mammalian effects

The reported 96-hour LC50 values for three aquatic species were 50 mg/L for the dab (Limanda limanda), a marine fish; 27-125 mg/L for the bluegill (Lepomis macrochirus), a freshwater fish; and 150 mg/L for the tidewater silverside (Menidia beryllina), an estuarine fish. The 48-hour LC50 value for a freshwater cladoceran, the water flea (Daphnia magna), was 35.2 mg/L.

In an embryo-larval test with the fathead minnow (Pimephales promelas), no adverse effects were observed at test concentrations up to 3.4 mg/L (USEPA/OWRS, 1980).

D. Short-term assays

Carbon tetrachloride was not mutagenic in bacteria, but at almost lethaldoses, it was mutagenic in yeast (WHO/IARC, 1982). It did not induce chromosomal damage in cultured rat liver epithelial cells, and did not induce unscheduled DNA synthesis in the hepatocytes of rats exposed in vivo. Studies of DNA binding were limited. No data were available on human mutagenicity or chromosomal damage.

VI. Exposure Standards and Recommendations

A. Air

1. Immediately Dangerous to Life or Health (IDLH): 300 ppm, with the notation "occupational carcinogen" (Sittig, 1981)

2. OSHA standards: (1974 values cited in USEPA/OWRS, 1980)

   TLV-TWA (8 hr): 10 ppm (65 mg/m^3)
   Ceiling: 25 ppm (162.5 mg/m^3
   Peak: 200 ppm (1,300 mg/m^3) for 5 minutes in a 4-hour period

3. NIOSH recommendations (USDHEW/NIOSH, 1976):

   Ceiling (1 hr): 2 ppm (12.6 mg/m^3)

4. ACGIH (1980):

   TLV-TWA (8 hr): 5 ppm (30 mg/m^3) - "suspect carcinogen"
   STEL (15 min): 20 ppm (125 mg/m^3)

Note: As of 1984, ACGIH (1984) has deleted the STEL value until further acute toxicological and industrial hygiene experience become available.

B. Water

The following ambient water quality criteria are cited by EPA (USEPA/OWRS, 1980):

To protect human health from potential carcinogenic effects: Preferably zero. An increase in the human lifetime cancer risk by $1x10^{-5}$ due to ingestion of contaminated water and aquatic organisms is estimated to occur at a carbon tetrachloride concentration of 4.0 ug/L.

To protect freshwater aquatic life: 35.2 mg/L (acute toxicity)
No available data on chronic toxicity

To protect marine aquatic life: 50.0 mg/L (acute toxicity)
No available data on chronic toxicity

C. Other

FDA has banned the sale of carbon tetrachloride in any product used in the home, including fire extinguishers. Its use as a solvent (spot remover) was banned in the U.S. by FDA in 1970 (WHO/IARC, 1979).

References for Carbon Tetrachloride Synopsis

ACGIH SEE American Conference of Governmental Industrial Hygienists, Inc.

American Conference of Governmental Industrial Hygienists, Inc. (1980) Documentation of the threshold limit values. 4th ed. Cincinnati: The Conference. 486p.

American Conference of Governmental Industrial Hygienists, Inc. (1984) Supplemental documentation [for documentation of the threshold limit values]. Cincinnati: The Conference. 61p.

Bouwer, E.J.; McCarty, P.L. (1983a) Transformations of 1- and 2-carbon halogenated aliphatic organic compounds under methanogenic conditions. Appl. Environ. Microbiol. 45:1286-1294.

Bouwer, E.J.; McCarty, P.L. (1983b) Transformations of halogenated organic compounds under denitrification conditions. Appl. Environ. Microbiol. 45: 1295-1299.

Casarett SEE Casarett and Doull's toxicology: the basic science of poisons

Casarett and Doull's toxicology: the basic science of poisons. (1980) 2nd

edition. Doull, J.; Klaassen, C.D.; Amdur, M.O.; eds. New York: Macmillan Publishing Co. 733p.
Clark, C.S.; Meyer, C.R.; Gartside, P.S.; Majeti, V.A.; Specker, B.; Balistreri, W.F.; Elia, V.J. (1982) Environmental health survey of drinking water contamination by leachate from a pesticide waste dump in Hardeman County, Tennessee. Arch. Environ. Health 37:9-18,125.
CRC SEE CRC handbook of chemistry and physics
CRC handbook of chemistry and physics. (1983) 64th ed. Weast, R.C.; Astle, M.J.; Berger, W.H.; eds. Boca Raton, FL: CRC Press, Inc. 2302p.
Kirsch-Volders, M., ed. (1984) Mutagenicity, carcinogenicity, and teratogenicity of industrial pollutants. New York: Plenum Press. 336p.
National Research Council. (1977) Drinking water and health. [Vol. 1]. Washington, DC: National Academy Press. 939p.
National Research Council. (1983) Drinking water and health. Vol. 5. Washington, DC: National Academy Press. 157p.
NRC SEE National Research Council
Sittig, M. (1981) Handbook of toxic and hazardous chemicals. Park Ridge, NJ: Noyes Publications. 729p.
TDB II SEE Toxicology data bank II
Toxicology data bank II. (1984) [Enhanced version of data base]. Carbon tetrachloride, TDB no. 53. (Last edited March 7, 1984). Bethesda, MD: U.S. Department of Health and Human Services, National Library of Medicine, Specialized Information Services. Magnetic disk.
U.S. Department of Health, Education, and Welfare; National Institute for Occupational Safety and Health. (1976) NIOSH revised recommended carbon tetrachloride standard. 7p. Available from: National Technical Information Service, Springfield, VA.
USDHEW/NIOSH SEE U.S. Department of Health, Education, and Welfare; National Institute for Occupational Safety and Health
U.S. Environmental Protection Agency, Office of Water Regulations and Standards. (1980) Ambient water quality criteria for carbon tetrachloride. EPA-440/5-80-026. Washington, DC: The Office. 123p. Available from: National Technical Information Service, Springfield, VA.
USEPA/OWRS SEE U.S. Environmental Protection Agency, Office of Water Regulations and Standards
Verschueren, K. (1983) Handbook of environmental data on organic chemicals. 2nd ed. New York: Van Nostrand Reinhold Company, Inc. 1310p.
WHO/IARC SEE World Health Organization, International Agency for Research on Cancer
Wilson, J.T.; McNabb, J.F. (1983) Biological transformation of organic pollutants in groundwater. Trans. Am. Geophys. Union (Eos) 64:505-507. Reprinted by U.S. Environmental Protection Agency as EPA-600/J-83-074. Available from: National Technical Information Service, Springfield, VA.
World Health Organization, International Agency for Research on Cancer. (1979) IARC monographs on the evaluation of the carcinogenic risk of chemicals to humans. Vol. 20. Some halogenated hydrocarbons. Lyon, France: The Agency. 609p.
World Health Organization, International Agency for Research on Cancer. (1982) IARC monographs on the evaluation of the carcinogenic risk of chemicals to man. Supplement 4 to Vol. 1-29. Chemicals, industrial processes and industries associated with cancer in humans. Lyon, France: The Agency. 292p.

Chloroform

Information abstracted from TDB II, 1984 (No. 56) unless otherwise specified

I. Chemical Identification

A. CAS Registry No.: 67-66-3

B. Synonyms:

Trichloromethane

C. Molecular formula: $CHCl_3$

D. Structural formula:

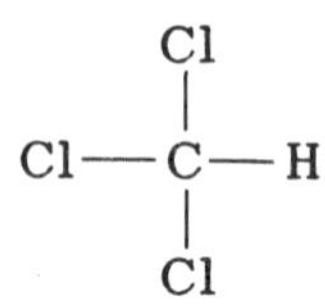

II. Physical and Chemical Properties

A. Molecular weight: 119.39

B. Physical state (color, form): Clear, colorless, mobile liquid

C. Odor: Sweet

D. Melting point: -63.5 deg C

E. Boiling point: 61.7 deg C

F. Solubilities:

Water: 8220 mg/L at 20 deg C (WHO/IARC, 1979); 1 ml/200 ml water at 25 deg C

Organic solvents: Soluble in ether, carbon disulfide, alcohol, benzene, and petroleum ether

G. Octanol/water partition coefficient (log P): 1.97 (Verschueren, 1983)

H. Density: 1.483 g/ml at 20 deg C

I. Dissociation constant: --

J. Viscosity: 0.58 cp at 20 deg C (CRC, 1983)

K. Surface tension (liquid interfacial tension with air): 27.1 mN/m (dynes/cm) at 20 deg C

L. Vapor pressure: 100 mm Hg at 10.4 deg C; 200 mm Hg at 25.9 deg C (WHO/IARC, 1979)

M. Vapor density: 4.12

N. Relative evaporation rate: Volatilizes from water with half-life of a few hours.

O. Spectral properties: Maximum absorption at less than 200 nm (vapor) (WHO/IARC, 1979)

III. Manufacturing/Use Information

Chloroform production in the U.S. in 1982 was 400 million lb. It is used as a chemical intermediate, especially in the synthesis of fluorocarbons for use in refrigerants, and as a general solvent for fats, oils, rubber, adhesives, and pesticides. It is registered for use in the U.S. as an insecticidal fumigant on stored grains.

IV. Environmental Fate/Exposure Potential

A. Potential for release from chemical disposal sites

Chloroform is very volatile; when spilled on land, it would be expected to evaporate rapidly due to its high vapor pressure. It is poorly adsorbed to soils having low organic carbon content, such as subsoils, clay, or aquatic sediment, and will leach into groundwater where it will reside for long periods of time. Chloroform will evaporate readily from water but the evaporation rate depends on mixing conditions. Modeling studies predicted a volatilization half-life of 36 hours in a river, 40 hours in a pond, and 9-10 days in a lake.

B. Biological uptake/accumulation

Chloroform has little or no potential for bioaccumulation because of its low log P. Monitoring data for marine organisms suggest only minimum chloroform uptake, with a reported bioconcentration factor of less than 1 for four species of fish.

C. Persistence

Chloroform is soluble in water. It has a negligible rate of hydrolysis and reportedly is not subject to photodegradation in aquatic systems. Chloroform is persistent in the atmosphere; its major reaction in the atmosphere is with photochemically produced hydroxyl radicals and it has a long half-life of 80 days. Chloroform is even less reactive in photochemical smog situations. Slow degradation can reportedly occur in groundwater in the presence of acclimated microbial populations. In laboratory studies simulating methanogenic groundwater conditions, chloroform was readily degraded after a lag period of acclimation (Bouwer and McCarty, 1983). There is no strong evidence for aerobic biodegradation.

D. Measured concentrations

Groundwater: Contaminated wells in New York and New Jersey showed chloroform concentrations from 67-490 ug/L. An EPA survey of finished drinking waters found a 70.3% occurrence in groundwater supplies.

Surface water: Sampling surveys from the Ohio and Delaware rivers measured chloroform concentrations ranging from 0.1-22 ug/L. Concentrations measured in Lakes Michigan, Erie, and Huron ranged from 1-30 ug/L. Concentrations in rivers near heavily industrialized areas ranged from 1-120 ug/L.

Drinking water: A survey of several U.S. cities reported chloroform concentrations ranging from 1-301 ug/L, with the values highest in summer and lowest in winter. In drinking-water wells in New York and New Jersey, concentrations were 67-490 ug/L. The EPA found chloroform in water supplies of 80 cities, with values ranging from less than 0.3-311 ug/L (Sittig, 1981).

Air: Average atmospheric concentrations in several U.S. areas were as follows: rural, 0.04 ppb; urban/suburban, 0.072 ppb. In eleven highly industrialized U.S. locations, the concentration range was 0-10.9 ppb, whereas in 10 U.S. cities randomly sampled, average concentrations ranged from 0.032-0.703 ppb.

Soil/Sediment: Chloroform was not detected in river sediment at an industrial location; none was detected in river sediment near Baltimore.

Food: Surveys have measured the presence of chloroform in several food groups, reporting values of 9-178 ppb in soft drinks; 56 ppb in butter; 15-17 ppb in cheese; and 34 ppb in mayonnaise.

E. Human environmental exposure

The major human exposure to any chloroform from chemical disposal sites would occur via ingestion of drinking water from chloroform-contaminated groundwater. (Drinking water disinfected by chlorination is a major source of ingested chloroform because chlorine is substituted onto the organic compounds present in the raw water.) Inhalation could also be a significant route of exposure as vapors reportedly may be transported long distances. Average daily intake was estimated in TDB II (1984) to be 0.0071 mg from air, and 0.064-0.136 mg from water. There are limited data on food intake. Measured occupational inhalation exposures ranged from 23-128 ppm in a plant manufacturing medicinals, and 7-170 ppm in a plant manufacturing film.

Body burden: In the vicinity of Love Canal in Niagara Falls, New York, median values measured in individuals were: breath, 26 ug/m^3; blood, 1.6 ng/ml; and urine, 860 ng/L.

V. Health Effects

A. Human health effects

Acute responses from atmospheric exposure to various concentrations of chloroform have been reported as: fainting sensation and vomiting from 4096 ppm; dizziness and salivation after a few minutes at 1475 ppm; dizziness, increased intracranial pressure, and nausea in 7 minutes; aftereffects of fatigue and headache for several hours from 1024 ppm. Chloroform exposure by inhalation has been fatal to humans, with rapid death attributable to respiratory failure and cardiac arrest, whereas delayed effects may include liver and kidney damage. Oral doses of 30 ml (44.6 g) and 100 ml (148.3 g) have been observed to produce severe nonfatal poisonings in humans, whereas ingestion of 200 ml (296.6 g) was fatal to adults (NRC, 1980). A splash of chloroform in the eye causes immediate burning pain, tearing, and reddening of the conjunctiva; the corneal epithelium is usually injured and may be irreversibly damaged.

Exposure to concentrations below anesthetic levels typically causes inebriation and excitation, changing to central nervous system depression. In workers chronically exposed to chloroform for 1-4 years, liver necrosis, hepatomegaly, and hepatitis have been detected. There is no conclusive evidence that chloroform is carcinogenic to humans.

Although chloroform crosses the placenta rapidly and enters fetal circulation, no data were available on developmental effects.

Absorption, Distribution, Metabolism, Excretion:

Chloroform can be absorbed from the gastrointestinal tract and to some extent through the skin, but the primary route is via the lungs. Chloroform is rapidly absorbed and distributed to all organs, with a relatively high concentration in nervous tissue. Chloroform is excreted rapidly, mainly by exhalation as an unchanged compound. A difference in the elimination rates between males and females was observed and explained by the affinity of chloroform for adipose tissue, which predominates in women (Kirsh-Volders, 1984). There is strong evidence that phosgene is formed as a metabolite and may be responsible for the liver and kidney toxicity of chloroform.

B. Non-human mammalian health effects

The effects of acute oral exposure in rats and mice are reportedly similar to those in humans; however, toxicity appears to be dependent upon species, sex, age, and strain. The oral LD50 for mice is 1120 mg/kg, and for rats, 800 mg/kg. In rats, low oral doses led to liver and kidney damage similar to that in humans.

Repeated oral chloroform administrations of 100 to 500 mg/kg to mice and rats has been shown to produce benign and malignant liver neoplasms in mice and malignant kidney tumors in male rats, whereas female rats were reportedly more susceptible to induction of hepatic tumors. An evaluation by IARC (WHO/IARC, 1979) indicated that "there is sufficient evidence that chloroform is carcinogenic in mice and rats."

Experiments have shown that when pregnant rats were exposed by inhalation, all doses from 30-300 ppm produced fetal malformations and retarded fetal development. Similar embryotoxic and developmental effects were reported in mice and rabbits given low oral doses during pregnancy. Although the evidence indicates that chloroform is fetotoxic and embryotoxic, it is apparently not highly teratogenic in the animal species that have been studied.

The pattern of absorption, distribution, and excretion in animals is comparable to that in humans.

C. Non-mammalian effects

The EPA (USEPA/OWRS, 1980) reviewed the aquatic toxicity data available for chloroform and reported a 48-hour LC50 value of 28,900 ug/L for the water flea (Daphnia magna), and 96-hour LC50 values of 43,800 ug/L for rainbow trout (Salmo gairdneri) and 115,000 ug/L for bluegill (Lepomis macrochirus). The 96-hour LC50 for the only marine organism tested, pink shrimp (Penaeus duorarum), was 81,500 ug/L. All data reported were from static tests with nominal (unmeasured) concentrations. In a 14-day exposure with radiolabeled chloroform, the bioconcentration factor for bluegill was 6 and the tissue half-life was less than 1 day, suggesting that residues of chloroform would not be an environmental hazard to consumers of aquatic organisms.

D. Short-term assays

There was no evidence that chloroform binds to DNA. Assays with Salmonella typhimurium and Escherichia coli K-12 showed chloroform to be nonmutagenic. It was inactive in a number of other short-term tests, including mutagenicity tests in Drosophila melanogaster, and those for chromosomal anomalies in human lymphocytes (WHO/IARC, 1982). One report on yeast indicated that chloroform induced mutagenesis and chromosomal aberrations in cells of Saccharomyces cerevisiae.

VI. Exposure Standards and Recommendations

A. Air

1. Immediately Dangerous to Life or Health (IDLH): 1000 ppm (Sittig, 1981)

2. OSHA standards:

Ceiling (10 min): 50 ppm (244 mg/m^3) (limit set for toxic effects other than cancer) (USDHHS/PHS, 1983)

3. NIOSH recommendations:

Ceiling (1 hr): 2 ppm (9.8 mg/m^3) (1976 value cited in USDHHS/CDC, 1983)

4. ACGIH (1980):

TLV-TWA (8 hr): 10 ppm (50 mg/m^3)
STEL (15 min): 50 ppm (225 mg/m^3)

Note: As of 1984, ACGIH (1984) has deleted the STEL value until further acute toxicological data and industrial hygiene experience become available.

B. Water

The following ambient water quality criteria are cited by EPA (USEPA/OWRS, 1980):

To protect human health from potential carcinogenic effects: Preferably zero. An increase in the human lifetime cancer risk by $1x10^{-5}$ due to ingestion of contaminated drinking water and fish and shellfish is estimated to occur at a chloroform concentration of 1.9 ug/L.

To protect freshwater aquatic life: 28.9 mg/L (acute)
1.2 mg/L (chronic)

To protect marine aquatic life: No value set due to insufficient data

Safe Drinking Water Act (SDWA):

EPA limits maximum contaminant level to 100 ug/L (USDHHS/PHS, 1983) 7-day SNARL = 3.2 mg/L; 24-hour SNARL = 22 mg/L (NRC, 1980)

C. Other

FDA banned the use of chloroform as an ingredient in human drug and cosmetic products as of July 29, 1976. EPA has banned the use of chloroform in foods (USDHHS/PHS, 1983).

References for Chloroform Synopsis

ACGIH SEE American Conference of Governmental Industrial Hygienists, Inc.

American Conference of Governmental Industrial Hygienists, Inc. (1980) Documentation of the threshold limit values. 4th ed. Cincinnati: The Conference. 486p.

American Conference of Governmental Industrial Hygienists, Inc. (1984) Supplemental documentation [for documentation of the threshold limit values]. Cincinnati: The Conference. 61p.

Bouwer, E.J.; McCarty, P.L. (1983) Transformations of 1- and 2-carbon halogenated aliphatic organic compounds under methanogenic conditions. Appl. Environ. Microbiol. 45:1286-1294.

CRC SEE CRC handbook of chemistry and physics

CRC handbook of chemistry and physics. (1983) 64th ed. Weast, R.C.; Astle, M.J.; Berger, W.H.; eds. Boca Raton, FL: CRC Press, Inc. 2302p.

Kirsch-Volders, M., editor. (1984) Mutagenicity, carcinogenicity, and teratogenicity of industrial pollutants. New York: Plenum Press. 336p.

National Research Council. (1980) Drinking water and health. Vol. 3. Washington, DC: National Academy Press. 415p.

NRC SEE National Research Council

Sittig, M. (1981) Handbook of toxic and hazardous chemicals. Park Ridge, NJ: Noyes Publications. 729p.

TDB II SEE Toxicology data bank II

Toxicology data bank II. (1984) [Enhanced version of data base]. Chloroform, TDB no. 56. (Last edited April 17, 1984). Bethesda, MD: U.S. Department of Health and Human Services, National Library of Medicine, Specialized Information Services. Magnetic-tape reels.

U.S. Department of Health and Human Services, Centers for Disease Control. (1983) NIOSH recommendations for occupational health standards. Morb. Mortal. Wkly. Rep. 32:(supplement)3S-22S.

U.S. Department of Health and Human Services, Public Health Service. (1983) Third annual report on carcinogens: summary. Research Triangle Park, NC: U.S. Department of Health and Human Services, National Toxicology Program Public Information Office. 229p.

USDHHS/CDC SEE U.S. Department of Health and Human Services, Centers for Disease Control

USDHHS/PHS SEE U.S. Department of Health and Human Services, Public Health Service

U.S. Environmental Protection Agency, Office of Water Regulations and Standards. (1980) Ambient water quality criteria for chloroform. EPA-440/5-80-033. Washington, DC: The Office. 61p. Available from: National Technical Information Service, Springfield, VA.

USEPA/OWRS SEE U.S. Environmental Protection Agency, Office of Water Regulations and Standards

Verschueren, K. (1983) Handbook of environmental data on organic chemicals. 2nd ed. New York: Van Nostrand Reinhold Company, Inc. 1310p.

WHO/IARC SEE World Health Organization, International Agency for Research on Cancer World Health Organization, International Agency for Research on Cancer. (1979) IARC monographs on the evaluation of the carcinogenic risk of chemicals to humans. Vol. 20. Some halogenated

hydrocarbons. Lyon, France: The Agency. 609p.
World Health Organization, International Agency for Research on Cancer. (1982) IARC monographs on the evaluation of the carcinogenic risk of chemicals to man. Supplement 4 to Vol. 1-29. Chemicals, industrial processes and industries associated with cancer in humans. Lyon, France: The Agency. 292p.

Hexachlorocyclopentadiene

Information abstracted from TDB II, 1984 (No. 4011) unless otherwise specified

I. Chemical Identification

A. CAS Registry No.: 77-47-4

B. Synonyms:

C-56
HEX
Perchlorocyclopentadiene

C. Molecular formula: C_5Cl_6

D. Structural formula:

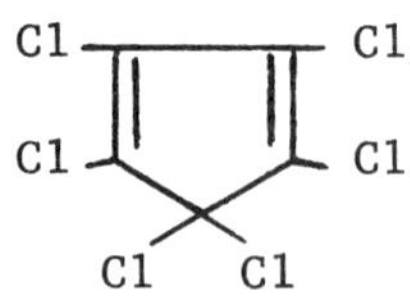

II. Physical and Chemical Properties

A. Molecular weight: 272.7

B. Physical state (color, form): Yellow-green liquid

C. Odor: Pungent

D. Melting point: -9 deg C

E. Boiling point: 239 deg C

F. Solubilities:

Water: 0.8 mg/L at 25 deg C

Organic solvents: Miscible with hexane (Bell et al., 1979)

G. Octanol/water partition coefficient (log P): 5.04 ± 0.04 (Wolfe et al., 1982)

H. Density: 1.702 g/ml at 25 deg C

I. Dissociation constant: --

J. Viscosity: --

K. Surface tension: --

L. Vapor pressure: 0.080 mm Hg at 25 deg C; 1 mm Hg at 60 deg C

M. Vapor density: 9.4 (air = 1)

N. Relative evaporation rate: --

O. Spectral properties: Maximum absorption at 323 nm (heptane)

III. Manufacturing/Use Information

Probable production of hexachlorocyclopentadiene (HEX) in the United States exceeded 908 kg (1,998 lb) in 1975 and 4,540 kg (9,988 lb) in 1977. As there are only two producers, actual data on production are confidential. The principal use of HEX is as a chemical intermediate in the manufacture of dyes, pharmaceuticals, resins, fire retardants, and insecticides, such as chlordane, dieldrin, and aldrin.

IV. Environmental Fate/Exposure Potential

A. Potential for release from chemical disposal sites

The high octanol/water partition coefficient of HEX suggests that it would be tightly bound to soil organic carbon and would not migrate in the soil compartment (Kenaga, 1980), and that it would be immobile with respect to leaching (Wolfe et al., 1982). Laboratory studies show that evaporation rates of HEX from sand, loam, and humus were 0.83, 0.33, and 0.14% for the first hour, and 0.23, 0.11, and 0.05% for the second hour, respectively; similar studies on HEX volatilization from water yielded rates of 4.7-8.8% for the first hour and 0.6-1.1% for the second hour (Kilzer et al., 1979).

B. Biological uptake/Bioaccumulation

The bioaccumulation/bioconcentration/biomagnification potential of HEX appears to be substantial based on its high log P (Wolfe et al., 1982). However, bioaccumulation factors from a short-term model ecosystem study indicate only a moderate accumulation potential for algae, snails, mosquito larvae, and mosquitofish, and no substantial biomagnification from algae to snails or from mosquito larvae to mosquitofish (Lu et al., 1975). In a 30-day continuous-flow bioassay with constant levels of HEX, steady-state bioconcentration factors of 29 (Veith et al., 1979) and less than 11 (Spehar et al., 1979) were reported for the fathead minnow (*Pimephales promelas*).

C. Persistence

HEX is predicted to be relatively immobile in soil, based on its log P and the rating scheme of Briggs (1973). Volatilization is likely to occur primarily at the soil surface and is inversely related to both the organic-matter levels and water-holding capacity of the soil (Kilzer et al., 1979). Leaching of HEX into groundwater should be very limited, and chemical hydrolysis and microbial metabolism are expected to reduce environmental levels (Rieck, 1977a [cited by USEPA/OHEA, 1984]; Rieck, 1977b [cited by USEPA/OHEA, 1984]; Thuma et al., 1978).

In shallow water a few centimeters deep, HEX has a photolytic half-life of approximately 0.2 hour (Butz et al., 1982; Wolfe et al., 1982); in deeper water where photolysis would not occur because of limited light penetration, hydrolysis and biodegradation should become the key degradative processes. The hydrolytic half-life of HEX is several days and is not strongly affected by pH in the range of 5-9, by salinity, or by suspended solids (Wolfe et al., 1982; Yu and Atallah, 1977 [cited by USEPA/OHEA, 1984]). HEX is known to volatilize from water (Kilzer et al., 1979; Weber, 1979 [cited by USEPA/OHEA, 1984]), but volatilization may be limited by diffusion, particularly in waters that are not well mixed, and by sorption onto sediments.

In the atmosphere, HEX does not seem to persist; its tropospheric residence time has been estimated to be approximately 5 hours, with photolysis and reaction with hydroxyl radicals and ozone being the key degradative processes (Cupitt, 1980).

D. Measured concentrations

Drinking water: In samples of raw and finished drinking waters from an Ottawa water treatment plant, HEX was not detected in the raw water, but the finished drinking water showed levels of 0.057-0.10 ug/L, suggesting that HEX may have been introduced during the treatment process (Benoit and Williams, 1981).

E. Human environmental exposure

The limited uses of HEX are not expected to result in significant residue levels in humans or in the environment, except among factory and other workers, such as those employed in sewage treatment plants who may be directly exposed.

V. Health Effects

A. Human health effects

HEX is highly toxic whether ingested, inhaled, or absorbed by the skin, and is capable of causing death. At 0.33 ppm, there is a pronounced and unmistakable pungent odor. The highly irritating nature of the compound and its lacrimatory activity make prolonged exposure intolerable (ACGIH, 1980).

In March 1977, a large volume of industrial HEX was illegally dumped into the municipal sewage system of Louisville, Kentucky, and 145 workers in the treatment plant were acutely exposed to the compound for a period of 2 days to 2 weeks. An evaluation revealed that 85 (59%) workers had experienced eye irritation, 65 (45%) had headaches, and 39 (27%) had throat irritation. Symptoms occurred throughout the plant, but the highest incidence was in primary sewage treatment areas. Medical examination of 41 employees 3 days after the plant was closed showed proteinuria and elevation of serum lactate dehydrogenase levels, but these findings were no longer present 3 weeks later.

Information on chronic effects, teratogenicity, and carcinogenicity resulting from exposure to HEX have not been located in the literature.

Absorption, Distribution, Metabolism, Excretion:

No information available for humans.

B. Non-human mammalian health effects

Acute exposure to HEX vapors caused lacrimation, salivation, and gasping respiration; tremors occurred at high concentrations. Diffuse degenerative changes were observed in the brain, heart, liver, adrenal glands, and kidneys. Severe pulmonary edema and hyperemia, and acute necrotizing bronchitis and bronchiolitis demonstrated the severity of irritation; incidence and severity were dependent on dose. Even at the lowest experimental concentration of 0.15 ppm, there were some degenerative changes in liver and kidneys of all species, and microscopic evidence of pulmonary irritation in mice.

The oral LD50 was reported to be 430 mg/kg for mice (Patty's, 1981), and 113 mg/kg for rats. A single oral dose of 420-620 mg/kg of the 93.3% compound was lethal to rats and rabbits. A dermal application of 430-610 mg/kg was lethal to rabbits.

Laboratory animals were chronically exposed 150 times to HEX vapors (89.5% purity) for 7 hours/day over a period of 216 days; guinea pigs, rabbits, and rats survived at 0.15 ppm (1.7 mg/m^3), but 4 of 5 mice died during the prolonged intermittent exposure. At a higher vapor concentration, that is, 0.34 ppm, for 7 hours/day, 5 days/week, none of the mice or rats survived 20 exposures; 4 of 6 rabbits died by the 25th exposure, but the guinea pigs survived the entire 30 exposures.

No teratogenic effects were observed in pregnant mice and rabbits fed by gavage at doses up to 75 mg/kg/day during active organogenesis. However, toxic effects on reproduction and development were noted in rabbits given 975 mg/kg orally between the 6th and 18th day of pregnancy.

Absorption, Distribution, Metabolism, Excretion:

In rats given oral and inhaled doses of HEX, excretion was primarily in feces and urine (Dorough, 1980 [cited by USEPA/OHEA, 1984]; Yu and Atallah, 1981 [cited by USEPA/OHEA, 1984]); very little of the administered HEX was lost in expired air. Metabolic pathways of HEX are not known; no studies to date have identified metabolites (USEPA/OHEA, 1984).

C. Non-mammalian effects

For the fathead minnow (Pimephales promelas), the reported 96-hour and 30-day LC50 values were 0.007 mg/L and 0.0067 mg/L, respectively, for both larvae and early juvenile stages. The 96-hour LC50 for the mysid shrimp (Mysidopsis bahia) was 0.007 (0.006-0.008) mg/L (USEPA/OWRS, 1980). Other LC50 data are available for freshwater and marine fishes and invertebrates, but none of these studies reported measured concentrations, only nominal levels.

D. Short-term assays

HEX was not mutagenic in either Salmonella typhimurium or Escherichia coli when incubated with mouse liver microsomes.

VI. Exposure Standards and Recommendations

A. Air

1. Immediately Dangerous to Life or Health (IDLH): --

2. OSHA standards: --

3. NIOSH recommendations: --

4. ACGIH (1980):

TLV-TWA (8 hr): 0.01 ppm (0.11 mg/m^3)
STEL (15 min): 0.03 ppm (0.33 mg/m^3)

Note: As of 1984, ACGIH (1984) has deleted the STEL value until further acute toxicological data and industrial hygiene experience become available.

B. Water

The following ambient water quality criteria are cited by EPA (USEPA/OWRS, 1980):

To protect human health: 206 ug/L

To protect freshwater aquatic life: 0.007 mg/L (acute toxicity)
0.0052 mg/L (chronic toxicity)

To protect marine aquatic life: 0.007 mg/L (acute toxicity) No available data on chronic toxicity

References for Hexachlorocyclopentadiene Synopsis

ACGIH SEE American Conference of Governmental Industrial Hygienists, Inc.

American Conference of Governmental Industrial Hygienists, Inc. (1980) Documentation of the threshold limit values. 4th ed. Cincinnati: The Conference. 486p.

American Conference of Governmental Industrial Hygienists, Inc. (1984) Supplemental documentation [for documentation of the threshold limit values]. Cincinnati: The Conference. 61p.

Bell, M.A.; Ewing, R.A.; Lutz, G.A. (1979) Reviews of the environmental effects of pollutants. XII. Hexachlorocyclopentadiene. Prepared by Battelle Columbus Laboratories, Columbus OH. EPA-600/1-78-047. Cincinnati, OH: U.S. Environmental Protection Agency, Health Effects Research Laboratory. 81p. Available from: Environmental Criteria and Assessment Office, U.S. Environmental Protection Agency, Cincinnati, OH.

Benoit, F.M.; Williams, D.T. (1981) Determination of hexachlorocyclopentadiene at the nanogram per liter level in drinking water. Bull. Environ. Contam. Toxicol. 27:303-308.

Briggs, G.G. (1973) Simple relationship between soil adsorption of organic chemicals and their octanol/water partition coefficients. In: British Crop Protection Council. Proceedings of the Seventh British Insecticide and Fungicide Conference, Nov. 19-22, Brighton, England. Vol. 1. London: The Council. p. 83-86.

Butz, R.G.; Yu, C.C.; Atallah, Y.H. (1982) Photolysis of hexachlorocyclopentadiene in water. Ecotoxicol. Environ. Saf. 6:347-357.

Cupitt, L.T. (1980) Fate of toxic and hazardous materials in the air environment. Prepared by U.S. Environmental Protection Agency, Environmental Sciences Research Laboratory, Research Triangle Park, NC. EPA-600/3-80-084. 28p. Available from: National Technical Information Service, Springfield, VA.

Dorough, H.W. (1980) Disposition of 14C-hexachlorocyclopentadiene (C56) in rats following inhalation exposure. Prepared for Velsicol Chemical Corporation, Chicago, IL. 27p. [Information cited is from USEPA/OHEA (1984)].

Kenaga, E.E. (1980) Predicted bioconcentration factors and soil sorption coefficients of pesticides and other chemicals. Ecotoxicol. Environ. Saf. 4:26-38.

Kilzer, L.; Scheunert, I.; Geyer, H.; Klein, W.; Korte, F. (1979) Laboratory screening of the volatilization rates of organic chemicals from water and soil. Chemosphere 8:751-761.

Lu, P.Y.; Metcalf, R.L.; Williams, J.W. (1975) Evaluation of environmental distribution and fate of hexachlorocyclopentadiene, chlordene, heptachlor, and heptachlor epoxide in a laboratory model ecosystem. J. Agric. Food Chem. 23:967-973.

Patty's SEE Patty's industrial hygiene and toxicology
Patty's industrial hygiene and toxicology. (1981) 3rd rev. ed. Vol. 2. Clayton, G.D.; Clayton, F.E.; eds. New York: John Wiley & Sons. 5112p.
Rieck, C.E. (1977a) Soil metabolism of 14C-hexachlorocyclopentadiene. Prepared by the University of Kentucky for Velsicol Chemical Corporation, Chicago, IL. [Information cited is from USEPA/OHEA (1984)].
Rieck, C.E. (1977b) Volatile products of 14C-hexachlorocyclopentadiene. Prepared by the University of Kentucky for Velsicol Chemical Corporation, Chicago, IL. [Information cited is from USEPA/OHEA (1984)].
Spehar, R.L.; Veith, G.D.; DeFoe, D.L.; Bergstedt, B.V. (1979) Toxicity and bioaccumulation of hexachlorocyclopentadiene, hexachloronorbornadiene and heptachloronorbornene in larval and early juvenile fathead minnows, Pimephales promelas. Bull. Environ. Contam. Toxicol. 21:576-583.
TDB II SEE Toxicology data bank II
Thuma, N.K.; O'Neill, P.E.; Brownlee, S.G.; Valentine, R.S. (1978) Biodegradation of spilled hazardous materials. In: Information Transfer, Inc. Control of hazardous materials spills. Rockville, MD: Information Transfer Inc. p. 217-220.
Toxicology data bank II. (1984) [Enhanced version of data base]. Hexachlorocyclopentadiene, TDB no. 4011. (Last edited June 5, 1984). Bethesda, MD: U.S. Department of Health and Human Services, National Library of Medicine, Specialized Information Services. Magnetic-tape reels.
U.S. Environmental Protection Agency, Office of Health and Environmental Assessment. (1984) Health assessment document for hexachlorocyclopentadiene. EPA-600/8-84-001F. Cincinnati, OH: The Office. 159p. Available from: Center for Environmental Research Information, U.S. Environmental Protection Agency, Cincinnati, OH.
U.S. Environmental Protection Agency, Office of Water Regulations and Standards. (1980) Ambient water quality criteria for hexachlorocyclopentadiene. EPA-440/5-80-055. Washington, DC: The Office. 92p. Available from: National Technical Information Service, Springfield, VA.
USEPA/OHEA SEE U.S. Environmental Protection Agency, Office of Health and Environmental Assessment
USEPA/OWRS SEE U.S. Environmental Protection Agency, Office of Water Regulations and Standards
Veith, G.D.; DeFoe, D.L.; Bergstedt, B.V. (1979) Measuring and estimating the bioconcentration factor of chemicals in fish. J. Fish. Res. Board Can. 36:1040-1048.
Weber, J.B. (1979) Adsorption of HEX by Cape Fear loam soil. Prepared by North Carolina State University for Velsicol Chemical Corporation, Chicago, IL. [Information cited is from USEPA/OHEA (1984)].
Wolfe, N.L.; Zepp, R.G.; Schlotzhauer, P.; Sink, M. (1982) Transformation pathways of hexachlorocyclopentadiene in the aquatic environment. Chemosphere 2:91-101.
Yu, C.C.; Atallah, Y.H. (1977) HEX hydrolysis at various pH and temperatures. Laboratory report. Project no. 482428, report no. 8. Prepared by Velsicol Chemical Corporation, Chicago, IL. [Information cited is from USEPA/OHEA (1984)].

Methyl Ethyl Ketone

Information abstracted from TDB II, 1984 (No. 99) unless otherwise specified

I. Chemical Identification

A. CAS Registry No.: 78-93-3

B. Synonyms:

2-Butanone
MEK
Methyl acetone

C. Molecular formula: C_4H_8O

D. Structural formula:

$$CH_3 - \overset{\overset{\displaystyle O}{\|}}{C} - CH_2 - CH_3$$

II. Physical and Chemical Properties

A. Molecular weight: 72.10

B. Physical state (color, form): Colorless liquid

C. Odor: Acetone-like (moderately sharp, fragrant, mint-like)

D. Melting point: -86.3 deg C

E. Boiling point: 79.6 deg C

F. Solubilities:

Water: 353,000 mg/L at 10 deg C; 190,000 mg/L at 90 deg C (Verschueren, 1983)

Organic solvents: Soluble in all proportions in acetone; miscible with alcohol, ether, benzene

G. Octanol/water partition coefficient (log P): 0.26 (Verschueren, 1983)

H. Density: 0.805 g/ml at 20 deg C

I. Dissociation constant: --

J. Viscosity: --

K. Surface tension: --

L. Vapor pressure: 70.6 mm Hg at 20 deg C

M. Vapor density: 2.41 (air = 1)

N. Relative evaporation rate: 2.7 (ether = 1); half-life in water is 3-12 days

O. Spectral properties: Maximum absorption at 273.5 nm (alcohol)

III. Manufacturing/Use Information

United States production of methyl ethyl ketone (MEK) in 1982 was 390 million kg (858 million lb). Its major uses are as a solvent for various coatings; in the manufacture of colorless synthetic resins, smokeless powder, artificial leather, synthetic rubber, lubricating oils, and ink; in the surface coating, and lacquer and varnish industries; and in pharmaceuticals and cosmetics, hardwood pulping, paint removers, cements and adhesives, organic synthesis, cleaning fluids, and fabric coating.

IV. Environmental Fate/Exposure Potential

A. Potential for release from chemical disposal sites

Land spills of MEK will partially leach into the ground and partially evaporate into the atmosphere. Its high vapor pressure indicates that volatilization from soil will be rapid, and its low partition coefficient indicates that soil absorption will be low. At surface impoundment disposal sites, MEK would be expected to evaporate into the atmosphere, based on estimated half-lives of 3 and 12 days in rivers and lakes, respectively.

B. Biological uptake/accumulation

Bioconcentration will not be significant because MEK has a very low octanol/water partition coefficient. It biodegrades slowly (removal times of 5-10 days) in surface water. Biodegradability studies in anaerobic systems suggest that, in groundwater, MEK may degrade slowly after a long acclimation period (about 1 week).

C. Persistence

The fate of MEK in soil is unknown, but it can be presumed that it will volatilize rapidly. It is highly degradable in aerobic systems. Degradation is completed in about 5-10 days, somewhat longer in salt water. In a river die-away test, the removal rate of MEK was 0.6 ppm/hour. Degradation in anaerobic systems requires acclimation of about 1 week.

In water surface layers, photolysis is possible because ultraviolet radiation reaching these layers can be absorbed by MEK; actual photolysis rates have not been determined. Hydrolysis and photooxidation in water will not be significant processes.

The dominant atmospheric degradation process of MEK is its reaction with photochemically produced hydroxyl radicals; the photodegradation half-life of MEK is 2.3 days. Acetaldehyde is the primary product of this reaction. Under simulated smog conditions, reactivity is somewhat faster with degradation rates ranging from 1.5% in 1 hour to 33% in 6.5 hours. Although the atmosphere "filters out" light of wavelengths shorter than 290 nm, some direct photodegradation of MEK may be possible because its absorption peak is at 273.5 nm.

D. Measured concentrations

Surface water: MEK was detected at 23 ug/L in 1 of the 204 sites sampled in 14 heavily industrialized U.S. river basins. MEK was detected but not quantified in the Black Warrior River in Tuscaloosa, Alabama. Concentrations detected in the Strait of Florida and in the Eastern Mediterranean Sea ranged from 7-16 ug/L.

Drinking water: In a Federal survey of groundwater supplies, MEK was detected at less than 5% of the sampling locations. It was detected but not quantified in drinking water supplies having varied sources in 7 cities with different types of pollutant sources.

Air: Air samples were negative for MEK or below the 0.1 ppb detection limit in various rural, urban, and suburban areas. However, during moderate to severe photochemical pollution episodes from September to October 1980 in the Los Angeles area, the MEK concentrations were 0-14 ppb. In 33 samples from industrialized areas of New Jersey, the range was 0.01-1.9 ppb with a median of 0.064 ppb. In the vicinity of a chemical reclamation plant, the MEK concentration was 94,000 ppb. In clouds over California, the concentration was 0-470 ppb, and in fog or ice fog, 0-trace amounts. In gasoline engine exhaust it was less than 100-1000 ppb, and in cigarette smoke, 50,000 ppb.

Effluent/Sediment: Though not detected in the sediment of a brackish industrial U.S. river, the MEK concentration of wastewater going into the river was 8,000-20,000 ug/L. It was detected in trench leachate from two low-level radiation disposal sites.

Food: MEK has been reported in several foods as follows: Swiss cheese, 0.3 ppm; cream, 0.154-0.177 ppm; milk, 0.077-0.79 ppm; milk fat, 8 ppm; southern peas, 0.074-0.390 ppm. It is a natural component of some foods, including roasted barley, bread, honey, chicken, oranges, black tea, and rum.

E. Human environmental exposure

The principal routes of human exposure to MEK are inhalation and skin contact in occupational settings, especially where it is used as a solvent for coatings. Exposure also occurs during photochemical smog episodes in which the average daily air intake is 0-0.823 mg, assuming 0-14 ppb MEK in the atmosphere. MEK is a natural component of some foods, so ingestion of those food items is a source of exposure.

Body burden: MEK has been detected in 5 of 8 samples of mothers' milk from 4 U.S. urban areas.

V. Health Effects

A. Human health effects

MEK has a low toxicity following both acute and chronic exposures. High atmospheric concentrations of 350 ppm are acutely irritating to the eyes, nose, and throat. If splashed in the eyes, it may produce painful irritation and corneal injury (Patty's, 1981).

Prolonged exposure may produce central nervous system depression and narcosis. Prolonged dermal contact may defat the skin and produce dermatitis. Workers chronically exposed to 300-600 ppm via inhalation and dermal contact complained of numbness of the fingers and arms. Workers exposed to 300-500 ppm experienced vomiting and nausea, and at lower concentrations, headaches and throat irritation. At a coated fabrics plant, MEK was implicated in toxic neuropathies diagnosed in workers, but direct causal evidence was lacking (Patty's, 1981). There is no information available on the carcinogenicity, mutagenicity, or teratogenicity potential of MEK in humans.

Absorption, Distribution, Metabolism, Excretion:

In humans, about 75% of inhaled MEK is absorbed by the pulmonary route (Patty's, 1981). Only 30-40% was eliminated unchanged in expired air; its metabolites are eliminated in urine as conjugates of glucuronic acid. MEK absorbed through the skin may be partially excreted via expired air. MEK may be found in normal human urine, possibly because of its dietary origin. MEK has a specific effect on proteins and can modify enzyme functions; thus, clonal lines of human neuroblastoma can be used to assay toxic effects of MEK because of their extreme sensitivity.

B. Non-human mammalian health effects

The acute oral LD50 for rats has been reported to be 3,400 mg/kg. In inhalation studies, rats exposed to 2,000 ppm MEK for 2 hours were apparently unaffected; 2-hour exposure to 4,000 ppm caused the death of 4 of the 6 animals. Guinea pigs that inhaled a 1% concentration (10,000 ppm) for a few minutes were not affected after 1 hour except for irritation of the eyes and nose; after 4-5 hours, CNS depression occurred.

Chronic exposure to 1,500 ppm of MEK for 7-9 weeks caused elevated plasma cholinesterase levels in mice, rats, and chickens. Sprague-Dawley rats inhaling 800 ppm for 4 weeks showed increased liver weights and liver-to-body weight ratios. Baboons exposed to MEK for 24 hours/day for 7 days showed slowed response times, but accuracy was minimally affected.

Pregnant female rats inhaling 3,000 ppm for 7 hours/day from the 6th to 15th days of gestation evidenced maternal toxicity by decreased weight gain and increased water consumption; fetotoxicity was evidenced by increased incidence of two minor skeletal variants. No information was available on carcinogenicity of MEK.

Absorption, Distribution, Metabolism, Excretion:

After rats were given a single oral dose of MEK, metabolites in the blood included 2-butanol, 3-hydroxy-2-butanol, and 2,3-butanediol, as well as the parent compound. Based on an experiment with intraperitoneally injected guinea pigs, the serum half-life of MEK was estimated to be 270 minutes.

C. Non-mammalian effects

Aquatic toxicity is reported by RTECS as follows:
96-hour TLm: greater than 1,000 ppm.

D. Short-term assays

Mutagenicity screenings of MEK in Escherichia coli WP2 and Salmonella typhimurium strains TA1535, TA1536, TA1537, and TA1538 were negative.

VI. Exposure Standards and Recommendations

A. Air

1. Immediately Dangerous to Life or Health (IDLH): 3,000 ppm (Sittig, 1981)

2. OSHA standards:

TLV-TWA (8 hr): 200 ppm (590 mg/m^3) (1978 value cited in USDHHS/CDC, 1983)

3. NIOSH recommendations:

TLV-TWA (10 hr): 200 ppm (590 mg/m^3) (value cited in USDHEW/NIOSH, 1978)

4. ACGIH (1980):

TLV-TWA (10 hr): 200 ppm (590 mg/m^3)
STEL (15 min): 300 ppm (885 mg/m^3)

B. Water

1-day SNARL = 7.5 mg/L; 10-day SNARL = 0.75 mg/L

References for Methyl Ethyl Ketone Synopsis

ACGIH SEE American Conference of Governmental Industrial Hygienists, Inc.

American Conference of Governmental Industrial Hygienists, Inc. (1980) Documentation of the threshold limit values. 4th ed. Cincinnati: The Conference. 486p.

Patty's SEE Patty's industrial hygiene and toxicology

Patty's industrial hygiene and toxicology. (1981) 3rd rev. ed. Vol. 2. Clayton, G.D.; Clayton, F.E.; eds. New York: John Wiley & Sons. 5112p.

Sittig, M. (1981) Handbook of toxic and hazardous chemicals. Park Ridge, NJ: Noyes Publications. 729p.

TDB II SEE Toxicology data bank II

Toxicology data bank II. (1984) [Enhanced version of data base]. Methyl ethyl ketone, TDB no. 99. (Last edited June 13, 1984). Bethesda, MD: U.S. Department of Health and Human Services, National Library of Medicine, Specialized Information Services. Magnetic-tape reels.

U.S. Department of Health and Human Services, Centers for Disease Control. (1983) NIOSH recommendations for occupational health standards. Morb. Mortal. Wkly. Rep. 32:(supplement)3S-22S.

U.S. Department of Health, Education, and Welfare; National Institute for Occupational Safety and Health. (1978) Criteria for a recommended standard occupational exposure to ketones. Cincinnati, OH: The Institute. 244p. Available from National Technical Information Service, Springfield, VA.

USDHEW/NIOSH SEE U.S. Department of Health, Education, and Welfare; National Institute for Occupational Safety and Health

USDHHS/CDC SEE U.S. Department of Health and Human Services, Centers for Disease Control

Verschueren, K. (1983) Handbook of environmental data on organic chemicals. 2nd ed. New York: Van Nostrand Reinhold Company, Inc. 1310p.

Naphthalene

Information abstracted from TDB II, 1984 (No. 184) unless otherwise specified

I. Chemical Identification

A. CAS Registry No.: 91-20-3

B. Synonyms:

Naphthalin
Naphthene
Tar camphor
White tar

C. Molecular formula: $C_{10}H_8$

D. Structural formula:

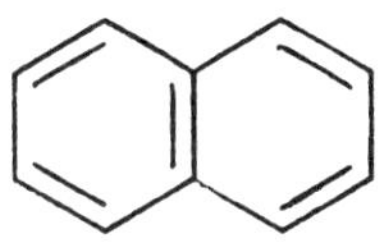

II. Physical and Chemical Properties

A. Molecular weight: 128.16

B. Physical state (color, form): White, crystalline flakes or solid

C. Odor: Strong coal tar odor

D. Melting point: 80.2 deg C

E. Boiling point: 217.9 deg C

F. Solubilities:

Water: 20 $\pm$ 2 mg/L at 22 deg C in seawater; 31-34 mg/L at 25 deg C in distilled water (Verschueren, 1983)

Organic solvents: Very soluble in methanol, ethanol, benzene, toluene, olive oil, turpentine, chloroform, carbon tetrachloride, carbon disulfide, ether, hydronaphthalenes, fixed and volatile oils, 1,2-dichloromethane; soluble in acetic acid, ethylene dichloride.

G. Octanol/water partition coefficient (log P): 3.01-3.59

H. Density: 1.162 g/ml at 20 deg C (Merck index, 1983)

I. Dissociation constant: --

J. Viscosity: --

K. Surface tension: --

L. Vapor pressure: 1 mm Hg at 53 deg C (Verschueren, 1983)

M. Vapor density: 4.42

N. Relative evaporation rate: Based on evaporation rate of 0.096 m/hour, calculated half-life in water at 25 deg C and 1 m depth is 7.15 hours (Verschueren, 1983)

O. Spectral properties: Maximum absorption 221 nm, 275.5 nm, 286 nm, and 311 nm (in alcohol)

III. Manufacturing/Use Information

United States production of naphthalene in 1982 was 317 million kg (699 million lb). It is used extensively as a raw material and intermediate in the chemical, plastics, and dye industries, as a moth repellent in the form of balls or flakes, as an air freshener, and as a surface-active agent (Patty's, 1981). It is also used in the manufacture of phthalic and anthranilic acids, naphthols, naphthylamines, sulfonic acid, hydronaphthalenes, synthetic resins, lampblack, and smokeless powder. Additionally, it is used to manufacture insecticides, fungicides, lacquers, and varnishes, and to preserve wood and other materials; in medicine, it has been applied as an antiseptic, anthelmintic, and dusting powder in skin diseases.

IV. Environmental Fate/Exposure Potential

A. Potential for release from chemical disposal sites

Naphthalene is soluble in water at only 20 mg/L, it is volatile, and it sublimes appreciably at temperatures above 80 deg C (Merck index, 1983). Therefore, it is assumed that spills or releases onto soil would evaporate. Its water solubility value suggests a potential for leaching from chemical disposal sites into groundwater. At 30 ug/L in estuarine waters, 0.7% naphthalene is adsorbed onto particles after 3 hours (Verschueren, 1983).

B. Biological uptake/accumulation

Naphthalene has a moderate potential for bioaccumulation based on its limited biodegradation and moderately high log P values. The steady-state bioconcentration factor for a marine copepod (*Eurytemora affinis*), was reported to be 350.

C. Persistence

Naphthalene is reported to be rapidly decomposed by a soil organism. In a laboratory study simulating anoxic denitrification conditions, there was no observed transformation of naphthalene (Bouwer and McCarty, 1983). Its biological oxygen demand (BOD) for 6 days was reported to be 59.5% of the calculated theoretical oxygen demand (ThOD), but two other sources reported that the BOD was 0% of the ThOD, suggesting that naphthalene was not biodegradable under the test conditions.

D. Measured concentrations

Water: Concentrations of naphthalene in groundwater and treated surface water were 0.001-0.025 ug/L; in ambient water, up to 2.0 ug/L; in sewage plant effluents, up to 2.2 ug/L; and in drinking-water supplies, up to 1.4 ug/L.

Air: Atmospheric concentration data reported for naphthalene were too variable to draw any general conclusions. Ambient air concentrations of 0.00006 ug/m^3 and 0.00035 ug/m^3 were reported in a small town and urban area, respectively. Levels at a coke oven ranged from 11.35 to 1,120 ug/m^3. The reported urban air concentrations during winter in seven selected U.S. cities ranged from 27,600 to 146,000 ug/m^3. These data from seven cities are suspect because they bracket the ACGIH short-term exposure level (STEL) of 75,000 ug/m^3 (ACGIH, 1980).

E. Human environmental exposure

Human exposure to naphthalene from chemical disposal sites could occur by breathing vapors and by deposition of particulate matter on the skin. Naphthalene may be ingested by drinking water from contaminated groundwater wells or surface waters.

V. Health Effects

A. Human health effects

Acute exposures to naphthalene vapors may cause eye and respiratory tract irritation, headache, nausea, and profuse sweating, depending on the duration of the exposure concentration; optic neuritis has also been observed (Patty's, 1981). Other symptoms of toxicity are confusion and excitement, as well as dysuria and hematuria. Skin irritation may result from dermal contact; cataracts and ocular irritation have been reported. In infants, skin contact has caused rashes and systemic poisoning (Patty's, 1981). Acute oral toxicity is enhanced when naphthalene is ingested in combination with fats, which facilitates its absorption and resulting systemic events (Patty's, 1981). Symptoms of oral toxicity are abdominal cramps with nausea, vomiting, and diarrhea; listlessness; coma with or without convulsions; irritation of the urinary bladder, and brown or black urine with or without albumin and casts; hemolysis, particularly in persons with red cell glucose-6-phosphate dehydrogenase deficiency; anemia, leukocytosis, fever, hemoglobinuria, jaundice, renal insufficiency, and liver function disturbances; and death in young infants from acute renal failure or kernicterus (i.e., acute neural symptoms associated with high levels of bilirubin in the blood). The probable oral lethal dose is estimated to be 5-15 g of naphthalene for an adult, 2-3 g for a child (Patty's, 1981).

Chronic exposure to naphthalene vapor or dust may result in corneal ulceration, cataracts, and lens opacity, and when inhaled may produce malaise, headache, and vomiting. Repeated contact may cause erythema and dermatitis, especially in hypersensitive individuals. The chronic effects of repeated ingestion are tremors, restlessness, extreme apprehension, and hallucinations, but the effects subside in a few days (Patty's, 1981).

Hemolytic anemia has been reported in a newborn whose mother ingested naphthalene during pregnancy. There is no evidence that the substance is mutagenic or carcinogenic.

Absorption, Distribution, Metabolism, Excretion:

Naphthalene is rapidly absorbed by the human system when inhaled, but more slowly through the intact skin and by ingestion. Cutaneous absorption in infants is increased when baby oil is used. Apparently, conversion to the hemolytic agents, α- and β-naphthol and α- and β-naphthoquinone, is rapid in adults and very slow in newborns. The naphthols are partially excreted as glucuronides. Naphthalene per se is nonhemolytic; the activity of the four metabolites decreases in the order listed above. The most active metabolite is probably naphthalene-1,2-oxide, which may rapidly convert to hydroxy- or oxy-derivatives, unless the system is glucose-6-phosphate dehydrogenase or glutathione deficient (Patty's, 1981).

B. Non-human mammalian health effects

The oral LD50 value for rats is 1,780 mg/kg; acute oral administration produced only mild eye effects. Rabbits fed naphthalene exhibited browning of the lenses and eye tumors, inhibition of the ciliary body and of ascorbic acid transport across the blood-aqueous fluid barrier, and development of cataracts. Dogs administered 3 g of naphthalene, simulating mothball proportions, developed distemper-like attacks and moderate anemia (Patty's, 1981).

Ten days ingestion of naphthalene by rats resulted in significant increases in liver weight, lipid peroxidation, and aniline hydroxylase activity. Rabbits administered 1,000 mg/kg/day developed changes in the lenses, which within 2 weeks produced mature cataracts, the biochemical basis of which were shown to be related to 1,2-dihydro-1,2-dihydroxynaphthalene, a liver metabolite of naphthalene.

C. Non-mammalian effects

The reported 96-hour LC50 values for aquatic species were 150 mg/L for the sunfish and mosquitofish, 2.3 mg/L for the rainbow trout, 4.9 mg/L for the fathead minnow, 3.8 mg/L for a polychaete, Neanthes arenaceodentata, and 199 mg/L for the Pacific oyster (Crassostrea gigas). The 48-hour LC50 value for the water flea (Daphnia magna) was reported to be 8.57 mg/L.

Absorption, Distribution, Metabolism, Excretion:

In several fish species, Gillichthys mirabilis, Oligocottus maculosus, and Citharichthys stigmaeus, naphthalene was metabolized to 1,2-dihydro-1,2-dihydroxynaphthalene and excreted in the urine (Patty's, 1981). Naphthalene was absorbed by common marine mussels, Mytilus edulis, and released again in unchanged form. In the rainbow trout after an 8-hour exposure to 0.005 mg/L, most tissues contained naphthalene concentrations 20 to 100 times greater than the water levels, with greatest concentration in liver tissue; half-life in fat was less than 24 hours. In marine animals that had accumulated naphthalene, 2 to 60 days (depending on the species) were required for its release from tissues and return to undetectable levels following transfer of the animals to oil-free seawater.

D. Short-term assays

Naphthalene was not mutagenic in Salmonella typhimurium with and without Arochlor 1254-induced hamster liver microsome activation; in S. typhimurium strains TA100, TA1535, TA1537, and TA98 with rat S-9 microsomal activation; and in S. typhimurium strain G-46 and Escherichia coli strain K12 after mouse S-9 activation. No carcinogenic activity was observed in an in vitro rat embryo cell/Rauscher leukemia virus test system at doses up to 100 mg/L, or in an in vitro mouse embryo cell/AKR leukemia virus test system at doses ranging from 50-100,000 ug/L. Cell transformation did not occur in a marine mammary gland organ culture system at naphthalene concentrations up to 1,000 mg/L.

VI. Exposure Standards and Recommendations

A. Air

1. Immediately Dangerous to Life or Health (IDLH): 500 ppm (Sittig, 1981)

2. OSHA standards: (Value cited in USEPA/OWRS, 1980)

   TLV-TWA: 10 ppm (50 mg/m^3)

3. NIOSH recommendations: --

4. ACGIH (1980):

TLV-TWA (10 hr): 10 ppm (50 mg/m^3)
STEL (15 min): 15 ppm (75 mg/m^3)

B. Water

The following ambient water quality criteria are cited by EPA (USEPA/OWRS, 1980):

To protect human health from effects due to ingestion of contaminated water and aquatic organisms: No value set due to insufficient data.

To protect freshwater aquatic life: 2.3 mg/L (acute)
0.62 mg/L (chronic)

To protect marine aquatic life: 2.35 mg/L (acute toxicity)
No available data on chronic toxicity

References for Naphthalene Synopsis

ACGIH SEE American Conference of Governmental Industrial Hygienists, Inc.

American Conference of Governmental Industrial Hygienists, Inc. (1980) Documentation of the threshold limit values. 4th ed. Cincinnati: The Conference. 486p.

Bouwer, E.J.; McCarty, P.L. (1983) Transformations of halogenated organic compounds under denitrification conditions. Appl. Environ. Microbiol. 45: 1295-1299.

Merck index. (1983) 10th ed. Windholz, M.; Budavari, S.; Blumetti, R.F.; Otterbein, E.S.; eds. Rahway, NJ: Merck & Co., Inc. 2154p.

Patty's SEE Patty's industrial hygiene and toxicology

Patty's industrial hygiene and toxicology. (1981) 3rd rev. ed. Vol. 2. Clayton, G.D.; Clayton, F.E.; eds. New York: John Wiley & Sons. 5112p.

Sittig, M. (1981) Handbook of toxic and hazardous chemicals. Park Ridge, NJ: Noyes Publications. 729p.

TDB II SEE Toxicology data bank II

Toxicology data bank II. (1984) [Enhanced version of data base]. Naphthalene, TDB no. 184. (Last edited June 12, 1984). Bethesda, MD: U.S. Department of Health and Human Services, National Library of Medicine, Specialized Information Services. Magnetic-tape reels.

U.S. Environmental Protection Agency, Office of Water Regulations and Standards. (1980) Ambient water quality criteria for naphthalene. EPA-440/5-80-059. Washington, DC: The Office. 67p. Available from: National Technical Information Service, Springfield, VA.

USEPA/OWRS SEE U.S. Environmental Protection Agency, Office of Water Regulations and Standards

Verschueren, K. (1983) Handbook of environmental data on organic chemicals. 2nd ed. New York: Van Nostrand Reinhold Company, Inc. 1310p.

Pentachlorophenol (PCP)

Information abstracted from TDB II, 1984 (No. 894) unless otherwise specified

I. Chemical Identification

A. CAS Registry No.: 87-86-5

B. Synonyms:

PCP
2,3,4,5,6-Pentachlorophenol

C. Molecular formula: C_6HCl_5O

D. Structural formula:

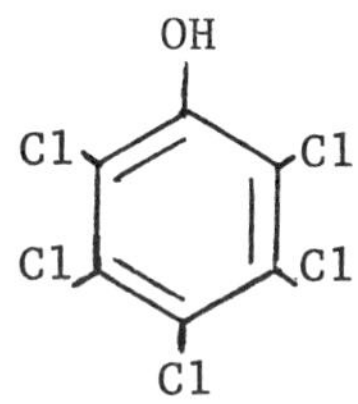

II. Physical and Chemical Properties

A. Molecular weight: 266.35

B. Physical state (color, form):

Pure: White crystalline solid
Technical grade: Dark greyish powder or flakes

C. Odor: Phenolic; very pungent odor when hot

D. Melting point: 190 deg C

E. Boiling point: 309 deg C

F. Solubilities:

Water: 20 mg/L at 30 deg C; 14 mg/L at 20 deg C (Verschueren, 1983)

Organic solvents: Soluble in most organic solvents

G. Octanol/water partition coefficient (log P): 5.01 (Verschueren, 1983)

H. Density: 1.978 g/ml at 22 deg C

I. Dissociation constant: --

J. Viscosity: --

K. Surface tension: --

L. Vapor pressure: 0.12 mm Hg at 100 deg C, 0.00011 mm Hg at 20 deg C (Patty's, 1981)

M. Vapor density: 9.20 (air = 1) (Verschueren, 1983)

N. Relative evaporation rate: --

O. Spectral properties: Maximum absorption at 300.5 and 308 nm (in alcohol)

P. Other: Decomposes at 310 deg C

III. Manufacturing/Use Information

Domestic production was estimated by the Stanford Research Institute (SRI) to be 46 million lb/year in 1975; USITC (1981) reported U.S. production to be 46.8 million lb in 1980. The major uses of PCP are as a wood preservative, pre-harvest defoliant, waterproofing agent for plywood and fiberboard, and termite extermination agent. Technical PCP has been reported to contain several impurities including chlorinated dibenzo-p-dioxins and polychlorinated dibenzofurans. The octachloro-dibenzo-p-dioxin content is typically 500-1,500 ppm (Verschueren, 1983). PCP has gained increasing use as a fungicide, herbicide, and insecticide since the early 1930's (Ahlborg and Thunberg, 1980).

IV. Environmental Fate/Exposure Potential

A. Potential for release from chemical disposal sites

PCP is only slightly soluble in water and it is relatively non-volatile, as evidenced by its low vapor pressure. Spills or releases of the pure compound onto the soil will not evaporate readily. PCP will readily sorb to soil or organic matter and might be washed off into surface waters by rainfall. Its water solubility value suggests a potential for leaching from waste disposal sites into groundwater.

B. Biological uptake/accumulation

PCP has a high potential for bioaccumulation because of its high log P. It is degraded very slowly. Any PCP entering surface waters may be absorbed by aquatic organisms; fish have been reported to accumulate PCP and its degradation products rapidly from water, with bioconcentration factors ranging from 500-8000 times that of the surrounding water (Ahlborg and Thunberg, 1980).

C. Persistence

Pentachlorophenol may undergo environmental degradation by photolysis in which octachlorodibenzo-p-dioxin may be produced; biodegradation by microorganisms may also occur. PCP applied to rice fields was dechlorinated by microorganisms with several chlorinated phenols as products. The decomposition rate of PCP in soil suspensions was reported to be more than 72 days for complete disappearance (Verschueren, 1983). Increases in pH as well as increases in temperature will stimulate degradation by photochemical reactions. Ahlborg and Thunberg (1980) reported that little or no degradation would occur in submerged soils in the absence of light and oxygen; however, Guthrie et al. (1984) reported almost complete biodegradation during anaerobic digestion of sewage sludge solids, but only after an acclimation period involving very gradual increases in PCP concentrations.

D. Measured concentrations

Groundwater: PCP was present at less than 1-2 ug/L in groundwater at 30-500 m from "dumping grounds" (Verschueren, 1983).

Surface water: The reported maximum levels of PCP were as follows: 0.7 ug/L in river samples; 0.01 ug/L in lake samples; and 10,000 ug/L in creek water containing industrial discharges (WHO/IARC, 1979).

Drinking water: The highest concentration reported in drinking water samples collected from 3 Oregon cities was 4.4 ug/L (Verschueren, 1983).

E. Human environmental exposure

Human exposure to any PCP from chemical disposal sites could occur by dermal contact with wastes at the disposal site, by bathing in lakes or streams containing PCP, or possibly by inhalation of wind-borne PCP dust. A special hazard would be burning of waste materials previously treated with PCP, since air-borne fly ash (including polychlorinated dibenzofurans and octachlorodibenzo-p-dioxins) could be disseminated into the atmosphere (Ahlborg and Thunberg, 1980). Because of PCP's low solubility and high log P, only minimal amounts would be expected to leach into groundwater. Average daily intake from drinking water has been calculated to be 0.003 mg/kg body weight per day (NRC, 1977).

Food: For the period 1965-1970, the average daily intake of PCP in foods was 1.67 ug/day; in 1973 it dropped to 0.725 ug/day, and rose to 0.76 ug/day in 1974 (WHO/IARC, 1979). PCP has been detected in lettuce grown in soil treated with PCP.

Occupational exposure: Air samples from 25 factories using PCP as a wood preservative showed that the average worker was exposed to 0.013 mg PCP/m^3 air, with a range of 0.004-1.00 mg/m^3 (WHO/IARC, 1979).

Body burden: Urine levels in workers in a wood-treatment factory ranged from 0.12-9.68 mg/L. PCP has been found in 85% of the urine samples collected from the general population, with a maximum level of 0.193 mg/L, and a mean level of 0.0063 mg/L. PCP has also been detected in human blood plasma, seminal fluid, and fingernails of non-occupationally exposed persons (WHO/IARC, 1979).

V. Health Effects

A. Human health effects

Acute exposure by ingestion causes respiratory depression and a decrease in blood pressure and urinary output; fever; motor weakness; and possible collapse with convulsions and death. Other effects from ingestion may be lung, liver, and kidney damage. The human oral lethal dose is estimated to be 50-500 mg/kg, which is between 1 teaspoon and 1 oz for a 70 kg (150 lb) person. The minimum lethal dose for humans is estimated to be 29 mg/kg (NRC, 1977). The most important acute effect of inhalation is cardiovascular depression with accompanying heart failure. Other effects from acute exposure may include visual damage, acute inflammation of conjunctiva, and corneal opacity. Dusts are particularly irritating to the eyes and nose in concentrations greater than 1 mg/m^3. Some irritation of the nose may occur at 0.3 mg/m^3.

Extended periods of exposure to PCP resulted in persistent chloracne and disorders of the nervous system and liver (WHO/IARC, 1979).

IARC (WHO/IARC, 1979) stated that the available data do not permit an evaluation of the carcinogenicity of PCP. IARC (WHO/IARC, 1982) later stated that evidence of carcinogenicity of PCP to humans and animals is inadequate.

Absorption, Distribution, Metabolism, Excretion:

PCP can be absorbed in the human via all routes, and unless renal and liver functions are impaired, PCP is rapidly eliminated from the blood and tissues by these routes. Approximately 90% of the doses in man were eliminated with urine and feces within 7 days. PCP has been shown to bind to albumin and to human plasma. The high capacity for protein binding of PCP in humans may explain the slow excretion of the small part of a dose remaining after the initial rapid elimination (Ahlborg and Thunberg, 1980).

B. Non-human mammalian health effects

The oral LD50's for PCP are: 120-140 mg/kg body weight for mice; 146 and 175 mg/kg for male and female rats, respectively; 100-130 mg/kg for rabbits; and 150-200 mg/kg for dogs (NRC, 1977). For rats, the LD50 by skin application is 320-330 mg/kg, and the LD50 by inhalation of aerosol is 11.7 mg/kg (WHO/IARC, 1979).

Information on subchronic and chronic effects of PCP in animals was limited. However, NRC (1977) stated that any studies of chronic effects of PCP must take into consideration the degree of contamination of PCP with chlorinated dimers, especially dibenzo-p-dioxins and dibenzofurans. Although information on developmental effects was minimal, NRC (1977) reported that commercial and pure PCP was embryotoxic and fetotoxic in rats and hamsters at doses of 15 mg/kg and above, and that high doses (less than 5 mg/kg/day) of PCP have been shown to be teratogenic in rats and hamsters when administered during susceptible days of gestation.

Absorption, Distribution, Metabolism, Excretion:

PCP is primarily excreted in urine. It is found in plasma protein, liver, and kidney, and is reported to inactivate respiratory enzymes and cause gross damage to mitochondria in animals.

C. Non-mammalian effects

The reported 96-hour LC50 values for freshwater fish species are as follows: 0.032 mg/L for the bluegill; 0.052 mg/L for the rainbow trout; 0.068 mg/L for both the channel catfish and chinook salmon; and 0.205 mg/L for the fathead minnow. In tests with birds that were 10-20 days old, the 96-hour oral LC50 values were 3,400 ppm in the bobwhite; 4,331 ppm in the ring-necked pheasant; 4,500 ppm in mallard ducks; and 5,204 ppm in Japanese quail.

D. Short-term assays

PCP was shown to be mutagenic in Saccharomyces cerevisiae. In a mammalian spot test for genetic alterations in somatic cells of mice, PCP was mutagenic and produced some chromosomal aberrations. PCP proved negative in the sex-linked level test in Drosophila (NRC, 1977).

VI. Exposure Standards and Recommendations

A. Air

1. Immediately Dangerous to Life or Health (IDLH): 13.5 ppm (150 mg/m^3 (Sittig, 1981)

2. OSHA standards:

   TLV-TWA: 0.045 ppm (0.5 mg/m^3) in air (1974 value as cited in TDB II, 1984)

3. NIOSH recommendations: --

4. ACGIH (1980):

TLV-TWA (10 hr): 0.045 ppm (0.5 mg/m^3)
STEL (15 min): 0.14 ppm (1.5 mg/m^3)

Note: As of 1984, ACGIH (1984) has deleted the STEL value until further acute toxicological data and industrial hygienc experience become available.

B. Water:

The following ambient water criteria are cited by EPA (USEPA/-OWRS, 1980):

To protect human health: 1,010 ug/L

To protect freshwater aquatic life: 0.055 mg/L (acute)
0.0032 mg/L (chronic)

To protect marine aquatic life: 0.0530 mg/L (acute)
0.0340 mg/L (chronic)

No-adverse-effect-level in drinking water: 0.021 mg/L (NRC, 1977)

C. Other:

The EPA (USEPA, 1984) announced in the Federal Register its intent to cancel the registration of wood preservative products containing PCP unless a number of modifications and restrictions were made in its use. The basis for concern is PCP's potential for teratogenicity and fetotoxicity. The cancellation of the registration is still pending.

References for Pentachlorophenol Synopsis

ACGIH SEE American Conference of Governmental Industrial Hygienists, Inc.

Ahlborg, U.G.; Thunberg, T.M. (1980) Chlorinated phenols: occurrence, toxicity, metabolism, and environmental impact. CRC Crit. Rev. Toxicol. 7(1):1-35.

American Conference of Governmental Industrial Hygienists, Inc. (1980) Documentation of the threshold limit values. 4th ed. Cincinnati: The Conference. 486p.

American Conference of Governmental Industrial Hygienists, Inc. (1984) Supplemental documentation [for documentation of the threshold limit values]. Cincinnati: The Conference. 61p.

Guthrie, M.A.; Kirsch, E.J.; Wukasch, R.F.; Grady, C.P.L., Jr. (1984) Pentachlorophenol biodegradation. II. Anaerobic. Water Res. 18:451-461.

National Research Council. (1977) Drinking water and health. [Vol. 1]. Washington, DC: National Academy Press. 939p.

NRC SEE National Research Council

Patty's SEE Patty's industrial hygiene and toxicology

Patty's industrial hygiene and toxicology. (1981) 3rd rev. ed. Vol. 2. Clayton, G.D.; Clayton, F.E.; eds. New York: John Wiley & Sons. 5112p.

Sittig, M. (1981) Handbook of toxic and hazardous chemicals. Park Ridge, NJ: Noyes Publications. 729p.

TDB II SEE Toxicology data bank II

Toxicology data bank II. (1984) [Enhanced version of data base]. Pentachlorophenol, TDB no. 894. (Last edited October 28, 1983). Bethesda, MD: U.S. Department of Health and Human Services, National Library of Medicine, Specialized Information Services. Magnetic-tape reels.

U.S. Environmental Protection Agency. (1984) Creosote, pentachlorophenol, and inorganic arsenicals; intent to cancel registrations of pesticide products containing creosote, pentachlorophenol (including its salts) and the inorganic arsenicals; determination concluding the rebuttable presumption against registration of the wood preservative uses of pesticide products; availability of position document. Fed. Regist. 49:28666-28689.

U.S. Environmental Protection Agency, Office of Water Regulations and Standards. (1980) Ambient water quality criteria for pentachlorophenol. EPA-440/5-80-065. Washington, DC: The Office. 89p. Available from: National Technical Information Service, Springfield, VA.

USEPA SEE U.S. Environmental Protection Agency

USEPA/OWRS SEE U.S. Environmental Protection Agency, Office of Water Regulations and Standards

U.S. International Trade Commission. (1981) Synthetic organic chemicals: United States production and sales, 1980. USITC publication 1183. Washington, DC: U.S. Government Printing Office. 327p.

USITC SEE U.S. International Trade Commission

Verschueren, K. (1983) Handbook of environmental data on organic chemicals. 2nd ed. New York: Van Nostrand Reinhold Company, Inc. 1310p.

WHO/IARC SEE World Health Organization, International Agency for Research on Cancer

World Health Organization, International Agency for Research on Cancer. (1979) IARC monographs on the evaluation of the carcinogenic risk of chemicals to humans. Vol. 20. Some halogenated hydrocarbons. Lyon, France: The Agency. 609p.

World Health Organization, International Agency for Research on Cancer. (1982) IARC monographs on the evaluation of the carcinogenic risk of chemicals to man. Supplement 4 to Vol. 1-29. Chemicals, industrial processes and industries associated with cancer in humans. Lyon, France: The Agency. 292p.

Phenol

Information abstracted from TDB II, 1984 (No. 113) unless otherwise specified

I. Chemical Identification

A. CAS Registry No.: 108-95-2

B. Synonyms:

Carbolic acid
Hydroxybenzene
Phenic acid
Phenyl alcohol

C. Molecular formula: C_6H_6O

D. Structural formula:

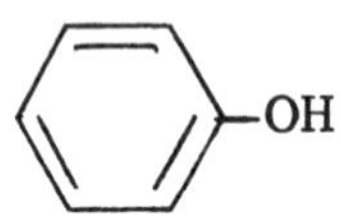

II. Physical and Chemical Properties

A. Molecular weight: 94.11

B. Physical state: Colorless or white crystals

C. Odor: Distinct, somewhat sickeningly sweet and acrid

D. Melting point: 43 deg C

E. Boiling point: 182 deg C

F. Solubilities:

Water: 93,000 mg/L at 25 deg C

Organic solvents: Very soluble in alcohol, chloroform, ether, glycerol, petrolatum, carbon disulfide, volatile and fixed oils, aqueous alkali hydroxides

G. Octanol/water partition coefficient (log P): 1.46

H. Density: 1.072 g/ml at 20 deg C

I. Dissociation constant:

Pure compound: Ka = 1.058 at 41 deg C

Aqueous solution: pKa = 10.0 at 25 deg C

J. Viscosity: --

K. Surface tension (liquid interfacial tension with air): 40.9 mN/m (dynes/cm) at 20 deg C

L. Vapor pressure: 0.3513 mm Hg at 25 deg C

M. Vapor density: 3.24 (air = 1)

N. Relative evaporation rate: Estimated evaporation half-life (aqueous) is 3.2 months.

O. Spectral properties: Maximum absorption at 218.5 nm, 266 nm (shoulder) 271 nm, 276 nm (shoulder) (in alcohol)

III. Manufacturing/Use Information

Phenol production in the U.S. in 1981 was 2.58 billion lb. Phenol is used as a monomer in phenolic resins, as an intermediate in chemical synthesis, as a solvent in the petroleum industry, and as a general disinfectant.

IV. Environmental Fate/Exposure Potential

A. Potential for release from chemical disposal sites

Phenol is very soluble in water, and experimental evidence indicates it is poorly adsorbed on soils, silty loam, or aquatic sediments. The vapor pressure of the pure compound is moderate, but phenol has a low evaporation rate from aqueous solutions.

B. Biological uptake/accumulation

Phenol has a low potential for bioaccumulation because it is readily biodegraded and has a low log P.

C. Persistence

Phenol is readily biodegraded by micro-organisms (except at very high concentrations) to carbon dioxide in aerobic systems, and to methane plus carbon dioxide in anaerobic systems. After a surface spill, phenol was reported to have persisted in groundwater for 19 months (Delfino and Dube, 1976). Photolysis should occur in aqueous systems but is of minor importance compared with biodegradation. Phenol released into the atmosphere may degrade by reaction with hydroxyl radicals, ozone, and nitrogen oxides, as well as by direct photolysis, because phenol exhibits some absorption of light at wavelengths of 290-330 nm.

D. Measured concentrations

In a sampling survey of the Delaware River, 2-4 ug/L phenol was detected in the winter. Phenol was not detected in the summer, attributed to more rapid biodegradation (Sheldon and Hites, 1978).

E. Human environmental exposure

Human exposure to phenol released from chemical disposal sites could occur via ingestion of drinking water and by dermal contact during bathing and showering. Phenol is unlikely to volatilize from a chemical disposal site, therefore, inhalation is not a likely route of exposure from this source.

V. Health Effects

A. Human health effects

Phenol is rapidly absorbed by all routes, including dermal contact, and results in systemic poisoning. It is caustic to tissues and is acutely toxic with a probable oral lethal dose of 50-500 mg/kg. The prominent acute effects are severe tissue burns due to direct contact, and CNS depression, often resulting in coma and death. Other effects from acute exposure may include cardiovascular depression; pulmonary distress; hepatic, renal, and gastrointestinal effects; and pigmentary changes.

There is limited information regarding the effects of chronic exposure to phenol. The carcinogenicity of phenol has not been evaluated by IARC (WHO/IARC, 1983), and phenol was not included in the Third Annual Report on Carcinogens published by the National Toxicology Program (USDHHS/PHS, 1983).

Absorption, Distribution, Metabolism, Excretion:

The principal route of phenol excretion is renal, as both the unchanged compound and its conjugated metabolites, phenylsulfate and phenylglucuronide. The half-life of phenol in the body is approximately 3.5 hours.

B. Non-human mammalian health effects

Acute toxic effects are similar to those of humans. Corneal opacification in rabbits and encephalopathies in Sprague-Dawley rats have also been characterized. LD50's for most animals, using various routes of exposure, range from 100-850 mg/kg. The cat is particularly sensitive to phenol, perhaps owing to the fact that, unlike most other species (including humans), one of its major phenol metabolites is 1,4-dihydroxybenzene. Prolonged administration may cause renal damage. The carcinogenic potential of phenol is uncertain but it has induced papillomas and carcinomas in mice.

C. Non-mammalian effects

The reported 96-hour LC50 values for freshwater fish species range from a low of approximately 6 mg/L for the bluegill (*Lepomis macrochirus*), to an intermediate value of approximately 30 mg/L for the fathead minnow (*Pimephales promelas*), to a high of approximately 45 mg/L for goldfish (*Carassius auratus*). For marine organisms, Verschueren (1983) reported 96-hour LC50 values of 9 mg/L for a fish, the goby (*Gobius minutus*), and 25 mg/L for a species of shrimp (*Crangon crangon*).

D. Short-term assays

Phenol has been shown to be mutagenic to *Salmonella typhimurium*, *Escherichia coli*, and *Drosophila*.

VI. Exposure Standards and Recommendations

A. Air

1. Immediately Dangerous to Life or Health (IDLH): 100 ppm (Sittig, 1981)

2. OSHA standards:

TLV-TWA (8 hr): 5 ppm (20 mg/m^3) (1976 value cited in USDHHS/CDC, 1983)

3. NIOSH recommendations: (1976 values as cited in ACGIH, 1980)

TLV-TWA (10 hr): 5 ppm (20 mg/m^3)
Ceiling (15 min): 15 ppm (60 mg/m^3)

4. ACGIH (1980):

TLV-TWA (10 hr): 5 ppm (20 mg/m^3)
STEL (15 min): 10 ppm (39 mg/m^3)

B. Water

The following ambient water quality criteria are cited by EPA (USEPA/OWRS, 1980):

To protect human health from effects due to ingestion of contaminated water and aquatic organisms: 3.5 mg/L

To protect freshwater aquatic life: 10.2 mg/L (acute)
2.56 mg/L (chronic)

To protect marine aquatic life:	5.8 mg/L (acute toxicity) No available data on chronic toxicity

References for Phenol Synopsis

ACGIH SEE American Conference of Governmental Industrial Hygienists, Inc.

American Conference of Governmental Industrial Hygienists, Inc. (1980) Documentation of the threshold limit values. 4th ed. Cincinnati: The Conference. 486p.

Delfino, J.; Dube, D.J. (1976) Persistent contamination of ground water by phenol. J. Environ. Sci. Health A11:345-355.

Sheldon, L.S.; Hites, R.A. (1978) Organic compounds in the Delaware River. Environ. Sci. Technol. 12:1188-1194.

Sittig, M. (1981) Handbook of toxic and hazardous chemicals. Park Ridge, NJ: Noyes Publications. 729p.

TDB II SEE Toxicology data bank II

Toxicology data bank II. (1984) [Enhanced version of data base]. Phenol, TDB no. 113. (Last edited April 5, 1984). Bethesda, MD: U.S. Department of Health and Human Services, National Library of Medicine, Specialized Information Services. Magnetic-tape reels.

U.S. Department of Health and Human Services, Centers for Disease Control. (1983) NIOSH recommendations for occupational health standards. Morb. Mortal. Wkly. Rep. 32:(supplement)3S-22S.

U.S. Department of Health and Human Services, Public Health Service. (1983) Third annual report on carcinogens: summary. Research Triangle Park, NC: U.S. Department of Health and Human Services, National Toxicology Program Public Information Office. 229p.

USDHHS/CDC SEE U.S. Department of Health and Human Services, Centers for Disease Control

USDHHS/PHS SEE U.S. Department of Health and Human Services, Public Health Service

U.S. Environmental Protection Agency, Office of Water Regulations and Standards. (1980) Ambient water quality criteria for phenol. EPA-440/5-80-066. Washington, DC: The Office. 87p. Available from: National Technical Information Service, Springfield, VA.

USEPA/OWRS SEE U.S. Environmental Protection Agency, Office of Water Regulations and Standards

Verschueren, K. (1983) Handbook of environmental data on organic chemicals. 2nd ed. New York: Van Nostrand Reinhold Company, Inc. 1310p.

WHO/IARC SEE World Health Organization, International Agency for Research on Cancer

World Health Organization, International Agency for Research on Cancer. (1983) IARC monographs on the evaluation of the carcinogenic risk of chemicals to humans. Vol. 31. Some food additives, feed additives and naturally occurring substances. Lyon, France: The Agency. 314p.

2,3,7,8-Tetrachlorodibenzo-p-dioxin

Information abstracted from TDB II, 1984 (No. 4151) unless otherwise specified

I. Chemical Identification

A. CAS Registry No.: 1746-01-6

B. Synonyms:

Dioxin
2,3,7,8-TCDD
TCDD

C. Molecular formula: $C_2H_4Cl_4O_2$

D. Structural formula:

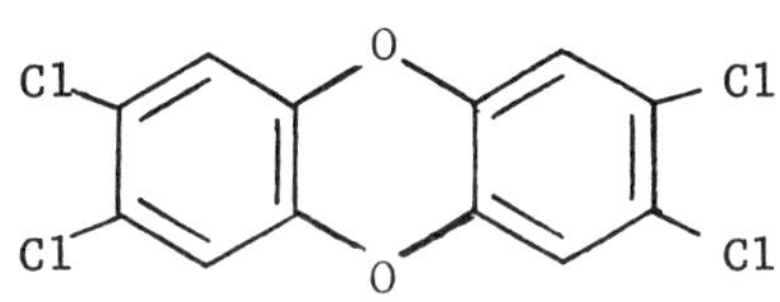

II. Physical and Chemical Properties

A. Molecular weight: 321.96

B. Physical state (color, form): Colorless needles

C. Odor: --

D. Melting point: 305-306 deg C

E. Boiling point: --

F. Solubilities:

Water: 0.2 ug/L (0.2 ppb)

Organic solvents: Soluble in ortho-dichlorobenzene, chlorobenzene, benzene, chloroform, acetone, n-octanol, methanol, lard oil.

G. Octanol/water partition coefficient (log P): 6.15 (USEPA/OWRS, 1984)

H. Density: --

I. Dissociation constant: --

J. Viscosity: --

K. Surface tension: --

L. Vapor pressure: 1.7 x 10^{-6} mm Hg at 25 deg C (NRCC, 1981)

M. Vapor density: --

N. Relative evaporation rate: Henry's law constant (estimated): 2.1 x 10^{-3} (atm)(m^3)/mol

O. Spectral properties: Maximum absorption at 248 nm, 310 nm (chloroform)

III. Manufacturing/Use Information

2,3,7,8-Tetrachlorodibenzo-p-dioxin (TCDD) is not manufactured commercially, but is formed during the production of 2,4,5-trichlorophenol by the alkaline hydrolysis of 1,2,4,5-tetrachlorobenzene, and as such TCDD has no uses; it is an inadvertent contaminant in herbicide precursors and thus in the herbicides themselves, such as the widely used 2,4,5-trichlorophenoxyacetic acid, more familiarly known as 2,4,5-T (Sittig, 1981). The 2,4,5-T produced commercially prior to 1965 contained as much as 30 mg/kg or more TCDD, but 2,4,5-T containing less than 0.05 mg/kg is now produced in several countries and is available in commercial quantities. Furthermore, techniques have been developed for producing 2,4,5-T containing levels of TCDD below the 0.02 mg/kg limit of detection.

IV. Environmental Fate/Exposure Potential

A. Potential for release from chemical disposal sites

Experimental data indicate that TCDD is highly sorbed to sediments and biota, not surprising considering its low water solubility and high octanol/water partition coefficient (Isensee and Jones, 1975; Ward and Matsumura, 1978). It tends to remain on or near the surface of soils; though vertical movement of TCDD is negligible, horizontal transfer of eroded soil by water and movement of contaminated airborne dust particles may occur (Helling et al., 1973; USEPA/OWRS, 1984). Its lateral displacement by soil erosion and runoff would be enhanced in areas of heavy rainfall (USEPA/OWRS, 1984). In view of the low water solubility of TCDD, vertical distribution could be due to volatilization through the air/soil interface or solvation of TCDD by organic solvents (NRCC, 1981).

B. Biological uptake/accumulation

Calculated and experimental results show that TCDD will bioaccumulate in aquatic organisms, but total amounts are related to its initial concentrations in aquatic phase and are significantly affected by the physical form -- sorbed or in solution -- in which TCDD occurs in the environment (Philipp et al., 1981). Using the measured octanol/water partition coefficient in various equations, bioconcentration factors ranging from 3,000 to 68,000 can be predicted (USEPA/OWRS, 1984). The best experimental values derived from equilibrium concentrations (reached in 7 to 15 days) in tested tissues are as follows: at an initial aquatic concentration of 0.05-1300 ppt of TCDD, the bioconcentration factor for the mosquitofish (Gambusia affinis) was 4,850; for the water flea (Daphnia magna), 7,070; for the snail (Physa sp.), 2,095; and for the alga (Oedogonium cardiacum), 2,075 (Isensee, 1978).

C. Persistence

TCDD is immobile in all soils, is not translocated in plants, and has a half-life in soil of about 1 year. In silty clay loam in Hagerstown, and in sand in Lakeland, both cities in Maryland that receive up to 100 mg TCDD/kg, amounts recovered after 1 year were 71% and 56%, respectively. Application of TCDD combined with formulated 2,4,5-T to soil, glass plates, and plant leaves resulted in losses of 15, 60, and 100%, respectively, of the applied doses after 6 hours of exposure to sunlight; the data suggested that the rate of degradation is faster at lower concentrations than at higher concentrations of TCDD. Microbial degradation of TCDD is reported to be relatively rare in nature; of approximately 100 microbial strains tested, only five degraded TCDD. The photodecomposition of TCDD, on wet or dry soil under artificial and natural sunlight, was found to be negligible (Crosby et al., 1971).

The observed half-life of TCDD in sediment-containing Wisconsin lake waters was found to be 550-590 days; in lake water alone, approximately 70% of the TCDD remained after 589 days (Ward and Matsumura, 1977). In an outdoor pond initially dosed with 53.7 ppt of radioactive TCDD, 49.7 and 29.4% remained after 12 and 25 months, respectively (Tsushimoto et al., 1982). Though TCDD in a pure state is photochemically stable, it will photolyze in sunlight in the presence of a hydrogen-donating substrate (Crosby and Wong, 1977). The volatilization half-life for TCDD has been estimated to be 5.5 and 12 years from pond and lake water, respectively (NRCC, 1981).

D. Measured concentrations

Drinking water: No information on TCDD levels in drinking water was available.

Effluent/Leachate: Two of 8 analyzed industrial wastewater samples from a trichlorophenol manufacturing plant site contained TCDD concentrations of 17 and 100 pg/g (VanNess et al., 1980). The concentration of TCDD in a leachate sample from a waste disposal site near Jacksonville, Arkansas, averaged 14 ppt with a range of 0-47,000 ppt (Thibodeaux, 1983). Two untreated leachate samples from the Love Canal chemical disposal site in New York showed a concentration of 1,560 ppt for TCDD (USEPA/ORD, 1982).

Air: The average atmospheric concentration of TCDD in two air particulate samples collected near the Jacksonville, Arkansas, chemical disposal site was 1,100 ppt (Thibodeaux, 1983). At the Love Canal area, TCDD was not detectable within detection limits of 1-20 ppt (USEPA/ORD, 1982).

Soil/Sediment: The TCDD concentration in soil near the chemical disposal site at Jacksonville, Arkansas, ranged from not detected to 2.9 ppb, and in sediments from a nearby pond and creek ranged from not detected to 22.1 ppb (Thibodeaux, 1983). Total tetrachlorodibenzo-p-dioxin was detected at concentrations up to 6.7 ppb in soils collected from Love Canal, but specific 2,3,7,8-TCDD concentrations were not reported. The 2,3,7,8-TCDD concentrations ranged from not detected to 672 ppb in sediments from Love Canal storm sewers, from not detected to 9,570 ppb in sediments from sump (USEPA/ORD, 1982), and from 0.9 to 312 ppb in sediments from storm sewers and creeks near the disposal site (Smith et al., 1983).

Food: Levels of TCDD in edible fish from Lake Ontario ranged from 0.002-0.162 ug/kg (NRCC, 1981), and from the Tittabawassee, Grand, and Saginaw Rivers, Lake Michigan, and Saginaw Bay ranged from 0.004-0.695 ug/kg (USEPA/OWRS, 1984).

E. Human environmental exposure

Probable routes of human exposure are inhalation of vapors and skin contact (Sittig, 1981) as a result of accidental TCDD contamination in industrial plants, and at municipal incinerators and chemical disposal sites. The major risk to humans may come from eating contaminated bottom-feeding fish and shellfish (USEPA/OWRS, 1984).

V. Health Effects

A. Human health effects

Symptoms of acute exposure to materials containing TCDD are nausea and vomiting, headache, and signs of irritation to the eyes, skin, and respiratory tract. Initially, a chemical-burn type of cutaneous reaction will occur, followed by chloracne after several days to weeks (Taylor, 1979), which may resolve in a matter of months or may persist for up to 15 years in severe cases.

Chronic exposure to TCDD may occur for chemical industry workers exposed to low levels of the contaminant on a daily basis during the manufacture of 2,4,5-T and pentachlorophenol. The first symptom usually noted in chronic exposure is chloracne; other reported effects include porphyria cutanea tarda, uroporphyrinuria, abnormal liver tests and liver enlargement, severe neurasthenia and depressive syndrome, peripheral neuropathy, and increased blood cholesterol and total lipids.

IARC (WHO/IARC, 1982) reported the evidence for human carcinogenicity of TCDD to be inadequate because there is no report of human exposure to TCDD alone. Several case-comparison studies have shown high relative risks for soft-tissue sarcoma and for lymphoma in association with exposure to phenoxyacetic acids or chlorophenols, in which TCDD was a likely contaminant; in several cohort studies, the number of cancer deaths was higher than expected in men exposed to TCDD during the manufacture or use of 2,4,5-trichlorophenol and/or 2,4,5-T (WHO/-IARC, 1982). It is not possible to attribute with certainty any positive finding to TCDD as a potential teratogen or cause of abortion in humans.

In one incident, waste oil contaminated with TCDD was sprayed on a riding arena for dust control. A 6-year-old girl played in the arena soil and then showed symptoms which included headache, nose bleeds, diarrhea, lethargy, hemorrhagic cystitis, and focal pyelonephritis. Three other children and one adult complained of skin lesions consistent with symptoms of chloracne.

B. Non-human mammalian health effects

The acute oral LD50 of TCDD ranged from 0.6 ug/kg for male guinea pigs (Schwetz et al., 1973) to 5,051 ug/kg for male hamsters (Henck et al., 1981). The oral LD50 for rats was 22 and 45 ug/kg for males and females, respectively, 114 ug/kg for mice, and 115 ug/kg for rabbits; dogs were less sensitive than rabbits (WHO/IARC, 1977). The percutaneous LD50 for rabbits was 275 ug/kg. Following a single lethal dose of TCDD, death may be delayed several weeks; in rats and guinea pigs, mortality was similar when TCDD was given as a single dose or as several divided daily or weekly doses within 4-5 weeks, indicating a cumulative effect (WHO/IARC, 1977). The most common symptoms after lethal exposure were weight loss, often characterized as "wasting away," and thymic atrophy. Although liver damage was not observed in the guinea pig, the most sensitive species to TCDD, it was extensive in rats and mice (Gupta et al., 1973).

Female rhesus monkeys fed 0.5 ug TCDD/kg/day developed anemia, severe pancytopenia, widespread hemorrhage related to marked thrombocytopenia, altered estradiol and progesterone levels, and reproductive difficulties including inability to conceive and maintain pregnancies. Death occurred in 5 of 8 animals between the 7th and 12th month of the experiment. Necropsy showed hypocellularity of bone marrow and lymph nodes as well as hypertrophy, hyperplasia, and metaplasia of epithelium, bile and pancreatic ducts, salivary glands, and palpebral conjunctivae (membranes lining the eyelids).

TCDD was a weak initiator of cancers in a two-stage mouse skin tumor model, and when applied to the backs of Swiss-Webster mice, an increase in integumentary fibrosarcomas was also observed. TCDD by gavage twice weekly for 2 years at 0.5 ug/kg/week caused hepatocellular carcinomas in female rats, follicular cell adenomas or carcinomas of the thyroid in male rats, and hepatocellular adenomas in male mice. At 2.0 ug/kg/week it produced hepatocellular adenomas or carcinomas and follicular cell adenomas of the thyroid in female mice.

TCDD is embryotoxic and teratogenic in mice in repeated or single doses of as little as 1-10 ug/kg administered during day 6-15 of pregnancy, causing increased frequencies of cleft palate and kidney abnormalities (WHO/IARC, 1977). In rats, embryolethal effects occur under experimental conditions; kidney anomalies, intestinal hemorrhages, and general edema can be produced in fetuses (WHO/IARC, 1977). Ingestion of 0.01 ug TCCD/kg/day resulted in decreased litter size, gestation survival, and neonatal survival and growth in rats; at dose levels of 0.01 and 0.1 ug/kg/day, reproductive capacity was affected through 3 successive generations.

Absorption, Distribution, Metabolism, Excretion:

Exposure to TCDD might occur by inhalation or by dermal or gastrointestinal absorption; inhalation exposure, however, has not been studied. After a single oral dose of 50 ug/kg, rats excreted almost 30% in feces during the first 48 hours and 1-2%/day thereafter; a total of 53.2% was eliminated in feces, 13.2% in urine, and 3.2% in expired air; TCDD accumulated primarily in the liver and fat, at levels 10 times that in other tissues. TCDD administered orally 5 times/week for 7 weeks also was excreted primarily via the feces, but 3-18% of the cumulative dose was found in the urine. Rats given daily oral doses of 15 ug 14C-TCCD/kg revealed biliary 14C activity, and incubation of bile with beta-glucuronidase resulted in extraction of glucuronide conjugates of 14C-TCDD; TCDD was slowly metabolized in the liver to polar metabolites and excreted in bile. Another feeding study led to a two to threefold increase in liver microsomal cytochrome P-450 levels and changes in liver microsomal metabolism with some hydroxylase activities increasing three to fivefold.

C. Non-mammalian effects

Most of the available data on the effects of TCDD to aquatic life are from studies involving freshwater microcosms. The data do not allow calculations of acute values, but some useful information can be derived. For example, 96-hour LC50 acute toxicity values would be greater than 0.01 ug/L for embryos of northern pike (*Esox lucius*), and embryos and yolk-sac fry of rainbow trout (*Salmo gairdneri*); greater than 0.1 ug/L for juvenile rainbow trout; greater than 0.2 ug/L for an oligochaete (*Paranais* sp.), and larvae of a mosquito (*Aedes aegypti*); greater than 0.24 ug/L for fingerling channel catfish (*Ictalurus punctatus*); greater than 1.0 ug/L for the coho salmon (*Oncorhynchus kisutch*); greater than 1.3 ug/L for the water flea (*Daphnia magna*), and a snail (*Physal* sp.); and greater than 10 ug/L for the guppy (*Poecilia reticulata*) (USEPA/OWRS, 1984). Based on effects caused by 96-hour exposures, it can be concluded that TCDD would be chronically toxic to rainbow trout at 0.001 ug/L, and to northern pike, channel catfish, coho salmon, and mosquitofish at 0.01 ug/L (USEPA/OWRS, 1984).

D. Short-term assays

TCDD increased the reversion frequency to streptomycin independence in Escherichia coli SD-4 at a concentration of 2 ug/ml. Early reports that frameshift mutations in Salmonella typhimurium were induced by toxic concentrations of TCDD have not been confirmed. Inhibition of mitosis and chromosomal abnormalities were observed in endosperm cells of Haemanthur katherinae Baker treated with 0.2 or 1.0 ug TCCD/L. In a sex-linked recessive lethal bioassay with Drosophila melanogaster, TCDD gave negative results.

VI. Exposure Standards and Recommendations

A. Air

There are no numerical limits for permissible exposure in air; in view of the effects of TCDD, all contact should be avoided (Sittig, 1981).

1. Immediately Dangerous to Life or Health (IDLH): --

2. OSHA standards: --

3. NIOSH recommendations: --

4. ACGIH: --

B. Water:

The following ambient water quality criteria are cited by EPA (USEPA/OWRS, 1984):

To protect human health from potential carcinogenic effects: Preferably zero. An increase in the lifetime cancer risk by 1×10^{-5} due to ingestion of contaminated water and aquatic organisms is estimated to occur at a TCDD concentration of 1.3×10^{-7} ug/L.

To protect freshwater and marine aquatic life: Measured bioconcentration factors (BCF) range from 390-13,000; assuming a BCF of 5,000, an ambient water concentration of greater than 0.00001 ug/L should result in concentrations in edible freshwater and marine fish and shellfish that exceed FDA advisory levels (USEPA/OWRS, 1984).

References for 2,3,7,8-Tetrachlorodibenzo-p-dioxin Synopsis

Crosby, D.G.; Wong, A.S. (1977) Environmental degradation of 2,3,7,8-tetrachlorodibenzo-p-dioxin (TCDD). Science 195:1337-1338.

Crosby, D.G.; Wong, A.S.; Plimmer, J.R.; Woolson, E.A. (1971) Photodecomposition of chlorinated dibenzo-p-dioxins. Science 173:748-749.

Gupta, B.N.; Vos, J.G.; Moore, J.A.; Zinkl, J.G.; Bullock, B.C. (1973) Pathologic effects of 2,3,7,8-tetrachlorodibenzo-p-dioxin in laboratory animals. Environ. Health Perspect. 5:125-140.

Helling, C.S.; Isensee, A.R.; Woolson, E.A.; Ensor, P.D.J.; Jones, G.E.; Plimmer, J.R.; Kearney, P.C. (1973) Chlorodioxins in pesticides, soils, and plants. J. Environ. Qual. 2:171-177.

Henck, J.M.; New, M.A.; Kociba, R.J.; Rao, K.S. (1981) 2,3,7,8-tetrachlorodibenzo-p-dioxin: acute oral toxicity in hamsters. Toxicol. Appl. Pharmacol. 59:405-407.

Isensee, A.R. (1978) Bioaccumulation of 2,3,7,8-tetrachlorodibenzo-para-dioxin. Ecol. Bull. 27:255-262.

Isensee, A.R.; Jones, G.E. (1975) Distribution of 2,3,7,8-tetrachlorodibenzo-p-dioxin (TCDD) in an aquatic model ecosystem. Environ. Sci. Technol. 9:668-672.

National Research Council of Canada. Associate Committee on Scientific Criteria for Environmental Quality. (1981) Polychlorinated dibenzo-p-dioxins: criteria for their effects on man and his environment. Publication NRCC No. 18574 of the Environmental Secretariat. 251p. Available from: The Council, Ottawa.

NRCC SEE National Research Council of Canada

Philipp, M.; Krasnobajew, V.; Zeyer, J.; Hutter, R. (1981) Fate of TCDD in microbial cultures and in soil under laboratory conditions. In: Leisinger, T.; Hutter, R.; Cook, A.M.; Nuesch, J.; eds. Microbial degradation of xenobiotics and recalcitrant compounds. FEMS Symposium No. 12. London: Academic Press. p. 221-233.

Schwetz, B.A.; Norris, J.M.; Sparschu, G.L.; Rowe, V.K.; Gehring, P.J.; Emerson, J.L.; Gerbig, C.G. (1973) Toxicology of chlorinated dibenzo-p-dioxins. Environ. Health Perspect. 5:87-99.

Sittig, M. (1981) Handbook of toxic and hazardous chemicals. Park Ridge, NJ: Noyes Publications. 729p.

Smith, R.M.; O'Keefe, P.W.; Aldous, K.M.; Hilker, D.R.; O'Brien, J.E. (1983) 2,3,7,8-tetrachlorodibenzo-p-dioxin in sediment samples from Love Canal storm sewers and creeks. Environ. Sci. Technol. 17:6-10.

Taylor, J.S. (1979) Environmental chloracne: update and overview. Ann. N.Y. Acad. Sci. 320:295-307.

TDB II SEE Toxicology data bank II

Thibodeaux, L.J. (1983) Offsite transport of 2,3,7,8-tetrachlorodibenzo-p-dioxin from a production disposal facility. In: Keith, L.F., ed. Chlorinated dioxins and dibenzofurans in the total environment. Vol. I. Stoneham, MA: Butterworth Medical and Scientific Publishers. p. 75-86.

Toxicology data bank II. (1984) [Enhanced version of data base]. 2,3,7,8-Tetrachlorodibenzo-p-dioxin, TDB no. 4151. (Last edited March 7, 1984). Bethesda, MD: U.S. Department of Health and Human Services, National Library of Medicine, Specialized Information Services. Magnetic-tape reels.

Tsushimoto, G.; Matsumura, F.; Sago, R. (1982) Fate of 2,3,7,8-tetrachlorodibenzo-p-dioxin (TCDD) in an outdoor pond and in model aquatic ecosystems. Environ. Toxicol. Chem. 1:61-68.

U.S. Environmental Protection Agency, Office of Research and Development. (1982) Environmental monitoring at Love Canal. Vol. 1. EPA-600/4-82-030a. Washington, DC: The Office. 319p.

U.S. Environmental Protection Agency, Office of Water Regulations and Standards. (1984) Ambient water quality criteria for 2,3,7,8-tetrachlorodibenzo-p-dioxin. EPA 440/5-84-007. Washington, DC: The Office. 267p.

USEPA/ORD SEE U.S. Environmental Protection Agency, Office of Research and Development

USEPA/OWRS SEE U.S. Environmental Protection Agency, Office of Water Regulations and Standards

VanNess, G.F.; Solch, J.G.; Taylor, M.L.; Tiernan, T.O. (1980) Tetrachlorodibenzo-p-dioxins in chemical wastes, aqueous effluents and soils. Chemosphere 9:553-563.

Ward, C.T.; Matsumura, F. (1977) Fate of 2,4,5-T contaminant, 2,3,7,8-tetrachlorodibenzo-p-dioxin (TCDD) in aquatic environments. Prepared for the U.S. Department of the Interior, Office of Water Research and Technology by the University of Wisconsin, Madison Water Resources Center. Technical report WIS WRC 77-01. 22p. Available from: National Technical Information Service, Springfield, VA.

Ward, C.T.; Matsumura, F. (1978) Fate of 2,4,5-T contaminant, 2,3,7,8-tetrachlorodibenzo-p-dioxin (TCDD) in a model aquatic environment. Arch. Environ. Toxicol. 7:349-357.

WHO/IARC SEE World Health Organization, International Agency for Research on Cancer

World Health Organization, International Agency for Research on Cancer. (1977) IARC monographs on the evaluation of the carcinogenic risk of chemicals to humans. Vol. 15. Some fumigants, the herbicides 2,4-D and 2,4,5-T ,chlorinated dibenzodioxins and miscellaneous industrial chemicals. Lyon, France: The Agency. 354p.

World Health Organization, International Agency for Research on Cancer. (1982) IARC monographs on the evaluation of the carcinogenic risk of chemicals to man. Supplement 4 to Vol. 1-29. Chemicals, industrial processes and industries associated with cancer in humans. Lyon, France: The Agency. 292p.

1,1,1-Trichloroethane

Information abstracted from TDB II, 1984 (No. 157) unless otherwise specified

I. Chemical Identification

A. CAS Registry No.: 71-55-6

B. Synonyms:

Chlorothene
Methyl chloroform

C. Molecular formula: $C_2H_3Cl_3$

D. Structural formula:

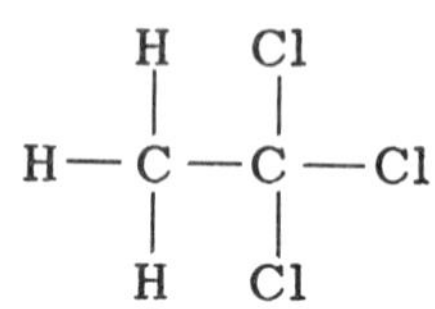

II. Physical and Chemical Properties

A. Molecular weight: 133.42

B. Physical state (color, form): Colorless liquid

C. Odor: Chloroform-like; sweetish

D. Melting point: -30.4 deg C

E. Boiling point: 74.1 deg C at 760 mm Hg

F. Solubilities:

Water: 4,400 mg/L at 20 deg C

Organic solvents: Soluble in acetone, benzene, methanol, carbon tetrachloride, ether, chloroform

G. Octanol/water partition coefficient (log P): 2.49

H. Density: 1.339 g/ml at 20 deg C (CRC, 1983)

I. Dissociation constant: --

J. Viscosity: 1.2 cp at 20 deg C (CRC, 1983)

K. Surface tension (liquid interfacial tension with air): 25.4 mN/m (dynes/cm) at 20 deg C

L. Vapor pressure: 100 mm Hg at 20 deg C, 155 mm Hg at 30 deg C (Verschueren, 1983); 127 mm Hg at 25 deg C

M. Vapor density: 4.63 (air = 1)

N. Relative evaporation rate: 12.8 (butyl acetate = 1); 1 (carbon tetrachloride = 1)

O. Spectral properties: --

III. Manufacturing/Use Information

United States production of 1,1,1-trichloroethane in 1981 was 614 million lb. Its major uses are as a solvent for cold cleaning of metals and other materials, for vapor degreasing, and for natural and synthetic resins, oils, waxes, tar, and alkaloids. It is also used as a chemical intermediate, a dry cleaning agent, a vapor-pressure depressant in aerosols, and for cleaning plastic molds. Formerly, it was used with ethylene gas for degreening citrus fruits and post-harvest fumigation of strawberries, but its registered agricultural uses have been canceled.

IV. Environmental Fate/Exposure Potential

A. Potential for release from chemical disposal sites

Due to the high vapor pressure of 1,1,1-trichloroethane, spills on land will rapidly evaporate into the atmosphere. It passes rapidly through soil into groundwater, suggesting that it is not adsorbed strongly by soils. Its primary loss from water is by evaporation, with a volatilization half-life of hours to several weeks, depending on wind and mixing conditions. In the atmosphere, the compound is fairly stable and may be transported long distances; a portion is returned to earth in rainwater.

B. Biological uptake/accumulation

1,1,1-Trichloroethane has a log P of 2.49, suggesting a limited tendency to bioconcentrate. Based on limited experimental data, the common logarithm bioconcentration factor is 1.0 in fish.

C. Persistence

The high vapor pressure of 1,1,1-trichloroethane suggests its rapid volatilization from soil and water. There is slow, or no, degradation in soils. Under acclimated conditions, slow degradation has been observed in loamy sand. In both anaerobic and aerated water, slow degradation may occur, but again, acclimation is necessary. In laboratory studies simulating methanogenic conditions in groundwater, 1,1,1-trichloroethane exhibited a lag period for acclimation of the microbial populations, after which nearly complete biodegradation occurred (Bouwer and McCarty, 1983a); under anoxic denitrification conditions, there was no observed transformation of 1,1,1-trichloroethane (Bouwer and McCarty, 1983b). In seawater, a half-life of 9 months has been determined experimentally, vinylidene chloride being the degradation product; no degradation has been found to occur in river water. Loss from an experimental mesocosm is entirely due to evaporation and ranges from a half-life of 24 days in spring to 11 days in winter. Hydrolysis is not a significant degradation process, having a half-life of approximately 6 months; photodegradation does not occur in water.

In the troposphere, 1,1,1-trichloroethane does not absorb light at wavelengths above 290 nm; thus, degradation by direct photolysis is insignificant. It does react slowly with hydroxyl radicals and has a half-life of 0.5-2.2 years; photo-oxidation products include phosgene, chlorine, hydrochloric acid, and carbon monoxide. The rate of degradation is increased by the presence of chlorine radicals and nitrogen oxides. About 15% of 1,1,1-trichloroethane drifts into the stratosphere, where it is rapidly degraded by photodissociation. As a result of its substantial release to air and its slow degradation, the chemical is increasing in the atmosphere by 12-17% a year.

D. Measured concentrations

Groundwater: In 13 U.S. cities, 23% of the raw groundwater was positive for 1,1,1-trichloroethane, with a median of 1.1 ug/L and maximum of 13 ug/L.

Surface water: In 105 U.S. cities, 12% of the raw surface water was positive, with a 0.2 ug/L median and 1.2 ug/L maximum. Mean values at industrial sites were greater than 10 ug/L, with maximum values as high as 334 ug/L.

Drinking water: In 133 U.S. cities with finished surface water, the median concentration was 0.4 ug/L and the maximum was 3.3 ug/L. In 23 cities with finished groundwater, 22% of the samples were positive, with a 2.1 ug/L median value and 3.0 maximum. Contaminated drinking-water wells in New York, New Jersey, Connecticut, and Maine contained 950-5,440 ug/L.

Precipitation: The concentration in rainwater was 0.0081 ug/L in La Jolla, California, and was 0.0069 ug/L in West Los Angeles on March 26, 1982. In snow, it was 0.0062 ug/L in southern California, 0.0006 ug/L in central California, and 0.027 ug/L in Alaska.

Air: Average atmospheric concentrations in various U.S. areas for 1977-1980 were reported to be 110 ppt with a range of 60-156 ppt in rural/remote sites; 420 ppt with a maximum of 700-8,000 ppt in urban/suburban areas, where 20% of the samples may be positive; and 1,200 ppt in source-dominated areas. Although maximum values are usually under 10 ppb, a value of 111 ppb has been reported in New Jersey.

Soil/Sediment: Near factories where 1,1,1-trichloroethane was used, the concentration in soil was 0.06-0.94 ppb, and in up- and down-stream sediment it was 0.039-2.6 ppb. The average background concentration in St. Francis National Forest was 0.42 ppb in soil and 0.45 ppb in sediment.

Food: The following concentrations of 1,1,1-trichloroethane have been reported: oils and fats, 0.005-0.010 mg/kg; fruits and vegetables, 0.001-0.004 mg/kg; and meat, tea, and bread, 0.002-0.007 mg/kg.

E. Human environmental exposure

The major human exposure to 1,1,1-trichloroethane from chemical disposal sites would occur via ingestion of drinking water from contaminated groundwater sources. Inhalation could also be a significant route of exposure as the chemical is reported to be transported long distances (it has been detected at the South Pole).

Average daily intake: The estimated daily drinking water intake of 1,1,1-trichloroethane from surface water sources is 0.8 ug, and from groundwater sources is 4.2 ug. The estimated intake from air is 12.2 ug in rural areas, 46.5 ug in urban/suburban areas, and 133.0 ug in source-dominated areas.

Body burden: Values measured in 8 subjects revealed concentrations of 0.0016-0.024 mg/kg in body fat and less than 0.0051 mg/kg in various organs. A sample of residents from Love Canal in Niagara Falls, New York, showed the following median and maximum concentrations, respectively: breath, 290 and 2,800 ng/m^3; blood, 0.85 and 2.0 ng/ml; and urine, 80 and 180 ng/L. 1,1,1-Trichloroethane was detected in 8 of 8 samples of mothers' milk from 4 U.S. urban areas.

V. Health Effects

A. Human health effects

The major acute effect of 1,1,1-trichloroethane absorption from the lungs or gastrointestinal tract is central nervous system depression proportional to the amount absorbed. After recovery from central nervous system depression, mild transient liver and kidney dysfunction may occur. The symptoms of acute overdosage by inhalation are as follows: neurologic -- dizziness, fainting, and coma; gastrointestinal -- nausea and vomiting; cardiovascular -- hypotension and cardiac arrhythmias that may result from myocardial sensitization; hepatic -- possible transient increase in liver enzymes. 1,1,1-Trichloroethane is an anesthetic, and as such is capable of causing death when inhaled at concentrations in excess of 14,000-15,000 ppm (ACGIH, 1980). Contact with the eyes results in chemosis and hyperemia. Burns of the esophagus have been noted after ingestion of large amounts of the compound.

Chronic dermal exposure to 1,1,1-trichloroethane results in transient erythema and defatting action, as well as possible skin vesications. Psychophysiological functions, such as subject reaction time, perceptual speed, and manual dexterity, are adversely affected by inhalation exposure to an average concentration of 350 ppm; a linear relationship was noted between the concentration in alveolar air and arterial blood.

The chemical cannot be classified as to human carcinogenicity according to IARC criteria. No epidemiologic carcinogenicity studies are available (USEPA/OHEA, 1984). No clinical reports associate maternal exposure with congenital malformations in offspring (USEPA/OHEA, 1984).

Absorption, Distribution, Metabolism, Excretion:

1,1,1-Trichloroethane is rapidly absorbed by ingestion, inhalation, and by direct skin contact. The biological half-life is reported as 8.7 ± 1.8 hours. Approximately 90% of the compound is excreted unchanged by the lungs over an 8-day period. However, to a limited extent (less than 6%), it is metabolized to trichloroacetic acid and trichloroethanol which are excreted in urine over a period of approximately 12 days. These urinary metabolites may be used as a biological indicator of exposure to 1,1,1-trichloroethane.

B. Non-human mammalian health effects

Acute oral LD50 values have been reported to be 0.75 g/kg for dogs, 5.66 g/kg for rabbits, 9.47 g/kg for guinea pigs, 10.3 g/kg for rats, and 11.24 g/kg for mice. Acute inhalation LC50 values for the rat are 24,000, 18,000, and 14,000 ppm at 1, 3, and 7 hours, respectively; for the mouse the LC50 is 13,500 ppm at 10 hours. A dose-dependent biphasic decline in arterial blood pressure and cardiovascular depression was observed in dogs exposed to 10,000-40,000 ppm (54-216 g/m^3) of 1,1,1-trichloroethane per minute.

Among rats exposed 7 hours/day, 5 days/week for 1-3 months, inhalation at 10,000 ppm caused staggering gait and weakness in 10 minutes and caused irregular respiration and semiconsciousness by 3 hours. Rats given 0.75 and 1.5 g/kg/day orally from 7-117 weeks of age developed nasal discharge, wheezing, urine staining on abdominal fur, and a hunched appearance, but tumors observed were not attributable to 1,1,1-trichloroethane. At blood levels greater than 75 ppm, a so-called positional nystagmus, indicating central vestibulo-oculomotor disturbances, was observed in rabbits. In two lifetime studies involving inhalation-exposed rats and orally-exposed mice, no carcinogenic effects could be demonstrated (Patty's, 1981). However, a more recent study in which mice were administered by gavage 1.5 or 3.0 g/kg of the compound 5 days/week for 2 years showed an increased incidence of hepatocellular carcinomas; rats were not similarly affected (NRC, 1983).

Inhalation studies on mice and rats (Patty's, 1981) and a multigenerational reproduction study in which mice were orally exposed to 1,1,1-trichloroethane revealed no fetal toxicity or teratogenicity. The limited data indicate that 1,1,1-trichloroethane is not teratogenic in mice or rats (NRC, 1983).

Absorption, Distribution, Metabolism, Excretion:

In rats, more than 98% of the absorbed dose was rapidly expired unchanged, 0.5% was converted to CO_2, and much of the remainder was excreted in the urine as the glucuronide of 2,2,2-trichloroethanol. When rats were exposed to 500 ppm in air, 6 hours/day for 5 days, concentrations in peritoneal fat were much greater than in other organs. After inhalation by mice, 1,1,1-trichloroethane concentrated primarily in the liver and to a lesser extent in the brain and kidney. Dogs inhaling 700, 1,500, and 2,000 ppm showed a cumulative uptake of 27.3, 44.7, and 71.2 mg/kg, respectively, after 1 hour, at which time excretion was 66-77% at all three doses and the mean ratio of total absorption value to total inspired value was 14%.

C. Non-mammalian effects

Aquatic toxicity is reported by RTECS as follows: 96-hour TLm: 10-100 ppm. The reported 96-hour LC50 values for the fathead minnow (Pimephales promelas) were 52.8 mg/L flow-through and 105 mg/L static (Verschueren, 1983).

D. Short-term Assays

Mutagenic activities were indicated when 1,1,1-trichloroethane was tested by the Salmonella/microsome test, the BASC test on Drosophila, and the micronucleus test on mouse bone marrow. It was mutagenic in the Ames Salmonella assay in strains TA1535 and TA100 (NRC, 1983). Tests with the yeast Saccharomyces cerevisiae were negative for mutagenicity.

VI. Exposure Standards and Recommendations

A. Air

1. Immediately Dangerous to Life or Health (IDLH): 1,000 ppm (Sittig, 1981)

2. OSHA standards:

   TLV-TWA (8 hr): 350 ppm (1,910 mg/m^3) (1976 value cited in USDHHS/CDC, 1983)

3. NIOSH recommendations:

Ceiling (15 min): 350 ppm (1,910 mg/m^3) (value cited in USDHHS/CDC, 1983)

4. ACGIH (1980):

TLV-TWA (8 hr): 350 ppm (1,910 mg/m^3)
STEL (15 min): 450 ppm (approximately 2,450 mg/m^3)

B. Water

The following ambient water quality criteria are cited by EPA (USEPA/OWRS, 1980):

To protect human health from effects due to ingestion of contaminated water and aquatic organisms: 18.4 mg/L.

To protect freshwater aquatic life: 18.0 mg/L (acute toxicity)
No data available on chronic toxicity

To protect marine aquatic life: 31.2 mg/L (acute toxicity)
No data available on chronic toxicity

References for 1,1,1-Trichloroethane Synopsis

ACGIH SEE American Conference of Governmental Industrial Hygienists, Inc.

American Conference of Governmental Industrial Hygienists, Inc. (1980) Documentation of the threshold limit values. 4th ed. Cincinnati: The Conference. 486p.

Bouwer, E.J.; McCarty, P.L. (1983a) Transformations of 1- and 2-carbon halogenated aliphatic organic compounds under methanogenic conditions. Appl. Environ. Microbiol. 45:1286-1294.

Bouwer, E.J.; McCarty, P.L. (1983b) Transformations of halogenated organic compounds under denitrification conditions. Appl. Environ. Microbiol. 45: 1295-1299.

CRC SEE CRC handbook of chemistry and physics

CRC handbook of chemistry and physics. (1983) 64th ed. Weast, R.C.; Astle, M.J.; Berger, W.H.; eds. Boca Raton, FL: CRC Press, Inc. 2302p.

National Research Council. (1983) Drinking water and health. Vol. 5. Washington, DC: National Academy Press. 157p.

NRC SEE National Research Council

Patty's SEE Patty's industrial hygiene and toxicology

Patty's industrial hygiene and toxicology. (1981) 3rd rev. ed. Vol. 2. Clayton, G.D.; Clayton, F.E.; eds. New York: John Wiley & Sons. 5112p.

Sittig, M. (1981) Handbook of toxic and hazardous chemicals. Park Ridge, NJ: Noyes Publications. 729p.

TDB II SEE Toxicology data bank II

Toxicology data bank II. (1984) [Enhanced version of data base]. 1,1,1-Trichloroethane, TDB no. 157. (Last edited April 11, 1984). Bethesda, MD: U.S. Department of Health and Human Services, National Library of Medicine, Specialized Information Services. Magnetic-tape reels.

U.S. Department of Health and Human Services, Centers for Disease Control. (1983) NIOSH recommendations for occupational health standards. Morb. Mortal. Wkly. Rep. 32:(supplement)3S-22S.

USDHHS/CDC SEE U.S. Department of Health and Human Services, Centers for Disease Control.

U.S. Environmental Protection Agency, Office of Health and Environmental Assessment. (1984) Health assessment document for 1,1,1-trichloroethane (methyl chloroform). EPA-600/8-82-003F. Research Triangle Park, NC: The Office. 164p.

U.S. Environmental Protection Agency, Office of Water Regulations and Standards. (1980) Ambient water quality for chlorinated ethanes. EPA-440/5-80-029. Washington, DC: The Office. 136p. Available from: National Technical Information Service, Washington, DC.

USEPA/OHEA SEE U.S. Environmental Protection Agency, Office of Health and Environmental Assessment

USEPA/OWRS SEE U.S. Environmental Protection Agency, Office of Water Regulations and Standards

Verschueren, K. (1983) Handbook of environmental data on organic chemicals. 2nd ed. New York: Van Nostrand Reinhold Company, Inc. 1310p.

Trichloroethylene (TCE)

Information abstracted from TDB II, 1984 (No. 133) unless otherwise specified

I. Chemical Identification

A. CAS Registry No.: 79-01-6

B. Synonyms:

Acetylene trichloride
1-Chloro-2,2-dichloroethylene
Trichloroethene
1,1,2-Trichloroethylene

C. Molecular formula: C_2HCl_3

D. Structural formula:

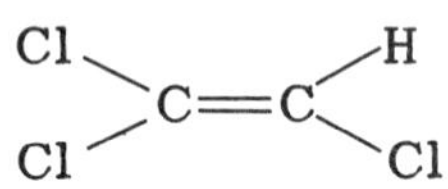

II. Physical and Chemical Properties

A. Molecular weight: 131.40

B. Physical state (color, form): Clear, colorless, or blue liquid

C. Odor: Chloroform-like, sweetish

D. Melting point: -73 deg C

E. Boiling point: 87 deg C

F. Solubilities:

Water: 1,110 mg/L at 25 deg C (Dilling, 1977)

Organic solvents: Soluble in chloroform, acetone; soluble in all proportions in alcohol, ether

G. Octanol/water partition coefficient (log P): 2.29

H. Density: 1.465 g/mL at 20 deg C

I. Dissociation constant: --

J. Viscosity: --

K. Surface tension (liquid interfacial tension with air): 29.3 mN/m (dynes/cm) at 20 deg C

L. Vapor pressure: 57.8 mm Hg at 20 deg C

M. Vapor density: 4.54 (air = 1)

N. Relative evaporation rate: Estimated half-life of minutes to hours, depending on turbulence

O. Spectral properties: Maximum absorption at less than 200 nm (vapor)

P. Other: Slowly decomposed with formation of HCl by light in the presence of moisture

III. Manufacturing/Use Information

Trichloroethylene (TCE) production in the U.S. in 1981 was 258 million lb. Its major uses are: as an industrial solvent in degreasing; in dry cleaning; in the manufacture of organic chemicals and pharmaceuticals; and as a refrigerant. Its use is prohibited in some states. It was formerly used as an extractant in food-processing (decaffeinating agent for coffee), and as an anesthetic.

IV. Environmental Fate/Exposure Potential

A. Potential for release from chemical disposal sites

Spills or releases of TCE to soil will evaporate rapidly due to TCE's high vapor pressure. It appears to be fairly stable in soil. TCE is reported to leach into groundwater, because TCE is readily soluble in water (1,110 mg/L) and is highly soluble in organic solvents.

B. Biological uptake/accumulation

Monitoring data of marine organisms suggest moderately low bioconcentration factors of 2-25 times that of surrounding seawater. Bioconcentration factors of 17-39 have been reported in bluegill sunfish and rainbow trout.

C. Persistence

TCE will be lost from aqueous solutions primarily by evaporation, with estimated half-life values ranging from minutes to hours, depending upon turbulence. The half-life for TCE in surface waters ranges from 3 hours in rapidly moving streams to 10 days or longer in ponds and lakes. Biodegradation, hydrolysis, and photooxidation in surface waters are extremely slow by comparison. The low adsorption coefficient of TCE indicates ready transport through soil and low adsorption to sediment. In laboratory studies simulating methanogenic groundwater conditions, TCE was biodegraded, although at a slow rate (Bouwer and McCarty, 1983). TCE released to the atmosphere is relatively reactive under smog conditions, with 50% degradation reported in 1 to 3.5 hours. Atmospheric residence time, based upon reaction with hydroxyl radical, is reported as 5 days, with formation of phosgene, dichloroacetyl chloride, and formyl chloride.

D. Measured concentrations

Chemical disposal sites: In the vicinity of Love Canal in Niagara Falls, New York, 7 of 9 samples of drinking water were reported to contain TCE, with concentrations of 0.01-0.25 ug/L. In the same area, 2 of 3 air samples contained TCE at concentrations of 1.6 and 3.4 ug/L, and home basement levels were estimated to be 0.83 ug/L. At a chemical disposal site in Edison, New Jersey, the TCE level in air was reported to range from a trace to 61 ug/L.

Groundwater: Dyksen and Hess (1982) reported TCE to be the most frequently detected chemical in a summary of state monitoring surveys, and it was found in the highest concentration of toxic organic chemicals studied; 28% of wells sampled were positive, with a maximum concentration reported to be 35,000 ug/L.

Surface water: Concentrations of TCE reported for industrial rivers in the U.S. ranged from 1-24 ug/L, with 188 ug/L reported in Lake Erie. TCE was reportedly the third most frequently detected compound in the Ohio River.

Drinking water: TCE was reported in 28 of 113 U.S. public water supplies, with a mean concentration of 2.1 ug/L.

Food: TCE has been detected in beverages, dairy products, fruits, vegetables, oils, and fats in concentrations ranging from 0.02 to 60.0 ug/kg. In marine fish, the range of measured concentrations is 40-1,100 ug/kg in the flesh and 0.66-20.0 ug/kg in the liver.

E. Human environmental exposure

Human exposure to TCE from chemical disposal sites could occur via ingestion of drinking water, by inhalation of TCE transported from disposal sites, or to a limited extent by direct contact with soil-bound wastes at the disposal site. Workers in degreasing plants may have high levels of exposure due to inhalation of vapors or absorption through the skin. Average daily intake has been reported in TDB II (1984) to be 11-33 ug from air and 2-20 ug from water. There are insufficient data on average daily intake from food.

Body burden: Analyses of body burdens in residents living near Love Canal in Niagara Falls, New York, have shown traces of TCE in 4 of 9 breath samples, and blood analyses have ranged from 0.09 to 2.50 ppb in 6 of 9 positive samples; the range reported for urine samples is

0.040-0.550 ppb in 9 of 9 positive samples. TCE has been detected in mothers' milk from 4 U.S. urban areas, with 8 of 8 samples positive.

V. Health Effects

A. Human health effects

There are accounts of poisoning of individuals who ingested TCE. The signs and symptoms of such acute oral exposures were reported by NRC (1980) to be principally gastrointestinal distress, narcosis, and occasional cardiac abnormalities. Only rarely are cases of hepatorenal involvement reported. The estimated human oral LD50 has been reported to be in the range from 50 to 500 mg/kg of body weight. In acute inhalation exposures, rapid coma may ensue with eventual death from hepatic or renal failure. An occasional sudden death suggests ventricular fibrillation. Signs of central nervous system depression, hepatic or renal failure, and critical cardiac events may be more common after acute exposures via inhalation than by ingestion. It is reported that these severe toxic effects may be seen after an inhalation exposure of 2,000 ppm over a period of 60 minutes.

Subchronic exposure to TCE vapors can cause irritation of the mucous membranes with resultant conjunctivitis and rhinitis. There have been reports of headache and dizziness in workers exposed to an air concentration averaging 10 ppm, and complaints of eye and throat irritation after exposures to 100-200 ppm.

There is little documented information on the chronic effects of human exposure to TCE, except that the defatting action of continued contact of TCE with the skin may cause dermatitis. In one study of serum enzymes in male volunteers, no changes in enzyme levels were observed after repeated exposures to vapors equivalent to an 8-hour maximum concentration of 50 ppm (NRC, 1983). Based on reports of chronic exposure of humans to TCE in occupational settings, there has been no report definitely establishing liver injury.

Information on the carcinogenic and teratogenic potential of TCE in humans is limited. TCE readily crosses the placenta and reportedly has appeared in the fetus within two minutes. According to IARC (WHO/-IARC, 1982), three studies showed no excess of cancers in workers exposed to TCE, but the results were considered inconclusive.

Absorption, Distribution, Metabolism, Excretion:

The major route of absorption of TCE is through the lungs, but moderate amounts can be absorbed through the intact skin and the gastrointestinal tract. Although a small portion of absorbed TCE is exhaled unchanged, a substantial amount is metabolized, primarily to trichloroethanol and trichloroacetic acid. The principle route of TCE excretion is renal. Retention of TCE is reported to account for a little more than half of the TCE inhaled. Most of the TCE is metabolized and the metabolites are eliminated by urinary excretion within the first few hours after exposure; some metabolites bind irreversibly to proteins, DNA, and RNA. TCE is quite soluble in lipids; thus, uptake is greater in adipose tissue than in muscle.

B. Non-human mammalian health effects

The acute oral LD50 in rats is 4,920 mg/kg; in mice it is approximately half this dose. The toxic effects of acute inhalation exposure to TCE are relatively similar to those in humans, including depression of the central nervous system and inhibition of cardiac function. Moderate degeneration of brain, lungs, kidneys, and spleen were shown in some experimental animals after acute inhalation exposure. The LD50 for rabbits treated percutaneously with TCE was 29,000 mg/kg.

Long-term studies of a variety of species of laboratory animals indicate that TCE may be no more hazardous upon chronic inhalation than it is with acute exposures; several species were exposed to TCE vapor 7 hr/day, 5 days/week, for 6 months and "maximum no-effect" concentrations were reported to range from 100 to 400 ppm (NRC, 1980). Slight evidence of adverse effects was observed as an increase in mortality of male mice exposed to high doses. Cats and guinea pigs showed some liver degeneration after inhalation exposures of 1,000 ppm daily over 10 months. Soybean meal was reported to be toxic when fed to cattle after it was extracted and processed with TCE.

TCE was reported to produce hepatomas in mice after long-term oral administration. Inhalation of TCE was associated with an increased incidence of lymphomas in female mice, but not in rats or hamsters. A modest increase was reported in the incidence of hepatocellular carcinomas in male mice exposed to high vapor concentrations, appearing to be a concentration-related incidence of tumor induction. No liver tumors were observed in rats. Although NRC (1980) reported that binding of TCE to mouse and rat microsomal proteins has been demonstrated, it concluded that TCE is apparently no more potentially carcinogenic with long-term exposure than with acute or short-term exposure. The NRC (1983) data indicate that TCE does not have significant teratogenic potential in mice and rats.

C. Non-mammalian effects

The reported 96-hour LC50 values for freshwater fish species were 66.8 mg/L for the bluegill (Lepomis macrochirus) and 40.7 mg/L for the fathead minnow (Pimephales promelas). For marine organisms, the 96-hour LC50 for grass shrimp (Palaemonetes pugio) was 2 mg/L and 20 mg/L for the sheepshead minnow (Cyprinodon variegatus).

D. Short-term assays

In studies reported by NRC (1983), TCE was found to be nonmutagenic in the Ames Salmonella assay but was mutagenic when tested in Saccharomyces cerevisiae D7. In a mouse host-mediated assay, TCE induced point mutations and gene conversions in yeast recovered from mouse liver and kidneys. It did not induce dominant lethal mutations in mice or rats, but positive results have been reported with a mouse in vivo somatic mutation assay for coat color mutants (WHO/IARC, 1982).

VI. Exposure Standards and Recommendations

A. Air

1. Immediately Dangerous to Life or Health (IDLH): 1,000 ppm (Sittig, 1981)

2. OSHA standards:

   TLV-TWA (8 hr): 100 ppm (535 mg/m^3)
   Ceiling: 200 ppm
   Peak: 300 ppm for 5 minutes in a 2-hour period (1972 value cited in USDOL/OSHA, 1978)

3. NIOSH recommendations: (1978 values cited in USDHHS/CDC, 1983)

   TLV-TWA (8 hr): 100 ppm
   Peak (10 min): 150 ppm

4. ACGIH (1982):

   TLV-TWA (10 hr): 50 ppm (270 mg/m^3)
   STEL (15 min): 200 ppm (1,080 mg/m^3)

B. Water

The following ambient water quality criteria are cited by EPA (USEPA/OWRS, 1980):

To protect human health from potential carcinogenic effects: Preferably a concentration of zero. An increase in the lifetime cancer risk by $1x10^{-5}$ due to ingestion of contaminated water and aquatic organisms is estimated to occur at 27 ug/L.

To protect freshwater aquatic life: 45 mg/L (acute toxicity)
No available data on chronic toxicity

To protect marine aquatic life: 2 mg/L (acute toxicity)
No available data on chronic toxicity

24-hour SNARL for TCE contamination of drinking water: 105 mg/L (NRC, 1980)

C. Other:

FDA has proposed regulations prohibiting the use of trichloroethylene in the processing of food, or human and animal drugs (WHO/-IARC, 1979).

References for Trichloroethylene Synopsis

ACGIH SEE American Conference of Governmental Industrial Hygienists, Inc.

American Conference of Governmental Industrial Hygienists, Inc. (1982) Supplemental documentation [for documentation of the threshold limit values]. Cincinnati: The Conference. 40p.

Bouwer, E.J.; McCarty, P.L. (1983) Transformations of 1- and 2-carbon halogenated aliphatic organic compounds under methanogenic conditions. Appl. Environ. Microbiol. 45:1286-1294.

Dilling, W.L. (1977) Interphase transfer processes. II. Evaporation rates of chloromethanes, ethanes, ethylenes, propanes, and propylenes from dilute aqueous solutions: comparisons with theoretical predictions. Environ. Sci. Technol. 11:405-409.

Dyksen, J.E.; Hess, A.F., III. (1982) Alternatives for controlling organics in groundwater supplies. J. Am. Water Works Assoc. 74:394-403.

National Research Council. (1980) Drinking water and health. Vol. 3. Washington, DC: National Academy Press. 415p.

National Research Council. (1983) Drinking water and health. Vol. 5. Washington, DC: National Academy Press. 157p.

NRC SEE National Research Council

Sittig, M. (1981) Handbook of toxic and hazardous chemicals. Park Ridge, NJ: Noyes Publications. 729p.

TDB II SEE Toxicology data bank II

Toxicology data bank II. (1984) [Enhanced version of data base]. Trichloroethylene, TDB no. 133. (Last edited April 5, 1984). Bethesda, MD: U.S. Department of Health and Human Services, National Library of Medicine, Specialized Information Services. Magnetic-tape reels.

U.S. Department of Health and Human Services, Centers for Disease Control. (1983) NIOSH recommendations for occupational health standards. Morb. Mortal. Wkly. Rep. 32:(supplement)3S-22S.

U.S. Department of Labor, Occupational Safety and Health Administration. (1978) Occupational safety and health: general industry standards. Rev. Nov. 7. Washington, DC: U.S. Department of Labor, Occupational Safety and Health Administration. 820p. Available from: U.S. Government Printing Office, Washington, DC.

USDHHS/CDC SEE U.S. Department of Health and Human Services, Centers for Disease Control

USDOL/OSHA SEE U.S. Department of Labor, Occupational Safety and Health Administration

U.S. Environmental Protection Agency, Office of Water Regulations and Standards. (1980) Ambient water quality criteria for trichloroethylene. EPA-440/5-80-077. Washington, DC: The Office. 58p. Available from: National

Technical Information Service, Springfield, VA.
USEPA/OWRS SEE U.S. Environmental Protection Agency, Office of Water Regulations and Standards
WHO/IARC SEE World Health Organization, International Agency for Research on Cancer
World Health Organization, International Agency for Research on Cancer. (1979) IARC monographs on the evaluation of the carcinogenic risk of chemicals to humans. Vol. 20. Some halogenated hydrocarbons. Lyon, France: The Agency. 609p.
World Health Organization, International Agency for Research on Cancer. (1982) IARC monographs on the evaluation of the carcinogenic risk of chemicals to man. Supplement 4 to Vol. 1-29. Chemicals, industrial processes and industries associated with cancer in humans. Lyon, France: The Agency. 292p.

Appendix B

Study Participants

EXECUTIVE SCIENTIFIC PANEL
on
HEALTH ASPECTS OF THE DISPOSAL OF WASTE CHEMICALS

Chairman

Joe W. Grisham, M.D.
Professor and Chairman
Department of Pathology
University of North Carolina
School of Medicine
Chapel Hill, North Carolina 27514

Martin Alexander, Ph.D.
Professor, Soil Microbiology
Department of Agronomy
708 Bradfield Hall
Cornell University
Ithaca, New York 14853

Morton Corn, Ph.D.
Professor and Director
Environmental Health Engineering
The Johns Hopkins University
School of Hygiene and
Public Health
615 North Wolfe St.
Baltimore, Maryland 21205

Robert E. Anderson, M.D.
Professor and Chairman
Department of Pathology
University of New Mexico
School of Medicine
Albuquerque, New Mexico 87131

John E. Davies, M.D., M.P.H.
Chairman
Department of Epidemiology
and Public Health
University of Miami
School of Medicine
Medical Arts Bldg., Rm. 100
1550 N.W. 10th Ave.
Miami, Florida 33101

Arthur D. Bloom, M.D.
Professor
Department of Pediatrics,
Human Genetics and Development
Columbia University
College of Physicians
and Surgeons
New York, New York 10032

John Doull, M.D., Ph.D.
Professor
Department of Pharmacology,
Toxicology and Therapeutics
University of Kansas
Medical Center
Room 3026
Rainbow Blvd. at 39th St.
Kansas City, Kansas 66103

Virgil H. Freed, Ph.D.
Professor and Head
Department of Agricultural Chemistry
339 Weniger Hall
Oregon State University
Corvallis, Oregon 97331

Gabriel L. Plaa, Ph.D.
Professor of Pharmacology and
Vice Dean for Research and Graduate School
University of Montreal
Faculty of Medicine
P.O. Box 6207, Succ A
Montreal, Quebec
Canada H3C 3T7

Eugene V. Perrin, M.D.
Professor of Pathology
Children's Hospital of Michigan
Wayne State University
3901 Beaubien Blvd.
Detroit, Michigan 48201

Paul V. Roberts, Ph.D.
Professor
Department of Civil Engineering
Stanford University
Stanford, California 94305

Henry C. Pitot, M.D., Ph.D.
Professor of Oncology and Pathology
McArdle Laboratory for Cancer Research
University of Wisconsin
School of Medicine
450 North Randall Ave.
Madison, Wisconsin 53706

Reuel A. Stallones, M.D.,M.P.H.
Dean
School of Public Health
University of Texas
P.O. Box 20186
Houston, Texas 77225

EXPOSURE TASK GROUP

Paul L. Bishop, P.E., Ph.D.
Professor of Civil Engineering
236 Kingsbury Hall
Department of Civil Engineering
University of New Hampshire
Durham, New Hampshire 03824

Edward Bouwer, Ph.D.
Assistant Professor
Department of Civil Engineering
University of Houston
Houston, Texas 77004

John A. Cherry, Ph.D.
Professor and Director
Groundwater Research Institute
Department of Earth Sciences
University of Waterloo
Waterloo, Ontario
Canada N2L 3G1

C.S. Giam, Ph.D.
Professor of Chemistry
Department of Industrial
Environmental Health Sciences
Graduate School of Public Health
University of Pittsburgh
Pittsburgh, Pennsylvania 15261

Ronald Hites, Ph.D.
Professor of Public and Environmental
Affairs and Chemistry
Department of Chemistry
Indiana University
Bloomington, Indiana 47401

Donald MacKay, P.E., Ph.D.
Professor
Department of Chemical Engineering
and Applied Chemistry
University of Toronto
Toronto, Ontario
Canada M5S 1A4

Douglas M. Mackay, Ph.D.
Research Associate
Department of Civil Engineering
Terman Engineering Center
4-700 Stanford University
Stanford, California 94305-2179

Warren T. Piver, Ph.D.
Chemical Engineer
Office of Health Hazard Assessment
National Institute of Environmental
Health Sciences
P.O. Box 12233
Research Triangle Park
North Carolina 27709

P. Suresh Chandra Rao, Ph.D.
Associate Professor
Department of Soil Physics
2169 McCarty Hall
University of Florida
Gainesville, Florida 32611

Lynn P. Wallace, P.E., Ph.D.
Associate Professor of
Civil Engineering
Department of Civil Engineering
Box 33 CB
Brigham Young University
Provo, Utah 84602

John T. Wilson, Ph.D.
Research Microbiologist
U.S. Environmental Protection Agency
R.S. Kerr ERL
Box 1198
Ada, Oklahoma 74820

HEALTH EFFECTS TASK GROUP

S. Gaylen Bradley, Ph.D.
Dean, School of Basic Sciences
Medical College of Virginia
Richmond, Virginia 23298

Doyle G. Graham, M.D., Ph.D.
Associate Professor of Pathology
Department of Pathology
Duke University Medical Center
Durham, North Carolina 27710

Philip S. Guzelian, M.D.
Professor, Department of Medicine
Medical College of Virginia
MCV Station
Richmond, Virginia 23298

Paul M. Newberne, D.V.M., Ph.D.
Professor of Nutritional Pathology
Massachusetts Institute of Technology
E18-611
Cambridge, Massachusetts 02139

John R. Van Ryzin, Ph.D.
Professor, Division of Biostatistics
Columbia University
School of Public Health
New York, New York 10032-3799

EPIDEMIOLOGY WORKSHOP

Henry A. Anderson, M.D.
Chief, Environmental and Chronic
Disease Epidemiology
Wisconsin Division of Health
Madison, Wisconsin 53701

Patricia A. Buffler, Ph.D.
Associate Dean for Research and
Associate Professor of Epidemiology
University of Texas
School of Public Health
Houston, Texas 77225

Dale D. Chitwood, Ph.D.
Research Assistant Professor
Department of Oncology
University of Miami School of Medicine
Medical Arts Building
Miami, Florida 33136

Robert S. Levine, M.D.
Associate Professor and Director
Division of Clinical Preventive Medicine
Department of Epidemiology and
Public Health
University of Miami School of Medicine
Medical Arts Building
Miami, Florida 33136

Raymond R. Neutra, M.D., Dr.P.H.
Chief, Epidemiology Studies Section
California Department of Health
Services
Berkeley, California 94704

Carl Shy, M.D.
Assistant Professor
Department of Epidemiology
University of North Carolina
School of Public Health
Chapel Hill, North Carolina 27514

Duncan C. Thomas, Ph.D.
Associate Professor
Department of Preventive Medicine
University of Southern California
School of Medicine
Los Angeles, California 90033-9987

EPIDEMIOLOGY TASK GROUP

Elizabeth Barrett-Connor, Ph.D.
Associate Professor and Chief
Division of Epidemiology
Department of Community Medicine
University of California
La Jolla, California 92093

Peter Greenwald, M.D.
Director
Division of Cancer Prevention and Control
Building 31, Room 4-A32
National Institutes of Health
Bethesda, Maryland 20205

Philip Cole, M.D.
Professor and Head
Department of Epidemiology
School of Public Health
University of Alabama at Birmingham
University Station
Birmingham, Alabama 35294

Darwin Labarthe, M.D., Ph.D.
Professor of Epidemiology
University of Texas
School of Public Health
Houston, Texas 77225

Brian MacMahon, M.D.
Professor and Chairman
Department of Epidemiology
Harvard University
Boston, Massachusetts 02115

CONSULTING SCIENTISTS

Robert L. Brent, M.D., Ph.D.
Director
Stein Research Center
Jefferson Medical University
Philadelphia, Pennsylvania 19107

James F. Crow, Ph.D.
Professor
Department of Genetics
University of Wisconsin
Madison, Wisconsin 53706

James M. Davidson, Ph.D.
Assistant Dean for Research
Research Administration
Institute of Food and Agricultural Sciences
University of Florida
Gainesville, Florida 32611

Dr. Jan W. Huismans
Director
United Nations Environment Programme
Palais des Nations
CH-1211 Geneva 10
Switzerland

Evelyn M. Hurlburt, Ph.D.
Division of Nuclear Medicine
and Radiation Health Sciences
The Johns Hopkins University
School of Medicine
Baltimore, Maryland 21205

Rudolf J. Jaeger, Ph.D.
Professor of Toxicology
Institute of Environmental Medicine
New York University Medical Center
New York, New York 10016

Leo J. Kinlen
Professor
CRC Cancer Epidemiology Unit
University of Edinburgh
14 George Square
Edinburgh EH8 9JZ
Scotland

Norman W. Klein, Ph.D.
Professor
Department of Animal Genetics
University of Connecticut
Storrs, Connecticut 06268

Fumio Matsumura, Ph.D.
Professor and Director
Pesticide Research Center
Michigan State University
East Lansing, Michigan 48824

Perry L. McCarty, P.E., Sc.D.
Professor and Chairman
Department of Civil
Engineering
Stanford University
Stanford, California 94305-2179

Albert E. Munson, Ph.D.
Professor
Department of Pharmacology
Medical College of Virginia
Richmond, Virginia 23298

Herbert H. Schaumburg, M.D.
Professor
Department of Neurology
Albert Einstein College of Medicine
of Yeshiva University
Bronx, New York 10461

Carol T. Schnizlein, Ph.D.
Postdoctoral Fellow
Department of Microbiology
and Immunology
Medical College of Virginia
Richmond, Virginia 23298

Arthur C. Upton, M.D.
Professor and Chairman
Institute of Environmental Medicine
New York University
Medical Center
New York, New York 10016

Mr. J. Ian Waddington
Director
Environmental Health Service
World Health Organization
Regional Office for Europe
8, Scherfigsvej
DK-2100 Copenhagen 0
Denmark

Ralph C. Wands, M.S., C.I.H.
Ralph C. Wands and Associates, Inc.
5445 Mt. Corcoran Place
Burke, Virginia 22015

PROJECT STAFF

UNIVERSITIES ASSOCIATED FOR RESEARCH AND EDUCATION IN PATHOLOGY, INC.

Kenneth M. Endicott, M.D., Project Co-Director
Executive Officer, UAREP

Thomas A. Hill, M.P.H.I.H.
Associate Project Director/Project Coordinator

Jaclyn M. Cusick
Secretary

Deborah S. Fox
Administrative Assistant

Bentley C. Gregg, Ph.D.
Senior Staff Scientist

Richard W. Leukroth, Jr., M.S.
Associate Staff Scientist

LIFE SCIENCES RESEARCH OFFICE

Kenneth D. Fisher, Ph.D., Project Co-Director
Director, LSRO

Richard G. Allison, Ph.D.
Senior Staff Scientist

Philip L. Altman, M.S.
Senior Staff Scientist

Gloria J. Cole
Secretary

Louise S. Erlick, M.L.S.
Technical Literature Retrieval and Report Specialist

Caren L. Feldman, M.P.H.
Associate Staff Scientist

J. Elaine Huey
Technical Literature Retrieval and Report Specialist

Margaret D. Palmieri, M.S.
Technical Literature Specialist

Stephen H. Simpson
Technical Literature Assistant

Appendix C

Individuals and Organizations Responding to the Draft Report

INDIVIDUALS AND ORGANIZATIONS RESPONDING TO THE DRAFT REPORT

Public Meeting Participants

The Executive Scientific Panel on Health Aspects of the Disposal of Waste Chemicals held a Public Meeting on December 17, 1984. Requests were received from 3 organizations for an opportunity to make oral presentations. The following individuals participated in the presentations:

Esther H. Forbes
Fauquier Organized Residents
Guarding the Environment
Warrenton, Virginia

Amanda M. Phillips
Health Associate
Toxic Chemicals Program
Environmental Defense Fund
Washington, DC

Bruce W. Karrh, M.D.
Chemical Manufacturers Association
Washington, DC

Ellen K. Silbergeld, Ph.D.
Chairperson
Toxic Chemicals Program
Environmental Defense Fund
Washington, DC

Written Comments

The following organizations and individuals submitted written materials for consideration by the Executive Scientific Panel:

C. Scott Clark, Ph.D.
Associate Professor
Institute of Environmental Health
University of Cincinnati Medical Center
Cincinnati, Ohio

Geraldine V. Cox, Ph.D.
Vice President and Technical Director
Chemical Manufacturers Association
Washington, DC

Henry S. Cole, Ph.D.
Research Director
National Campaign Against
Toxic Hazards
Washington, DC

Janet S. Hathaway
Staff Attorney
Public Citizen's Congress Watch
Washington, DC

William F. O'Keefe
Vice President
American Petroleum Institute
Washington, DC

Timothy F. O'Leary
Associate Director
Health, Safety and Chemical Regulations
Chemical Manufacturers Association
Washington, DC

David Ozonoff, M.D.
Boston University School of Public Health
Boston, Massachusetts

Amanda M. Phillips
Health Associate
Toxic Chemicals Program
Environmental Defense Fund
Washington, DC

Charles Poole, M.P.H.
Associate Epidemiologist
Epidemiology Resources Inc.
and the
New England Epidemiology Institute
Chestnut Hill, Massachusetts

Kenneth J. Rothman, Dr.P.H.
Senior Epidemiologist
Epidemiology Resources Inc.
and the
New England Epidemiology Institute
Chestnut Hill, Massachusetts

Randal P. Schumacher
Director
Health, Safety and Chemical Regulations
Chemical Manufacturers Association
Washington, DC

Ellen K. Silbergeld, Ph.D.
Chairperson
Toxic Chemicals Program
Environmental Defense Fund
Washington, DC

Theodore R. Torkelson, Sc.D.
Chairman
Health and Science Committee
Halogenated Solvents Industry Alliance
Washington, DC

Index